Crime Scene to Court

The Essentials of Forensic Science
5th Edition

Crime Scene to Court

The Essentials of Forensic Science 5th Edition

Edited by

Niamh NicDaeid

Leverhulme Research Centre for Forensic Science,
University of Dundee, UK
Email: n.nicdaeid@dundee.ac.uk

and

Peter C. White

ReForensics, PO Box 1208, Lincoln, UK
Email: Lyric29P@aol.com

ROYAL SOCIETY OF CHEMISTRY

Print ISBN: 978-1-83916-800-0
PDF ISBN: 978-1-83767-224-0
EPUB ISBN: 978-1-83767-225-7

A catalogue record for this book is available from the British Library

The Royal Society of Chemistry is a charity, registered in England and Wales, Number 207890, and a company incorporated in England by Royal Charter (Registered No. RC000524), registered office: Burlington House, Piccadilly, London W1J 0BA, UK, Telephone: +44 (0)20 7437 8656.

Visit our website at books.rsc.org

Printed in the United Kingdom by CPI Group (UK) Ltd, Croydon, CR0 4YY, UK

Preface

It was 26 years ago when the first edition of *Crime Scene to Court – The Essentials of Forensic Science* was published by the Royal Society of Chemistry (RSC) to support my (PW) teaching at undergraduate and postgraduate levels as no suitable teaching text covering the UK legal system was available. At the outset it was deemed important that forensic practitioners should contribute chapters on their own particular discipline and written at a level suitable for students and the lay person. Apart from being adopted as a core forensic science textbook by many universities both in the UK and overseas, the book has since gained a much wider readership with it being used by crime scene officers, forensic scientists, police and the legal profession for both training and reference purposes. It has also gathered interest from the lay readership.

Such was the success of this book that the second edition was published in 2004. This provided the opportunity to increase the number of disciplines and monitor changes taking place in the delivery of a forensic science service in the UK. For similar reasons the third edition was published in 2010 and a fourth edition in 2016. Through each of these editions, authors were requested to retain as much as possible of their introductory and historical material so that readers did not have to refer back to previous editions or other publications. Additionally, Bibliographies were used as a means to provide a reader with more in depth and/or recent knowledge.

In 2022 the RSC discussed with me (PW) the possibility of a fifth edition. Readers of the fourth edition would have seen that I indicated that it would be my last edition as the editor and was to enjoy retirement. Having kept professional interests going and staying in

Crime Scene to Court: The Essentials of Forensic Science, 5th Edition
Edited by Niamh NicDaeid and Peter C. White

Published by the Royal Society of Chemistry, www.rsc.org

touch with colleagues, I believed that with on-going changes and challenges in providing a forensic science service still prevailing, there should be another edition and agreed to set up and edit this next edition. This required making some major changes due to a number of the previous authors retiring and finding their suitable replacements, a reorganisation of the sequence of the chapters and the inclusion of one new chapter namely, Forensic Collision Investigation by Mark Crouch. Inclusion of this chapter not only extends the range of forensic disciplines being covered but also shows students that a working knowledge and interest in physics and mathematics is required if wanting to pursue a career within this area of forensic science.

Some of the retiring authors have contributed to all the previous editions and I would like to thank them for all their time and support. Four chapters have had to be rewritten completely by the new authors and include, Fire Investigation by Iain Peck, Trace and Contact Evidence by Sarah Jacob, Body Fluids by Caroline Crawford and Claire Jarman and Forensic Examination of Documents and Handwriting by Anthony Stockton. There have also been new co-authors for Forensic Science – Rachel Bolton-King, Digital Forensics – Benjamin Donnachie and for Forensic Archaeology and Anthropology – Lucina Hackman. Again, as with previous editions all of these new authors are respected practitioners thus ensuring readers benefit from the knowledge and experience of their particular forensic discipline. I would also like to thank the other previous authors for continued support and updating of their chapters.

A comment made by a reviewer also prompted a change in the sequence of chapters in this edition so that they were more aligned with the route of crime scene investigation through laboratory analysis and attending court. Interestingly, this also resulted in a group of chapters that are concerned broadly with identification of an individual, an important role in the judicial process as with no suspect any legal procedure cannot be completed. There was some difficulty in grouping of the other chapters as most disciplines require both scene of crime investigation as well as laboratory analysis. These were therefore grouped according to their level of crime scene or laboratory activity.

Sadly, after putting all this in place and editing a couple of chapters, I was unable to continue my editing duties due to requiring a second major spinal operation and sought another editor to assist and take over completion of this edition. I am delighted that Professor Niamh NicDaeid agreed to undertake this task. Niamh started her forensic career with me in the Forensic Science Unit at the University of Strathclyde and became a member of my staff when I was appointed as the Director of the Unit. Having developed an excellent and extensive forensic science research record coupled with considerable case work experience, Niamh is now the Director of the Leverhulme Research Centre for Forensic Science at the University of Dundee. With funding from the Leverhulme Trust, the strategic aim of her work is to support the evidence presented in courts with robust scientific data and hence heads up a leading forensic science research centre which has many national and international contacts and collaborations – I therefore believe Niamh is an ideal person to take over as the editor.

Again, as identified in chapters here and in all previous editions, the provision of a forensic science service in the UK is still changing and developing. The definition of forensic science, who delivers it, quality control of working and reporting standards, sustainability of a forensic service, including research, are just some of the issues that are being reviewed. These issues are discussed in more detail in the introductory chapter and other authors have identified where these changes have affected their particular discipline. An additional factor has been living and working through Covid 19 with all industries, including the providers of the forensic science service, having to make changes to their working practices. Crime scene investigation, laboratory work and court procedures were all affected. As commented upon in some chapters this required new working practices and procedures to be implemented to investigate scenes, reduce backlogs in laboratory work and delays in court work. Any change of procedures was required to meet previous or modified operational standards and as now no longer under lock down, some of the these modified standards are still being adopted.

Looking back to when Home Office Laboratories and my previous employer, the Metropolitan Police Forensic Science Laboratory, were providing the majority of the forensic service as identified in the first edition of this book, it may appear that the provision of this service has become fragmented and more bureaucratic. There may be a modicum of truth in this statement if just considering the wide range of practitioners and disciplines that now contribute to and provide our current forensic science service and how they are managed. However, no matter who or how this service is delivered, it must meet public expectations, inspire confidence and strive to ensure no miscarriages of justice. Through the Forensic Science Regulator Act 2021 there are now statutory powers for any forensic science service to comply with the Regulator's Code of Practice and Conduct for Forensic Service Providers and Practitioners which should help to inspire confidence. Interestingly, in 2023, a Westminster commission had been set up to focus on risks posed by current provision that may contribute to miscarriages of justice. Hopefully there will be some positive outcomes and we hope they will inspire public confidence in the legal system and those who provide a forensic contribution. Perhaps readers will find out more if there were to be another edition of this book.

Finally, we would just like to thank all the authors and the RSC publishing team for all their help and support in making this fifth edition possible. We hope the book will continue to assist and be a benefit to undergraduate and postgraduate students in their forensic courses and anyone pursuing careers within the judicial system. In closing, PW would just like to thank Janet Freshwater at the RSC, with whom he has worked with since the first edition and most importantly to Niamh for taking over the editorship of this book.

Peter White
ReForensics, PO Box 1208, Lincoln, UK

Niamh NicDaeid
University of Dundee, UK

Abbreviations

6-MAM	6-Monoacetylmorphine
ABFA	American Board of Forensic Anthropology
ABPI	Association of the British Pharmaceutical Industry
ABS	Antilock braking system
ACM	Airbag control module
ACPO	UK Association of Chief Police Officers
ADD	Accumulated degree days
ADH	Alcohol dehydrogenase
ADH	Accumulated degree hours
AFIS	Automated Fingerprint Identification System
AFSP	Association of Forensic Science Providers
ALS	Alternative light sources
ANFO	Ammonium nitrate/fuel oil
ANPR	Automated numberplate recognition
AP	Acid phosphatase
ASQDE	American Society of Questioned Document Examiners
ATM	Automatic Teller Machine
AUP	Acceptable use policy
BMK	Benzenemethylketone
BPA	Bloodstain pattern analysis
BPA	Base pairs
BZP	*N*-Benzylpiperazine
CAI	Case assessment and interpretation
CAID	Child abuse image database
CAST	Centre of Applied Science and Technology
CBRN	Chemical, biological, radiological and nuclear
CHIS	Covert human intelligence source

Crime Scene to Court: The Essentials of Forensic Science, 5th Edition
Edited by Niamh NicDaeid and Peter C. White

Published by the Royal Society of Chemistry, www.rsc.org

CIfA	Chartered Institute for Archaeologists
CITES	Convention on International Trade in Endangered Species
CJS	Criminal justice system
CLSM	Confocal laser scanning microscopy
CMA	UK Computer Misuse Act
CN	Chloroacetophenone
COPFS	Crown Office Procurator Fiscal Service
COSHH	Control of Substances Hazardous to Health
CPR	Criminal procedure rules
CPS	Crown Prosecution Service
CRFP	Council for Registration of Forensic Practitioners
CS	*o*-Chlorobenzalmalononitrile
CSC	UK Cyber Security Council
CSC	Crime Scene Coordinator
CSC	*ortho*-Chlorobenzalmalononitrile
CSE	Crime scene examiners
CSFS	Chartered Society of Forensic Science
CSI	Crime scene investigators
DABFT	Diplomate of the American Board of Forensic Toxicology
DNA	Deoxyribonucleic acid
DPP	Director of Public Prosecutions
DSP	Dynamic Stability Programme
DSTL	Defence Science and Technology Laboratory
DVI	Disaster victim identification
EAFE	Association for Forensic Entomology
EDR	Event data recorder
ELISA	Enzyme-linked immunosorbent assay
EMCDDA	European Monitoring Centre for Drugs and Drug Addiction
EMIT	Enzyme multiplied immunoassay technique
ENFSI	European Network of Forensic Science Institutes
EPG	Electrophoretogram
ESC	Electronic stability control
ESDA	Electrostatic detection
ET-AAS	Electrothermal atomic absorption spectrometry
FAAS	Fame atomic absorption spectroscopy
FASE	Forensic Anthropology Society of Europe
FBI	Federal Bureau of Investigation
FCN	Forensic Capability Network

FEL	Forensic Explosives Laboratory
FID	Flame ionisation detectors
FPIA	Fluorescence polarisation immunoassay
FSI	Forensic Scene Investigator
FSNI	Forensic Science Northern Ireland
FSR	Forensic Science Regulator
FSS	Forensic Science Service
FTIR	Fourier transform infrared spectroscopy
GABA	gamma-Aminobutyric acid
GAM	General additive models
GC	Gas chromatography
GC–MS	Gas chromatography–mass spectrometry
TLC	Thin layer chromatography
GC–TEA	Gas chromatography–thermal energy analyser
GDPR	General data protection regulation
GPR	Ground penetrating radar
GRIM	Glass refractive index measurement
GSR	Gunshot residue
H_d	Defence hypothesis
HID	High intensity discharge
HMTD	Hexamethylene triperoxide diamine
HMX	Cyclotetramethylenetetranitramine
HOFSS	Home Office Forensic Science Service
H_p	Prosecution hypothesis
HPLC	High-performance liquid chromatography
HRMS	High resolution mass spectrometry
HVR	Hypervariable regions
IABPA	International Association of Bloodstain Pattern Analysts
IBIS	Integrated Ballistics Identification System
ICMP	International Commission for Missing Persons
ICP–MS	Inductively coupled plasma mass spectrometry
ICTY	International Tribunal for the Former Yugoslavia
IED	Improvised explosive device
IFSA	International Forensic Strategic Alliance
ILAC	International Laboratory Accreditation Cooperation
IOPC	Independent Office for Police Conduct
IPM	Integrated pest management
ISO	International Organization for Standardisation
KC	King's Counsel
KM	Kastle–Meyer

LC–MS	Liquid chromatography–mass spectroscopy
LC–MS–MS	Liquid chromatography tandem mass spectrometry
LCN	Low copy number
LCV	Leucocrystal violet
LDA	Lysergic acid diethylamide
LED	Light emitting diode
LMG	Leuco Malachite Green
LPM	Low power microscope
LR	Likelihood ratio
LRCFS	Leverhulme Research Centre for Forensic Science
MA	Methylamphetamine
MADLI	Matrix assisted laser desorption ionisation
MDA	Methylenedioxyamphetamine
MDA	Misuse of Drugs Act 1971
MDMA	Methylenedioxymethylamphetamine
MEOS	Microsomal enzyme oxidizing system
MIRS	Medium-infrared spectroscopy
MNI	Minimum number of individuals
MPDD	Missing Persons DNA Database
mRNA	Messenger RNA
mtDNA	Mitochondrial DNA
NABIS	National Ballistics Intelligence Service
NCA	National Crime Agency
NDNAD	National DNA database
NGS	Next generation sequencing
NIRS	Near-infrared spectroscopy
NIST	National Institute of Standards and Technology
NMR	Nuclear magnetic resonance
NPCC	National Police Chiefs Council
NPS	New/novel psychoactive substances
NTFS	New technology file system
OC	Oleoresin capsicum
OS	Operating systems
OSAC	Organization of Scientific Area Committees for Forensic Science
PACE	Police and Criminal Evidence Act 1984
PCAST	President's Council of Advisors on Science and Technology
PCR	Polymerase chain reaction
PCR-RFLP	Polymerase chain reaction-restriction fragment length polymorphism

PDMS	Polydimethylsiloxane
PDOF	Principle direction of force
PETN	Pentaerythritol tetranitrate
PGM	Phosphoglucomutase
PID	Photo ionisation detectors
PMI	Post-mortem interval
PMImin	Minimum post-mortem interval
POI	Point of impact
PolSA	Police Search Advisor
PPE	Personal protective equipment
PRT	Perception response time
PSNI	Police Service of Northern Ireland
PTPH	Plea and Trial and Preparation Hearings
QAA	Quality Assurance Agency
RAPD	Random amplification of polymorphic DNA
RDX	Cyclotrimethylenetrinitramine
RFU	Relative fluorescence units
RIA	Radioimmunoassay
SAR	Search and rescue
SCRAs	Synthetic cannabinoid receptor agonists
SEM	Scanning electron microscopy
SEM–EDS	Scanning electron microscopy with energy-dispersive spectroscopy
SFC	Supercritical fluid chromatography
SFR	Streamlined forensic report
SIMS	Secondary ionisation mass spectrometry
SNPs	Single nucleotide polymorphisms
SOCO	Scene of crimes officers
SPA	Scottish Police Authority
SPSA	Scottish Police Services Authority
SSD	Solid-state drives
SSM	Scientific Support Managers
STR	Short tandem repeat
SWSTAIN	Scientific Working Group on Bloodstain Pattern Analysis
TATP	Triacetone triperoxide
TCDO	Temporary class drug order
Tetryl	2,4,6-Trinitrophenylmethylnitramine
TFMPP	Trifluoromethylphenylpiperazine
TLC	Thin layer chromatography
TNT	2,4,6-Trinitrotoluene

UHPLC	Ultra high performance liquid chromatography
UKAS	United Kingdom Accreditation Service
UNODC	United Nations Office on Drugs and Crime
UV	Ultraviolet
VIN	Vehicle identification number
VPDD	Vulnerable Persons DNA Database
VRD	Victim recovery dogs
XRD	X-ray diffraction
XRF	X-ray fluorescence
YHRD	Y Chromosome Haplotype Reference Database
Δ^9-THC	Δ^9-Tetrahydrocannibinol

Contents

Crime Scene to Court: The Essentials of Forensic Science, 5th Edition
Edited by Niamh NicDaeid and Peter C. White

Published by the Royal Society of Chemistry, www.rsc.org

1 Forensic Science

Brian Rankin*[a] and Rachel Bolton-King*[b]

[a] Formerly Head of Centre for Forensic Investigation, Teesside University, Middleborough, Tees Valley, North East England; [b] School of Science & Technology, Nottingham Trent University, Clifton, Nottingham, England
*Emails: bwjrankin@gmail.com; rachel.bolton-king@ntu.ac.uk

1.1 Introduction

Forensic science is a fascinating subject, the science is serious but the term 'forensic' seems to give it a 'sexy' appeal. Forensic science practitioners (henceforth collectively referred to as forensic scientists within this book) soon discover when talking to the general public that many people have very high expectations of forensic science and what it can provide to an investigation and a court of law. This is mainly because of their fascination with the subject, often alluded to as the 'CSI effect', which may be a consequence of the proliferation of TV dramas such as CSI on the subject of crime scene investigation and forensic practices typically covering scene and laboratory investigations, digital technology, pathology and many other related subjects. The material in this chapter sets out to address these high expectations, by providing an explanation and understanding of forensic science, a discussion of its origins and how forensic science services (frequently referred to as forensic practice) have been developed over the years, and how it now operates in the United Kingdom (UK) within a broader international setting. The duties of the forensic scientist and how the required high standards of analysis, reporting

Crime Scene to Court: The Essentials of Forensic Science, 5th Edition
Edited by Niamh NicDaeid and Peter C. White

Published by the Royal Society of Chemistry, www.rsc.org

and behaviour are obtained, maintained and delivered also form important aspects of this chapter.

1.1.1 Forensic Science—A Definition

If one were to ask one hundred forensic scientists to define forensic science it is possible that one would receive one hundred different definitions, but it might be expected that among these the terms 'science' and 'legal process' or 'law' would dominate. This would rightly refer to their work at a scene or in a specialist unit or facility, such as a laboratory within the criminal justice system (CJS). Therefore, a high-level working definition is 'forensic science is science used for the purpose of the law'. Consequently, any subject or topic carried out using a scientific method for the resolution of legal disputes is considered forensic science. This broad definition covers criminal prosecutions in the widest sense including consumer and environmental protection, health and safety at work, and civil proceedings such as breach of contract and negligence. It also covers crime against both humans and non-humans, although the focus is often on the former. When the first UK Home Office Forensic Science Regulator (FSR) was appointed in 2008 the definition was reviewed and defined as "forensic science is any scientific and technical knowledge that is applied to the investigation of a crime and the evaluation of evidence to assist courts in resolving questions of fact in court". The current FSR, in his foreword for the 2022 Quality Assurance Agency (QAA) Benchmark statement for forensic science explained forensic science as "a critical part of the investigation of crime and the administration of justice, not only to identify offenders and provide expert evidence to the courts but it is one of the strongest safeguards against false allegation and wrongful conviction." However, the application and delivery of forensic science is always changing such that the term 'forensic practice', 'forensic science practice' and 'forensic science activities' are now being applied more widely to the use of science in the investigation of crime by the police and by the courts as evidence in resolving an issue in any subsequent trial. This practice will therefore typically require an individual or individuals to be apprehended as those responsible for the incident. Consequently, the definitions of forensic science will adapt to the times but still have the foundation of operating within a legal framework. During 2015 the Chief Scientist, Sir Mark Walport, published his second annual report entitled *Forensic Science and beyond: Authenticity, Provenance and*

Assurance. Interestingly, this report attempted to set forensic science in a much wider context rather than the traditional definition of the application of science within a legal framework.

This widening of the definition follows a general trend. In 2022 the Sydney declaration delivered a new definition, which interestingly excludes any specific reference to the word 'forensic' (relating to courts of law) and provides seven supporting principles for forensic science. Instead, forensic science is defined as a case-based (or multi case-based) research-orientated endeavour using the principles of science to study and understand traces – the remnants of past activities (such as an individuals' presence and actions) – through their detection, recognition, examination and interpretation to understand anomalous events of public interest *e.g.* crimes, litigations, security incidents). Regardless of the exact definition, the practice of forensic science will continue to evolve as part of some form of investigation to support justice.

The term 'forensic scientist' however is generally accepted as meaning someone who works in a laboratory or specialist facility carrying out examinations on various materials who ultimately can appear as an expert witness in a court of law. These materials typically being those recovered from a crime scene (a physical location or individual for example), or something relating to an investigation. The forensic scientist can also be someone who attends crime scenes as a specialist, often in more difficult and high profile cases, as part of a team of forensic practitioners. The term 'forensic practitioner' is therefore used as an alternative to forensic scientist and may be considered a more inclusive term to encompass all those involved within the investigative process from the crime scene to the court room.

Regardless of terminology, all those involved in the investigative process in England and Wales within the legal framework are subject to the Criminal Procedural Rules (CPR) and specifically Parts 19, 31 and 33 of those rules (these rules are also broadly adopted in Scotland and Northern Ireland by forensic scientists). The CPR and obligations of those who may ultimately be involved in disclosing and giving written or oral evidence in a court of law are covered in more detail in subsequent chapters. Those governed by the CPR include crime scene examiners, digital forensic examiners, fingerprint officers, and others as core within the investigative process, remembering that there are other equally important specialisms and experts available to advise investigators and courts such as forensic medical examiners, custody retention officers, *etc.* With the advance of digital technologies, those investigating the specialised area of 'digital forensics and cybercrime' are an ever-growing expertise. In recognition of the growing digital/cyber area the UK Government delivered the National Cyber Security Strategy 2016–2021. As a result of this strategy the UK Cyber Security Council (CSC) was created as an umbrella body for existing professional organisations in support of the challenges of cyber security and in 2021 the CSC was granted charitable status.

In the mind of the public, confusion can also exist between forensic scientists and those involved in forensic medicine (the latter sometimes being referred to as legal medicine). The forensic scientist, as defined above, can be involved in all types of criminal investigation but forensic medics in general restrict their activities to criminal and civil cases where a human body is involved. These are nearly always serious cases such as murder or rape and other sexual acts, and will require the participation of pathologists, police surgeons, forensic medical examiners or Sexual Assault Referral Centres (SARCs).

Although the term forensic practitioner seems to be more acceptable for the courts, more specific terms are still used and often preferred within the forensic science community. This lends itself to a much wider definition of forensic practice to incorporate the recovery, examination, and analysis of materials as well as incorporating the various technological changes all within the court environment. This was reflected in the previous definitions given by the FSR and others. However, the narrower definition is implied in the title of this book and the following chapters, which discuss the use of science in the investigation of offences against organisations, the individual and property including cyber-attacks, murder, violent assault, robbery, arson, breaking and entering, fraud, motoring offences, illicit drugs, poisoning, *etc.* This chapter will not attempt to resolve the ongoing debate of whether forensic science is a science, an art, both or neither and this chapter will not cover the probative value of contact trace materials such as fingerprints, footwear, tool marks or physical fits in the context of establishing a common source.

1.1.2 Historical Background

The origins of forensic science can be traced back to the 6th century, with legal medicine being practiced by the Chinese. During the next ten centuries, advances in both medical and scientific knowledge contributed to a considerable increase in the use of medical evidence in courts. Other types of scientific evidence did not start to evolve until the 18th and 19th centuries, a period during which much of our modern-day knowledge of chemistry was just starting to be developed. Toxicology, the study of poisons, emerged as one of the new forensic disciplines, and was highlighted by the work of Orfila in 1840 with his investigation into the death of a Frenchman, Monsieur Lafarge. Following examination of the internal organs from the exhumed body, Orfila testified on the basis of chemical tests that these contained arsenic, which was not a contamination from his laboratory or the cemetery earth. This evidence subsequently resulted in Madame Lafarge being charged with the murder of her husband, but more importantly raised the significant problem of contamination. As our technology advances and deals with smaller and smaller traces, contamination is an ongoing concern for any forensic scientist whether they are recovering, examining, analysing various materials or interpreting their findings. It is highly significant that the oral evidence to the court noted that the arsenic was related to the body and not the surrounding ground or laboratory. This could be argued as a form of quality assurance with an early approach to the evaluation and interpretation of findings with given alternatives *i.e.* the arsenic found in the body could have come from another source, for example the background level of arsenic in soil.

During the latter part of the 19th century there was also considerable interest in trying to identify a person, which today would be encapsulated within the term biometrics. One approach, studied by Alphonse Bertillon, was to record and compare facial and limb measurements from individuals. This proved to be unsuccessful because of the difficulty of obtaining accurate measurements. However, it was the first recorded attempt in a criminal investigation to use a classification system based on scientific measurement. Interestingly, and in accord with this principle, forensic scientists today use the results from a combination of analytical measurements to discriminate between groups or to compare samples. Hence, modern approaches have resulted in greater success using CCTV, speech recognition and facial recognition. A more successful development in human identification was to come from fingerprint examinations. Although Bertillon is reported to have used fingermarks/latent fingerprints (latent inferring something that cannot easily be seen by the naked eye) from a crime scene to solve a case, it was Sir William Herschel, a British civil servant in India, and Henry Faulds who were credited with performing most of the early investigations. Faulds, a Scottish physician, is also accredited with establishing the fact that fingerprints remain unchanged throughout an individual's life. However, it was not until 1901 when Sir Edward Henry devised a fingerprint classification scheme for cataloguing and retrieving prints, that the full potential of personal identification through fingerprint evidence could be used in forensic investigations.

The development in the evaluation and comparison of fingerprints is much debated and will no doubt continue. Within the UK, a specific number of points is no longer required for a comparison to result in an identification. This zero-point criteria has placed a greater individual responsibility on fingerprint officers in exercising their evaluation and interpretation. Probably the biggest UK inquiry into fingermark examination and interpretation was as the result of an initial prosecution, which was later over-turned during the Scottish Fingerprint Inquiry, which issued its report in 2011. However, broader inquiries into the scientific validity of forensic science have been undertaken in the United States of America (USA), in 2009 by the National Academy of Sciences and in 2016 by The President's Council of Advisors on Science and Technology (PCAST). Despite establishing the Organization of Scientific Area Committees for Forensic Science (OSAC) in 2014, who began to address the lack of discipline-specific standards in forensic science, the PCAST report continued to make strong recommendations for strengthening the robustness and validity of feature-comparison disciplines, such as latent fingermarks. Whilst these reports and their recommendations received strong criticism and objection from practicing forensic scientists across the globe, they have resulted in significant research growth, aiming to defend, scientifically underpin and quantify the error associated with the methods used in the pursuit of justice. In addition, research has started to evaluate the consistency and depth of reporting practices within and between forensic disciplines, which has further evidenced the need for change. Examples include articles published in *Science & Justice* by Bali *et al.* in 2020 and Sunde in 2021. Most recently (2022), Needham and colleagues demonstrated that fingerprint officers report more consistently when using a prompted approach such as an analysis form. However, forensic scientists and researchers such as Morrison (2022) highlight the need to move towards developing more objective systems that can compare and evaluate forensic evidence using machine learning and sufficiently large, relevant training populations; an example was published by Basu and colleagues in 2022 in the field of forensic firearm examination. The key to evaluation and interpretation being the need to have a balanced approach (with alternative propositions) and considering the likelihood of each of these alternatives to calculate the overall probability for the hypothesis (likelihood ratios).

Body fluid samples have also been found to contain information that can help to identify an individual. The progress made in this area has been dramatic and major advances have occurred within the past thirty years. Now we simply refer to 'DNA profiling' to cover the examination and amplification of the DNA from a variety of body fluids and body parts. However, until 1900 it was impossible to determine if a blood sample or stain was of human or animal origin, or to classify human blood into the four main groups: A, B, AB and O or various enzyme systems. These categorisation methods have now been superseded by the ability to differentiate between species and individuals using their DNA. Since the discovery of human DNA fingerprinting in 1985, research into the recovery, extraction and amplification of material to produce a DNA profile has seen massive advances in sensitivity and has had an instrumental impact on how forensic science has been used to support the investigation and prosecution of crime on a global scale. For example, DNA analysis can now differentiate between identical twins and can be used to identify human trafficking routes. Originally enough blood to cover a 2p coin was required to develop a DNA profile, but now DNA can be recovered by swabbing or extracting an area that has no visible staining but is thought to have been 'touched' and thus contain trace levels *i.e.* minute quantities of material. Hence the terms 'Low template DNA' and 'touch DNA'. However, the term 'trace DNA' is now deemed more appropriate to imply nano or pico amounts of DNA. As we no longer need to be able to see the (body fluid) stain by eye, it has been increasingly important to explain how that DNA came to be in that location, thereby considering where and when the contact was made and the surrounding circumstances.

In connection with 'contact', it is Edmund Locard who is attributed to the important basic overarching principle and succinct phrase, 'every contact leaves a trace'. Whilst this phrase is well known within the forensic science community, it does not fully encapsulate the detail provided within the relevant passage from his 1920 publication 'L'enquête criminelle et les methodes scientifique', which according to Roux and colleagues translates as:

> "The truth is that no one can act with the intensity required for criminal activity without leaving multiple signs of his/her passage (…). The types of evidence I want to show here are of two kinds: sometimes the criminal has left signs of their presence at the scene, sometimes through inverse action he/she has taken with them on their body or their clothes signs of their presence or of their actions. Left or taken, these marks are of extremely different sorts."

Although the examination of fingermarks or body fluids, which might be present in only trace amounts, may directly implicate an association of particular person with a victim or a crime scene, other types of trace evidence such as glass, paint, fire accelerants, gunshot residues, drugs, fibres, hair, pollen, digital evidence, *etc.*, can provide links that establish contact between objects and/or people involved with a victim or present at a crime scene.

It is interesting that today and noted above, the technology and sensitivity to detect smaller and smaller quantities, often sub-nanogram, does not necessarily implicate an individual but triggers the question as to how the traces of that individual got to where it was recovered from. Further it often raises the questions of secondary and tertiary transfer between people or objects, and even whether transfer could have occurred through the air. The ability to now analyse such a variety of materials stems from technological advances that have occurred particularly in the past thirty years. Many of the analytical techniques that have been devised offer impressive sensitivity and permit examination of minute quantities (traces) of material which cannot be observed directly by the human eye. To provide some indication of the amount of material being examined in these trace samples, think of a grain of sugar. This can be seen without any difficulty and weighs about 1 milligram (1 mg), (*i.e.* one thousandth of a gram or 1×10^{-3} g). Now consider one millionth of this quantity, which is 1 nanogram (1 ng or 1×10^{-9} g). This amount of sugar cannot be seen, but quantities as small as this can be detected by many analytical techniques. Even lower detection limits (picograms or 1×10^{-12} g) can be obtained routinely with some instrumental methods and techniques. Although this is beneficial, extreme caution is required in the evaluation and interpretation at every stage of any investigation and its subsequent analysis to ensure a result is genuine and not due to contamination or any other artefact. The correct interpretation of a result can then be used to advise the investigator and support the court of law to administer justice. An incorrect result or interpretation may lead to an unresolved case, no conviction or a miscarriage of justice.

Rapid developments in computer technology have also played an important role in the advancement of forensic science and practice. Apart from their use in manufacturing instruments and producing analytical data, computers permit the storage of very large amounts of information that can be searched very quickly. With increased computer capacity has come the establishment of local, national and international databases. For example, DNA (recovered from body fluids and biological materials), fingermarks/prints and footwear marks, *etc.* The purpose of these databases is to help in the identification of an individual or items associated with an individual as well as supporting the overall evaluation of findings. Databases can save a tremendous amount of time and effort in an investigation and are beneficial both to the police in following their enquiries and to forensic scientists identifying emerging trends and providing evaluative evidence and information to the courts. However welcome databases may be to any investigation, their use needs to be balanced against ethical issues surrounding how the information is gathered and what consents have been provided as well as consideration of individuals' rights. For example, the retention of DNA profiles of individuals not convicted of a crime was debated in the European court, resulting in a review of how long samples and profiles could be retained on the national DNA database (NDNAD). Other forensic disciplines, such as fingerprints and firearms, also have national databases in the UK and elsewhere, known as IDENT1 (or AFIS outside of the UK) and Integrated Ballistics Identification System (IBIS) respectively. Such databases can facilitate more effective intelligence gathering and the investigation of crime on a potentially global scale.

There are other interesting debates in relation to databases including who creates and owns a database, who contributes, what information is required to be placed within the datasets of the database, who cleans and curates the data, how is data stored and transported and with whom can this information be shared? Further, how useful will the data be to an investigator, evaluator or interpreter of the information? This is particularly a challenge in England and Wales where forensic science provision is currently split between numerous public and private providers, and the forensic science marketplace is extensive and likely unsustainable, making the provision and implementation of national databases in this jurisdiction for areas such as footwear, fibres and drugs very difficult. In addition, access to and usage of data stored by private companies in England and Wales and internationally may create further challenges. This is not necessarily the case in Scotland or Northern Ireland where, in each jurisdiction, there is a single forensic science provider funded by the devolved governments. Concerns have increasingly been raised about the acquisition, use and destruction of biometric data within the CJS including, more recently, the use of genealogy data to support criminal investigations. As a result, in 2018 the Home Office Biometrics Strategy was published and later the Home Office Biometrics and Forensics Ethics Group and Home Office Age Estimation Science

Advisory Committee were formed. Similarly, a Biometrics Commissioner was appointed in Scotland following the passing of the Scottish Biometrics Commissioner Act 2020 in which a Code of Practice was created to promote the adoption of appropriate legal and ethical practices relating to biometric data, its use and retention.

As previously mentioned, obvious advances in technology relate to the area of telecommunication. With increasing use of personal computers and digital devices, fixed and mobile, a whole new, ever-growing area of forensic science has emerged, which will be covered in greater detail in subsequent chapters.

1.1.3 Forensic Science in England and Wales

Before specialist laboratories were established, police forces in many parts of the world relied on scientific assistance from people who, through their occupations, were able to provide the expertise required. The expertise being established by having a specific detailed knowledge greater than what one would expect a random person to have. Without a centralised system, knowing whom to approach was a problem for investigators. This resulted initially in the creation of formalised institutions which were almost invariably established as parts of universities or hospitals. Over time, these were restricted to examining and providing expertise in a limited number of forensic science disciplines. As a result, police forces took the step of developing their own forensic science laboratories. Europe took the lead in this development, with the first police forensic laboratory being opened in 1910 in Lyons, France. Thereafter, police laboratories appeared in Germany (Dresden, 1915), Austria (Vienna, 1923) and other countries including Holland, Finland and Sweden, with these last three all coming into service in 1925. It was not until 1923 that the Los Angeles Police Department set up its own forensic science laboratory in the USA. The reason for this change was the failure to obtain an indictment in a case as a result of improper handling of evidence before laboratory examination. Many other police departments across America followed this lead, with the FBI (Federal Bureau of Investigation) laboratory opening in 1932. Interestingly, the first police forensic science laboratory in the UK was not established until 1935, when the Metropolitan Police Laboratory sited at Hendon was opened. How this laboratory started is a fascinating history. It all arose from the unofficial efforts of a constable, Cyril Cuthbertson, who was interested in medicine and criminalistics (in the UK this term is usually associated with the examination of physical evidence such as footwear marks, but in the USA it has a much wider meaning and covers most of the activities undertaken in a forensic science laboratory) and became involved in applying scientific tests that helped his police colleagues in their investigations. Following his examination of a document and his attendance at court as a witness, praise for his testimony and skills soon filtered back to Scotland Yard. The Police Commissioner, Lord Trenchard, took a considerable interest in this matter as he could see the benefits of a laboratory dedicated to his police force. As a consequence, over a period of time, he persistently engaged the Home Office over this matter and this eventually paid off. The success of this laboratory resulted in the Home Office sanctioning the development of their own forensic laboratories and a research laboratory, under the banner of the Home Office Forensic Science Service (HOFSS), providing regional laboratories for police forces in all areas of England and Wales. These laboratories were all financed from central and local government funds until 1991 when the HOFSS became The Forensic Science Service (FSS), an executive agency of the Home Office. Executive Agencies were created to try and operate at 'arm's length' from the government with the longer view of being self-financed and separate from government. In 1996 the Metropolitan Police Forensic Science Laboratory in London became part of the FSS.

Wherever possible, facilities were provided locally but the corporate FSS structure allowed the concentration of specialist expertise in particular laboratories so that a comprehensive service was available across England and Wales. Research for the FSS was carried out at one of the Birmingham sites. In 1999, with the introduction of agency status came the ability of the FSS to charge 'customers' (primarily but not exclusively the police) for the facilities offered on a contract, case by case, item by item, or hourly rate basis depending on the circumstances. Hence agency status for the FSS permitted the provision of services to any customer in the UK or overseas. These facilities were also available to the defence in criminal cases. Where work was performed for both prosecution and defence the work from each would be conducted at different laboratories and client confidentiality would be maintained. Other agencies which were formerly part of government departments and offered forensic science facilities were the Laboratory of the Government Chemist (LGC), particularly in the area of drugs and documents, and the Defence Science and Technology Laboratory (DSTL), formerly the Defence Evaluation and Research Agency, operating under the Forensic Explosives Laboratory (FEL) in respect of explosives.

With the closure of the FSS in 2012 the primary customer, the police, needed to ensure delivery of scientific evidence across the whole investigative process. This transition was supported by a program called Transforming Forensics which was police led and funded by the Home Office. This support continues through a group called the Forensic Capability Network (FCN), which emerged in 2020. Other significant changes arose from the government's desire to level the playing field and not have it dominated by any one provider such as the FSS. Additional companies were formed by amalgamation, such as LGC and Forensic Alliance to become LGC Forensics (now Eurofins), or the setting up new larger companies, such as Key Forensic Services Limited and Cellmark, or other smaller companies and sole traders. Consequently, many of the bigger organisations came together in 2013 to form a body called The Association of Forensic Science Providers (AFSP), which aims to be an independent representative body facilitating the effective delivery of justice and promoting public confidence in forensic science (AFSP also had representation from Scotland, Northern Ireland and the Republic of Ireland forensic science laboratories). In addition, the number of non-commercial police in-house services began to rise. These major changes resulted in a competitive market for the provision of scientific services to the police and the justice system in England and Wales where the provision of services being contracted by means of a rather complex procurement process. Interestingly, England and Wales is often referred to as having a fully commercial forensic science provision. However, the interpretation and significance of this depends on one's definition of forensic science and its practice.

There are also hidden dangers in a totally 'privatised' forensic science service. For example, it could be argued that commercial pressures and competition could lead to compromised standards. Constraints on budgets could also restrict the amount of work requested, the material submitted, the analytical work to be performed and the timeliness of reporting results, not to mention how research is funded in a relatively immature market. Research and development is the life blood of any developing industry and is particularly important within forensic science to keep at the cutting edge and in front of the perpetrators of crime. The danger with this scenario is that these restrictions could prevent the forensic scientist from providing 'best evidence' based on a holistic approach to the investigation and reach a conclusion, which might either provide a court with stronger evidence to support a prosecution, or equally show that the accused could not have perpetrated a criminal act.

Therefore, it is essential that these dangers are identified and appropriate controls are put into place. The key overriding philosophy is to provide scientific services from reliable witnesses, which subsequently provides assurance to the justice system and confidence to the public. The significance here from the courts' perspective being the word 'reliable'. Again, this can be followed up with the CPR as noted earlier.

In England and Wales, crime scene investigation, fingerprint recovery and enhancement are primarily undertaken within police forces *i.e.* public sector. A considerable amount of other forensic examinations, footwear, digital, *etc.* are also carried out within police forces. The commercial providers supply wide ranging forensic science services including drug analysis and identification, DNA profiling and other areas of criminalistics involving various analytical procedures and contact traces such as glass, fibres, fire investigation, *etc.* Therefore, considerable debate surrounds the totality of commercialisation of forensic provision within England and Wales – as there remains clearly a public and private sector provision.

It seems inevitable for there to be a continued debate regarding public and private provision of forensic science services in England and Wales. Increasing areas of forensic science work is performed within police forces with the expectation that the other commercial organisations will take up some of the slack. Crime scene and fingerprint/fingermark comparisons, have always been undertaken within the police forces in England and Wales. However, whilst digital forensic units are now commonplace in police forces, numerous private companies also exist. In addition, whilst NABIS (National Ballistics Intelligence Service) was set up as an independent national policing organisation to investigate gun crime in 2008, it is hosted by West Midlands Police in England and collectively funded by all UK police forces and other partner law enforcement agencies. While forensic science provision in England and Wales is largely provided by commercial providers and police forces, there will always be a need for some private input depending on the specific requirements of an investigation, especially for niche areas such as speaker recognition, microbes and pollen.

Since the demise of the Council for Registration of Forensic Practitioners (CRFP) and the FSS coupled with all the changes within public, private, learned society and professional bodies, there has never been a clearly communicated strategy for forensic science in England and Wales, while the devolved Nations (Scotland and Northern Ireland) have fared better. In England and Wales, this lack of an overarching strategy and vision hampers the totally integrated investigative process from scene to court. In 2015, the Home Office delivered its strategy for forensic science, in which it referred to a national approach to forensic science delivery in the criminal justice sector. However, in 2019 the House of Lords Science and Technology Committee published its review of forensic science in England and Wales which continued to voice concerns for the sustainability of forensic science in this jurisdiction. Following this review, the FCN was established in 2020 to provide national support for forensic science activities taking place within police forces in England and Wales. Although the government issued its digital forensic strategy in 2020, the Government's Response to the House of Lords Science and Technology Committee's report on forensic science in 2021 continued to acknowledge that more needed to be done in England and Wales to support the sustainability and viability of forensic science provision. Supporting this endeavour in England is the work provided by the Forensic Science Data Laboratory and the Aston Institute for Forensic Linguistics, which were established in 2019 with grant funding from UK Research and Innovation (UKRI) to develop empirically validated methods for evaluating forensic evidence. Therefore, there have been some efforts to increase funding for forensic science activities. In addition, the National Police Chief's Council appointed its first Police Chief Scientific Advisor, Professor Paul Taylor, who aims to keep UK policing at the forefront of best practice and connect policing to expertise in science and technology.

1.1.4 Forensic Science in Scotland and Northern Ireland

While there is still a desire and commitment to achieving an integrated approach to forensic science in the UK, the forensic science provision is separately managed within Northern Ireland and Scotland where justice is a devolved responsibility. Differences in their governance and legal systems means their forensic science provisions are currently publicly funded through government agencies. In Northern Ireland similarly, most forensic science services are provided by Forensic Science Northern Ireland (FSNI) with some provided by the Police Service of Northern Ireland (PSNI).

In Scotland, forensic science facilities were originally provided by independent laboratories in Aberdeen, Dundee, Edinburgh, and Glasgow, each associated with, but independent of, specific police forces. In 2007 the Scottish Police Services Authority (SPSA) was created to provide a corporate service to the police forces for Scotland and effectively created one forensic science provision for the country providing a more integrated scientific service to the police. SPSA was replaced, in 2013, by a single agency the Scottish Police Authority (SPA) with responsibilities for Police Scotland and an independent forensic science service. In 2016 the Leverhulme Research Centre for Forensic Science was established in Scotland at the University of Dundee with grant funding from the Leverhulme Trust and with the strategic aim to support the evidence presented in the Courts with robust scientific data.

1.1.5 Other Forensic Science Provision in the UK

Although the laboratories and organisations referred to above can generally be regarded as the 'official' ones, there are a wide range of practitioners and practices throughout the UK delivering an independent forensic service to clients. These include university departments, public analysts, fire and rescue services, civil investigation companies, large and small practices and individual practitioners, who are collectively referred to as either small and medium enterprises (SME) or sole traders. Although these may undertake prosecution work, they have a particular role in working with lawyers retained by a defendant in a criminal case to explore the strengths and weaknesses of scientific evidence tendered by the prosecution. This may include the laboratory examination of original or new material in a case, but it will usually involve an evaluation of the results obtained by the original scientist(s) and the interpretation offered. The latter may require modification in the light of further information provided by the client or discovered by the retained expert. Funding for such defence examinations is generally sought from the legal aid board, although their resources have been declining over recent years. Whether a public (*e.g.* police force or government funded forensic science provider) or commercial organisation, all those practicing forensic science should be seeking appropriate internationally recognised validation of their quality control and management practices. Across the forensic science domain this usually means gaining accreditation of their practices through ISO 17025 (for laboratory practice) and ISO 17020 (for scene practice) accreditation standards, regardless of their size (see Section 1.4).

The introduction of charging for scientific services has therefore meant that private laboratories and other institutions are all competing for custom from the police or in offering a service for the defence. Unfortunately, whilst many of these laboratories had sought quality control and accreditation of their procedures and facilities, there was no recognised system of accreditation or regulation of the forensic science profession in any part of the UK. This meant that any organisation could offer and supply forensic science services whether or not it had the technical competence and experience. This position changed in 2008 with the introduction of a forensic science regulator (FSR) and a tendering process (in England and Wales) for scientific services.

1.1.6 Development in Crime Scene Investigation

Another historical development in forensic science was the introduction of civilian scene of crimes officers (SOCOs) into police forces. SOCOs (also known as crime scene examiners, (CSE), crime scene investigators (CSI) or forensic investigators) carry out the searching of crime scenes resulting in the collection and packaging of various materials that may provide potential directions for the investigation. Contrary to common belief, it is nowadays quite rare in England and Wales for a forensic laboratory-based scientist to attend a scene (although they attend more regularly in Scotland and Northern Ireland) because the introduction of CSEs into the investigative process was intended to reduce the amount of time forensic scientists were being called away from their laboratory work.

Over the years the importance of CSE/CSI and associated development of forensic strategies at crime scenes together with the need for collection, packaging and transportation of material of potential evidential value has been increasingly recognised. In earlier days this would have been carried out by a detective or a forensic scientist, but it is now usually performed by specialists who have received relevant training in all aspects of crime scene examination including latent fingermarks, evidential traces and recording practices, such as photography. In Scotland and Northern Ireland CSE/CSI training is provided nationally by SPA forensic services and PSNI, respectively. In England and Wales the College of Policing provide a national training programme for delivery by licensed forces; forensic courses are no longer delivered at their national facility at Harperley Hall, Durham. Such professionalism should ensure that the integrity of items received by the forensic scientist for examination will not be disputed. Another consequence of this civilianisation was the introduction of Scientific Support Managers (SSM) to each police force. Their responsibility is to oversee the management of forensic scientific services for the police force in terms of fingermark identification, photography and forensic science work. This has extended to include various technology advances such as laser scanning, digital work and CCTV. The overall role of the SSM depends on the size of the individual police force and, latterly, where various neighbouring police forces have been amalgamated, the individual SSM roles have been replaced with one overall Scientific Support Director or Regional Director.

Over the years there have been various qualifications within the forensic science arena. *Via* Harperley Hall, the police offered various diplomas and then foundation degrees in crime scene and fingerprints. There were also several diplomas offered by the Forensic Science Society, FSSoc, (now the Chartered Society of Forensic Science (CSFS)). However, these are no longer available.

Academically, the first forensic science qualification in the UK was a Master of Science degree in forensic science provided by the University of Strathclyde in Scotland in 1967. This was followed by a Master of Science in forensic science at Kings College London established in 1985. These qualifications were followed by a wave of undergraduate forensic degrees beginning in around 2000 and master degrees developing around 2004, which unfortunately varied in quality resulting in various critical comments from the practitioner communities. These comments were taken on board by the FSSoc, which developed an educational accreditation scheme to ensure forensic science and academic quality. In this scheme, universities apply against component standards (forensic science, forensic investigation, anthropology, archaeology and digital forensics) to achieve either educational accreditation (full undergraduate and postgraduate degrees), recognition (less than full degrees) or endorsement (essentially post-16 education or foundation year).

The UK Quality Code for Higher Education (Quality Code) sets out the expectations and practices that all UK higher education are required to meet. The Quality Assurance Agency for higher education (QAA) sits alongside the Quality Code to help providers develop courses and refine curricula. The QAA have recognised forensic science as a subject in its own right and a benchmark statement on forensic science was produced for undergraduate and postgraduate awards incorporating the CSFS component standards and this statement underwent significant revision in 2022 to reflect the recent changes.

With the changing marketplace for forensic science, the commercialisation and several high profile miscarriages of justice there were calls for greater regulation. The first full-time permanent post of the Forensic Science Regulator (FSR) was created by the UK government in 2008 following a recommendation from the House of Commons Science and Technology Committee. The FSR sits within the Home Office and operates independently to ensure that quality standards apply across all forensic science services to the criminal justice service in England and Wales and are adopted (but are not mandated) by Scotland and Northern Ireland. This involves the setting, maintaining and monitoring of quality standards. Consequently, the FSR makes an important contribution to the continuing development and delivery of forensic science practice to the CJS. This is achieved *via* compliance with the Regulator's Code of Practice (often referred to as the code). Initially this compliance was voluntary, but statutory powers for the FSR to enforce scientific quality standards within forensic science have been enacted by government through the Forensic Science Regulator Act 2021.

The FSR is supported by an advisory council and numerous specialist groups, who develop and deliver the standards to the Regulator for ratification. Since 2008 there have been three Regulators; Mr Andrew Rennison (2008–2014), Dr Gill Tully (2014–2021) and Mr Gary Pugh (2021–present). The role to set, maintain and monitor quality standards within the investigative process remains at the heart of the role, although with the number of forensic providers increasing and expanding especially within police forces this makes the FSR role even more critical.

Historically, the forensic science work performed in the UK has for many years been regarded very highly throughout the world for its standards, innovation and integrity. However, the implications of the change in marketplace, number of forensic providers and funding for research has had some impact on its ability to deliver an optimal service in England and Wales specifically. It is vital that sufficient monitoring and review is provided with resulting actions taken where necessary, to continue to preserve the global reputation of forensic science in England and Wales and, more importantly, to ensure that it is not responsible for miscarriages of justice. In 2023, an all party parliamentary group (APPG) Westminster commission was established to investigate forensic science in England and Wales with a specific focus on the risk posed by the current provision in this jurisdiction to contribute to miscarriages of justice, and is expected to report in 2025 or 2026.

As an overview to this section, Table 1.1 details some key dates of change that have taken place within the UK landscape since the

Table 1.1 A list of key dates of change within UK forensic science provision.

Year	Key change
1991	HOFSS becomes Executive Agency – charging
1994	Using Forensic Science Effectively – FSS/ACPO
1995	NDNAD implemented
1996	FSS and MPFSL merge LGC privatised Forensic Alliance Ltd.
1998	CRFP established
1999	FSS becomes Trading fund
2003	McFarland Report on FSS and two main recommendations (1) Separation of NDNAD from FSS (2) Creation of a Government Owned Company (GovCo) and then a Public Private Partnership (PPP)
2004	FSSoc becomes Professional body Skills for Justice established
2005	LGC acquire Forensic Alliance – LGC Forensics FSS becomes GovCo
2006	Orchid Forensics (Cellmark)
2007	Interim and then permanent Forensic Science Regulator
2008	First FSR appointed NABIS established
2009	CRFP goes into demise
2011	Scottish Fingerprint Inquiry
2012	Closure of FSS (select committee) and NPIA Requirement for ISO 17020/17025 accreditation initiated New Professional Body – College of Policing created
2013	Royal Charter for FSSoc to 'The Chartered Society of Forensic Sciences' (CSFS)
2014	Launch of CSFS OSAC established New FSR appointed
2016	Home Office issued 'Strategy for Forensic Science' PCAST Report published
2017	House of Lords Science and Technology Select Committee inquiry established
2018	SPA published Forensic 2026 Strategy Home Office Biometrics Strategy published
2019	House of Lords Science and Technology Select Committee report – Forensic science and the criminal justice system: a blueprint for change published
2020	Transforming Forensics published Digital Forensic Science Strategy FCN established Scottish Biometrics Commissioner Act 2020 with Code of Practice
2021	New National Policing Chief Scientific Advisor appointed Sydney declaration published online New FSR appointed FSR Act 2021
2023	FSR Code of Practice approved by Parliament FCN Forensic Science Data Service possibly launched

early 90s. However, another change worthy of highlighting has been the increased partnership and collaboration between academia, forensic science providers and forensic science units within police forces and industry. While some universities and police forces have worked and/or collaborated together informally for many years the PCAST's call for greater research underpinning forensic science was a key driver for the formalisation of some of these partnerships. The Staffordshire Forensic Partnership, Lancashire Forensic Science Academy and Forensic Innovation Centre are all good examples. The Leverhulme Research Centre for Forensic Science in Scotland as part of its work, catalysed a partnership between the National Academies of Science (Royal Society and Royal Society of Edinburgh) and the senior judiciary across the UK to develop judicial primers. These are texts written in simple prose developed in partnership between the judiciary, forensic scientists and the academic community to provide concise yet accurate explanations of the science behind forensic evidence together with any limitations when applied in casework. A range of judicial primers are freely available from the Royal Society's website and include statistics in legal proceedings, collision investigation, forensic DNA analysis, ballistics, forensic gait analysis, forensic anthropology and fire investigation. Such a collaborative approach is world-leading and a pivotal step forward in connecting science and the law. Both the Royal Society and the Royal Society of Edinburgh also now hold regular events for scientists and the judiciary.

A need was also identified to increase UK practitioners' access to forensic research and in 2020 a proposal was developed by Bolton-King and colleagues to create a new platform to specifically share unpublished (non-peer reviewed) forensic research and journal pre-prints. The collaboration and sharing of forensic science work between academia, forensic science providers and policing has been supported by the FCN for the last few years. With the challenges faced particularly in digital forensics, the FCN have also built upon and expanded the proposed repository and platform, securing Home Office funding in 2023 to explore the commissioning of a new forensic science data service. If launched, this service aims to provide quality-assured and easily accessible data, reports and tools to those working in the sector.

An increase in open access peer reviewed literature across forensic science journals is also providing access to research and data for forensic scientists.

Further to these national efforts, the International Forensic Strategic Alliance (IFSA) established in 2004, supports the connection of the six existing regional forensic science networks and three strategic partners (Interpol, United Nations Office on Drugs and Crime (UNODC) and the Leverhulme Research Centre for Forensic Science), setting minimum required standards for forensic practice on a global level. IFSA also produced a Research and Innovation Position Statement at its meeting in Scotland in 2021, which was leveraged by the European Network of Forensic Science Institutes (ENFSI) to influence Horizon Europe calls and provide funding for forensic science research and innovation across Europe. The IFSA Research and Innovation Position Statement was also used in discussions with UKRI to align funding in the UK to the global direction of travel.

Other significant developments that support the sharing and advancing of forensic science through international collaboration have included:

- Interpol's Forensic Science Managers Symposium publications of peer review papers summarising relevant research within a range of forensic science areas.
- Establishing the Scientific Advisory Board for the International Criminal Court who provide the Office of the Prosecutor with updates and advice on advances in technology, scientific methods and procedures in forensic science.
- Launching the 2030 Agenda for Sustainable Development and implementing 17 Sustainable Development Goals (SDG), for which the UNODC supports Member States in their efforts to make the world safer from drugs and crime.
- Specifically aligning international research outputs and educational learning outcomes to SDG 16 (Peace, Justice and Strong Institutions) and 4 (Quality Education) within forensic science and wider criminal justice sector.
- Increasing input from forensic scientists to support the humanitarian work provided by charities and non-government organisations such as the International Committee of the Red Cross to ensure the protection and dignified treatment of the dead.

1.2 When Is Forensic Science Required?

A police officer investigating an incident will seek clarification of four issues:

(1) Has a crime been committed?
(2) If so, who is responsible?

(3) If the responsible person has been traced, is there enough evidence to charge the person and support a prosecution?
(4) Is the crime connected with any other previous crimes?

This clarification is seldom the sole responsibility of one officer and the investigation and court trial will reveal the involvement of specialist police officers, civilian staff, lawyers and scientists. Overall, forensic practice can be expected to contribute to the clarification of all four issues. One very important attribute of a forensic scientist is their ability to accept that they work for the court of law and consequently their approach should not be biased. In other words, they are able to consider information from various sources, review it and provide an independent unbiased evaluation and interpretation, *i.e.* the information evaluated and interpreted has not been subject to cognitive bias (Section 1.6) in particular.

1.2.1 Has a Crime Been Committed?

In most cases there may be little doubt that a crime has been committed, but there are a number of occasions when only a scientific examination of items can inform the investigator that this is the case. For example, the alleged possession of an illicit drug will require identification of the seized material. Similarly, to support an offence of driving under the influence of drink or drugs, a blood or breath sample taken from a motorist will require an accurate analysis not only to establish that alcohol or a drug is present, but that any alcohol or drug present exceeds a permitted level. The presence of semen on a vaginal swab from an under-age girl is evidence of illegal sexual activity. Similarly, the demonstration of toxic levels of a poison in tissues removed at post-mortem from a body of an individual believed to have died from natural causes will be a strong indication of a crime or suicide. Doubts as to the authenticity of a document may be resolved by scientific examination and provide evidence of fraud.

1.2.2 Who Is Responsible?

If a latent fingermark is developed and recovered from a crime scene and the criminal's fingerprints are already in a database, the person potentially responsible for that crime may soon be identified. Similarly, the existence of a database of DNA profiles may enable initial identification of an offender who has bled at the scene of violence or who has left other body fluids in a sexual assault. Although specific identification of an offender may not be provided by scientific examination, useful investigative leads can be produced which will enable the investigator to close certain lines of enquiry and follow others potentially leading to a perpetrator.

1.2.3 Is the Suspect Responsible?

Irrespective of any support received from the forensic scientist, the usual diligent police investigator often produces a suspect and the investigator will look to the forensic scientist to provide corroborative or exculpatory evidence to enable a charge, or not, to be made, and may assist the court in deciding whether a defendant is guilty or not guilty. The scientific examination will normally be directed towards two aspects:

(1) Examination of material left on the victim or at the crime scene which is characteristic of the suspect.
(2) Examination of the clothing and property of the suspect for the presence of material characteristic of the victim or the crime scene.

1.2.3.1 Materials Characteristic of the Suspect

The biological, physical or chemical characteristics of materials found on the victim or at a crime scene can help to confirm the identity of a suspect and/or provide evidence of their involvement or presence at the crime scene. Blood, semen, saliva, fingerprints, hair and teeth are all characteristics of an individual. Finding fibres from clothing or the characteristic pattern of the soles of shoes worn by a suspect may provide evidence of their involvement, as can materials found that may be associated with their particular occupation. The characteristics of a vehicle used in the crime, such as oil drips, tyre marks or tachograph may indicate where the vehicle had been parked or driven over ground near to or at the crime scene. Paint, glass or plastic from the vehicle after a collision may help to identify the particular vehicle and hence the owner who could become a suspect. The characteristic marks that may arise from weapons, tools or other items used in committing a crime, *e.g.* a knife in a stabbing, a screwdriver used in forcing an entry or a firearm used in a robbery, especially if found on a suspect, could provide further evidence of a suspect's association with a crime. Botanical evidence such as pollen may provide links between crime scenes and suspects; finally, even insects found on decayed bodies can assist the forensic scientist and the courts. Investigation of mobile phones or other digital related software are common and can help show communications which may have taken place between individuals as well as locating where these devices may have been. Clearly it is very unlikely that all these possibilities will be realised in a single case, but knowing the circumstances surrounding a case and taking into account previous experiences, the forensic scientist should be in a position to exploit their skills to the benefit of the investigator and the courts.

1.2.3.2 Materials Characteristic of the Crime Scene or Victim

The crime scene could be an individual, a building or an outdoor area. Any search of a building or outdoor area usually yields materials that are characteristic of the particular location, for example:

(1) Domestic premises—external and internal painting, external and internal glass, furnishings, crockery and glassware, *etc.*
(2) Commercial premises—as for domestic premises, plus process materials.
(3) External scenes such as gardens, waste ground and fields—soil, vegetation and miscellaneous debris.

Where the scene involves a living or deceased victim, biological material and clothing discussed above for the suspect will also apply as will digital devices.

1.3 Duties of the Forensic Scientist

Having established when the service of a forensic scientist might be required, we can now identify their duties as follows:

(1) To attend a crime scene (which may be rarely, depending on jurisdiction), examine material collected or submitted to a forensic science laboratory to provide information previously unknown or to corroborate or exclude information already available.
(2) To provide the results of any examination in a report that will enable the investigator to identify an offender or corroborate/exclude other evidence in order to facilitate the preparation of a case for presentation to a court.

(3) To indicate wherever appropriate, whether there is any association between the marks or materials recovered at the crime scene to those recovered in any previous cases.
(4) To present written and/or verbal evidence, based on evaluation and interpretation of their findings within the context of the given case to a court of law to enable it to deliver justice by reaching an appropriate decision as to guilt or otherwise of a defendant.

Under the adversarial system of trial used in the UK, USA and many other parts of the world the individual forensic scientists may be regarded as, and claimed to be, an independent witness for the court. However, they may not always be regarded in this way by the courts and public. It is therefore essential for forensic scientists to be able to demonstrate competence, impartiality and integrity by attention to issues such as the following:

(1) Forensic scientists should only give evidence on work carried out personally or under their direct supervision and within their area(s) of subject expertise. Although, one of the privileges bestowed on an expert witness is that they can interpret factual evidence given by another witness under oath or affirmation in the light of scientific findings and knowledge.
(2) Where scientific examinations are relied on for legal purposes, the methods used should be based on established and accepted scientific principles, validated and, preferably, published in reputable peer reviewed scientific literature, so that they can be scrutinised by the scientific community at large.

Where the scientific findings require interpretation, the basis of any interpretation should be available to the scientific community.

It is important to recognise that the responsibilities of individual forensic scientists are personal and not corporate. Thus, in giving evidence they are completely and solely responsible for their own experimental results and for the facts and opinions expressed. However, the corporate environment should be a supportive structure providing appropriate training and continual professional development, standardised methods and procedures, evaluation of performance and a quality management system (QMS). Attention to the latter can be a real source of assurance to the individual forensic scientist, the CJS and the public at large. Further, given the corporate environment mentioned above it is critical that the forensic scientists are also empowered to innovate and be creative in order to adapt to changing trace materials and technology.

1.4 Quality in Forensic Science

There are many definitions of quality, but for our purposes, within forensic science, the International Organization for Standardisation (ISO) is appropriate which defines quality as the 'degree to which a set of inherent characteristics fulfils a requirement' (modified from ISO 9000:2015, 3.6.2 in ISO 21043-1:2018).

The ISO 17025 standard was adapted specifically to give guidance to forensic laboratories on both quality management and the technical requirements for the practices and procedures in operation. This standard can be considered the technical complement to ISO 9000. As a result, any organisation that satisfies the requirements of ISO 17025 will also meet the intent of ISO 9000 requirements; the reverse is not true, however. The ultimate 'customer' to be satisfied is the court, and it would expect there to be a total QMS in place that will ensure the integrity of material examined by the scientist, the examination carried out and the reliability of any scientific evidence presented. It is expected that the 17025 standard will primarily cover the laboratory work. However, the investigative process is not restricted to the laboratory and the scene is equally and arguably more important. The relevant ISO standard here is 17020. These standards focus on the procedures and standards at the scene and within the laboratory as outlined later (Section 1.5) and dates for their implementation in England and Wales have been established by the FSR. A parallel argument exists alongside these standards, which are those of the individual practitioner's competency and professional judgement. Many argue that these latter standards are incorporated within the ISO standards, but the question still arises as to how the individual becomes competent with the ability to make professional judgements (Section 1.6).

1.4.1 Quality at the Crime Scene—Laboratory Chain

The quality control system must clearly extend outside the forensic science laboratory environment and places a responsibility on everyone involved in an investigation to maintain a 'chain of custody', from crime scene to court. It is also argued that the standard starts upon initial reporting of the incident. In fact, in terms of the

Table 1.2 The crime scene-forensic science laboratory links in the chain of custody to ensure quality in this procedure.

Link	Category	Comment
1	Preservation of the scene	This can be difficult in the early stages, particularly when injury or hazard is involved but the police must establish access control as soon as possible. Thereafter access must be restricted to those who can make a real contribution to the investigation.
2	Search for material of potential evidential value	This must be systematic with careful records kept of the location of all material collected.
3	Packaging and labelling of collected material	This must ensure that the material arrives at the laboratory as far as possible in the condition in which it is collected and that it can be related to the source.
4	Storage and transmission to the laboratory	Preservation of the condition is a priority *e.g.* refrigeration may be appropriate.

materials used to 'bag and tag' items it is equally important that they have been produced under a QMS to ensure the quality—particularly with items which may contain minute quantities of trace material such as DNA—hence the term 'DNA free' recovery items (*e.g.* swabs) and packaging. A more appropriate expression would be 'chain of integrity', since the court will need to know not only the identity of the links in the chain but also their behaviour, as illustrated in Table 1.2.

Even with this brief consideration it can be noted that apart from knowing the identity of the 'custodians' of items at each stage, the court will need to be assured of their awareness of the consequences of any deficiency in the processes in which they are involved. For this reason, in many investigations most of the process is conducted by specialist personnel with occasional assistance from laboratory-based forensic scientists. Everyone involved needs to protect the items from the twin problems of deterioration and contamination. The latter is vital when a suspect has been arrested, since all possible contact between items from two sources must be prevented with proof that the appropriate actions have been taken.

1.4.2 Laboratory Quality Procedures

Clearly, if the integrity of the articles received by the laboratory, either as a commercial forensic science laboratory or an in-house police scientific laboratory, has been maintained, the responsibility is then transferred to the forensic scientist. The methods used for the examination of various evidential materials are detailed in subsequent chapters, but certain principles apply to all examinations:

(1) Prevention of contamination is a prime requirement, particularly as such small amounts of material can be examined and characterised. The practitioner must be able to demonstrate that the procedures used have prevented the adventitious transfer of evidential material between two sources including themselves.
(2) Security of all items must be assured by recording the names of all individuals having contact with them. This is usually achieved by signing an attached label. Leaving items unattended in the forensic science laboratory must also be avoided.
(3) Careful permanent, contemporaneous records should be kept at each stage of the laboratory examination to avoid any possibility of confusion by assigning results to the wrong item.
(4) All the procedures and methods used by the forensic scientist should be fully documented and any decisions justified. These are often referred to as Standard Operational Procedures (SOPs) and Standard Methods. Forensic science providers should normally have a comprehensive system detailing procedures to be followed and methods to be used. One of the important points here is to recognise that because it's written down it does not need to be followed religiously if there's a better way. One must always be aware of continual improvement and follow the SOP for improvements. However, the final source of assurance is the competence and integrity of individuals.

1.5 Accreditation of Forensic Science Facilities

Over many years there has been an increasing call for third party accreditation of forensic science laboratories, in common with many other scientific laboratories and industrial organisations. In the USA this call has been met by the American Society of Crime Laboratory Directors through their Laboratory Accreditation Board. Under the auspices of this board the organisation, staffing and facilities of a laboratory are subjected to evaluation and on-site inspection before accreditation. A full re-inspection is carried out every 5 years. A good proportion of the many forensic science laboratories in the USA have been accredited by this process, and some in other parts of the world, especially Europe, Asia and Australia through alternative international accreditation bodies which may also have signed up to the ILAC (International Laboratory Accreditation Cooperation) Agreement.

Given the smaller number of laboratories in the UK it has proved more convenient to use a well-established system of accreditation applicable to all laboratories offering testing services to a client. The United Kingdom Accreditation Service (UKAS) is recognised by the government as the body for accrediting all types of laboratories and in this role UKAS established a number of standards, all of which have now been subsumed under the standards previously mentioned; ISO/IEC 17020 (Scene work) and ISO/IEC 17025 (Laboratory work and associated management) and ISO 9000. However, with a truly integrated investigative process it could be argued that one standard covering the whole process would be more appropriate. All of these processes are being internationalised through the ILAC organisation. Clearly, the wider recognition will be of great value to laboratories engaged in work supporting trade across international boundaries. However, given the increasingly international nature of crime, the concept of all forensic laboratories working to a common high standard has its attractions.

The accreditation process involves a series of steps. The applicant organisation, SME or sole trader submits documentation, including its quality manual, to UKAS who then assign a Technical Officer and a Lead Assessor to be responsible for advising whether the forensic provider should be accredited. This advice will be based on a pre-assessment visit for informal discussion and a broad review of the quality system followed by a formal inspection of the provider by an assessment team. At the end of the inspection the team will discuss any non-compliance found and agree the appropriate corrective action. If the assessment team is satisfied with the corrective actions, UKAS will review all the evidence and decide whether to accredit. Following accreditation, the provider will be subjected to regular surveillance and re-assessment visits to ensure that standards are being maintained.

Forensic scientists are required to tackle a wide variety of scientific analytical problems, many of which have no commercial analogue. This means that widely publicised and used methods such as those of the British Standards Institution may not be an option. As previously mentioned (Section 1.1), the call for forensic scientists to develop the necessary degree of objectivity in the method applied to a particular problem requires one to consider the definition of an objective method or test. A former Forensic Science Working Group that took part in formulating standards for forensic science spent much time in defining what was meant by an objective test. Their definition was that an objective test is one which, having been documented and validated, is under control so that it can be demonstrated that all appropriately trained staff will obtain the same results within defined limits.

Objective tests are controlled by:

(1) Documentation of the test.
(2) Validation of the test.
(3) Training and authorisation of staff.
(4) Maintenance of equipment.

and where appropriate by:

(1) Calibration of equipment.
(2) Use of appropriate reference materials.
(3) Provision of guidance for interpretation.
(4) Checking of results.
(5) Testing staff proficiency and continual staff development.
(6) Recording of equipment/test performance and measurement of uncertainty.
(7) Consider latest innovations and technology for future applications.

For a sole trader, SME or large organisation to become accredited the individual forensic scientists performing any method need to be deemed competent and proficient. However, these two terms are closely related, much debated and often confused. In summary, competency refers to an individual's continual ability to successfully carry out a 'task' to a particular standard. Proficiency relates to the high degree of skill and expertise to carry out a task. Both terms involve the application of knowledge and skill, with competency being the essential skills that are continually required for a certain task and proficiency implying the mastery of them. The aspect of personal accountability is further explored in Section 1.6.

1.6 Personal Accountability in Forensic Science

The ultimate role of the forensic scientist is the presentation, either written as a witness statement or report, or verbal direct to the court as a professional or expert witness (skilled witness in Scotland). In fulfilling this role, the witness is completely and solely accountable for the work carried out, experimental results presented and for the facts or opinions expressed. If appearing as an expert witness then it is the court which decides if the witness is to be accepted as an expert and therefore permitted to express an opinion as opposed to the provision of fact. The expert witness is viewed by the court as rather special in that they are permitted to express an opinion. There has been much debate surrounding the role of the expert witness specifically from the law commission. Expert opinions must be justified to the court, often in the face of fierce cross-examination, and the witness cannot shelter behind the laboratory manual or base an opinion on a consensus or majority vote. This requires the witness to be a professional in the best sense of the word: that is, to have an initial developed competence which is continuously maintained and advanced (*i.e.* proficient) together with a powerful sense of balance, honesty and integrity. Clearly an employing organisation has a responsibility in this respect, but there may be additional assurance to the court through qualifications, experiences, expertise and memberships – such as professional body membership. The latter would seek to provide evidence of competence with a code of conduct and advice on professional behaviour. Given the nature of forensic science a range of appropriate professional bodies exist.

During the late 1990s the need for a 'professional body' was recognised and resulted in a register of competent forensic scientists being developed. This was held by the Council for the Registration of Forensic Practitioners (CRFP), an independent regulatory body to promote public confidence in forensic practice in the UK. Initially the introduction of the CRFP was viewed as a major step forward for the forensic science profession in the UK. However, the criteria used for registration were not necessarily 'standards' in the accepted meaning of the word and on 31 March 2009 the CRFP ceased operation, primarily due to lack of funding and unsuccessful attempts to become self-funded.

With the demise of the CRFP, the question was then raised about how to replace it. The FSR issued a consultative paper reviewing the options for the accreditation of forensic scientists. The outcome, although not necessarily covering all forensic scientists within the investigative process, was to have all forensic science laboratory functions accredited to an ISO standard. This would be carried out by UKAS, which would not only cover the practices and procedures but also look at a proportion of individuals' cases, competency, proficiency and training records. However, there still remained a gap to be filled for the independent specialist expert not necessarily employed by a main forensic provider. A further consequence of the

Table 1.3 Professional standards of competence in forensic science.

Unit	Element
1. Prepare to carry out examination	1.1 Determine case requirements 1.2 Establish the integrity of items and samples 1.3 Inspect items and samples submitted for examination
2. Examine items and samples	2.1 Monitor and maintain integrity of items and samples 2.2 Identify and recover potential evidence 2.3 Determine examinations to be undertaken 2.4 Carry out examinations 2.5 Produce laboratory notes and records
3. Undertake specialist scene examination	3.1 Establish the requirements for the investigation 3.2 Prepare to examine the scene of the incident 3.3 Examine the scene of the incident 3.4 Carry out site surveys and tests
4. Interpret findings	4.1 Collate results of examinations 4.2 Interpret examination findings
5. Report findings	5.1 Produce report 5.2 Participate in pre-trial consultation 5.3 Present oral evidence to courts and inquiries

loss of the CRFP was the loss of the register. There are other registers held by various bodies, but the overall criteria to be accepted on the register varies and a national gap remains in this regard.

Another body that was involved with drawing up standards of competence (Table 1.3) for forensic scientists in the UK was SEMTA (Science, Engineering and Manufacturing Technologies, and formerly the Forensic Science Sector Committee of the Science, Technology and Mathematics Council) and Skills for Justice, two of the government Sector Skills Councils. Skills for Justice represents employers in the Justice sector including the Fire and Rescue sector. It was formed in 2004 and has the overarching mission to help organisations and individuals within these sectors to maintain a safe and just society. It does this through workforce skills and influencing policy at all levels to ensure it takes each sector's workforce into account. To date, its greatest impact on the Justice and Fire and Rescue sectors could arguably be its review and 2007 updating of the National Occupational Standards (NOS) for forensic science, which had previously been owned by SEMTA. Skills for Justice in consultation with industry practitioners have reviewed and developed NOS in three areas—Crime

Scene (evidence recovery from the crime, CSI, SOCOs, *etc.*), Forensic Identification (fingerprints, footwear, marks, DNA, *etc.*) and Forensic Laboratories (conducting forensic science laboratory investigations and analysis). The Fire and Rescue Service became a subscriber to Skills for Justice in 2009 and NOS are also available for this area. The NOS specify the standards of performance which practitioners are expected to achieve in their work and the knowledge and skills which they need to perform effectively, *i.e.* they describe competent performance in terms of the outcomes of an individual's work. Skills for Justice remain the custodian of the NOS which are normally reviewed every three years or when the employer instigates a review.

As introduced earlier, it is important to have an overall QMS which can ensure practices and procedures. However, individual competency is integral to a successful profession and Table 1.4 shows how this can be implemented. Having described standards of competence it then becomes necessary to develop a strategy for assessing scientists against

Table 1.4 An illustrative 5 point QMS.

	Type of test	Area of competence tested
1	**A case review.** Between line manager and practitioner to assess what was done by practitioner, decision making processes, what could have been done better, review of forensic strategy, overall outcome of items taken for analysis, *etc.*	Knowledge & understanding
2	**Simulated competence test.** A simulated competence check providing the practitioner with a scenario with known outcomes and based on the relevant National Occupational Standards (NOS). This can include 'blind trials' where known items are entered into the workflow unknown to the practitioner.	Knowledge, understanding & skills
3	**Observation (site visit).** Line manager observes the practitioner undertake an agreed number of tasks to assess competence against agreed NOS.	Knowledge, understanding & skills
4	**Knowledge test.** Practitioner sits a written knowledge test which is marked by a competent peer or line manager to ensure practitioner has knowledge required to perform role competently.	Knowledge & understanding
5	**Dip-sampling.** A set number of 'work products' are sampled randomly every month to ensure consistency in competent delivery of work tasks.	Consistency of performance over time

such standards and such assessment strategies are presently being developed.

To further support the continual development of individuals' professional competence and to promote the development and dissemination of new scientific and investigative methods within the sector, a range of forensic science journals are published internationally. This includes *Science & Justice*, (formerly the *Journal of the Forensic Science Society*, 1960–1994) published by the CSFS. In a similar way to the University of Durham (in collaboration with the predecessor to the College of Policing), the CSFS once offered practitioners professional and postgraduate diplomas. The Society's diplomas also offered a stepping-stone to a master's degree with the option of an additional research project at either the University of Strathclyde or Staffordshire University. Although both the Durham and CSFS diplomas have now ceased, the CSFS introduced Certificates of Professional Competence (CPC) in a growing range of disciplines. Like the previous diplomas, there is a strong element of practical assessment, and the overall model includes the initial application, online questions in the applicant's specialised area as well as questions on the legal framework within which they work. This is followed by a practical assessment based on the typical work carried out by the practitioner and a supporting statement or report. The CSFS gained its Royal Charter in 2013 enabling it to offer 'Chartered' status to its practitioners. The diversity of offerings provided by the CSFS further support the ISO standards in relation to individual competence and professional judgement.

Due to the breadth of disciplines that span forensic science activities, it is important to recognise that professional membership and educational accreditation are also sought from other professional bodies who are actively involved in developing and nationally implementing changes in quality standards, policy and nationally endorsed codes of practice within forensic science. Examples include the Royal Societies of Chemistry and Biology, Royal Anthropological Institute, Chartered Institute for Archaeologists, UK Association of Fire Investigators and the Institute of Fire Engineers. Another Society within the investigative process was The Fingerprint Society (FPS), founded in 1974 as a learned Society for fingerprint examiners. Its aim was to 'advance the study and application of fingerprints and to facilitate the cooperation among persons interested in the field of personal identification'. Unfortunately, the Fingerprint Society proved unviable and merged with the CSFS in 2016.

A final area of interest and importance to forensic science practice over recent years has been that of cognitive bias. This is a situation when you are unaware that you are being influenced in making your decisions such that you may not make a personal, independent, honest decision. As way of example, consider the scenario where a stamped-in engine or chassis mark on a car or motor bike has been deliberately erased. The process to recover the number appears to be successful and some digits are being recovered by chemical treatment. During the quality assurance process you are asked check the digits and whether you can see the '8'? The very fact that the digit '8' was mentioned means you are unduly influenced as it may also be a '3' or a '5' depending on the style of the numbers. In fact, it might not be a number at all and may be a letter. Being provided with some information this can be termed confirmation bias creating a cognitive influence and bias. The forensic scientist must be aware of such situations and ask the 'checker' the appropriate questions, and the 'checker' must also ensure they provide an independent review.

Other forms of bias include contextual bias, meaning that an individual has been influenced by others and therefore their decision may not be fully their own and they do not necessarily appreciate the impact of the other information. This has been investigated extensively by a range of authors across numerus forensic disciplines and can occur at many stages during an investigation. Hence, anyone involved must be fully aware of the impact and minimise bias whilst performing their duties. Some of the authors discuss issues of bias within their own particular discipline in this book. However, for a fuller understanding of this topic, readers are referred to other chapters in this book where the precautions and comments are relevant and transferable to many other forensic disciplines.

1.7 Conclusion

The discussion in this chapter has aimed to provide the reader with a background understanding of forensic science and the role of forensic scientists across the UK. It also covers how they achieve professional status and how the word 'forensic' implies working and interacting with the legal process. Working as a forensic scientist can be physically, emotionally and intellectually demanding, but also very satisfying and personally rewarding. The succeeding chapters will show why this is so.

Further Reading

I. E. Dror, Cognitive neuroscience in forensic science: understanding and utilizing the human element, *Philos. Trans. R. Soc. London, Ser. B*, 2015, **370**(1674), 20140255.

Encyclopedia of Forensic Sciences, ed. M. M. Houck, Elsevier, Amsterdam, 3rd edn, 2022.

G. S. Morrison, Advancing a paradigm shift in evaluation of forensic evidence: The rise of forensic data science, *Forensic Sci. Int. Synergy*, 2022, **5**, 100270.

C. Roux, R. Bucht, F. Crispino, P. De Forest, C. Lennard, P. Margot, M. D. Miranda, N. NicDaeid, O. Ribaux, A. Ross and S. Willis, The Sydney Declaration – revisiting the essence of forensic science through its fundamental principles, *Forensic Sci Int.*, 2022, **332**, 111182.

Science and Technology Select Committee, *Forensic science and the criminal justice system: a blueprint for change*, Parliament, London, 2019, https://publications.parliament.uk/pa/ld201719/ldselect/ldsctech/333/33302.htm.

2 The Crime Scene

Mark Butler

Teesside University, School of Health & Life Sciences, Tees Valley, Middlesbrough TS1 3BX, UK
Email: M.butler@tees.ac.uk

2.1 Introduction

Forensic science is a serious and fascinating subject, even the term *forensic* is used in our language to denote rigor, quality and the level of detail expended on an activity. Without doubt its existence is enthralling to many and in media and entertainment platforms it is often presented with glamorous appeal. Forensic practitioners sometimes discover when talking to the public, that people have unrealistically high expectations of forensic science and what it can provide to an investigation or its impact in the court room. This fascination with the subject and those that practise it, has been cited as the 'CSI effect', arguably a consequence of the proliferation of TV dramas and gaming, either as a primary or secondary aspect of the production. The mood now is that while the CSI effect is a feature the forensic practitioner should be mindful of, it has perhaps been inflated and is much less of an influence than initially perceived. Moreover, the forensic practitioner should skilfully emphasise the boundaries or limitations of the interpretation. Forensic science often works in the realm of probability, in essence a spectrum in terms of how much support can be offered to an opinion or hypothesis. It does

Crime Scene to Court: The Essentials of Forensic Science, 5th Edition
Edited by Niamh NicDaeid and Peter C. White

Published by the Royal Society of Chemistry, www.rsc.org

not always offer binary answers. The material in this chapter sets out to address the governance for high expectations for those that operate at the crime scene.

This chapter focuses on the work of the United Kingdom police Crime Scene Investigator however it is worth acknowledging that aspects of the role to varying degrees are also performed in other disciplines for example Armed Forces, Fire Service and Environmental Health.

Crime Scene Investigation in UK policing has seen some of the most remarkable changes in the last few years notably in quality governance and inter force cooperation. Irrespective of technological advances there have always been some core skills that have continuously been vital to the operation of crime scene investigation. The ability for the examiner to communicate, appreciate what tools, protocols are necessary for forensic evidence and intelligence to be managed. These have remained unchanged for decades. This chapter unpacks the role and title of the CSI, exploring its many early identities before discussing the training and education requirements to perform this technical and scientific work.

It is worth noting at this juncture that there is a certain rhythm to crime scene examination, a way of working. Where possible this chapter adheres to this order of scene processing: visual examination, photography, forensic recovery, and fingerprint enhancement, before illuminating how real-world operations require examiners to sometimes break away from this model.

2.2 Myths in Crime Scene Examination: CSI Effect

This section addresses a couple of important myths that seem to occupy students (in particular), at the beginning of their studies in forensic investigation and no doubt members of the public too. Its presentation now forms a prelude to a detailed explanation of the CSI's role.

2.2.1 "Homicide CSI"

The Crime Scene Investigator has a varied and complex role to play in modern British policing. The range of scenes they attend is extensive; the very same individuals fingerprinting a vehicle that has been broken into or a home burglary often attend more serious crimes such as sexual assaults and murder. The clothing they wear and the precautions they take are proportionate to the offence type and the tasks they have been assigned. No special CSI squad exists ready on standby for a murder investigation, the truth is that the bulk of a CSI's work will be attending routine crime often called volume and priority crime. It is possible to say that volume and priority crime would encompass for example: burglary, street robbery and vehicle related crime. It is the frequency by which these crimes occur that denote the term volume and depending on the evidence presented at these scenes can also include antisocial behaviour and some assaults. These early definitions are important as some Crime Scene Investigators may only be employed to deal with this level of crime type.

2.2.2 "Is Only the Coroner Allowed to Handle the Deceased?"

This seems to be a rather popular myth and is of course untrue! In fact, at a death investigation, the coroner is highly unlikely to be present. It is the role of the CSI who recovers fragile forensic evidence from the deceased and packages the body ready for transportation to the mortuary. The CSI may be called upon to attend stressful situations involving cadavers, for example cutting down the dead body in scenes of hanging, or recovery of victims and body parts in disaster work. The information the CSI obtains will no doubt be needed by the coroner, but the hands-on recording and recovery no matter how demanding or unpleasant, is performed by the CSI. Northern Ireland legal procedures work similarly to those in England and Wales; however, it is important to recognise that practitioners in Scotland have a separate system. Here the Procurator Fiscal performs a similar role to that of coroner in cases of death investigation, however the legal reach of the Fiscal is wider and more diverse. That said the operational work of the CSI in Scotland will be similar to that outlined above and below.

2.2.3 Blurred Lines

The boundaries between the work of the CSI and forensic scientist are not always easy to see and popular television dramas have done little to be authentic to reality not just in the limitations of forensic science but equally who does what! A classic example is that in some television productions the role of the pathologist has expanded and tipped into the realm of protagonist, engaged in investigative duties significantly beyond the already complex and demanding work of establishing cause of death.

In simple terms the CSI attends crime scenes to *record, enhance and recover* forensic evidence. Pathologists attend the scenes of suspicious deaths and perform in the UK what is known as a post-mortem at the mortuary: an examination to determine how the deceased died on behalf of the coroner.

Other areas of forensic investigation should be delivered using multiple agencies and specialists, such as forensic: archaeologists, anthropologists, chemists, biologists, and entomologists where appropriate. A CSI and pathologist may be part of a case conference discussing blood pattern distribution however the lead and ultimate presentation in court should be performed by a Forensic Biologist. The important aspect to appreciate is that each specialist should assist but step back from reporting on areas where they are not qualified to do so. It will often be the role of the Crime Scene Manger to facilitate this process. This comparison between fact and fiction may seem to expose a rather limited role, when actually it is highly complex with significant cognitive demands, possessing many levels or subshells depending on the context and requirements of the case. The CSI must therefore have a tremendous appreciation for the needs of other specialists, evidence types and sequential processes to maximise forensic opportunities and dissolve any tensions between each of these participants.

2.3 Defining the CSI Role

2.3.1 Professionalism, Character and Approach

Some effort has now been made to scrub back the layers of mis-selling of the CSI role, however there is general acknowledgement that it takes a certain kind of individual to be a Crime Scene Investigator. Many role profiles will reference a person with highly developed observational skills, an enquiring mind, meticulous note taking, patience, an ability to deal with difficult and stressful situations and an aptitude to work predominantly on one's own often at unsociable hours. These could be with other professionals too, for example: police, fire service, forensic service providers, hospital staff, Coroner's Office, utility companies, insurance companies and armed forces – in conditions which could be cramped, dangerous as well as visually and psychologically demanding to absorb. What is not always explicit in role profiles is the importance of character; illustrating tolerance, human understanding and in particular humour, the last of these not merely confined to anecdotal comments but recognised in research as playing a part in good performance.

These softer skills have always been necessary for crime scene investigation and crime prevention work, notably when eliciting information from victims who are distressed, vulnerable, confused or even deeply frustrated and angry.

The operational name of the CSI has indeed undergone changes, with previous incarnations being: Scientific Support Aid/Officer, Scenes of Crime Officer leading to Crime Scene Examiner/Investigator, Forensic Scene Investigator (FSI) and for those employed predominantly for volume crime work, Volume Crime Scene Investigator/Examiner. The evolution in title from a Supporter, an Officer to an Investigator can also be tracked in police and forensic doctrines. The intelligence gathering features of the role have been emphasised as playing a crucial part in the investigative process, recognising the cognitive skills needed to perform the many functions demanded of CSIs. Pushing the capabilities that the role has to offer, expanding it beyond recording and collecting trace evidence, has been a major theme in the last two decades. Moreover, police forces using the term Forensic Investigator (or FSI) is interesting as it captures the additional scenes investigators attend that may not be categorised as a crime, but due to the sensitive nature of the incident, the scene still requires forensic processing until full understanding of the events can be formally accepted, for example, Sudden Infant Death Syndrome.

It is worth emphasising that this evolution in role title has not been adopted by all police forces, nor has it been uniformly implemented at the same time, which demonstrates that despite recent improvements in cooperative working, differences still exist. These differences are compounded, as there are also diverse philosophies about how the role is viewed or performed. CSIs traditionally report fact and Forensic Scientists report opinion. The reality is that this is not quite as binary as first presented. What is true is that the CSI is a *professional witness*, reporting on fact first and foremost but should feel comfortable expanding on their knowledge and offering their thoughts provided these are bounded and supported by experience and training.

2.4 Training

2.4.1 National Training, Individual Force Training, Higher Education

Over the last 3 decades CSI training has seen major change, a chronology is detailed in the 4th Edition. Today police forces largely deliver their own scenes of crime training *via* a licensed product from the College of Policing. There are two main products, the first being Stage 1, Volume and Priority Crime, and Stage 2 Serious and Complex Crime. There are also additional and advanced products for the management and coordination of crime scenes. Some police forces have more advanced resources and so regional training centres have now emerged.

There are several Higher Education (HE) providers that have crime scene or forensic investigation within their degree programmes, some of these are being taught by former practitioners. For those seeking out quality in this domain it is advisable to refer to the Chartered Society of Forensic Sciences (CSFS) web site Accredited Course Search for a listing of HE providers with CSFS accreditation.

The CSFS plays an important part in Continuing Professional Development for its members and does this in a variety of ways, for example conferences and access to research material.

2.5 Health and Safety, Terminology and Jargon

Crime scenes can be hazardous, even a volume crime scene such as burglary is likely to involve items strewn onto the floor making moving around difficult, if nothing else needing to walk precariously to avoid damaging evidence or indeed the owner's property.

There are times when the CSI must access hard-to-reach places such as flat roof extensions, to evaluate footwear and instrument marks left by the offender during their entry or escape. Being able to assess these hazards and the likelihood of injury are critical skills for the CSI. These are especially important in complex environments such as fire scenes where hazards are not obvious or require additional monitoring for example: structures, air quality or reignition. The CSI needs to be aware of the Health and Safety at Work Act (1974). There is also the Management of Health and Safety at Work Regulations (1999) as well as much information from the Health and Safety Executive (www.hse.gov.uk) on lone working, documentation on the web site is very clear on who is responsible and what kinds of risk assessment are needed. Working alone and maintaining safety is very important at crime scenes, and invariably involves having a good grasp of the basics, such as good personal protective equipment (PPE) *e.g.*, nitrile protective gloves, face mask, over-shoes, and scene suit. At more dangerous scenes examiners should have at their disposal, hard hat, goggles/visor, face mask (appropriate filters), steel toe and sole lined boots, Kevlar gloves, proper lighting (and additional backup), spare batteries, two methods of communication coupled with a plan of entry and movement in the examination space. Specific officers with additional training in CBRN scenes (chemical, biological, radiological and nuclear) should have more advanced forms of protective technology to keep the examiner safe.

Police forces will also monitor examiners in routine health checks including lung function especially as fingerprint powders are frequently used. It is the examiners responsibility to ensure they are up to date with any immunisations or similar inoculations. One point of health and safety often missed in textbooks is mental health. CSIs should regard monitoring of this component of their health with as much importance as staying up to date with boosters or having the correct PPE. Attending occupational health debriefings post incident along with routine monitoring should be accepted and regarded with the same commitment as a vehicle check to ensure all equipment is present and working. Crime scenes can be hostile environments, however the daily grind of run-of-the mill offences and dealing with victims of crime every day can have an accumulative effect. This can be magnified with the increase in documentation and protocols in recent years, irregular working hours and insufficient capacity in staffing numbers. Stress in the workplace is a real phenomenon. Proper management and a supportive culture will go a long way to manage and prevent illness of this type.

2.6 Supervision, Coordination and Management

CSIs are generally supervised by a 'Senior CSI/Crime Scene Operations Manager', this individual may deal with authorising leave, performance indicators as well as taking on a guidance role at serious and major crime scenes.

Performance and quality enhancement has become highly significant as police forces respond to financial challenges, and the *cost-of-living* crisis. There is also a strong desire to move beyond only measuring frequency of scenes attended by CSIs and the number of forensic samples recovered with positive outcomes as a metric for good performance. Strategies to reduce repeat and near repeat victimisation and measuring crime by the amount of harm it causes and not simply a frequency count has gathered momentum, and the CSI has a serious role to play in this harm focussed policing.

It is also important to comment that crime scene management is not always performed by someone with permanent supervision authority but instead by a highly capable and experienced CSI who takes on the role when directed to, usually for serious crime examination duties. Senior CSIs/Crime Scene Operation Managers may be required to be a Crime Scene Coordinator, managing multiple crime scenes; creating strategies, gathering intelligence, and ensuring the major crime team and Senior Investigator Officer are being appraised of what is being recorded or what specialists are needed as well as dealing with tasks and actions that come into the Major Incident Room.

2.6.1 Defining the Scene

One of the most difficult aspects of crime scene management is perhaps not an obvious one to the uninitiated. Crime scene investigation and preservation can be intrusive, even to those who are not victims. Intrusion can be in the form of questions, limited or forbidden access to a public space or even a person's home. The size, scope and complexity of the scene can be difficult to define on arrival. It may range from a street scene, a fight outside a nightclub, to even a transport disaster with multiple casualties. The list of serious and major crime types would be exhausting to record. However, every CSI needs the skills to manage the scene until supervision arrives. The 'golden hour' is often commented upon in medicine. This same phraseology can be paralleled to crime scene examination with anecdotal comments suggesting a further drive towards the 'platinum fifteen' minutes. The work performed in the first fifteen minutes by those arriving at the scene, are arguably the most crucial. At this point the crime scene is at its most vulnerable, it should be anticipated that medical staff, fire service or police officers dispatched to assist will all impact on the forensic evidence in some form. The decisions at this stage to maximise the forensic potential are possibly the most important. Decisions and actions to rescue, save lives, hold persons or vehicles nearby, close roads are all likely to displace or corrupt the evidence at the scene from its original location or condition.

Moreover, decisions on fixed points for initial cordons (if at all) may have been decided not by highly skilled forensic investigators but instead by uniformed police officers doing their very best in what is likely to be a difficult, dynamic, and dangerous situation. The actions of these officers should be reviewed immediately by the first attending forensic practitioner. It is inexcusable, to leave this task for supervising scene managers who have yet to arrive, nor is it acceptable to give up simply because the scene is no longer perfectly sterile. No matter how careful emergency service personnel are, spent firearm cartridges will get kicked, blood showing intricate patterns with the potential to inform what happened could be smudged and fibres displaced from the injured to emergency service personnel uniforms. The key role here for the CSI is to manage, salvage and engage in their own forensic evidence rescue. Early telephone calls and actions to recover uniforms and footwear for elimination will go a long way to mitigate or regain lost potential. At the scene the old maxim, *Confirm, Cordon, Secure and Control* still stands today and should be unpacked by those first on the scene to examine it. Confirm what has happened and where, check to see cordons are in place and are appropriate based on the initial assessment. Ensure *via* police commanders that access under the cordon is controlled and restricted to those who have a genuine need to proceed. Police barrier tape is not enough; someone needs to begin logging those coming in and out of the scene.

The first CSI on the scene should also make assessments to decide on what additional resources are needed that have not been brought to the scene, this should be communicated to the Crime Scene Manager (CSM) as soon as possible. Early notice puts the CSM in a better position to keep some staff on duty and gear up deployment of other examiners to assist. In addition, any new information such as the strong possibility that additional scenes might materialise should also be passed on. It might be the case that the limited number of examiners available, are kept for possible forensic examinations of these addresses. The CSM will be cognisant that using staff to attend one location prevents those very staff from attending any other locations for fear of contaminating them. This is perhaps one of the biggest frustrations observed in TV dramas. The investigator needs time and opportunity to decontaminate equipment, shower, change clothes, *etc.*

Figure 2.1 shows the cordoning off of the area, the setting up of the crime scene tent to cover and protect the scene and stepping plates for the personnel to reach the scene, reduce contamination and

Figure 2.1 Simulated case showing the reverse view of the crime scene tent with stepping plates (in front of hedge), cordon tapes and general surround.

preserve the area. A photographic record of the items found at the scene and their numbered location markers is shown in Figure 2.2.

Establishing and maintaining the cordon can be immensely problematic depending on the scene. Consider a sexual assault in the stairwell to a block of flats. Placing an outer cordon some distance from the flats may be easy but invariably there will also be an inner cordon around the primary scene. These cordons need to be guarded with those passing under them being logged in and out. For forensic or police staff this is easy to appreciate but what happens to the residents of the property, those that want to return or leave or have visitors? General visitors can perhaps come back another day but a community nurse that needs to change the dressing covering an ulcer of a diabetic patient needs to be sifted out of those that do not need to enter the building. It is dealing with events such as the one outlined where we begin to appreciate the dynamic and complex nature of defining the crime scene and those who require access.

Enforcing the same level of restricted access at volume crimes is rare. Police forces in these cases do their best to ensure businesses and schools are up and running as soon as possible. Not allowing victims of routine crime to get back to some state of normality quickly, only prolongs the discomfort and distress victims face. In most instances a telephone call and early preparation can alleviate tension or difficulty. Providing advice on what to look for before

Figure 2.2 Photograph (with fill-in flash) of a simulated scene with exhibits and numbered location markers.

items are disturbed can be crucial as well as how to handle items in such a way to minimise or hopefully omit destruction to any DNA, fingerprint, footwear or particulate trace evidence. It is clear already that communication and problem solving are skills in high demand by CSIs.

2.7 Visual Examination and Note Taking

CSIs will indeed need to be methodical in their search and examination of forensic material and may work with other specialist units, for example a search team led by a Police Search Advisor (PolSA). A police search team will have additional equipment to search in for example a dense undergrowth or even under floorboards. It is not uncommon for

CSIs to remain and assist these search teams, usually by undertaking photographic duties and providing further forensic advice.

The visual examination is the most important phase; it happens predominantly at the beginning and may be a distinct stage of its own, however in truth it never stops. Even as the CSI begins the fingerprint examination invariably the last element, they might still observe forensic material like fibres or blood, which in an ideal world would have been recovered first. The visual examination is also an early point in the process to think clearly and logically about a strategy on how access might need to be established including what additional resources are needed: people as well as equipment. The CSI must always be able to justify their actions. To have sound evidence/intelligence-based reasons on why they did the things they did. This becomes especially important for the Crime Scene Manager at major crime scenes but is equally valid a comment for volume or routine crime examination. These actions if not justified might come under scrutiny in the courtroom, a formal review by another police force or the Independent Office for Police Conduct (IOPC).

It is also an opportunity for the CSI to make notes either on a tablet, laptop or hardcopy, this is a crucial time as these notes are often referred to as 'contemporaneous', that is, 'notes made at the time' of the examination. The CSI could be asked to attend court many months or even years from when the examination was carried out. These notes serve as the best record for ascertaining information on exhibit location, recovery times and actions taken. They are supplemented with photographic images and a sketch. Being a gifted artist is not a requirement to be a CSI however the investigator should be able to draw plans (not necessarily to scale) with measurements to two fixed points using the coordinate system of measurement. There are many different methods of sketching and recording accurately such as triangulation to three fixed points, use of angles and arcs, however simple approaches of 'across and up' are arguably best understood and most adopted.

Even in daylight or bright conditions the CSI will use a light source when searching for evidence, especially when searching for fingerprints. Today CSIs use a variety of light sources of different wavelengths and filters to improve the contrast between the evidence and the surface it is on, making it easier to see, but also to visualise evidence that may be difficult or impossible to see with a naked eye. On this point blood absorbs light, for it to be best seen on dark surfaces, it first needs to be stained using chemicals, a fact missed in many popular TV dramas.

2.7.1 Legal Considerations for Civilian Crime Scene Investigators

The vast majority of CSIs in the UK are civilians employed by police forces. The Police Reform Act 2002 c.30 Part 4, Chapter 1, Section 38 discusses the powers of police authority employees. For more easily digestible material on the subject the Explanatory Notes section Part 2: Investigating Officers paragraph 208, neatly and clearly identifies Crime Scene Investigators (described as a Scenes of Crime Officer in the explanatory notes) as being covered by the legislation. These legal powers are important and allow a CSI to be effective in their duties. An example is the ability to search and seize property for further forensic analysis.

The Chief Officer (often referred to as Chief Constable) gives the CSI the legal authority to enable them to do their work and earlier reference to these legal powers may be found in the Police and Criminal Evidence Act 1984 (PACE). Different powers and procedures exist in Scotland and Northern Ireland.

2.8 Photography

2.8.1 Telling a Story

Taking good quality photographs is as much about art as it is science. The ability to understand the behaviour of light and how it travels through a lens is fundamental but great photography requires much more; it necessitates a need to understand people. Images should tell a story; the viewer should be led through the crime scene in some sensible order when the finished album is produced. Photographs of places and property are often taken from all four corners, capturing all the sides of the subject as well as safeguarding that the ceiling and floor (if a confined space) are not missed. The CSI should also where possible ensure that their equipment or indeed themselves are not captured in reflective surfaces such as mirrors, polished metal or similar as this causes much distraction for those viewing the final images. Doors should be marked so that their original location can be reset as the CSI photographs behind these structures. CSIs could be asked to photograph anything from: a site of antisocial behaviour, fire scenes, recovered property, stolen vehicles or vehicles used in crime, burglary scenes, drug paraphernalia/cannabis factory, locations of assault and those assaulted, cadavers, digital devices or scenes involving high economic crime as well as landscape flora/fauna. In addition to those listed, specific evidence types that require 1:1

capture notably instrument marks, tyre marks, fingermarks, footwear marks and sometimes blood distribution fall under the remit for most CSIs. Here the CSI is required to take a photograph with a scale and label to allow the final image to be produced to actual size. This photographic work extends to capturing findings during the post-mortem however, some of the most challenging images to take are from those who have suffered abuse, domestic violence, and common/sexual assault. The CSI needs to be sensitive, explain what is going to happen, be mindful of personal space and the intrusive feeling a camera lens and flash evoke. Thoughtful CSIs work with social workers, parents, and guardians; introducing themselves first even if the subject is a child. The CSI who thinks ahead may bring along colouring books, stickers, or inexpensive toys for children to play with, helping the child to be quiet and calm before they photograph the injuries. These can be anything from footwear transfer marks, grip mark bruises (contusions), general cuts, bites, lacerations, abrasions and burns on the skin. Many of these will also require 1 : 1 photography. Other instances of injury photography could require the CSI to attend hospitals and liaise with medical staff. The CSI is not trained to remove bandages even for minor wounds for fear of introducing an infection, and so preparation and negotiation before-hand of appropriate times to carry out this work within busy wards is crucial. This role should be of equal importance to photographing a crime scene.

2.8.2 Cameras, Specialist Photography, Drones

The skill set of the CSI in this area needs to be highly advanced. Being able to use all the manual controls on a modern camera is essential as the investigator thinks about flash techniques to fill in shadow, controlling the depth of field by changing the lens aperture, to carefully selecting shutter speeds for long exposures such as night-time photography or when using high intensity light sources to fluoresce a mark. An example of the use of oblique lighting to highlight and photograph a Vehicle Identification Number (VIN) is shown in Figure 2.3.

Drones (unmanned aerial vehicles) are also being used by UK (United Kingdom) police forces, some of which have specific forensic search and image capture capability. Whilst the general consensus is that these aircraft offer less disturbance to evidence on the ground than conventional methods notably a helicopter; the CSI should still inform drone pilots of the fragile nature of trace evidence *e.g.*, fibres, plant material, wet blood. This may be directly from the aircraft's downdraught or from the dust or water it might displace, with the

Figure 2.3 Use of oblique lighting to highlight and facilitate the photography of a Vehicle Identification Number (VIN).

potential of covering up evidence including footwear and fingermarks. The additional opportunities these technologies bring however are worth considering; here imaging has the potential to allow post review measurements and reconstruction of the scene.

2.9 Forensic Recovery and Continued Preservation

2.9.1 Recovery

CSIs need to afford equal consideration on how items are to be handled, recovered and packaged. These considerations will also be dependent on who they are working with and the nature of the crime. That said, in general terms disposable sterile tweezers and forceps are used by examiners to handle evidence before it is packaged and during subsequent laboratory examinations. For example, cigarette ends will often be gripped from the burnt end rather than the filter. The CSI must also consider their own safety as well as avoiding destroying trace evidence such as fingerprints or DNA. The recovery of sharps and used syringes are examples when the primary focus should be on confidence during the recovery and handling stage. These are instances when tweezers are avoided, the examiner whilst wearing appropriate gloves should be committed and firm when handling items like this rather than gingerly gripping edges for fear of destroying trace material.

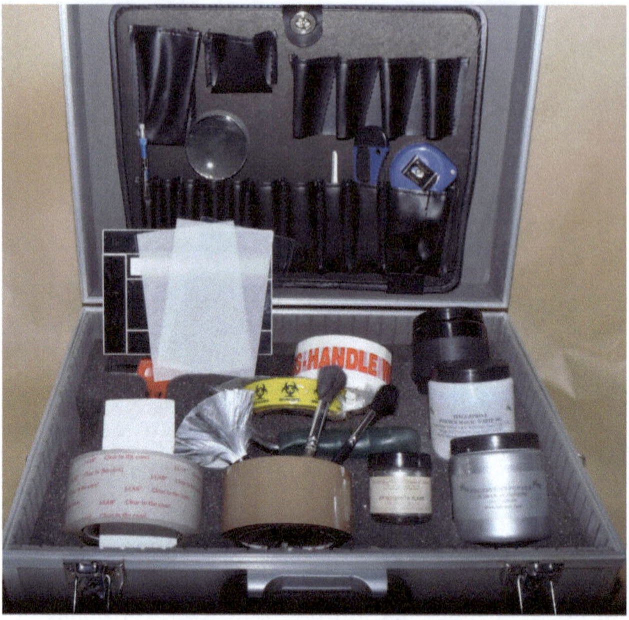

Figure 2.4 A CSIs case: J-lar and precut lifting tape with fingerprint template and acetate sheets. Parcel tape, Zephyr and animal hair fingerprints brushes, aluminium powder, black and white powder, magnetic powder and warning tape.

To recover evidence from a scene the investigator will carry their own personal case of equipment, such as swabs and mini-tapes for the recovery of DNA material, fingerprint powders and brushes as illustrated in Figure 2.4. Due to the variety of surfaces on which fingerprints are found it is necessary to carry a range of powders to aid their recovery. Readers are referred to Section 2.12 for a more detailed discussion on the recovery of fingerprints.

2.9.2 Packaging

Crime Scene Investigators traditionally use an array of packaging equipment, for example brown paper bags of all sizes, from small envelope styles to those with the ability to house a double mattress. Plastic bags are usually for non-organic items or those not requiring fingerprint treatments. Packaging blood-stained clothing, plant material or even footwear in plastic for extended periods can result in the items producing mould growth, hampering further forensic opportunities. Nylon bags can be used in combination with plastic to recover fire debris samples; these ensure volatile hydrocarbons from a potential fuel source like petrol are kept contained. Weapon tubes are used for sharp items such as knives, screwdrivers, and syringes. Depending on the item, extra protection is advocated with a metal disc in the inside to ensure the item does not penetrate the plastic base, an easy error if the examiner or police officer over tightens the screw thread! Packaging in the form of boxes with a clear window covering is needed for: weapons, firearms, footwear gelatine lifts/casts or even documentation.

Documentation in general may need to be examined for indented writing (electrostatic imaging) and ensuring that no pressure or additional folds are placed on the paper is crucial. Some of these boxes are ready made, others not, and it should be a skill of a competent practitioner to construct suitable packaging for preserving the integrity of any evidence from being lost as well as shielding the exhibit from any contaminants getting in. Items for fingerprint treatment should not be allowed to rattle around, instead the packaging should be made to fit or appropriate restraining options implemented to prevent friction and the rubbing away of any latent fingermarks. The condition of the object might just be as valuable as the forensic evidence contained within or on it; it is for this reason that cigarette ends are packaged in a polypot or universal container, as are spent cartridges and bullets.

2.9.3 Sealing, Labelling, Continuity

Each packaging type will have similar methods for sealing and security, for example brown paper bags have the opening corners and ends folded over twice and sealed, along with any remaining opening side edges or glued strips. Each taped edge is signed over with the examiners signature using a permanent marker. The process extends to signing over seals when evidence is placed inside a weapons tube, polypots, or universal container (tape to tape signature seal for paper wraps). This is a security measure that affords confidence on who did the packaging, and any tampering will result in a broken signature seal. At a convenient place on the packaging a series of important details are written, referred to as minimum details, these are: exhibit number, date, location, description, name and signature. This information also appears on the Criminal Justice Act exhibit label, which is either fixed or printed on the packaging. This label has an additional important feature in that it records who has handled this item irrespective of duration. Modern computer tracking systems use a system of barcodes or tags however traditional methods of exhibit labels to establish continuity are still used. It is imperative that rapid identification of who has handled a forensic item, and when they did so from the moment it was found to when it is presented in court is

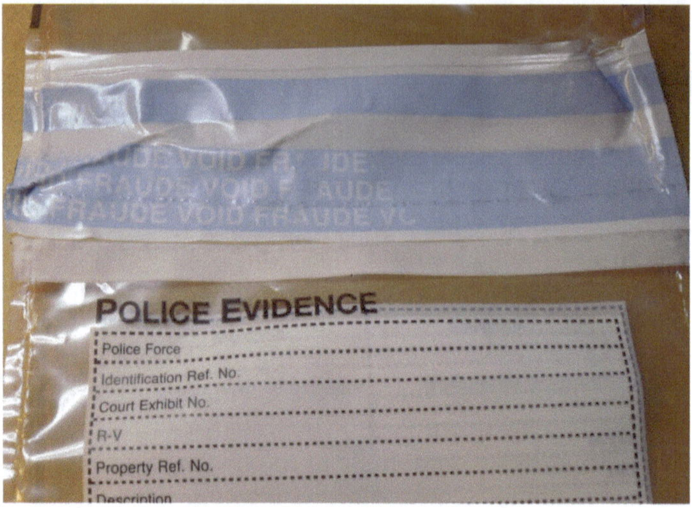

Figure 2.5 Tamper evident bag with 'VOID' label along sealed edge showing unauthorised access attempt.

made transparent. This is often referred to as continuity or the evidential chain. Finally, if any attempt has been made to tamper with or try and open a sealed bag, "VOID" will be seen, as shown in Figure 2.5.

2.9.4 Drying Clothing

Police forces have sophisticated drying cabinets capable of retaining particulate material falling off garments, as well as having design features to ensure adequate washing and cleaning down between uses. Processes for use are now well ingrained in quality management standard operating procedures.

2.10 Enhancement

The majority of police forces in the UK will have their own chemical enhancement laboratory or at least share these facilities; these units focus on mostly using chemical treatments to enhance fingermarks (or footwear marks) that are not available for the CSI to use at the crime scene. All police chemical laboratories now have ISO accreditation. It is important to recognise that they function not solely for the submissions by CSIs but also fraud or economic crime officers and may include samples such as document examination and drug paraphernalia. CSIs at all levels are under increasing pressure to deal with as much enhancement at the crime scene as possible. Examiners need to therefore be skilled in appreciating what fingerprint brush and powder combination works best on any given surface. Guidance to help Chemical Mark Enhancement Technicians and CSIs has always existed in some format. The Home Office Scientific Development Branch were once the providers of this information. This branch later became integrated with CAST (Centre of Applied Science and Technology), and now CAST has become integrated with DSTL (Defence Science and Technology Laboratory). This lineage is important to appreciate as it helps to understand the context should one come across different references relating to chemical treatments either in research literature or indeed practice, especially the sequential treatments. What is fundamental is that every CSI should be aware of these sequential processes and what opportunities for further fingermark enhancement are available, to maximise prospects for the investigation and support the investigating officer. Moreover, being mindful of the condition the original item was in, for example saturated by rain, may affect what treatments are possible and time spent examining the article. It may also be reasonable to ask the chemical enhancement lab to attend a crime scene; this will invariably be a serious incident such as a murder scene or drug factory. Here they will perform some of the enhancement on site due to the sheer scale of the operation.

2.11 Quality Management

2.11.1 ISO 17020 and 17025

In previous years, the majority of police forces, forensic examination and chemical mark enhancement laboratories did not have these standards, in fact it was only ISO9001 for fingerprint work that was the mainstay within the sector. Some but not all practitioners had the previously accredited CRFP (Council of Registered Forensic Practitioners) status, or the Diploma in Scientific Support Skills (later Crime Scene Examination). There were no standards other than those personally adopted as good practice such as operation manuals for using equipment and how resources should be left after use. It was not unusual for CSIs to be asked to quickly fingerprint small objects whilst sat at their desk doing paperwork, or even briefly test/screen items for blood or carry out swabbing on drinks cans. For more serious offences, items were examined in a laboratory. Good practice was adopted by using brown paper sheets to cover surfaces prior to examination, personal protective equipment and cleaning products were used, but that was largely it. The forensic space was denoted by little more than the sign on the door rather than through accreditation. Today's police forces forensic examination rooms and chemical mark enhancement units have become standardised through ISO 17025.

ISO 17025 specifically focuses on the competency testing and calibration within laboratories. The United Kingdom Accreditation Service (UKAS) is the recognised body that assesses against these international standards. The assessment is also designed to check the competency of the organisation and those practising forensic investigation. Sampled rechecking of exhibits is now in place.

As an adjunct to above, the Forensic Science Regulator has also provided extensive document support for helping forensic practitioners recognise the many guises of cognitive bias. In essence, thinking about the systems, processes, interpretation, and decision making that happen in forensic practice. Focusing on the many sides of quality management, all of which need to be tackled. Through the Forensic Science Regulator Act 2021, this role has become more authoritative with a legal footing and in a much stronger position to assist in building better practices and monitoring compliance.

There are many other aspects that ISO 17025 covers such as handling and storage of exhibits, but fundamentally it checks the operational credibility of the laboratory and those using it. This shores up through regulation and accreditation the question as to whether this valuable forensic resource is indeed fit for purpose.

It may have been noticed that the information above does not capture one of the most important features of good forensic investigation and that is at the crime scene itself, notably a proper and strict crime scene examination policy. It is here that if mistakes or improper practices are done such as: coughing, sneezing, talking over items for DNA analysis, inadequate packaging, improper enhancement methods, then no matter how professionally the end destination forensic testing facility is operated, the evidence is clearly compromised. ISO 17020 aims to address the inspection element of the role, with language pointing to consistency, competency, impartiality, and achieving public trust in the work undertaken. The dates of when specific codes are to come into effect are published in newsletters by the Forensic Regulator. These documents are a useful resource for anyone wishing to gain a better understanding of how accreditation works and what specific international standards are aligned to crime scene work. A deadline of October 2022 was given to police forces (extended due to Covid-19) to acquire this accreditation. At the time of writing September 2022 there is still some way to go for some police forces to achieve this accreditation, however it is accurate to say they have made huge steps and are strongly working towards it.

2.12 Fingerprinting

2.12.1 Brush and Powder Combinations

Traditional brushes that examiners predominantly use are fibreglass or are of the animal hair variety, the latter most commonly being squirrel. Animal hair brushes are used exclusively for non-metallic granular powder, this powder is spherical in construction and suitable on semi-porous surfaces, such as a weathered windowsill. These powders are favoured by some practitioners even for smooth surfaces such as uPVC window frames. It should be appreciated, that these examiners are highly skilled and aware of the need to control such an unstable powder during the application process. Fingermarks enhanced by these powders are photographed first and if appropriate subsequently recovered with a gelatine lifter.

Fibreglass brush designs are usually reserved for applying flake powder; perhaps the most famous and cost effective of which is aluminium powder. Powders like this are sensitive to moisture content; their structure, flat by design, makes them highly stable and suitable for lifting. An example of a fingermark lift from a car window is shown in Figure 2.6.

Magneta flake is another metallic powder that is highly regarded among examiners simply due to its versatility on surface types. Its application is different to that mentioned above in that it is applied using a magnetic wand. Manufacturing statements comment that the sensitivity of this powder is imparted not just in the milling process but also through its amino acid coating.

This is by no means an exhaustive list as the application of fluorescent powders and high intensity lights at the crime scene could easily warrant a chapter on its own.

2.13 Impact of Covid-19

The impact of Covid-19 changed many forensic practice procedures between 2020 and 2022, indeed new ways of working needed to be

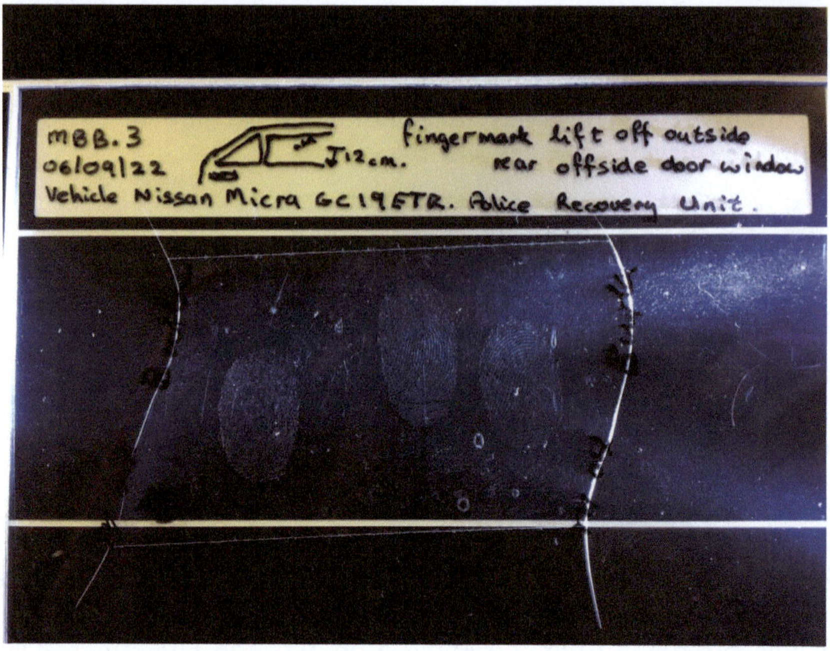

Figure 2.6 Fingermark lift on a template board annotated with the exhibit number, date, location, description from where the mark was lifted, along with a directional arrow pointing to a fixed point enabling the orientation of the mark to be established.

implemented immediately. This period saw CSIs not attend certain scenes if anyone in the household tested positive for Covid-19 or showed symptoms. Advice needed to be given to help homeowners and businesses respond by prioritising and securing evidence, placing it to one-side until a point in time it was safe for the CSI to deal with it. Simple tasks such as taking elimination fingerprints from people who had a legitimate reason to have their fingermarks at the crime scene location was halted to reduce the spread of the virus. Thankfully there is now a gradual return to providing a crime scene investigation service that resembles that experienced pre-pandemic.

2.14 Specialisms

Some CSIs develop their skills further in specific areas for example: fire investigation, blood pattern analysis, mass disaster and victim identification, archaeology, anthropology, entomology and more. All these topics are provided in an introductory or awareness format during the training a CSI gets. However, these specific standalone courses often run by private practitioners and training providers enable the CSI to go deeper into the topic and be a specific person of contact within the police force on the matter. Whilst caution should be given and noted that it *would not be appropriate* for these individuals to be pressured into providing expert testimony, they nevertheless offer much intelligence value to an investigation. Moreover, the role these advanced practitioners play impacts on the design and structure of policy and 'go-to' documents when specific expertise is difficult to source immediately. For example, an officer recently returning from an entomology course should be expected to update other staff, any existing departmental literature, or even create a policy on the subject. These instructions would cover contact details of experts, what samples to take as well as preparation for recording and preserving insect life where appropriate.

2.15 Closing the Scene

At some point when the forensic examination has concluded the scene will need to be closed and officially handed over to another agency or released back to the owner/local authority. For serious crimes the Crime Scene Manager (CSM) should record when this activity occurs.

The official closing of serious crime scenes requires authorisation from the Senior Investigating Officer (SIO). The decision, date and time is also logged in their major incident file. It is not unusual for a final walk through at the scene between the SIO and CSM or Coordinator, reviewing all crime scene actions to ensure and check that absolutely everything that should have been done has been implemented. Moreover, any outstanding actions should have been communicated to the examiners with the result of these recorded by the incident team. The decision to close a scene down should be done with the same degree of formality, as it was to establish it.

2.16 Giving Evidence in Court

All crime scene staff, whether they are administration officers receiving and submitting exhibits, chemical enhancement technicians, fingerprint experts, or CSIs, could be called upon to give evidence at court. This is because the court needs to satisfy itself that the evidence was properly handled, processed, or correctly identified and stored.

CSIs rarely attend court to give evidence, invariably their statements are simply accepted. However, that does not mean they are not given advanced skills in this area or asked to maintain this skill. CSIs need to understand the ways of the Criminal Justice System, the different court procedures and what is expected of them. It is a skill to communicate the evidence and intelligence obtained at a crime scene and explain complex processes and jargon to an audience who may have never heard these terms before. In addition to this, the defence's legal team will ask some tough questions designed to test the evidence collected. The CSI must be calm in these situations and comment on what they know to be true. More often than not the CSI reports on facts, providing the court on the details of the scene, what was observed, where evidence was and how it was recovered or transported. Whilst this forensic presentation may be primarily factual, the court does appreciate the expertise of the CSI and there are times when a CSI will provide further useful opinion-based information to the court. These instances may be rare, but it is important to note their existence and that this presentation of evidence may be in both written and oral form. For example, CSIs might be asked to discuss the most likely point of entry of a burglary, explaining which scenes may be linked due to the *Modus Operandi* of the offender. Ultimately it will be for the court to decide on who provides expert evidence.

2.17 Conclusion

This chapter explored the role of the CSI, focusing not simply on the technical skills needed but also the softer elements the job demands to be effective and professional. Complex issues such as cordoning, recording, and preserving evidence at the crime scene were also discussed and that people from different services will always impact on the forensic evidence to some degree. Here the CSI needs to be able to manage, adapt and keep pushing for control of the crime scene, appreciating that the initial stages are incredibly dynamic until protocols have been setup, understood and respected.

Different agencies, specialisms and training were key themes presented and that these need to be managed to avoid conflict. Although this chapter was clear in how scenes are to be controlled with much debate on cordons, one last point should be recognised: early conversations with specialists are imperative and that the information from these conversations could impact significantly on the preservation of evidence. For example, consider allowing members of the public to lay flowers near to the crime scene, an action that signifies mourning and paying respect. Now study the implications should the SIO and CSM wish to employ the services of a forensic botanist. No matter how well implemented, a police cordon will not stop the transference of pollen. This is the plant biological equivalent, of sneezing all over the evidence that still needs to be processed for DNA analysis. The learning point here is that early communication is necessary in these highly active moments; taking the time out to get the best advice will ultimately make for a better more profitable crime scene examination.

Further Reading

M. Chowdhury, A broken system? Examining the perilous state of quality assurance in crime scene practice, *Sci. Justice*, 2021, **61**(5), 564–572.

G. Horsman, Decision support for first responders and digital device prioritisation, *Forensic Sci. Int.: Digital Invest.*, 2021, **38**, 301219.

M. Illes, P. Wilson and C. Bruce, Forensic epistemology: testing the reasoning skills of crime scene experts, *J. Can. Soc. Forensic Sci.*, 2019, **52**(4), 151–173.

B. A. Klentz, G. M. Winters and J. E. Chapman, The CSI Effect and the impact of DNA evidence on mock jurors and jury deliberations, *Psychol. Crime Law*, 2020, **26**(6), 552–570.

N. Marsh, *Forensic photography: a practitioner's guide*, John Wiley & Sons Inc., Hoboken, New Jersey, 2014.

D. Wyatt and D. Wilson-Kovacs, Understanding crime scene examination through an ethnographic lens, *WIREs Forensic Sci.*, 2019, **1**(6).

3 Forensic Ecology

Patricia Wiltshire

Department of Geography and Environment, University of Southampton, Southampton Road, Southampton, SO17 1BJ, UK
Email: patricia.wiltshire1@btinternet.com

3.1 Introduction

Ecology is the multidisciplinary study of organisms and their relationship with the environment. It is concerned with the distribution, abundance, and complexity, of both the biological community and its habitat – its physico-chemical surroundings. Any community together with its habitat constitutes an ecosystem and, whatever its scale, this is the functional unit for study. Forensic ecology really comes into its own when other lines of evidence, such as human DNA and fingerprints are lacking. Its power is a consequence of the extraordinary complexity of the natural world, such that no two sites sharing identical ecological profiles have yet to be found.

Ecology focusses on whole organisms, or fragments of them, and these provide proxy evidence of the ecosystems from which they are derived. While no practitioner can be an expert on all, the forensic ecologist needs some knowledge of animals, plants, fungi, bacteria, and other microbes, as well as a grasp of geology, soil science, and aquatic ecosystems. The effect on communities of ambient temperature, humidity, oxygen tension, airflow, and light intensity within habitats also need to be appreciated.

Crime Scene to Court: The Essentials of Forensic Science, 5th Edition
Edited by Niamh NicDaeid and Peter C. White

Published by the Royal Society of Chemistry, www.rsc.org

Although other related classes of evidence are important, it is botany (including palynology) and mycology that have become the lynchpins in forensic ecology. Both can provide pivotally important information at crime scenes, even identifying pre-meditation in grave digging and, through the recognition of anomalous growth of species, identifying the nature, and sometimes the timings, of disturbances.

Both are also valuable because of the ubiquity and copious amounts of proxy markers provided by plants and fungi. Their spores, pollen, seeds, and other living and dead parts act as proxy indicators of the species from which they are derived. Although soil can provide proxy evidence in its own right, it influences the plants on which animals and fungi ultimately depend. If plant proxies indicate soil type, then the geology, which is the ultimate source of mineral component of soil, is also indicated. Thus, proxy indicators of organisms can provide information on ambient conditions, vegetation, soil, and geology, and thus indicate geographical location; and because the responses of plants and fungi to environmental influences are broadly predictable, their proxies can also provide temporal, and even climatic information.

Proxies of organisms in soil, sediment, dust, water, vegetation, or on any palyniferous surface, can be transferred to clothing, footwear, vehicles, hair, skin, nails, gut, nasal passages, cadavers, and a wide range of other materials and objects. Such transfer can be primary, secondary, or even tertiary, depending on the physico-chemical characteristics of the evidence itself. Once retrieved, both macroscopic and microscopic proxies can provide both intelligence and/or probative evidence.

There is now considerable information on the effects of earthworms, other invertebrates, foxes, badgers, rodents, birds, vegetation, and fungi, on the crime scene, and what it contains. Much is also known about decomposition processes on the surface, within various soil and sediment types, and in still and flowing water. Conversely, research is being directed to the effects of buried corpses on surrounding materials, such as soils and sediments, and on the decomposer organisms they contain. At a crime scene, the forensic ecologist should be able to guide the investigating officer to the potential of specialisms other than their own and, for the maximum retrieval of pertinent environmental evidence, be familiar with other available research bases and scientists.

Most living things will be found where they perform best and, in addition to habitat preference, many other factors will influence where they are found. All species have particular needs and tolerances, and these will determine the niches they occupy, and reflect their local, regional, national, and even international distributions. Demanding species with narrow tolerance to environmental change (stenotopic organisms) will generally have restricted geographical ranges, and their presence in any forensic profile may be useful for the implication or elimination of a place as being relevant to an investigation. Less demanding (eurytopic) species may have wider spatial and geographical distributions, and so be found in a broad range of habitats, but their presence in a profile can still be of forensic value even though their information content invariably needs more nuanced interpretation.

Forensic ecology has been developing over about 30 years, and the greatest strides were made in the 1990s and early 2000s as cases were being analysed, and various databases were being established. There has always been a paucity of able ecologists with the necessary broad knowledge, field experience, microscopy skills, and robust personal qualities, to work in forensic investigation. As a consequence, forensic ecology is less well-known and understood, and there can still be confusion as to the nature and potential of the subject. This presents a dilemma for both investigating officers and the courts. Ecological knowledge takes time to acquire, and the most effective practitioners are those with many years' experience both in the laboratory and the field, and who understand ecosystem processes. Importantly, they also need to understand the limits and potentialities of the subject within the forensic context. No beginner should attempt to provide evidence for court scrutiny.

Some forensic analyses such as DNA sequencing, fibres, footmarks, gunshot residue, and fingerprints are well-established and have become routine. In terms of diversity of population, these involve the use of relatively small datasets, and competence can be achieved relatively quickly.

Forensic entomology is another frequently used discipline where expertise generally takes longer to attain but the techniques in that field are well described and rest on a knowledge of relatively few species. Botany, palynology, mycology, and phycology (study of algae) may involve thousands of species and they are not applied as frequently, primarily because they are not learned quickly and the number of cases where they are justified is relatively small. Here, expertise also relies less on standardised forensic techniques and sophisticated equipment, but more on depth of knowledge, experience, and the availability of extensive reference collections and databases. Progress in forensic ecology relies on the cumulative results of criminal and civil investigations rather than experimental data. It is also true that standardisation at laboratory, national, or international level is difficult to achieve because the skill base is individualistic. Processes can be standardised, but individual expertise cannot be formalised. Strict protocols for many areas of forensic ecology have not yet been formulated, but comprehensive ones for forensic palynology are now published, and ones for forensic mycology are in active preparation.

Soils, and the rocks and sediments giving rise to the inorganic fraction within them, can play a role, and there are efficient elemental and mineralogical techniques used for characterising inorganic trace evidence. In some circumstances, for example where synthetic materials are involved, these can be valuable, but the specificity of mineralogical assemblages within any landscape is not as great as those of biological ones although the combined analyses of inorganic and organic particulates should always be considered. The palynologist or mycologist will often carry out some soil analysis in the course of their work and will make routine comparisons on the nature of the matrix such as colour, texture, and organic content.

Some attempts have been made using soil biochemistry in forensic investigation but, for trace evidence, pollen, spores, some soil organisms, and even microfossils, have provided more reliable forensic evidence of higher specificity. A particular advantage of organic particulates is that the high counts of palynomorphs from large numbers of taxa (often to species level) means that an experienced practitioner can envisage the environment from which the profile was derived and, by eye, be confident of eliminating any irrelevant location with a high level of confidence. This can represent a huge saving in resources because, unlike in other kinds of environmental work, detailed analysis of every potential place may be unnecessary. For example, if quick scans of preparations from an exhibit yield only a heathland profile, then gardens or broadleaf woodland can be eliminated. Even microhabitats within ecosystems can be recognised so that specific locations can be highlighted. The biological profile of any community will have some universal elements typical of that community, but, when rare elements are present, high degrees of reliability can be achieved for positive recognition.

Palynomorphs (microscopic entities such as pollen and spores) are ubiquitous and can be retrieved from inside and outside buildings, from vegetation, soils, and very many objects and fabrics. Most carry a negative charge, and they can cling tenaciously to surfaces and fabrics by electrostatic forces. On numerous occasions they have provided the only probative forensic evidence and have contributed to cases involving:

(1) Estimation of body deposition time;
(2) Estimation of post-mortem interval;
(3) Demonstration of offender pathways and events at and around an offence;
(4) Demonstrating sequences of events at crime scenes;
(5) Linkage of people, objects, and places (trace evidence);
(6) Location of human remains and graves;
(7) Differentiation of a kill site from a deposition site;
(8) Establishment of cause of death;
(9) Challenging or confirming witness testimony;
(10) Elimination of irrelevant locations in an investigation;
(11) Gut content analysis.

While each ecologically based subject area will present its own specific kind of information, none should be treated as 'stand-alone,' nor be in competition with any another, but rather be regarded as contributing parts to a whole picture. Some skills will be more important in certain scenarios than others but, invariably, they are complementary. At the start of an enquiry, thought must be given to the potential contribution of different specialisms as well as the emphasis needed in a particular investigation.

3.2 Nature of the Evidence

There is a wide range of organisms that could, under some circumstances, be useful in the forensic context, yet many have not been used in casework. Table 3.1 summarises the main groups which can provide proxy indicators. Animals have been excluded from the table although, on occasion, fragments of animals (carapaces, exoskeletons, hair, and fur) have provided trace evidence and so should not be ignored.

3.2.1 Algae

Algae are photosynthetic, mainly aquatic, organisms with an enormous range of morphological, molecular, and reproductive variation. They are ubiquitous in fresh, brackish, and salt water, and if some moisture is periodically available, some can grow in a wide range of

Table 3.1 Simplified scheme showing some organisms that may be encountered in forensic botanical and mycological investigations.

		Plants		
			Seed bearing	
Algae[a]	Fungi[a]	Non-seed bearing	Gymnosperms (naked seeds)	Angiosperms (seeds enclosed in ovaries)
Macroscopic (*e.g.*, kelps) Microscopic (*e.g.*, diatoms and dinoflagellates)	Mushrooms Lichens Moulds Smuts Rusts Mildews Slime moulds Yeasts	Mosses Liverworts Hornworts Ferns Club mosses Horsetails	Conifers Seed ferns Ginkgo (maidenhair tree) Cycads Gnetophytes	Flowering plants (monocotyledons and dicotyledons)
These produce spores which germinate to form the next generation.			These produce pollen grains containing sperm cells which effect fertilisation after pollination	

[a] These categories of organisms are used in the colloquial sense as recent molecular data indicate that many "traditional" groups of organisms comprise ones that are now classified in different kingdoms.

terrestrial habitats, including on and in buildings. Large, macroscopic algae, such as the marine kelps, can achieve great size, but most algae are microscopic and only visible when their populations discolour water, or form surface growths. Some can be free-floating, forming tangled masses around submerged plants or other objects, whilst others are truly planktonic. They can colonise the hair and clothing of corpses in water and, in terrestrial situations, can often cause a "greening" of bone once it becomes de-fleshed and available to daylight.

Generally, the most valuable algae in forensic cases are diatoms. They are unicellular, photosynthetic, yellow-brown algae (Kingdom *Straminipila*[†]) and are distinguished by each cell having a durable, siliceous frustule over the cell wall. The frustule is composed of two valves which fit together like a lidded box. The valves are highly variable in morphology and surface patterning (Figure 3.1), and this allows relatively high precision in identification; a huge number of species has been recognised (approximately 100 000).

Diatoms are found in terrestrial and aquatic habitats. They colonise surfaces quickly and exhibit seasonal growth and, because of their sensitivity to water quality, many species are habitat specific. For these reasons they have been used as proxy indicators of environmental

[†] Sometimes as "*Stramenopila*", but *Straminipila* is the formal scientific name; *Chromista* is a synonym.

Figure 3.1 Examples of diatoms, illustrating their considerable variation in shape and surface sculpturing of the silicious frustule which aids identification. Images are not to scale.

conditions in many areas of science. Diatoms are useful as trace evidence, and they can be extracted from clothing and other objects, but they are most valuable where drowning is thought to be a cause of death. Specific assemblages may provide information on the depth where water was inhaled, and whether the water was salt, brackish, or marine. Assemblages of a number of species are invariably needed for interpretation and, to date, light microscopy is the most usual and efficient method of identification.

Other algae potentially useful to the forensic scientist are the *Chlorophyta* (green algae), a morphologically diverse and species-rich group, and the single-celled *Pyrrophyta* (dinoflagellates). Living dinoflagellates show great morphological variation but it is the fossil dinoflagellates that can be of most use in forensic investigations. The free-swimming organism eventually produces a robust cyst whose cell walls contain sporopollenin – the same substance found in the outer walls of pollen grains and plant spores. Such cysts became embedded into rock-forming sediments and remain as fossils. During soil formation the particles become released from the rock and cysts find their way into the soil matrix from which they can provide additional sources of trace evidence. Other fossil spores of unknown affinity are often referred to as acritarchs. These also show large variation in morphology and, again, provide interesting markers in soils and sediments.

3.2.2 Animals

Of particular interest in forensic ecology are the cells of testate amoebae as these are encased in a membrane which can survive the acid treatment used in preparing soil samples for microscopy. The outer wall (test) can be constructed of silica, calcium carbonate, or foreign agglutinated particles. Some testate amoebae are common in lakes, damp sediments, or peat while others, such as the Foraminifera, are marine. Testate amoebae fossils can weather out of rocks and find their way into soils and sediments and depending on their resistance to acid treatment, they may be found in palynological preparations and provide useful trace evidence. Other animal remains that have, on occasion, been relevant in forensic investigation are: Mollusca (*e.g.*, snails and slugs – even squid from gut contents); Ostracoda; Acari (mites); Insects (dipteran flies, lice, fleas, and others); Nematoda (round/eel worms); cysts or eggs of other animals (including birds). Feathers, hair/fur and animal bone, teeth, claws, and exoskeletons may also have forensic relevance. Mites and nematode eggs often become included in palynological samples. Large numbers of invertebrates are often observed in association with decomposing corpses and can give information on time of deposition, supplementing that available from flies and from fungi.

3.2.3 Fungi and Fungal-like Organisms

Mycology is the study of all kinds of fungi including mushrooms, lichens, moulds, smuts, rusts, mildews, slime moulds, and yeasts. They belong to the same major taxonomic group as animals and are not related to plants. Fungi can colonise and grow on many organic materials, from food and textiles to human remains, but also on fine organic films on paper, glass, concrete, plastic, and other inorganic and inanimate objects.

The world estimate for plants is about 400 000 species and most have been described and given scientific names but the global estimate for fungi is at least 3 million, and only about 150 000 species have been described and named. There are about 15 000 species of fungi recorded for Great Britain and Ireland, compared with only 2000 native plants. Further, around 40 fungi new to Britain are discovered annually, and about 2000 new fungal species globally are described each year. Unlike most plants, many fungi are microscopic and are easily missed, and with species new to science still being discovered, even in Europe, fungi are more difficult to identify than plants. There are very few mycologists available to identify those which prove to be most relevant in forensic investigations, especially microscopic species. Examples of some fungal spores are shown in Figure 3.2.

All true fungi, and their spores, have walls composed of chitin. This is a polymer which is both chemically and structurally robust and is also found in the exoskeleton of arthropods, including insects. Fungal spores are invariably microscopic and may be produced by sexual or

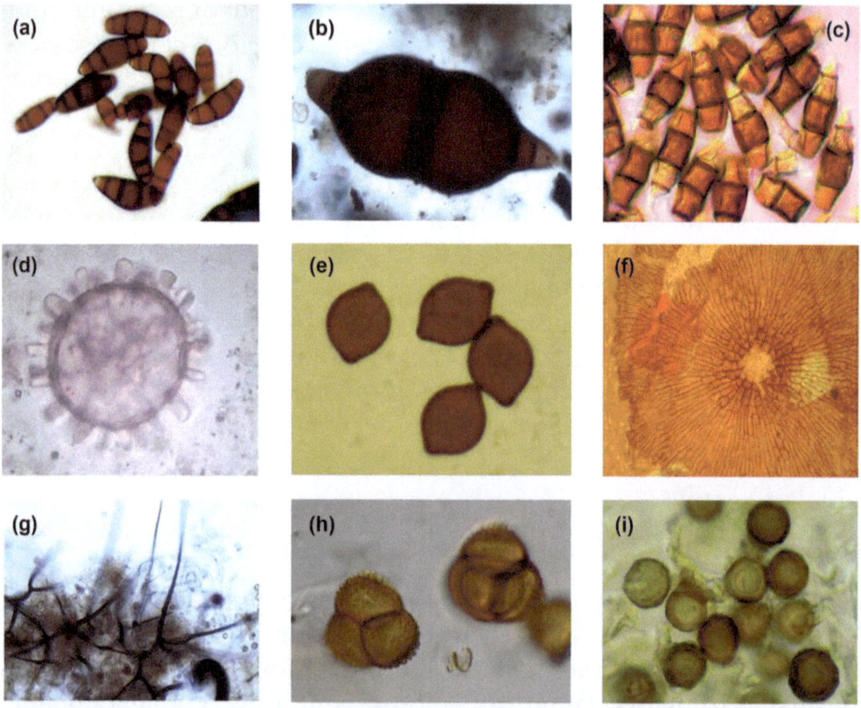

Figure 3.2 Examples of unusual fungal palynomorphs from recent forensic cases: (a) *Curvularia inaequalis*, (b) *Caryospora callicarpa*, (c) *Pestalotiopsis funerea*, (d) *Choiromyces melandriformis*, (e) *Melanospora zamiae*, (f) *Thecaphora frezzii*, (g) *Myxotrichum chartarum*, (h) *Microthyrium inconspicuum*, and (i) *Thermomyces lanuginosus*. Images are not to scale. Courtesy of Professor D. L. Hawksworth and Dr J. A. Webb.

asexual means; they are formed in various kinds of sporophores which may be very large, or themselves microscopic.

In the last fourteen years, fungi have proved to be of critical importance in many forensic criminal and civil cases, both solely and in conjunction with plant material. Due to habitat and geographical restriction, and host or substrate specificity, fungi are useful proxy environmental markers. Contrary to popular belief, most spores are rarely dispersed more than 1 cm to 1 m from the spore-producing body and, when combined with palynology, fungal spores and other fungal bodies provide excellent trace evidence. A fungal spore profile can be obtained along with one constructed from pollen and plant spores and, where appropriate expertise is available, fungal palynomorphs are now routinely analysed alongside pollen and plant spores in the same forensic samples and specimens. Fungi with more specific nutritional and growth requirements may be highly informative as their presence indicates specific conditions which may be location limited. For example, some fungal pathogens are highly host specific and can indicate the presence of a plant even in the absence of its pollen. Like plants, fungi have different ecological requirements, but their modes of spore production differ, and the taphonomic factors affecting their presence in any sample will differ from those operating on plant-derived evidence. This means that two distinct classes of evidence can be obtained from the same samples.

The "weed" fungi, such as species of *Alternaria*, *Aspergillus*, *Mucor*, *Penicillium*, and the less-common *Scopulariopsis brevicaulis*, are ubiquitous and very common in soils. They tend to be opportunistic colonisers of a wide range of substrates, including those provided by food, corpses, body fluids, and associated materials and objects, including clothing and soft furnishings. Even the slightest smear or stain can be sufficient to support fungal growth and be a source of valuable evidence. Temperature and humidity are important constraints on microbial growth, and these parameters are interdependent. Some specialised species of fungi (and bacteria) are thermophilic (up to 122 °C) or cryophilic (down to −20 °C) but the great majority have optima around 25–30 °C. Humidity can be measured as water activity (a_w) which is the partial vapour pressure of water in a solution divided by the vapour pressure of pure water at the same temperature. A value of 0 means there is no water and a value of 1 means there is saturation. It is a measure of unbound (available water) in a system. There is no microbial activity when a_w is <6.0; bacteria need an a_w of 0.91 and many fungi can grow between a_w 0.75 and a_w 0.9. This means that they can colonise mummified cadavers when all bacterial growth is inhibited.

Fungi are invariably ignored by forensic pathologists and yet cultures from hair, skin, and mucosa have provided accurate temporal evidence where post-mortem interval has been important. In many cases in the UK, fungal growth on a cadaver's skin has been shown rarely to establish until approximately seven days after death and thereafter the speed of colony growth depends on temperature and humidity in the ambient environment.

On occasion, a fungus may be identified (at least to genus) from photographic evidence, and published information on species' growth rates at different temperatures can be used to assess the time since fungal growth started. The crime scene officer should examine and photograph fungal colonies on any object or corpse, using macro-lens and graduated scale, and every colony should be given a reference. Photographic evaluation is not as accurate as physical examination and where interpretation is based on pictorial records, every caveat must be applied. Where many species colonise a cadaver, relative growth of colonies, especially where there is overgrowth one by another, can provide invaluable temporal information. Sampling should be the responsibility of a mycologist with laboratory facilities, and it is always preferable to obtain samples with sterile swabs for culturing onto various media. This will allow accurate estimates of growth rates under specific humidity and temperature regimes. Wherever possible, tapings should be made to facilitate identification. The temperature of the crime scene should be closely monitored over several days and, at outside sites, accumulated data should be compared with meteorological information in the same way as that for entomology. In addition, measurement of the ambient humidity as well as that at the interface of an object, or skin of a corpse, should be recorded.

There have been cases where fungi have grown in body fluids soaked into furnishings and fabrics, including bedding. These may be of adventive species falling out of the airspora, or even from growth out of a corpse onto fabrics or furniture, and into stains made by body fluids. Even tiny stains can be colonised and yield information about timings.

Where psychotropic and poisonous species have been involved, accurate identification of whole fungi and their spores have been helpful in demonstrating their use as "social drugs" or cases of poisoning. Analysis of gut contents can also confirm the ingestion of macroscopic pieces of mushrooms, as well as their spores. This allows identification of species containing mind-altering substances, as well as toxic ones.

Identification of fungi is difficult, but molecular techniques can sometimes provide accurate information when macroscopic material or spore masses are available.

3.2.4 Plants

Identification of plant material is of particular forensic scientific value especially as communities of plants (vegetation), individual plants, and parts of plants, play such a huge role in everyday life. They provide food, fuel, building materials, other resources, and essential medicines, yet some are highly toxic, and are the source of narcotics, hallucinogens, and poisons. Plants have predictable lifecycles (phases of which can often be linked to specific seasons), predictable growth responses, and to a great extent, predictable distributions.

Identification of whole plants, as well as isolated parts, fragments, and propagules, can eliminate or implicate certain plant species in criminal cases. Plant materials can provide important trace evidence, and can demonstrate food adulteration, origin of materials, and the consumption or use of poisonous and/or illegal plants and plant substances.

3.2.4.1 Non-seed-bearing Plants

These all produce spores and include Bryophyta (mosses), Hepatophyta (liverworts), Anthocerophyta (hornworts), Psilophyta (whisk ferns), Lycophyta (club mosses), Pteridophyta (ferns), and Sphenophyta (horsetails). Unlike the mosses, liverworts and hornworts, the ferns and horsetails have true roots and vascular systems, and many species have spores with sufficient morphological differentiation to allow identification, sometimes to species. Many non-seed-bearing plants are highly specific in habitat requirement so presence of their spores in forensic samples may help to characterise habitats. Fragments of leaves and spores of all these plants can be useful as trace evidence since many are highly habitat specific. For example, species in the genus *Riccia* (liverwort) are characteristic of fallow fields and field edges, and different species of *Sphagnum* moss can indicate varying water levels in acidic, wet places; on more than one occasion, both have helped provenance footwear and identify approach paths to crime scenes.

3.2.4.2 Seed-bearing Plants

There are two phyla of seed-bearing plants, the Gymnospermae (seeds are naked) and the Angiospermae (seeds are enclosed in an ovary).

Gymnosperms include Pteridospermophyta (seed ferns), Gnetophyta (joint firs and others), Ginkgophyta (maidenhair tree), Cycadophyta (cycads), and Coniferophyta (conifers). Most are woody trees, and members of the Coniferophyta are the most numerous and widespread of the gymnosperms.

The angiosperms (flowering plants) may be divided into two classes, the Dicotyledonae and Monocotyledonae. They range from tiny aquatic species that consist of a leaf of 2 mm in diameter and a single, tiny root, to huge trees over 100 m tall with trunk girths of 20 m or more – and from submerged aquatics to the desert cacti. They have specialised conducting and supporting cells in their stems, roots, and leaves, and a wide range of specialised cells and tissues can occur in all parts of the plant. They exhibit numerous diverse morphological features.

The definitive structure in angiosperms is the flower. The female part has one or more carpels, each of which consists of a stigma, style, and ovary as illustrated in Figure 3.3. The ovary produces one or more

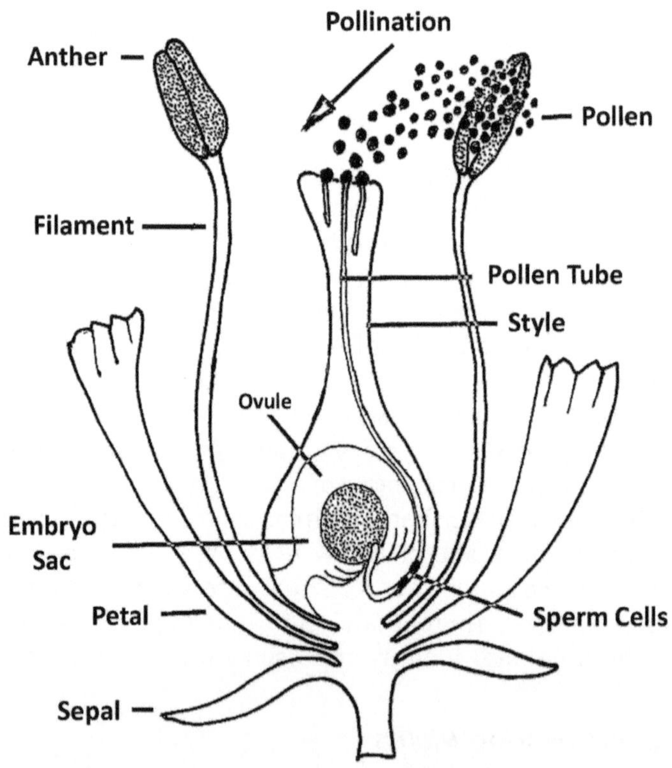

Figure 3.3 Stylised diagram of a flower illustrating the fertilisation process after pollination.

ova (sing. ovum) which each develops into a seed after fertilisation. Pollen grains are produced in the anthers and contain sperm cells; they effect fertilisation after successful pollination. The angiosperms produce the most diverse cells, tissues, and structures in the plant kingdom. They also produce a very wide range of fruiting structures which surround the ovary and developing seeds, and these have provided valuable trace evidence in many forensic cases, including poisonings, cases of malnutrition, and finding locations.

3.2.4.3 Wood, Dendrochronology, and Dendrochemistry

During the growing season, a woody plant will produce new xylem (wood) tissue just below the bark; this carries water and nutrients from the roots to the rest of the plant. Wood cells continue to form during favourable conditions and slow or cease during winter, or other unfavourable conditions. The transitions are marked in the wood structure and a cross-section shows a series of rings. Generally, the wider the ring, the more favourable are conditions for growth and narrow rings are produced when the plant is under stress or insect attack. In tropical plants, where conditions vary little, growth is more regular, and rings are not produced in the same way as in more temperate species. The wood of angiosperms is more complex than that of conifers, and several kinds of cell are produced. There are many variations in arrangement and kinds of cell in wood, and samples can often be identified accurately to genus but, if sufficient growth rings are present, events may also be timed.

Since growth rings are produced annually in both stems and roots, a remarkably precise estimate of the age of a piece of wood can be obtained for some species. Ring patterns for replicate wood samples from any geographic area with similar environmental conditions can be cross correlated to dated structures, and standard time curves for different species established. Specimens of unknown age can then be dated by comparison. This technique, dendrochronology, is used in a wide variety of disciplines such as archaeology, palaeobotany, climatology, and ecology, as well as dating and matching wooden objects from crime scenes or checking for forged antiques and illegally imported artefacts. Trees also retain pollutants within the wood as it is being laid down in the tree, and chemical analysis of tree rings (dendrochemistry) can help to resolve environmental disputes by demonstrating the source of contamination, and putting the episode in a precise chronological framework. One technique used in dendrochemistry is energy dispersive X-ray fluorescence (EDXRF), and it is possible to find pollution events by targeting elements such as sulphur and chlorine for fossil fuels, and lead for leaded petrol. Other metals are useful for indicators of mining and smelting.

Identification to the level of an individual wood source can, however, exceptionally be achieved by a physical fit (*e.g.*, a splinter of wood matching a gouge, fitting together ends of a broken twig), or some individual-specific phenotypic peculiarity.

3.3 Palynology

Palynology has been a well-established sub-discipline of botany with accepted conventions and analytical methods for over a hundred years. It is the study of any microscopic entity (palynomorph), which has been scattered away from source, although for most palynologists, palynomorphs are restricted to pollen grains and plant spores. Only the seed-bearing plants produce pollen, but the spores of all other groups of plant, and those of fungi have contributed to the versatility and usefulness of the science. It is used to great effect in palaeoecology (study of ancient environments), archaeology (study of historic and prehistoric man and his environment), aerobiology (detection and forecasting of seasonality of airborne allergens), melissopalynology (pollen content of honey); palaeobotany (study of ancient plants and their evolution), oil prospection (as markers for different geological layers), and for forensic investigation.

Palynology has also been used to support convictions and damage claims in many criminal and civil cases. In order to maximise the information provided by any sample, the forensic palynologist must identify everything as precisely as possible: pollen, plant and fungal spores, other plant and fungal remains, trichomes (plant hairs), phytoliths (silica cell inclusions), arthropod fragments, charred fragments, plant cells and tissues, and indeed any detrital material capable of identification. It is, therefore, imperative that the palynologist has had in-depth botanical, ecological, and preferably other biological training. But where necessary, they must also seek additional specialist advice to complement their personal expertise, for example in the identification of particulates such as fossil plant spores, microscopic animals, eggs, and cysts.

The spores and pollen grains of all plant groups have an inner wall of cellulose (intine) and an outer one (exine) in which is embedded a complex, chemically-robust, but mechanically-fragile polymer – sporopollenin. Like fungi, pollen and spores can persist for millions of years in rocks, and thousands of years in more recent sediments. They can also persist for many years in soil although they can be decomposed by microorganisms within weeks in well-aerated and bioactive (microbially rich) situations.

Pollen and plant spores vary greatly in diameter from 5 μm to 200 μm, with most falling between 25–60 μm, whilst most fungal spores range from 2 μm to 800 μm. The resolution in identification differs between different groups of plants and fungi. Although the spores of some species of mosses, liverworts, hornworts, ferns, club mosses, and horsetails have distinctive patterning of the outer wall, discrimination is relatively poor for these non-vascular plants. Higher resolution can be achieved for spores of ferns, but the range of morphology is still rather narrow. Spores may be trilete and variously sculptured, or bean-shaped (Figure 3.4), and sometimes enclosed in an outer, variously ornamented covering – the perine. There are, however, relatively few diagnostic features for spore identification when compared with the pollen of seed-bearing plants.

Identification of pollen is aided by the presence or absence of various ornamentation of the outer wall of the grain, as well as the number and organization of apertures such as pores and furrows. The grain may be smooth, or support various kinds of network, spines, rods, variously arranged mounds, and even combinations of them. Artificial intelligence and computer recognition of pollen and spores has been attempted but, to date, the results have been disappointing. It has been possible to identify a few assemblages of pollen grains, but the taxa recognised by machine are low in number, and the obstacle of background detritus, variable orientation, and preservation of grains, as well as intraspecific variation in morphology, have so far proved to be insurmountable.

A range of pollination methods have evolved in the seed-bearing plants, and this is reflected in the structures which produce and

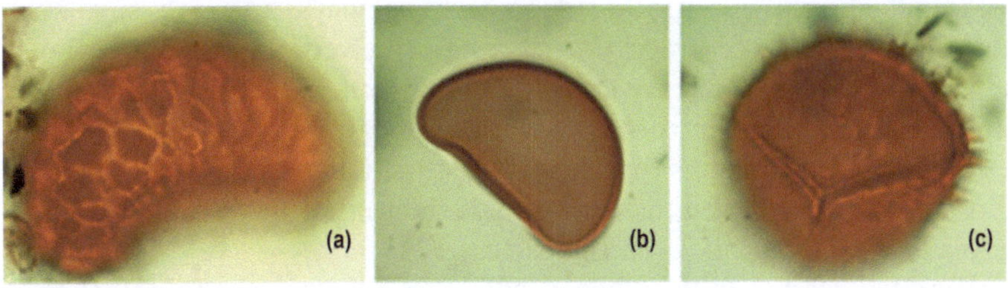

Figure 3.4 Light micrographs of spore-bearing plants (a) polypody fern, (b) a spore of a liverwort, *Riccia*, and (c) a fern spore which has lost its outer perine, making specific identification impossible. Images are not to scale. Courtesy of Dr J. A. Webb.

support the sexual organs. The gymnosperms mostly utilise the wind to carry their pollen from male cones to the ovules on female ones. Like the gymnosperms, angiosperms with wind pollination produce copious amounts of well dispersed pollen, and their flowers are generally devoid of colour, perfume, and nectar. Insect pollinated plants, however, invariably have invested resources into flowers of complex structure and have evolved to attract pollinators with colour, perfume, nectar, and often, specific shapes. They produce relatively small amounts of pollen as their vectors carry out direct pollination.

The pollen grains of many gymnosperms are more-or-less spherical while others have grains with characteristic air sacs which enhance buoyancy in dispersal, and yet others may have a hooked or curved papilla. Patterning on the air sacs, variation in wall thickness, grain size, and variation in sculpturing and patterning of the main body of the grain all provide diagnostic features for identification. However, when compared to angiosperms, there are relatively few diagnostic characters, and discrimination is relatively poor. Images of some gymnosperm and angiosperm pollen grains can be seen in Figure 3.5.

Flowering plants produce the most diverse ranges of pollen morphology, but the resolution in identification is variable depending on the taxonomic group to which the parent plant belongs. Some can be identified to species, others to genera, while others can only be assigned to family. Within a family, it is often possible to determine groups or types. For example, in the rose family (Rosaceae) many species cannot be differentiated from one another although some in *Prunus*-type (plum, sloe, cherry and others) can be differentiated from *Crataegus*-type (hawthorn, whitebeam and others), and some, such as *Dryas octopetala*, (mountain avens) or *Agrimonia eupatoria* (hemp agrimony) may be taken to species. In the forensic context, it is always wise to err on the side of caution and simply record the taxon as 'undifferentiated Rosaceae' if there is any hint of doubt in identification.

Fungal spores are produced in various ways and are similarly diverse depending on the group of fungi involved, and their methods of dispersal. Some are tiny cells with little outer differentiation, and often very abundant in the airspora. These can be almost impossible to identify, even to genus, but others are much larger and multi-cellular and/or ornamented and, based on their micro-morphology, lend themselves to identification (even to species).

Standard reference keys based solely on micro-morphological features are available to aid the identification of palynomorphs and many pictorial reference publications are available. Whilst there are several excellent texts on pollen identification, and much information

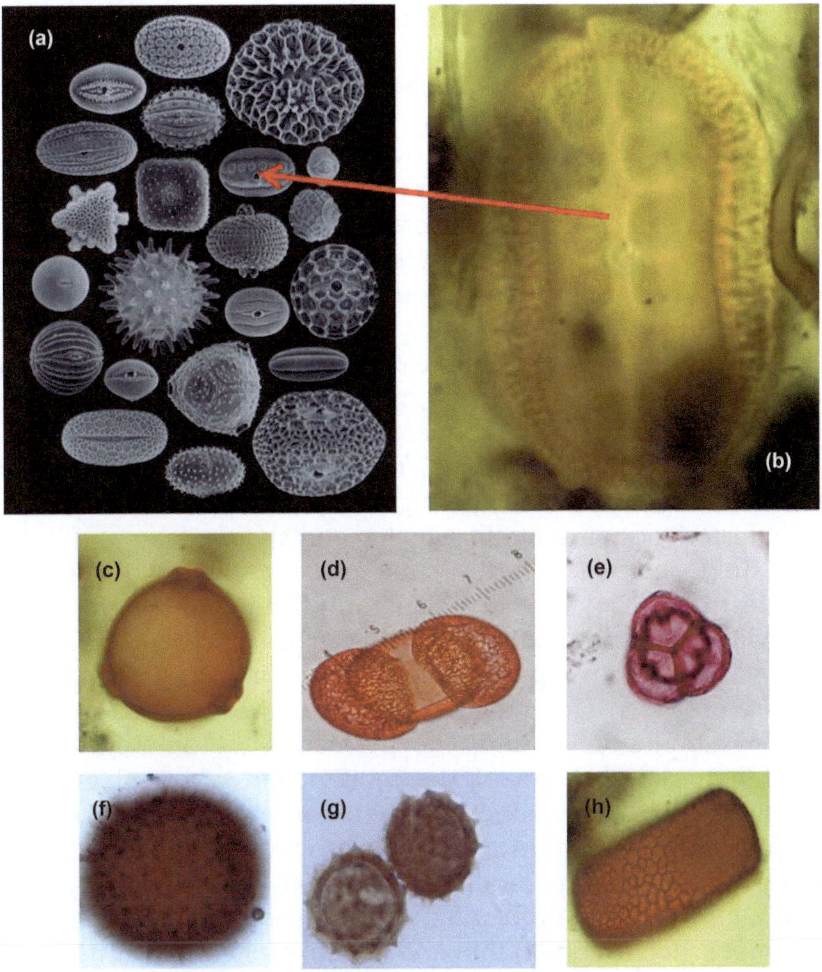

Figure 3.5 Examples of pollen grains showing variation in surface sculpturing, aggregation, and aperture arrangement. (a) SEM of pollen in the family Acanthaceae showing the large range in morphology and sculpturing in this plant family and (b) a light micrograph of a pollen grain of a shrimp plant, identified by the arrow as one of the Acanthaceae pollen grains in (a). The light micrographs of some other pollen grains are shown for (c) birch, (d) pine, (e) heather, (f) mallow, (g) aster and (h) balsam. Images are not to scale. Courtesy of: (a) Dr R. Scotland, (b) Dr J. A. Webb, (d) Professor V. M. Bryant.

is available on the internet, it is essential that the palynologist has access to a reliably identified and comprehensive reference collection and, because of inter- and intraspecific variation, does not rely only on pictures for identification. The recognition and classification of fungal spores is a highly specialised skill and, when working in a forensic context, palynologists should always seek the advice of a mycologist who possesses wide experience of macro- and micro-fungi and has access to reference material for direct comparison. It is bad practice to identify any palynomorph without available reference material, but particularly so in the case of fungal spores because of the similarity of many taxa. While there are no comprehensive texts for fungal spore identification comparable to those for pollen, there are many specialist, well-illustrated works dealing with particular fungal groups of which the specialist needs to be aware.

As well as the native flora of the country in which an investigation occurs, the forensic palynologist must also be familiar with exotic taxa. With the UK as an example, cases often involve gardens and parks planted with many species alien to the British flora. Again, there is no substitute for collecting reference material and learning the characteristics of unfamiliar pollen and spores.

In a single forensic case, a sample may consist of an assemblage of more than 200 kinds of palynomorph and, although a comprehensive physical database of pollen types is essential for comparison, the effective practitioner draws on a personal mental database of a very wide range of taxa in routine analysis. In most forensic samples in Britain, the percentage of unknown grains should not exceed 2–5% of the total pollen recorded although palynomorphs may be unidentifiable because of differential decay, crumpling, excessive thinning of the exine, or poor methods of extraction or slide preparation. On occasion, an exotic pollen grain or spore may elude identification, and this should then be described in as much detail as possible and photographed for future reference. If the percentage of unidentified taxa is high, the competence of the forensic practitioner should be questioned. Likewise, if no unidentifiable palynomorphs are reported, that may indicate over-confidence.

Scanning electron microscopy (SEM) has little application in routine forensic palynology and, indeed, research has shown that light microscopy is quicker and more efficient. Ordinary light and/or phase contrast microscopy is preferred, especially as correct identification often requires examination of the internal wall structure and, in the case of fungi, presence of septa and inter-connections. The SEM only allows the outer surface of the grain or spore to be seen. Many thousands of pollen grains and spores must be identified and counted in a forensic case and the sheer laboriousness of working with SEM precludes its use as it would be impossible to identify and count sufficient numbers of palynomorphs for construction of a palynological profile.

In forensic investigations, palynomorphs may be retrieved from a wide range of items and materials, including soils. For identification by light microscopy, specimens are subjected to harsh chemical treatment to remove humic acids, cellulose, silica, and any other background material likely to obscure the grains. The aim is to remove all except the sporopollenin and chitin making up the walls of pollen and spores. These chemical treatments preclude DNA analysis, and so molecular methods cannot help in identifying palynomorphs that are retrieved from exhibits in the standard way.

3.3.1 Palynological Taphonomy

Palynological taphonomy may be defined as being "all the factors that influence whether a palynomorph will be found at a certain place at a certain time". This is a complex matter and only salient factors are presented here. The mixture of pollen and spores in the air is known as the 'airspora' and it is envisaged as falling as 'pollen rain'. Pollen and spore production by plants, the concentration of palynomorphs in the airspora, and patterns of distribution, are varied. Palynologists continue to devote much effort into measuring these variables, and the literature contains many models of temporal and spatial variation which have been constructed to facilitate interpretation of complex palynological data. Such models, and those derived from experimental trials testing pollen movement, transfer, and retrieval, are of interest but as they relate only to specific sets of experimental parameters (models, which are based on relatively small data sets and are geographically limited), their results cannot be extrapolated to forensic cases without scrutiny. Whatever the nature of the airspora and its palynomorph load at any place and at any time, the only data of concern to a forensic palynologist are (1) those obtained from comparator samples from the crime scene or other pertinent place, and (2) what is retrieved from exhibits in a criminal investigation. Every forensic case is unique, and interpretation must be made on the basis of sound sampling, analysis, and interpretation of specific data relating to each investigation. The application of any theoretical model would inevitably introduce an element of doubt. Most instances of trace evidence comparison centre around the likelihood of a specific palynological profile having been transferred from a specific place. For example, what is the likelihood that this shoe contacted the ground at that place?

3.3.1.1 Production and Dispersal

Pollination is achieved mainly by wind and animal vectors (mostly insects) and examples of some plant species pollinated by wind and insect vectors are illustrated in Figure 3.6. Wind-pollinated flowers produce vast amounts of pollen to enhance the probability of locating a female. Flowers are often grouped into dense inflorescences, such as catkins, that can shake in the wind and maximise the movement caused by air currents. For example, a single catkin of *Alnus* (alder) can produce up to 6 million grains in one season.

Inevitably, wind-pollinated plants are generally well represented in the airspora while insect-pollinated and self-pollinated taxa tend to be under-represented, or not at all. Pollination by vectors leads to exchange of genetic material between individuals but, despite the advantages of out-crossing, more than half of all flowering plants in the temperate region regularly undergo self-pollination – even before the flower bud opens. This is a common phenomenon in more extreme environments where animal vectors are rare or unreliable, or where a given population is well-suited to a particular habitat and genetic diversity would not be particularly advantageous. Some plants

Figure 3.6 Top row shows examples of wind-pollinated flowers (a) hazel and (b) plantain. Bottom row are examples of insect-pollinated flowers (c) foxglove, (d) clover and (e) pansy. Images are not to scale.

such as *Urtica dioica* (nettle) and *Ilex aquifolium* (holly) are dioecious (sexes on separate plants). In some instances, there might be a large population at a site but, if largely female, they might not register in the pollen record. Some plants such as the bee pollinated *Trifolium repens* (white clover) may be common but their pollen is rare because of their low pollen production and poor dispersal. It may only be when the plant dies and falls to the ground that pollen becomes released from the parent.

Rare palynomorphs contribute to the specificity of any pollen/spore profile and have provided compelling evidence in cases but, even in those that are wind-pollinated, nearly all of the pollen produced will fall close to the parent.

3.3.1.2 Patterns of Distribution and Residuality

Analysis of thousands of surface samples has shown that palynomorph assemblages at ground level are highly heterogeneous and unpredictable, especially where people have manipulated the environment, as in towns and gardens. The patchiness of pollen and spore concentration in any sample taken from the ground is, amongst other things, a function of:

(1) Heterogeneity in the seasonal and annual pollen rain;
(2) Diversity of palynomorphs inherently present (as well as fossilized palynomorphs or from plants and fungi formerly present);
(3) Proximity of the site from pollen producing plants;
(4) Vulnerability or recalcitrance of palynomorphs to decomposition;
(5) Effects of physical barriers;
(6) Action of people and animals (including disturbance and mixing).

Any surface open to the pollen and spore rain will have an individual palynological signature although it might be similar in some respects to others in close proximity. Similarity declines with distance and so far, all sampled sites prove to have unique profiles.

In most situations, wherever decomposer microorganisms are present, most pollen and spores will eventually decay, and an experienced palynologist can sometimes differentiate between the current year and older residual pollen from the state of the exine. Some studies have also demonstrated variable resistance of certain palynomorphs to decomposition. *Alnus* (alder), *Tilia* (lime), *Polypodium* (polypody fern), and *Lycopodium* (clubmoss) have particularly robust pollen or spores which can remain for many years, while others such as Juncaceae (rushes) have pollen that decomposes very quickly indeed and are rarely found. Fungal spores may behave in a comparable way; those of many mushrooms and 'weedy' fungi with thin-walled spores are only infrequently found, while those of some asexual fungi producing thick-walled, multicelled spores, can persist long after the substrate on which they formed has decomposed.

There are few experimental studies on palynomorph residuality, but artificial situations in a laboratory rarely concur with what analyses of actual soils reveal. It is therefore unsound to extrapolate from a few limited published results, and generalisations should not be applied in real cases without further research. Palynomorphs will accumulate on surfaces and in the environment at large and can remain for varying amounts of time – often for years, even decades, depending on their ambient environment. Resistant palynomorphs can be over-represented because of past accumulation, and so a palynological profile may not reflect the present local vegetation as closely as might be expected but also include some of that which was formerly present.

Where pollen and fungal spore release is seasonally variable, as in Europe, prediction of periodicity in production and dispersal for various plant and fungal taxa can be useful, especially for allergy studies. Pollen and spore calendars are available for some areas, and these may have some value in the neighbourhood of monitoring sites; but they are imprecise, do not have wide applicability, and vary annually and from place to place, so they do not have the precision required for the interpretation of forensic data. Only the very broadest inferences may be made from them.

On occasion, a crime scene may involve a surface where palynomorphs are periodically removed; lawn or hedge clippings constitute the comparator sample, it will represent a palynological 'snapshot'. The profile from that sample will depend on the frequency of removal of the receiving surface (*e.g.*, lawn or hedge), and the nature of the subsequent pollen rain. Seasonality can be of great importance here, and the sample might not be representative of conditions at the time an offence was committed. Wherever possible, it is prudent to obtain comparator samples at the same time the following year.

In most cases, the comparator sample will be accumulated surface soil or plant litter, and these may have received pollen and spores for at least one year. This will make temporal estimates problematic as an accumulated mixture will mask the seasonal highs and lows of species abundance in the airspora and seasonality will be masked.

Where crime scenes involve bodies of water, much of the pollen and spore load in the surface sediments and muddy foreshore may be waterborne rather than airborne. These sites are capable of receiving pollen and spores from whole catchments of rivers or streams, and their profiles are often highly diverse containing many taxa exotic to the local or even national flora. At some crime scenes, the extreme heterogeneity of palynological profiles derived from artificial plantings can produce highly specific profiles of trace evidence and be valuable for linking objects and places.

The vulnerability of pollen and spores to palynological processing in the laboratory is often ignored. Although sporopollenin and chitin are resistant to the highly corrosive reagents used to dissolve background detritus in palynological samples, prolonged exposure can result in disappearance of the palynomorph. This is particularly true of very thin-walled palynomorphs; Juncaceae (rushes) and many of the less robust fungal spores, and delicate appendages, evidently fail to survive the process.

3.3.1.3 Transfer

Pollen and spores are present virtually everywhere. They are easily picked up from some surfaces but can stick tenaciously, or become deeply embedded, in others. They settle onto soils, plant litter, foliage, and inanimate objects, and are a component of household dust. They are breathed into nasal passages, and get trapped by clothing, hair, fur, feathers, and synthetic materials such as soft and hard plastics, and nylon. Soils and other palyniferous materials can be picked up by items such as footwear and transferred to vehicle pedals and other surfaces that have been contacted. There can also be secondary transfer from one item to another, although the pollen assemblage may be diluted in the process. Palynomorphs can adhere for considerable periods even when the mineral component of the soil has been shed during wear or usage.

It is often difficult to retrieve pollen and spores from surfaces such as rough wooden boards, coarse paper, and coarse fabrics. They can also be held by electrostatic forces and the tenacity with which they are held is variable. For example, pollen picked up by shiny leather shoes can be rubbed into a carpet, but it is not so easily transferred from the carpet to other objects. Textiles often act as pollen and spore traps, and it requires considerable effort to extract palynomorphs for analysis. Textiles offer interstices as well as electrostatic forces, but 'dressings' on fabrics reduce both effects. For example, dressed or starched cotton is much less tenacious than synthetic fleece, nylon, or woollen fabric, and the clay coating on high quality paper also reduces the adhesion of palynomorphs. Remarkably, articles washed by machine with strong detergents, or which have been subjected to dry cleaning, have yielded profiles of palynomorphs which could be related back to crime scenes. Washing and drycleaning machines appear to be relatively ineffective in removing pollen and spores from the weave of fabric and the interstices of washable footwear.

3.4 Forensic Science Applications

3.4.1 Forensic Palynology

Standard protocols for forensic palynological analysis are available, but the practitioner will still need to be inventive when obtaining samples from the huge array of items that might be presented for examination. Once analytical material has been obtained, the aim is to remove all background detritus by a sequence of corrosive reagents to produce an homogenised, concentrated assemblage of palynomorphs, stained and mounted on microscope slides. The mountant should ensure that the palynomorphs are fixed so that any that require re-examination can be found and glycerol jelly is the preferred mountant. Another popular mountant is silicone oil but this is unsuitable for forensic analysis because palynomorphs are mobile and cannot be relocated after examination. In the UK adversarial system, all samples must be available for scrutiny by the opposing faction and all pivotally important or contentious palynomorphs must be able to be relocated on a slide preparation.

During processing, every precaution must be taken against aerial and reagent contamination, and the laboratory should ideally have positive air pressure to prevent pollen drifting in draughts. Reagent blanks must be run alongside trials and trap slides arranged around the laboratory to check for aerial contamination.

Initially, each prepared slide is rapidly scanned. This gives information on pollen concentration and the state of preservation, as well as the main features of the community from which the sample was obtained. On occasion, the scanning procedure can provide rapid intelligence for police, and detailed analysis necessary for evidential purposes can follow if required. It is even possible to eliminate sites by just the background detritus remaining after processing, thus minimising cost for the investigating officer.

Analysis is typically carried out using high-power phase-contrast microscopy and any contentious palynomorph is checked against a properly provenanced and named reference material. Conventionally, a minimum of 300 palynomorphs is counted and considered to constitute a 'statistically sound sample'. This number has been broadly accepted by palaeoecologists who work mostly on peat and lake deposits and find that the law of diminishing returns operates after 300 palynomorphs have been identified and counted. In the forensic context no prescribed number is possible. Some samples may yield insufficient pollen and spores to achieve a sum of 300 and yet, the profile may still, on occasion, provide meaningful information. In most investigations, every effort is made to achieve much higher palynomorph counts to maximise the information gleaned from the sample; in some circumstances, counting may run to thousands of pollen grains and spores as this may be the only way to detect critically important rare pollen types or fungal spores.

When counting is finished, each slide is scanned further to pick up any "rare" taxa, although such grains cannot be included in the pollen sum. Analysis is context driven and depends on many factors pertaining to any one investigation. Any unusual or unknown palynomorphs are photographed for further reference, and their position is recorded on the slide by the use of an England Finder™; the location coordinates are important for when slides are made available to palynologists working for the opposing side.

The pollen and spore counts are expressed as percentages of the total sum. On occasion, where certain palynomorphs are over-abundant, others can appear to have disproportionately low representation. In this case, it is standard practice to present the data in two ways: with over-abundant taxa (a) included, and (b) excluded, from the pollen sum. Where samples are of the same material such as soil or faeces are to be compared, 'absolute' methods may be applied by counting in relation to known quantities of markers which are added to standard quantities of sample. This is infrequent, however, and it is invariably impossible to obtain comparable samples. For example, washings from turbinate bones can only be compared to grave soil by using relative counting as described above.

Due to the complexity of the data set, and the large numbers of variables, no suitable statistical technique has been universally accepted for application in forensic palynology. Where the data involve very many comparators and large numbers of exhibits, the dataset can be unwieldy. In such cases, multivariate statistics can be used to draw attention to certain relationships and patterning, but interpretation must be made using botanical criteria. Another approach is to use likelihood ratio analysis (Bayesian analysis) but, although it has recently been applied to palynological data, its value is questionable for forensic science casework. It is also difficult to provide any graphic representation of forensic palynological data, and results are usually provided to the court in tables of percentage data. On occasion, partial pollen diagrams similar to those used in palaeoecology can be constructed to emphasise certain aspects of profiles, but many datasets do not lend themselves to graphic representation.

3.4.2 Body Deposition Times and Post-mortem Intervals

Accurate estimates of post-mortem interval, or the length of time a corpse has lain at a deposition site, can facilitate the identification of unknown victims by narrowing the search in missing persons' registers. There are occasions where entomology may fail to give the required level of accuracy in post-mortem intervals or body deposition times. These are usually when a victim has been dead for a long period and cycles of the first waves of insects have been completed, or where insects have failed to gain any access to the body. In several recent cases, analysis of fungal growth rates on cadaver skin and bone, and on associated materials such as carpets and fabrics, have supplied accurate information.

If a victim has been in the same position for a few years, plant roots or stems growing over or through the remains are capable of being dated by dendrochronology, and they can provide a minimum time for deposition. In the shorter term, the relationship of the human remains to successive falls of leaf litter, growth of grasses, and other herbs can also provide time estimates. In one case where skeletonised remains were found in January, deposition time was accurately estimated from the continued growth of *Calystegia sepium* (bindweed) underneath the corpse, and the inhibition by body fluids of fruit-ripening in female *Urtica dioica* (nettle) plants. The experienced field ecologist simultaneously absorbs many types of evidence, and in some cases, it has been combined observations that have contributed to the accuracy of temporal estimates. Examples are: earthworm and molluscan activity; fungal growth; presence of certain micro-arthropods beneath the corpse and scattered bones; root, shoot, and leaf growth; fern life-cycle stage; recovery of cut and trampled stems; decomposition of fabrics; growth of algae and mosses; differential orientation, growth, and discolouration of lichens; and activity of voles and other rodents. Sequential palynological analysis of various bones, including those of the skull, has also provided useful information. In a case where a skeleton was found in February, it was possible to narrow the time of deposition to a window of about six weeks in the previous autumn from shoot growth patterns of bramble stems, and the presence of autumnal spores of a species of rust fungus growing on bramble leaves in the grave-fill.

3.4.3 Location of Clandestine Graves and Buried Remains

The elimination of sites for clandestine burials saves time and resources for search teams. Large areas of landscape can be eliminated by virtue of their geology, shallow soils, unfavourable topography, and relative access for vehicles and people. If the grave site is suspected, the plant ecologist can give information on the recent history of the vegetation and soils. Places that would have been inaccessible at the time of an offence, or ground that is now inaccessible but was formerly more open, can be identified. Surface vegetation can also indicate hydrology and sub-surface barriers, so unsuitable places are highlighted. Estimates of the ages of trees, fallen wood, thickets of brambles and nettles, ant hills, and other features can also help.

It is relatively easy to eliminate irrelevant places but more difficult to locate the burial or the corpse. Responses of surface vegetation and fungi (especially certain mushrooms) can sometimes highlight disturbed ground, and if grasses colonise the surface of a grave, they can be enhanced by the release of ammonia during deamination of body proteins. The nature of the herbaceous sward will depend on very many factors and can be highly unpredictable and subtle; its development does not necessarily follow a predictable pattern, but an experienced forensic ecologist can be sensitive to even slight anomalies.

If a suspect has been identified, palynological analysis of belongings can yield detailed information regarding the kinds of place contacted. From the assemblages and profiles retrieved from footwear, digging implements and/or vehicles, the experienced palynologist is able to visualise communities of vegetation, sometimes in remarkable detail. Assemblages of pollen grains, plant spores, and fungal spores can give information on: the vegetation; geology; soil type; soil fertility; altitude; hydrology; amount of leaf and woody litter on the ground; and degree of artificiality of the ecosystem. If the botanist/palynologist is unfamiliar with details of the vegetation of a part of the country that is relevant, a detailed description can be given to a local ecologist; this has resulted in the location of graves in several cases.

3.4.4 Differentiation of Kill Sites from Deposition Sites

Frequently, murder victims have been found in places other than where they were killed. If there are sufficient palynological differences between the two sites, and hair, clothing, or footwear of the victim contacted the ground at the kill site, it is usually possible to differentiate between them. Furthermore, the 'kill' site may be so distinctive that it can be suggested by reconstruction from palynological evidence. The same methods as those for finding clandestine graves can be employed (see above). Irrelevant places can be eliminated just by visual inspection since, if the plant communities do not resemble those retrieved from the palynological profile from the victim, they are unlikely to be relevant.

3.4.5 Cause of Death

Irrigation of the turbinate bones inside the skull releases palynomorphs that have been inhaled in life. Even after death when mucosal membranes have decomposed, palynomorphs can remain stuck to the bony complex. The mucosa of the nostrils should be ignored since pollen and spores there could represent contamination before and after death. Cases where the turbinates have been analysed have provided evidence to show that:

(1) A victim was likely to have drowned in bathwater rather than a natural body of water.
(2) Dust retrieved from a neonate indicated that, before being killed, it had drawn breath in the bag of peat in which it was found.
(3) A young child had possibly been suffocated by a pillow with a cotton pillowcase.
(4) A young man inhaled soil from the surface of his grave before being buried.
(5) A young girl had inhaled pollen from the bank of nettles on the approach path to her grave.

Causes of death can also be established from other palynological samples. For example, the analyses of swabs from wounds have shown the presence of pollen from the garden from which the "weapon" was obtained, whilst in another case an analysis of gut contents confirmed the presence of *Cannabis* pollen and *Psilocybe* (magic mushroom) spores.

3.4.6 Evidence of Contact (Trace Evidence)

Fragments of glass, paint, fibres, hair, and DNA are regarded as typical trace evidence samples. Pollen and spores provide superb trace evidence as they can also provide links between people, objects, and places. Offenders walk on soil, dust, vegetation, and leaf/woody litter, lean against buildings, sit on seats, climb across lichen and moss-covered walls and roofs, and brush against vegetation and other palyniferous surfaces. Invariably, these can be directly transferred to anything with which they come in contact.

The whole assemblage of both plant and palynomorph taxa must be considered when comparing and interpreting the palynological profiles of places and objects. Some taxa will generally be considered to be more meaningful than others. A high pollen-producing, easily dispersed, wind-pollinated plant such as *Pinus sylvestris* (Scots pine) is more likely to be picked up than an insect-pollinated plant such as *Salix cinerea* (grey willow) or self-pollinated one such as *Triticum aestivum* (wheat). In most places in Britain, a value of 5% for pine might certainly be less significant than the same value for willow or wheat. However, a high abundance of any palynomorph in a comparator sample may not be reflected on an exhibit because of subsequent wear and abrasion, or even attempts to remove trace evidence by an offender. Furthermore, except where a footprint or tyre mark are involved, the actual place contacted by an exhibit can never be known precisely, and comparator samples are taken from the 'most likely places of contact'.

The distribution of palynomorphs at any location is heterogeneous, and a single comparator sample will inevitably contain only a 'fragment' of the full palynological profile at a site. To gain a meaningful 'picture of place', as many spatially separated samples as possible should be obtained, and the results combined. Exhibits such as footwear and vehicles will also have only fragments of the full palynological profile. Because an offender would have walked about after the offence, there may be a combination of multiple fragments from one critical location on the footwear, and this can be compared with combined comparators. But taphonomy can be more complex.

In some circumstances, the palyniferous material, or accumulated palynomorphs themselves, can be secondarily transferred to other objects, possibly over several vehicles, or several places. Such subsequent secondary transfer will be diluted, and only a third order fragment may be retrieved, but there can still be important trace evidence in partial profiles if the original profile were distinctive enough. In two cases rare fungal spores survived in vehicles that had been burnt and provided critical evidence to the cases and, in another, the profile of the crime scene in a British woodland was retrieved from a vehicle that had driven across Europe.

There is unlikely ever to be a perfect 'match' between any two comparator samples, although the closer sampling sites are to one another, the more similar they will be. Likewise, because of fragmentary profiles and multiple depositions, an absolute match between comparator samples and exhibits can never be expected. The degree of confidence in determining whether there had been contact depends on the richness of the respective palynological profiles and is enhanced by the presence of rare taxa.

For links between exhibits and crime scenes to be acceptable to the court, there needs to be either a complex assemblage, with many points of similarity between place and object, or some exceptionally unusual or rare component. If both are present, palynological trace evidence can be very powerful. This makes the forensic palynologist a useful member of the investigative team, and in spite of all the caveats that apply, palynological profiles that can be distinctive enough to demonstrate convincing links between items and places have been repeatedly demonstrated.

3.5 The Crime Scene

The role of the botanical ecologist and mycologist can be pivotal at a crime scene, and if someone with appropriate skills is not consulted early in an investigation, critical information can be lost or compromised.

3.5.1 Inside Buildings and Vehicles

A building or vehicle may act as an incubator for insect and fungal growth and development. The value of forensic entomology is accepted when deaths occur indoors, but few realise that mycology may give equal, or even better, information under some circumstances. Indeed, fungal colony growth has given accurate post-mortem interval and other temporal estimates in several recent cases. Unlike flies, fungi are not affected by the dark, and will continue to grow and develop over a wide range of temperature. They can also grow on many materials – even glass, stone, and polished wood – if there is a film of nutrient available. Soil and dusty footprints found at indoor crime scenes can also yield botanical, mycological, and mineralogical trace evidence that might link offenders with specific and pertinent outdoor places. Palynological profiles of scenes inside buildings can also be compared with those from exhibits as footwear transfers palynomorphs to carpets and other furnishings, and even the presence of cut flowers might give specificity to an indoor crime scene.

3.5.2 Outside Buildings and Vehicles

Forensic ecology is mostly applied to outside scenes of crime, and it can provide intelligence, and aid the police in interviewing suspects. Ideally, the forensic ecologist should access the site as soon as the offence is reported, and evidence retrieval is maximised if scientists and police work according to a hierarchy for access. There are no fixed protocols, and each case must be treated individually. The following should be avoided:

(a) trampling;
(b) destruction of botanical/mycological evidence by cutting or raking;
(c) contamination by police, archaeologists, the pathologist, and other forensic scientists;
(d) setting up approach paths in inappropriate places;
(e) injudicious placing of generators, lights, and tents.

The use of tents and other coverings can result in ambient environmental conditions being greatly altered in a short time. The incubation effect can enhance the performance of some organisms, such as insects and fungi, but adversely affect plants. If a tent is unavoidable, one side should be left completely open to minimise the incubation effect which will affect information on timings.

There can be pressure to remove a corpse to the mortuary, but this should be resisted until an adequate and appropriate ecological survey can be conducted. Indeed, the place where the offence has been committed, or where a corpse has been deposited, should be protected, especially if vertical barriers to air flow, such as fences, hedges, and trees, are involved. The ground beneath the body may yield important temporal evidence, and palynological profiles from vertical elements can give information on sequences of events occurring during an offence.

In most cases at outside sites, plants and fungi continue to grow even in suboptimal weather conditions and compared to flies, they are relatively undisturbed by ambient temperature, humidity, wind speed, and light intensity. An event such as disturbance or deposition of a corpse may impact the growth and appearance of fungi and plants, but their changed performance can provide valuable clues to previous events, and the order in which they happened. Such subtle changes can be destroyed by inadvertent disturbance by investigating teams.

It is impossible to construct ecological protocols applicable to all scenarios, but several kinds of information can be provided by ecological techniques, and crime scene practice should facilitate them:

(1) Direction of offender approach and departure pathways;
(2) Timings and order of events;
(3) Traces of evidence transferred from offenders and victims, and traces transferred to them from the crime scene.

The first two can be evaluated only by inspection of the crime scene while trace evidence is obtained from samples collected at the scene and examined later. The precautionary principle should always be applied, and adequate and appropriate samples should always be collected even if they are not used subsequently.

3.5.3 Crime Scene Evaluation

After being briefed by the police, the ecologist should wear full protective clothing, including hair and foot covers and mask, and access the crime scene *via* the common approach path, keeping to foot plates where being used. A comprehensive photographic record should be obtained of vegetation at the crime scene, both in a panorama using 360° and close-up where necessary. Photographs should also be provided from any other place pertinent to the case such as the garden of a suspect's home. Subsequent visits should also be made to those places.

One of the first tasks of the ecologist may be to identify an offender approach path. Often, there would have been few options for access because of natural barriers (banks of bramble or nettle, deep streams, barbed wire fences, *etc.*) and, once these have been recognised, anomalies in surface litter, foliage, stems, and soil should be noted. Direction of access can be determined and sometimes timed by:

(a) orientation of lichens and other fungi;
(b) angles of broken plant stems;
(c) amount of etiolation (growth in partial or complete absence of light) in previously covered shoots and leaves;
(d) compensatory growth of specific plants;
(e) footprints in wet leaf litter and soil;
(f) discolouration of specific lichen species.

3.5.4 Vegetation Survey

It is essential to list as many plant species as possible at a scene of crime, and at any other site pertinent to the investigation. Since comparator samples will necessarily represent a portion of the place, plants growing there might not be represented in any palynological comparator profile, and yet it may be retrieved from a critical exhibit. It is important to know the potential range of taxa that could be found in a palynological preparation and, although sampling would be targeted in an attempt to include actual locations contacted by an offender, there is no guarantee that they would be sampled. The actual profile obtained from a range of comparator samples, combined with a list of all potential taxa that might have been transferred to an offender, is the best that can be achieved. Since some pollen is airborne, it is often necessary to extend the botanical survey to as wide an area as possible around a crime scene. Any other location pertinent to the investigation should also be visited, surveyed, and sampled.

This approach has the benefit of economy since, in many instances, places can be quickly eliminated from, or implicated in, an enquiry by comparison with their vegetation and the palynological profiles obtained from exhibits. Rapid scanning of preparations from exhibits will allow the palynologist to assess the feasibility of the item having contacted specific places. However, this can only be done effectively if the field survey has been carried out by the ecologist/botanist/palynologist personally responsible for the analysis and interpretation of the samples. Because of complexities of taphonomy, as a high priority, the analyst should always visit the crime scene and other places pertinent to an enquiry. If there is a lag between the offence and the ecologist being involved, it is often pertinent to survey the site on the anniversary of the crime.

3.5.5 Sampling from the Crime Scene

Often, the aim in a conventional ecological evaluation is to determine the abundance, distribution, and performance of various plant and animal taxa within the environment. Here, sampling methods are designed to maximise the kind of information required; the aim is to evaluate the environment in terms of what can be compared with items retrieved from items associated with an offender, or what can be useful for timing events. The sampling employed does not lend itself easily to statistical approaches partly because, due to constraints on time and resources, it has to be targeted and closely focused rather than random. Soil and/or vegetation will be taken to compare trace evidence from suspect(s) and the site, or to estimate timing of events. These "comparator" samples are not controls *sensu stricto* since the aim is to focus on the most likely places contacted by the offender; sampling is generally systematic along paths, and in any place likely to be pertinent.

Any sampling protocol will be a compromise, but every attempt should be made to obtain as comprehensive palynological profile as possible, and this involves multiple samples; the larger the sample area, the more accurate will be the assessment of the ecological status of the location. It has proved productive for each ground sample to be typically 30 × 30 cm. Another approach is to take pinches of surface soils over a prescribed plot although this precludes any information of surface variation in a place. Samples are homogenised in the laboratory and aliquots taken for analysis, each sub-sample representing an area of at least 900 cm^2. Because there is often limited opportunity for collection, and as a cover against accidental loss, it is advised to take more samples than considered necessary.

An offender footprint will provide ideal samples, but otherwise, only material that is likely to have been contacted should be collected. In most cases, the surface (and no deeper than 0.5 cm) should be lightly scraped, using a fresh sterilized teaspoon or spatula. Deeper sampling may be necessary in the case of soft surfaces where feet have sunk into the soil or mud. In the case of buried remains, there is a high likelihood of the offender having stood in the grave-fill, so it is important to obtain a large sample of excavated grave spoil for later homogenisation in the laboratory. Because soils change dramatically with depth in terms of organic content, biota, and chemistry, whole vertical grave profiles can be obtained with stainless steel rectangular 'monolith' holders. All samples can be stored in plastic bags and frozen, while litter or plant material is best placed in paper containers and kept dry to prevent fungal growth. Fresh plant material should be laid flat between sheets of absorbent paper (*e.g.*, clean newspaper) for later identification. Each sample should be labelled with: GPS reading; type of material; date; time; collector reference; operational name; and police force. Samples are then given to the exhibits officer for storage and transport to the laboratory as appropriate.

If a body of water is involved at the crime scene, samples should be obtained at the water's edge, the water itself at various depths, and the basal sediment at all locations deemed appropriate by investigating officers. Water and mud samples can be frozen for storage.

3.5.6 Samples from a Mortuary

The cadaver may provide the best comparator of all, and it should be appraised as a source of important botanical, mycological, and, occasionally, zoological evidence. Hair, skin, nails, and upper nasal passages can be sampled for palynomorphs and other evidence, and techniques have been devised to obtain the maximum evidence. If a corpse had been stored before burial, fungal growth on skin, bone, and items associated with the body, may provide information on the storage conditions, as well as post-mortem interval, or time since deposition.

Any botanical or mycological material retrieved by the pathologist should be given to the botanist/palynologist. Combing of hair must be followed by washing and centrifugation of the washings for palynological analysis. This task is best carried out by the botanist/palynologist.

Rather than accepting standard practice, the pathologist should be asked to retrieve the whole of the stomach contents. If a victim is thought to have survived any period after a fatal attack, or if poisoning is suspected, a request should be made for samples to be provided from several areas of the ileum (small intestine), ascending, transverse, and descending colon (large intestine), the rectum, and possibly any faeces in the gut. There are some data on the speed of transit of food through various parts of the gut for healthy individuals, so the presence in the lower gut of food items, known to have been eaten by a victim, can help provide temporal estimates pertinent to the case. Vomit found at a crime scene should also be sampled as it may be analysed in the same way as gut contents and, importantly, reveal the nature of the last meal of the deceased. The cellulose cell walls of plants render them indigestible, and they are usually well preserved, so it is often relatively easy to identify seeds, fruit tissues, leaf and stem fragments, and even root-crop food and fungi, but detailed knowledge of plant anatomy is crucial in gut content analysis since the material is often comminuted into plant tissues, and even to individual cells.

Starch grains (amyloplasts) are nearly always present in plant foods, and they may be seen intact inside cells, or floating freely in stomach fluids. Starch grain analysis can aid the identification of the plant with considerable precision, but digestive enzymes (amylases) in saliva and the stomach start to degrade them as soon as they are eaten. Partially degraded starch grains are frequently found in stomach fluid although very commonly, any starch seen is amorphous and gelatinous and this usually indicates some proprietary food as the source.

Identification of pollen and various kinds of spore can be particularly important in cases of poisoning and the use of recreational drugs such as *Cannabis*, opium poppy seeds (*Papaver somniferum*), magic mushrooms such as *Psilocybe semilanceata*, and even exotics such as South American *Psychotria viridis* one of the component plants of the hallucinogenic mixture Ayahuasca.

There is no absolute benchmark for the time taken for food to pass through the gut, but digestion of plant and animal tissue can be recognised, and it is sometimes possible to give an opinion that the food has been in the stomach for more or less than, a specific period. The estimates are crude, but they have proved useful in some investigations.

3.6 Databases

Databases, some comprehensive and many limited, have been constructed for a variety of manufactured materials such as footwear, fibres, paint, glass, and bullets, and practitioners' reference and population data are available for footwear patterns, fibres, manufactured glass, and gunshot residues, and there is even an injuries database of marks caused by impact with manufactured objects. Databases of biological material are not as easy to compile. DNA and fingerprint profiles are considered to be individualising, and there are national databases of human samples for comparison with unknowns. Analytical advances show that the DNA of monozygotic twins can now be differentiated, and the added complication of chimerism is becoming recognised. Methods and expertise have increased in recent years, and difficulties in the interpretation and evaluation of both DNA and fingerprint analysis have been highlighted, with fallibilities in both presented in court. Apart from application in wildlife crime where many species are studied, both DNA and fingerprint study involves just one species (*Homo sapiens*) whereas the forensic ecologist is faced with vast numbers of species, both micro- and macroscopic – both living and dead.

Comprehensive databases involving living or dead non-human organisms are fewer because of the sheer numbers involved and the fact that so many new species are classified each year. Nevertheless, there are databases covering the geographical distribution of plant, animal, fungal, and microbial species, and these can help broadly to identify habitat location.

An extensive literature is available to aid identification of plants (including diatoms and other algae), animals, fungi, other microbes, and fossils, and museum herbaria and fungaria, such as those at the Royal Botanic Gardens Kew, invariably hold reference collections of most species of plants and fungi, and most practitioners make significant efforts throughout their careers to collect and construct their own compendia of reference material.

Although the UK is well served with detailed databases, accessible through the UK National Biodiversity Network, there is better coverage for groups for which there is a long history of biological study, and in regions with long recording traditions. The Global Biodiversity Information Facility (GBIF) accumulates records of all organisms from anywhere that are submitted to it, and has inputs from many national recording centres, though quality control is something of an issue as information is mostly organism based and used primarily for purposes of conservation. In the United Kingdom, the *National Vegetation Classification: Plant Communities* (five volumes) is valuable, as well as records in the many County Floras, often researched and compiled by amateur botanists and naturalists. The Botanical Society of Britain and Ireland's *Atlas of the British and Irish Flora* is especially valuable in the UK because of its intense level of recording but, in many parts of the world, recording may be sparse or even absent. Although they can be useful for highlighting or eliminating pertinent areas, databases alone are of limited value for forensic investigation. Describing the kind of place from proxy indicators is sometimes easy for an experienced practitioner, but there can never be an 'easy fix' for the inexperienced by referring to a database.

Criminal and civil investigations themselves provide databases of information most valuable to forensic investigation but, invariably, the information resides in the archives of individual practitioners and dissemination of that information depends on their publishing case histories. In such publications, there will be examples of species assemblages, and the trace evidence they generate, with results where high likelihood of matches will be demonstrated between locations and any items that have contacted them. But usually, it is only possible to predict the kind of place from which an assemblage has originated though, with additional information and police intelligence, it has been possible to locate specific sites with considerable accuracy. Being able to describe the nature of a location from its proxy evidence is invaluable, but the acquisition of a database of individual places is untenable because each one will have an individual palynological profile.

3.7 Statistical Techniques

All results obtained in ecological investigations are probabilistic, and caveats must always be applied when presenting interpretations to the court. It is often assumed that there will be statistical techniques presented to aid interpretations, but their value is limited in the face of ecosystem heterogeneity and the sheer numbers of variables involved.

Generally, if the quantitative results from each of a large number of comparator samples are processed with those from an equally substantial number of exhibits, dimensionality-reduction techniques such as principal components analysis (PCA) and discriminant analysis (DA) are useful and can help focus attention on the most likely exhibits. To evaluate the dissimilarity/similarity of samples, other techniques such as the Bray Curtis dissimilarity index (erroneously termed distance measure), and true distance measures such as Euclidean, Minkowski, or Canberra distances have been used, as well as Bayesian Networking methods. Their relative performances have also been tested by log-LR cost (Cllr) calculation. The experienced palynologist may certainly use a Bayesian approach when formulating initial, mutually exclusive, hypotheses, but some forensic environmental scientists appear to have an over-reliance on theoretical approaches and, some erroneous conclusions have been published. It is unwise to rely mainly on such methods, and all final decisions should be based on personal expertise.

3.8 Experimentation

In nature, organisms, parts of organisms, and their physico-chemical environments, occur in a huge (unknowable) numbers of combinations – ecosystems can overlap and merge one into another, and there are transitional situations. Natural, semi-natural, and artificial habitats are highly heterogeneous, especially at the microscopic level, and experience has shown that it is possible to differentiate sites that are just a few metres apart. An experienced ecologist can easily identify, for example, an oak woodland from its proxy indicators – but no two oak woodlands, or even parts of the same woodland, are identical, and recognition of a specific oak woodland (or an area within it), requires detailed profiles of proxy indicators relating to the specific microsite. Even so, in this example, it is of great value early in an investigation for all other kinds of woodland to be eliminated. Effort has been devoted to constructing predictive models from proxy indicators such as minerals, pollen, and other markers but, for forensic purposes, no model can provide sufficient specificity at the level required for probative evidence.

Forensic ecology, and specifically palynology and soil science, have become extremely popular subjects in university courses as relatively small, convenient research projects can be designed to demonstrate some potential value of the discipline. Most academic papers are based on the results from single experimental trials, often using student-generated data and while the laudable aim may be to provide information upon which predictive models may be constructed, the notion that such experiments will be practically useful in forensic casework is flawed. It cannot be stressed strongly enough that, although these topics are useful to teach research methodology and approaches to problem-solving to students, the results generated will be insufficiently robust to furnish conclusive databases or models for actual cases unless performed in large scale.

Within an investigation, small research trials on, for example, pollen and spore taphonomy at a crime scene can be useful, and individual plant species can be tested to evaluate their responses to various stimuli to aid interpretation of observed phenomena. Even then, the number of biological and environmental variables encountered at any crime scene (indoors or outside) is so overwhelmingly large that, except in the broadest of terms, results from tests *within* investigations can never be extrapolated to other scenarios. It is impossible to reconstruct and replicate, in the field or the laboratory, the variables encountered at any crime scene.

An effective forensic ecologist is one who has worked on investigations and has accrued a large personal database of case histories. Such a database, in effect, reflects experiments in nature and can be a guide to avoid unnecessary work – but in every case prior assumptions must be minimised because of the individual nature of every place and every scenario. No individual would be able to replicate sufficient trials for any aspect of forensic ecology to enable more than broad predictions for intelligence purposes. It is of concern that results from so many preliminary experiments have been imbued with the gravitas afforded by publication itself. Information in papers is often taken as being accepted wisdom and quoted uncritically in further publications. The practitioner starting work in any case of serious criminal investigation should be wary and critical of forensic experimental literature in ecology and instead, rely on basic principles, methods, and databases – and be wary of cognitive bias at every stage of the work, particularly in the interpretation of results.

3.9 DNA Identification and Molecular Ecological Profiling

3.9.1 DNA Identification

Identification of plants and fungi by DNA sequencing has increasingly been used over the last 10 years or so, and there is now international consensus on the most appropriate choice of primers and sequences ("barcodes") for species identification in different groups. Identification below the species level, even to crop and other cultivated plant varieties, let alone individuals, is much more problematic except in well-researched cases, but this is likely to improve. In most forensic cases, plants and fungi have to be identified by their morphological characteristics. This is achieved by visual inspection and may require a low power lens, light microscope, or scanning electron microscopy; these require relatively simple preparation of the unknown for comparison with known reference material. However, sometimes, the state of the material makes identification by these visual techniques impossible, and then identification by analysis of their DNA offers an alternative.

Molecular techniques are of particular value where the identity of a species is suspected but cannot otherwise be confirmed. DNA sequencing can then sometimes confirm or eliminate its presence from fragmentary samples such as powders, or tablets. For example, DNA has confirmed the species identity of *Cannabis sativa* (hemp) in drug-related criminal cases, and of seeds of known plants such as *Solanum lycopersicum* (tomato) in faecal matter.

In forensic cases, it would sometimes be particularly useful to know whether an *individual* of a species is involved. As in humans, if the DNA of tissues from a particular cat, bird, deer, or cow, were compared with that of a *specific* animal, its identity may be confirmed. Also, as in humans, there must be reference material of the animal for comparison as well as an adequate DNA database. When considering identification of specific organisms, chimaeras must be borne in mind. Plants and animals (including humans) can possess more than one genome, with cells and tissues of differing genetic constitution being variously distributed throughout the body; this phenomenon is much more common than previously thought and problems of sampling for comparison with reference material can be envisaged.

The situation of plants and fungi is much more complex than that of most animals. A diploid plant will contain DNA from both parents so that plant tissue may be linked back to an individual if that individual is known, but many reproduce only by vegetative means and form clones of hundreds or even thousands of plants with an identical genetic constitution. Clones are also created by horticulturalists and agriculturalists, and commercial production means they can be widely dispersed geographically, with identical individuals spread nationally or even globally. Clones can be identified by molecular techniques, and this has been demonstrated for *Cannabis*. An example of the potential and limitation of this approach is shown in a recent molecular study of *Populus nigra* (black poplar) in Britain. It showed that there are at least fifteen clones of this species present, and that members of a single clone are dispersed throughout the country; one particular clone had genetically identical individuals in Cheshire, Yorkshire, and Essex. Added to the problem of cloning, many species regularly hybridize so the genetic profile can be complex. Large human DNA databases are necessary because the human global population runs into billions, and in the case of plants and fungi, each species would need a similarly large database; even then, the problem of cloning and hybridization would remain a problem.

It must also be remembered that the publicly available sequence databases, such as GenBank, do not include barcode sequences from all known species; in the case of the fungi, only around a third of the described species are represented – and in some genera different species have identical barcodes so different gene sequences have to be used to discriminate between them.

Palyniferous surfaces from which pollen could be easily retrieved, and so in a condition suitable for molecular comparison, rarely feature in forensic cases. Invariably, palynomorphs are enmeshed or embedded in fabrics, soils, or other matrices, and the methods used to retrieve them usually remove the cellular contents. DNA-based identification is also unfeasible for pollen and spores present in palynological preparations as the acids used in processing denature the DNA. If clothing or some other object picked up pollen/spore dust of a single taxon in sufficient quantity, or when dealing with a living fungal colony on an object, DNA identification might be possible. However, the dilemma of an adequate database remains (see above) and in most cases identification is more likely to be achieved and more quickly and reliably by microscopical examination.

3.9.2 Molecular Ecological Profiling

There is now a huge and rapidly expanding body of molecular work on ecological profiling. This can involve metabarcoding (comparison of short DNA sequences in a sample), metagenomics (comparison of total genomes in a sample), and also machine learning technologies using microbial DNA, in particular for provenancing and as trace evidence. There have been promising results in differentiating countries globally using dust and soil samples; but the challenges for these techniques in actual forensic investigation are great.

The basic aim is to obtain molecular signatures of environmental samples by extracting the DNA content from all the organisms in a sample using metabarcoding or metagenomics. Some of the sequences will match those of named species represented in the genetic databases, but invariably many will not – a consequence of the incompleteness of our knowledge (see above). In such studies, several samples are generally taken from an area of interest at a depth down to 5 or 10 cm, homogenized, and the DNA analysed as if from one spot. This is unhelpful if a link with items of footwear, for example, is sought as only the point of contact with the actual surface would be really informative.

The reproducibility of such studies is also crucial for forensic cases as the microbial communities being targeted are highly heterogeneous, ephemeral in nature, and are hugely responsive to ambient environmental factors which can change diurnally, or even hourly. There have been advances in differentiating microbial profiles from homes and offices, but the high diversity and temporal instability of such profiles means that they are not yet capable of producing reliable, probative evidence. There are also challenges in creating reliable and comprehensive databases to facilitate the interpretation of results, and the development of standard operating procedures is fraught with difficulty in spite of the enthusiasm and optimism of some authors.

There remains a question over reproducibility as the processing of such samples is costly. Sequencing from the same extract is exceptionally repeated, and this leaves issues open to question in any court situation. The methodologies are, however, continually being improved and the forensic ecologist should monitor them as a time may come when they will have more utility than is currently the case. For the moment, it seems that data captured by the human eye and computing within the brain are not yet redundant and will not be superseded for a considerable time.

3.10 Forensic Ecology and the Courts

The forensic ecologist/botanist/palynologist must observe forensic science protocols for:

(1) Keeping adequate records;
(2) Continuity of possession;
(3) Submission of reports;
(4) Making statements;
(5) Disclosure;
(6) Making required declaration of understanding;
(7) Providing indices of unused material.

Evidence from forensic ecological investigations has been tried and tested many times in the British Court system. It has been included successfully in cases which have resulted in conviction of offenders and, indeed, on occasion has provided the only forensic science evidence. Detailed protocols have now been published to aid the practitioner and lawyers but, unlike some areas of forensic science, practitioners do not come within the remit of any formal accreditation body. A consequence of this is that palynologists with limited botanical or ecological background and experience, and virtually no experience of working in the forensic context, can offer their services to the police. Those wishing to utilise the service of a forensic ecologist should, however, look at their academic qualifications, publications record, and also their standing in pertinent learned societies.

Some aspects of forensic ecology are easily tested. For example, estimations of post-mortem interval, or deposition times, can be confirmed by testimony of witnesses, or confession of an offender. If clandestine graves are located, then their very discovery verifies the data and, if the data were palynological, then by default, these would implicate the owner of the exhibit(s) from which the palynological information was obtained. Presentation of palynological data as trace evidence is more difficult. A jury composed of non-scientists has to be introduced to the concept of taphonomic variability and a plethora of unfamiliar terms and ideas. These data do not lend themselves easily to graphic representation, and the palynologist has to have an opportunity to explain, in some detail, the significance of the various values for palynomorph taxa. The palynological report must be written as clearly as possible to make it accessible to the non-specialist, but palynological data are rarely simple, and over-simplification can lead to spurious arguments. The best approach for dealing with ecological and palynological data would be for both prosecution and defence specialists to work together for a consensus to be presented to the court, with statements of what they agreed upon and what differences remained, but to date, this has rarely happened in the United Kingdom in criminal cases although the practice is well-established in the Civil Courts.

3.11 Conclusion

The usefulness of a forensic ecologist with both field and analytical experience is that once a profile of trace evidence has been obtained from an exhibit (whatever its nature) the forensic ecologist can eliminate or implicate locations by eye at minimal cost when combining botany, palynology, mycology, and a knowledge of soils. Even quick scans of preparations can inform the investigator whether the techniques will be useful to their case or not. This presents a very great advantage over other kinds of environmental analyses where comparison necessitates detailed analysis of all samples, both exhibits and comparators, before any comparison can be made. Confirmation of degree of similarity and specificity between places and objects can be obtained through more detailed analysis of comparator samples after irrelevant locations have been discounted. A major difference between pure research and forensic work is that the forensic practitioner must work with what is available rather than being able to conduct planned projects under controlled conditions.

What is especially gratifying is that in around the 300 or more cases in which forensic ecology and palynology (including mycology) have been used in the UK, the results have always been supported by other findings. Further, the results have often been so overwhelming and clear they have led to guilty pleas, so saving taxpayers money and the bereaved the stress of protracted court hearings.

Further Reading

R. F. Evert and S. E. Eichhorn, *Raven Biology of Plants*, W.H. Freeman and Company Publishers, New York, 8th edn, 2013.

D. L. Hawksworth and P. E. J. Wiltshire, Forensic mycology: the use of fungi in criminal investigations, *Forensic Sci. Int.*, 2011, **206**, 1–11.

D. L. Hawksworth and P. E. J. Wiltshire, Forensic Mycology: Current Perspectives, *Res. Rep. Forensic Med. Sci.*, 2015, **5**, 75–83.

Palynology: Principles and Applications, ed. J. Jansonius and D. C. McGregor, American Association of Stratigraphic Palynologists Foundation, Dallas, 1996, vol. 3.

P. E. J. Wiltshire, Protocols for Forensic Palynology, *Palynology*, 2016, **40**(1), 4–24.

P. E. J. Wiltshire, Forensic Botany, in *Encyclopedia of Forensic and Legal Medicine*, ed. J. Payne-James and R. W. Byard, Elsevier, Oxford, 2nd edn, 2016, vol. 2, pp. 520–527.

P. E. J. Wiltshire, D. L. Hawksworth, J. A. Webb and K. J. Edwards, Palynology and mycology provide separate classes of probative evidence from the same forensic samples: a rape case from southern England, *Forensic Sci. Int.*, 2014, **244**, 186–195.

P. E. J. Wiltshire, D. L. Hawksworth, J. A. Webb and K. J. Edwards, Two sources and two kinds of trace evidence: enhancing the links between clothing, footwear and crime scene, *Forensic Sci. Int.*, 2015, **254**, 231–242.

4 Bloodstain Pattern Analysis

Jo Millington

Spattered Ltd, UK
Email: jo@spattered.co.uk

4.1 Introduction

Bloodstain pattern analysis (BPA) is a discipline within forensic science that deals with the examination and identification of bloodstains and the interpretation and evaluation of those bloodstains to attempt to establish the activities that may have caused them. Because there is a direct requirement for blood, BPA is inextricably linked to violent crime. Although it can be powerfully applied within the context of crime scene reconstruction it also has relevance in the assessment of bloodstaining on items, for example clothing and weapons, that are examined in the laboratory.

Human observation forms a significant component of BPA and, as a consequence, it is widely considered to be subjective. However, the discipline is underpinned by sound scientific principles and a body of research that dates back to at least the late 19th century. Piotrowski's work for example, the 'Origin, shape, direction & distribution of the bloodstains following head wounds caused by blows', forms a solid foundation for the observation-based interpretation that can be applied by scientists today.

Whilst there is the potential for every bloodstain pattern to be unique, in practice there is a wide range of variability in patterns that

Crime Scene to Court: The Essentials of Forensic Science, 5th Edition
Edited by Niamh NicDaeid and Peter C. White

Published by the Royal Society of Chemistry, www.rsc.org

are encountered even if they have been caused by the same action. However, the application of the principles of fluid dynamics, in conjunction with experimental data and knowledge assimilated through experience can ensure that the identification and interpretation of bloodstain distributions is sufficiently objective to be used with confidence in the criminal justice system.

BPA can offer an insight into blood evidence that not only allows the activities associated with blood loss to be determined but critically, from the perspective of the court, can be probative when evaluating alternative scenarios that have been proposed: *he kicked versus he did not kick, but walked through the blood of the injured party*.

Although DNA profiling can be applied effectively to help determine from whom the blood could have originated, if the source of that blood is not challenged, the question 'how did the blood get there?' becomes relevant. This 'activity-level' proposition sits comfortably within the discipline of BPA and is an area where the evaluation of bloodstain distributions can be most effectively applied.

BPA may assist an investigation in a number of ways. It can be especially probative at scenes where assaults are known to have occurred and can be used to establish the relative positions of individuals as blood was being distributed, the possible sequence of events and whether blood had been subsequently disturbed – such as if cleaning had taken place. Pattern recognition can be particularly powerful if it allows blood that may have originated from different people to be identified. For example, if blood from an assailant can be differentiated, by inspection of its shape or location, from blood of the victim. In these cases the expedited DNA analysis of indicative stains can provide information that, in conjunction with BPA, can be extremely useful in the early stages of an investigation.

In addition to pattern recognition, by understanding how the physiological characteristics of blood influence stain formation, how blood interacts with surfaces and the factors that impact on how blood is distributed in various scenarios, it is possible to form an understanding of how blood might be expected to transfer when allegations are made. This is where BPA can add value to a case and it can be equally applied in the evaluation of the absence of blood, an area of BPA that is frequently under-utilised. It is a fallacy that an individual will be heavily stained with blood if they are involved in an assault. In fact, in many cases, it is the individuals that are not directly involved in an assault, but within proximity to it, that become stained more heavily than the assailant.

As with many forensic disciplines it is rare to be able to offer a definitive BPA opinion; however, by understanding the strengths and limitations (and communicating these properly) a robust conclusion is achievable and, when taken into account with evidence of other specialists, can be extremely probative.

The way in which BPA can contribute to an investigation undergoes somewhat of a change as the process evolves. At the start of an enquiry a scientist may attend a scene and examine bloodstaining with little or no other information. In these circumstances the bloodstain pattern analyst may be able to answer investigative questions that are 'crime-centred': *What has happened here? What caused the blood to be distributed? Where did an assault take place? Was a weapon used? Was the assailant injured? Have attempts been made to conceal or clean up blood? Are we looking for a body?*

As the investigation progresses and information emerges, scenarios might be put forward to explain the blood findings, by a suspect for example. In this mode BPA can be used to test propositions that relate to the core concerns of the prosecution and defence and here the scientist works evaluatively. In evaluation mode, the questions under consideration are usually 'defendant-centred': *Do the blood distribution findings provide support to establish whether or not the defendant punched* or *kicked* or *stamped upon the victim as alleged?* Answers to hypotheses such as these are usually of help to the court.

In relatively recent history a small handful of professional BPA organisations have emerged. This includes the International Association of Bloodstain Pattern Analysts (IABPA) that was founded in 1983 with the aims to encourage and promote BPA science, standardise scientific techniques, promote education and encourage research and disseminate information.

In 2002 the Federal Bureau of Investigation (FBI) established a Scientific Working Group on Bloodstain Pattern Analysis (SWGSTAIN) in order to facilitate collaboration between practitioners within forensic laboratories, law enforcement, private industry and academia. This group has been replaced by the Physics/Pattern Interpretation Scientific Area Committee within the Overseas Security Advisory Council (OSAC). OSAC was initiated by the National Institute of Standards and Technology (NIST) but this committee continues to develop standards and guidelines relating to the scientific detection and analysis of bloodstain patterns at crime scenes and on associated evidence. The importance of these ongoing collaborations cannot be overstated and they have led to the development of resources which address the substantive and operational issues within the field of BPA.

These include best practice guidelines in relation to recommended terminology, which encourages consistent communication within the discipline, along with an extensive repository of relevant literature and models for education and operational quality frameworks. In addition, these groups support and commission a wide range of research, not least of which has involved the development of an open-source database on blood dynamics captured using high-speed photography.

Best practice in BPA is further defined in the UK Forensic Science Regulator's Guidance on BPA which is updated regularly.

In addition to the broad literature that is available, this chapter aims to introduce the fundamentals that underpin BPA and illustrate how BPA can be used in the investigation of violent crime.

4.2 Bloodstain Pattern Classification and Terminology

It goes without saying that the development of a robust taxonomy is critical when identification is based on the recognition of features. BPA is no exception and over the years bloodstain pattern analysts have strived to develop a comprehensive classification system that can be applied to the broad spectrum of bloodstain patterns that are encountered. A number of schemes, which have been defined through research, observation and casework exist. All have the common aim to encourage reproducibility in the interpretation of bloodstain patterns. One such system utilises 'Objective Criteria' which facilitates an objective and consistent approach in the differentiation of bloodstain patterns and encourage the visual examination (and recording) of the following features:

- Shape of the whole pattern
- Alignment of bloodstains in respect of the whole pattern
- Alignment of bloodstains with regard to other stains
- Distribution of stain sizes
- Distribution of stains across the pattern (sometimes referred to as the density of stains)

Irrespective of the scheme that is adopted, the classification of bloodstains is fundamentally based on fluid (blood) dynamics and, on a basic level, whether the pattern is formed of blood drops.

The spatter group incorporates blood distributions that are composed of blood drops and examples include drip stains, impact, cast-off and projected patterns. The non-spatter group deals with blood distributions that are not composed of spatter and includes transfer stains (also referred to as contact stains), altered bloodstains and larger volume stains such as pools of blood. Patterns that exhibit a combination of spatter and non-spatter features are often sub-classified as complex or composite patterns.

If a bloodstain or pattern of bloodstains exhibits diagnostic features then it should be possible for a trained observer to assign that bloodstaining to one of the main spatter, non-spatter or composite groups. The features that allow bloodstains to be categorised accurately are outlined in the following sections.

4.2.1 Blood Spatter

Our understanding of blood spatter dynamics builds on a foundation of knowledge relating to how single drops of blood behave. This section primarily concerns the formation and properties of blood drops falling passively under the influence of gravity. Examples include blood dripping from wounds or weapons and the phenomenon that is associated with drip patterns where a liquid, at least one of which is blood, drips into another liquid causing the production of satellite staining, or secondary spatter. Consideration is then given to more complex spatter patterns which comprise multiple droplets of blood. The latter are synonymous with assault-related activities.

4.2.1.1 Single Drops of Blood

The formation of a blood drop is dependent on two competing forces: gravity and surface tension. Surface tension acts to retain blood at the dropping site and, when it is overcome by the force of gravity, blood is able to drip from the surface. Once formed and in free fall, a drop of blood will momentarily oscillate before forming a sphere. The sphere represents the drops stable state and it will not break up in flight unless it is acted upon by another force.

The volume of a blood drop is determined predominantly by the shape and size of the dropping site, although the rate of flow of blood and whether the surface is already wet with blood also play a part. If blood is present across a larger surface area, such as a fist, the surface tension of the blood tends to be greater than if the surface area is small, such as blood that has collected at the tip of a knife. As a consequence drops formed from the broader surface area tend to be larger. If the available surface is extremely small, for example blood collected on a single hair or fibre, then it may be impossible for gravity alone to overcome the surface tension of the blood and no drops will form.

Because a number of factors are at play, there is no standard drop volume and the resulting blood droplet can vary from as little as 15 microlitres (μL) up to 100 μL, or in some circumstances more.

4.2.1.2 Stains Resulting from Single Drops

The shape of a stain formed when blood lands onto a target will also depend on a number of factors, not least the angle at which the droplet impacts the surface (Figure 4.1). When a drop of blood falls perpendicularly (*i.e.* the trajectory of the droplet and the plane of the target are at 90° to each other) onto a perfectly smooth flat surface for example glass, the resulting stain, referred to as the drip stain, will be

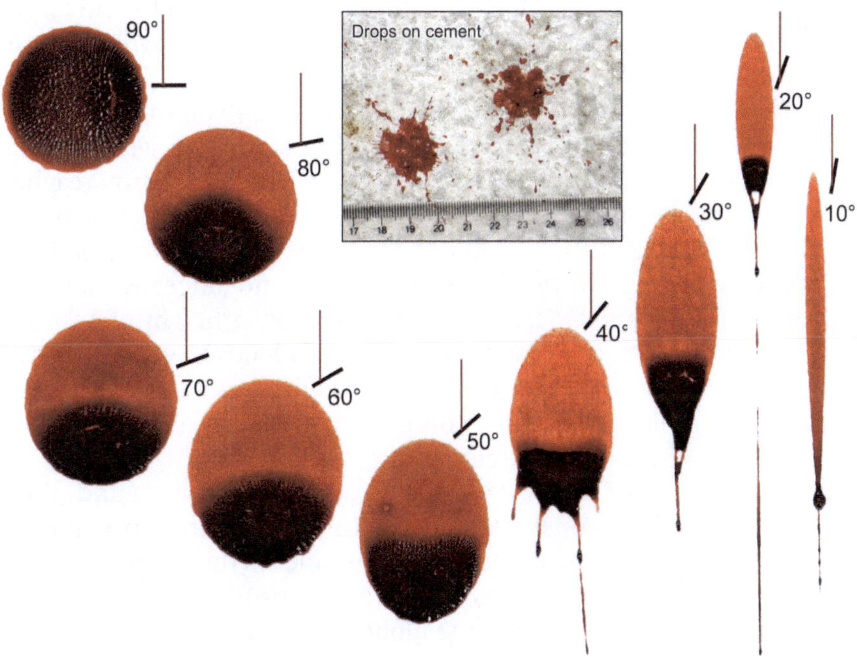

Figure 4.1 Bloodstains generated as a result of single drops of blood landing onto smooth card at angles of 10–90°. Stains impacting the target at angles of less than 40° tend to show pronounced directionality. An example of secondary stains caused due to wave cast-off can be seen underneath the stain formed at an impact angle of 20°. Inset bloodstains generated as a result of single drops landing onto cement block at angle of 90°.

circular with no distortion at its edges. As the roughness of the surface increases, such as with cement, the resulting stain is able to disrupt on contact to form edge characteristics such as spines of blood (Figure 4.1, inset). If the surface is sufficiently textured, such as with cement, the surface tension of the blood will be overcome dramatically on impact and secondary features will start to develop. These typically lead to the production of discrete satellite stains (sometimes referred to as secondary spatter) that radiate from the parent stain. Blood dripping repeatedly into liquid blood can generate very pronounced satellite staining and this type of drip pattern, sometimes referred to as 'blood into blood', is commonplace at scenes where an individual has been bleeding onto the floor, for example, in or near the same place.

If a drop of blood in free fall is considered in more detail, three variables affect the resulting size and shape of the bloodstain formed when it lands perpendicularly onto a target surface:

(1) The volume of the individual drop. If the drop height and target surface remain constant, the resulting stain size will increase with increasing volume.
(2) The distance the drop has fallen. If the drop volume and target surface remain constant, the resulting stain size will increase with increasing drop height up until the free-falling drop reaches terminal velocity. Terminal velocity is the point at which the effect of drag is equal to the force imposed on a falling body by gravity, in other words the blood droplet is no longer accelerating towards its target. This is achieved at a drop height of approximately 7 metres (m) for a 50 μL drop of blood. In general terms, it is not possible to predict drop height from the size of the resulting bloodstain *unless* other variables such as drop volume and the effect of the target surface are known.
(3) The nature of the target surface. If drop height and drop volume remain constant, larger stains will tend to form on non-absorbent surfaces as the entire volume of the drop is able to spread unimpeded across the surface. Absorbent surfaces tend to absorb a proportion of the blood and therefore smaller stains are produced.

As the target surface is more acutely angled in relation to the drop's flight path the resulting stain will elongate. Although the surface texture and blood volume continue to influence stain morphology, on an angled smooth surface, the resulting stain will develop into an ellipse which can show 'directionality' in the form of a tapered end. Directionality is a characteristic that can be used to establish the direction that the blood droplet was travelling at the moment it was deposited. As the directional angle of impact (the internal angle between the plane of the target surface and the path of the impacting blood drop) becomes more acute, the elongation becomes more pronounced. As the stain starts to develop the available blood volume decreases and this causes the overall stain to taper or close off. In some instances, residual blood can lift away from the target entirely to form secondary droplets and then satellite stains. This particular phenomenon is sometimes referred to as 'wave cast-off' and is typically associated with drops that have landed on a target at impact angles of less than 40°. The resulting satellite droplets can travel some considerable distance from the parent stain and produce stains that are usually in line with it but are excessively elongated. The interpretation of 'wave cast-off' can be particularly challenging if it is present on surfaces that are disconnected from the parent stain. For example, if a person drips blood from a nosebleed onto their trousers, satellite stains can be distributed across their footwear or onto the shoes of others. The resulting satellite stains can show features that are common to other pattern types and if they are examined in isolation have the potential to be misinterpreted.

A mathematical relationship exists between the maximum width and *elliptical* length of a directional bloodstain such that the ratio of these variables provides an estimate of the angle of impact (Figure 4.2). This relationship is relevant to single drip stains and can be applied very effectively in the reconstruction of complex spatter patterns.

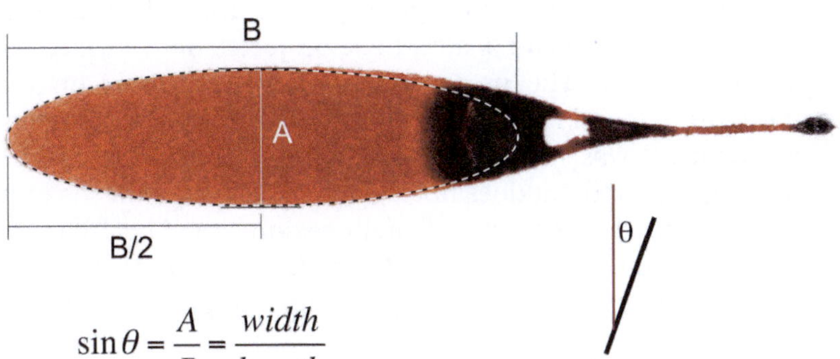

$$\sin\theta = \frac{A}{B} = \frac{width}{length}$$

Figure 4.2 Mathematical relationship between the elliptical length of a bloodstain and its maximum width, which can be used to calculate the angle of impact (θ).

4.2.1.3 Drip Trails

A trail of dripped bloodstains will inevitably develop if a bleeding person or an object sufficiently stained with blood moves. Drip trails can be of particular interest if they lead away from a crime scene and, if they persist, can indicate that an individual *e.g.* the perpetrator has been injured. Individual stains within a drip trail may exhibit edge characteristics that align with the prevailing direction of travel, although in reality this can be difficult to determine because the individual stains are usually generated under their own set of complex and variable dynamics. This is particularly the case if an individual is running or swinging their arm/hands for example.

4.2.1.4 Satellite Stains

Satellite staining or spatter is a secondary phenomenon caused when liquid drips into another liquid, in this context at least one of which is blood. The resulting drip pattern consists of small droplets of relatively uniform size, usually 1–2 millimetres (mm) in diameter, which are distributed at low energy around the parent stain. On a non-absorbent surface the resulting stains retain a dense, globular appearance and these can be particularly prominent when the stains are dry and viewed laterally. The distance travelled by satellite droplets depends on the volume of blood involved, the rate of drip, the dropping height and the topography of the target surface. Increases in each will generally result in a more random and energetic response. The limits of travel for blood dripping into blood experimentally have been shown to be approximately 1 m horizontally and 50 centimetres (cm) vertically, with the number of spatter stains decreasing away from the origin. In reality, this means that when an injured person drips blood onto the ground the potential exists for spatter to be projected onto surfaces and people in the vicinity. The presence of small bloodstains on their footwear (Figure 4.3) and lower clothing can therefore provide evidence that the individual was in close proximity, typically within a metre, when bleeding occurred but does not help in identifying how the injury was sustained. The identification of satellite stains must be progressed carefully because, on some surfaces, it can demonstrate features that are common to other pattern types, for example impact patterns.

4.2.1.5 Case Example

If blood flow from an injury is unchecked, the development of drip patterns can be extremely probative, none more so than drip stains

Figure 4.3 A drip pattern caused as a result of blood dripping into blood and a simulated representation of satellite spatter which can be deposited on the inner aspect of the footwear of an individual stood nearby.

that were observed at a murder scene where the victim had sustained multiple stab injuries. The deceased was found at home, lying on the floor of his hall. There was a moderate amount of bloodstaining close to the body, in the form of a blood pool, which was consistent with him having continued to bleed for some time whilst he was in the position that he was found. Of particular interest were a number of circular blood drops, or drips stains, in and amongst the staining that had developed from his wounds. Similar stains were also present on the uppermost surfaces of the clothing of the victim. In proximity to the deceased, these stains were almost indecipherable from other bloodstaining that had developed, but away from the body, they became clearer and a number appeared to have been distorted as a consequence of continued foot traffic whilst the blood had still been wet. Blood from the centre of these distorted stains had been removed by the tread of a shoe and had then been distributed throughout the scene in the form of partial footwear marks.

A sample of blood from the drip stains located on the body and also those observed some distance from the deceased were subjected to DNA profiling and an individual was nominated following a match on the National DNA Database (NDNAD). When the suspect was located he was found to have a serious laceration to his hand.

In addition to the strong evidential value of the presence of blood in the scene, which matched the suspect, the pattern of the bloodstains was probative. In this case, the blood distribution findings demonstrated that the individual had been present at the scene whilst injured and had bled onto the victim after he was on the floor; and that there had been subsequent foot traffic within the scene before the drops of blood had time to dry. Other examinations demonstrated a match between the suspect's footwear and the partial marks in blood that had been observed at the scene.

This case illustrates a number of important aspects of scene work and the preservation of evidence. Firstly, that the body of a victim forms an integral part of the scene and if it cannot be examined by a specialist *in situ* that a number of steps should be taken to preserve and record the body before it is removed. This should include preservation of the clothing, particularly if it is bloodstained, and removal of clothing should be considered so that potentially discriminating patterns, in addition to other evidence, are preserved before the body is transported to the mortuary. Frequently post-incident activities can obscure vital staining due to the continued loss or distribution of blood. Secondly, that no forensic discipline should operate in a vacuum. In this case, the targeted use of DNA profiling led to the identification of a potential perpetrator and in conjunction with bloodstain pattern analysis, and footwear mark comparison, provided a context which was critical to the investigation.

4.2.2 Impact Patterns

Spatter generated as a result of an impact is arguably the most common type of pattern encountered in casework and can be generated by a wide range of actions virtually all of which are attributed to violence. Kicking, stamping, beating, punching and shooting all have the potential to cause blood to spatter, although non-violent actions such as a fall, movement of an individual or medical intervention should not be discounted. Impact patterns are caused when an object strikes directly into liquid blood causing it to be displaced

Figure 4.4 Impact pattern comprising blood deposited on a horizontal smooth surface.

and break into small droplets of varying size. In general terms the greater the force the smaller the average size of the droplets (Figure 4.4). The droplets, which have variable trajectories and velocities, are dispersed radially from the impact site following the path of least resistance.

4.2.2.1 Factors Influencing the Generation of Impact Spatter

As with most bloodstain pattern types the resulting impact pattern is influenced by a number of factors which affect the distance travelled, direction and quantity of blood spattered due to the application of force. These include physical aspects such as the amount of blood available at the impact site, the shape, size and mass of the striking object, the relative position of surfaces on which spatter is deposited to the impact site and the force, speed and angle of the weapon/striking object at the moment of impact. By way of an example, beating with a large flat object such as a frying pan will probably cause blood spatter to be distributed at relatively low level with little or no blood being projected back onto the attacker because of the blocking effect of the weapon. In contrast, blood can be distributed more evenly following impacts with a hammer because the striking face exerts less of a constraint on the developing spatter.

If two blood droplets of different size have the same initial velocity, and trajectory, the larger droplet will travel further because it is more easily able to overcome the retarding effect of air resistance, which otherwise overwhelms the smaller droplet.

The individual spatter stains within an impact pattern are distributed randomly and exhibit a considerable variation in stain size, the majority of which are quite small, usually 3 mm or less, although stains as large as 10 mm and sometimes greater can be produced. As the distance from the impact site increases, the density or number of stains within the pattern decreases, in other words stains disperse as the distance from the origin increases. Spatter stains from a single impact will align at angles to each other and radiate from a common area of origin. Estimates of the three-dimensional area of origin can be calculated using methods that are often employed in crime scene reconstruction and these are considered later.

The classification of impact patterns has undergone somewhat of a change over the years and current thinking reflects developments that have been made in the understanding of blood dynamics in impact situations. In the 1970s, Professor Herbert MacDonell, working in the USA, classified impact patterns on the basis of the velocity of the impacting object, typically projectiles fired from low, medium and high velocity weapons. In adopting this approach, the interaction between blood and the impacting object at the impact site is largely overlooked. This is particularly limiting since it has been demonstrated that resulting stain size is not strictly a function of velocity and that the complex dynamics that develop at the area of contact between a weapon and the bloodstained surface actually have a greater effect on the range of drop sizes that develop than the velocity of the impacting object.

A more flexible approach in deciding what type of action might have caused an impact pattern is to base the interpretation on the size of stains within it. This is particularly useful because it is evident that there is an overlap in stain size within patterns that have resulted from impacts of greatly differing forces and mechanisms. In 1985, Terry Laber presented an alternate classification scheme that has since been widely adopted. This scheme, which is outlined below, focuses on resultant stain size and allows much greater flexibility in deciding what type of action might have caused a pattern.

Large: an impact pattern consisting of individual stains that are predominantly 6 mm or larger. Drip patterns typically comprise stains in this size range.

Medium: an impact pattern consisting of individual stains that are predominantly 2–6 mm in size. Cast-off bloodstaining is characteristic of this size range.

Fine: an impact pattern consisting of individual stains that are predominantly 2 mm or less. Impact spatter resulting from a medium-force impact such as a beating and high-velocity gunshot impacts can incorporate stains of this size.

Mist: an impact pattern consisting of individual stains that are predominantly less than 0.1 mm in diameter. Because of the pronounced influence of air resistance on these minute droplets, they are only able to travel short distances from the area of origin. Spatter generated as a result of high-force impacts, such as high-velocity gunshot, is characteristic of this size range.

4.2.2.2 Information That can be Derived from Impact Patterns

It is important to recognise that it is not necessarily the case that bloodstaining deposited on an item or clothing will be exclusively diagnostic of that article having been actively used in or worn during an assault. Furthermore, not all aspects of a continuing assault will result in blood loss or blood transfer, and it is entirely possible for an individual to be assaulted without blood being lost or transferred at that moment. Also, if a blow or repeated blows are delivered into separate areas, or areas which are not bloodstained, the potential for blood to transfer to the assailant, or weapon if used, can be significantly reduced, and it is possible for no blood to transfer during assault-related activities even when there are bleeding injuries. In scenarios where multiple blows are delivered it becomes more likely that the assailant will become bloodstained to some extent.

In all scenarios, if there is prolonged or close contact after injuries are sustained, or if activities take place which expose an individual to blood that has already been shed, the chance of blood being transferred will increase. Whether that blood is ultimately useful in determining what happened or who may have been involved will be dependent on the individual case circumstances and also the accounts that have been provided by the parties involved. The evaluation of patterns against the framework of the case circumstances will be discussed in more detail later.

The potential to provide significant information to an investigation exists if the bloodstain patterns that have been generated are interpreted accurately and impact patterns are no exception. Successful identification of impact patterns within a scene, can allow the site of an assault to be located and also allow the relative positions of items, and people, with respect to that impact site to be established.

In some instances, the distribution of blood might allow the nature of the impact and/or the type of weapon that caused the spatter to be proposed. Most importantly, the evaluation of impact patterns can provide an indication of the likelihood that bloodstaining would be seen on the assailant, or on other individuals that might have been present at the scene, and can allow the significance of bloodstains that might subsequently be found on their clothing and footwear to be determined.

Assaults involving kicking and stamping actions can generate a wide range of patterns, typically on footwear and trousers. Figure 4.5 shows a diagnostic kicking distribution on a shoe. The pattern demonstrates several key features. Firstly, the concentration of blood around the seams, stitching and in other recessed areas. This is indicative of blood having been pushed into these areas as a result of forceful contact. Secondly, the presence of spattered blood which emanates from the transfer (or 'contact') staining at the toe. In stamping scenarios, where forces are attributed to the sole or heel of a shoe, the resulting spatter may be projected up onto the shoe and onto the inside of the lower part of the trouser leg. The latter can be a significant finding and demonstrates the importance of examining

Figure 4.5 Impact pattern on a training shoe caused by multiple kicks into wet blood. The large area of transfer (contact) staining at the toe is associated with radiating impact spatter on the side of the shoe. Courtesy of A. Readfearn.

clothing in the context of how they would be orientated as worn rather than in isolation. The presence of transfer stains caused by hair, hairs embedded into the recesses of footwear or the presence of body tissue within the blood can increase the evidential significance of the findings.

A punching type action can produce equally diagnostic bloodstain patterns although the propensity for these to develop will be dependent on factors discussed previously such as the nature of the injury and the amount of blood available when punches are thrown. Punching directly into a source of wet blood, especially multiple times, will increase the chance of a 'classic' punching pattern developing and an examiner may see transfer staining and impact patterns around the cuff of a long-sleeved garment (Figure 4.6) and spatter continuing to the upper sleeve regions and across the chest.

Although the presence of blood on the hands of the assailant is extremely probative (Figure 4.6A and B), unless it is recorded at the time including digitally or on body worn video, it is easily lost by washing and only part of the pattern may be retained on the shirt which would ultimately be subjected to a laboratory examination (Figure 4.6C).

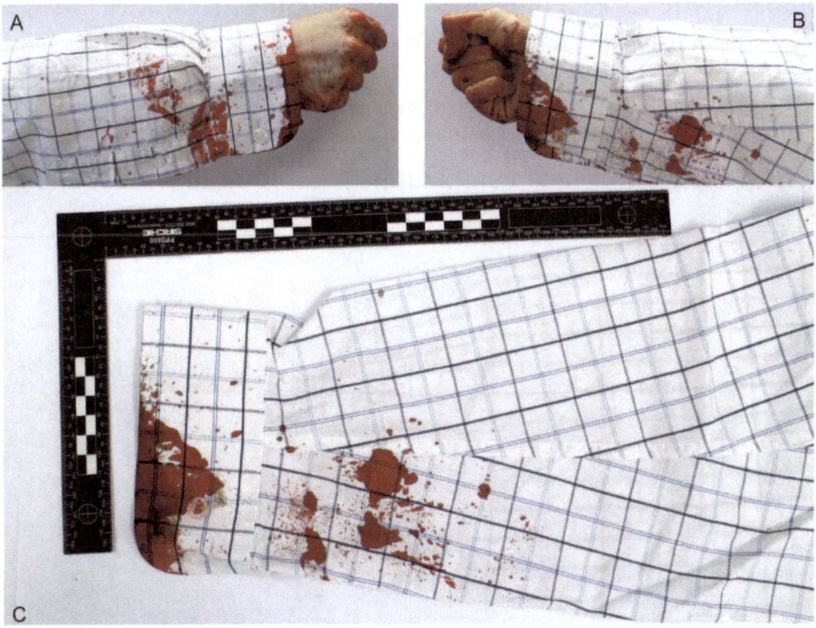

Figure 4.6 (A) and (B) A bloodstained fist and shirt cuff/sleeve caused by punching into a heavily bloodstained surface. (C) Cuff/sleeve of shirt as examined in the laboratory, now without the context of the hand. Courtesy of A. Readfearn.

Partial pattern capture, the dynamic nature of assaults, the scope for patterns to be subsequently disturbed through wear or as a result of cleaning, and the distorting effects of fabric on the resulting bloodstains are all factors that can make the interpretation of impact patterns, and other bloodstain types, on clothing (in particular) extremely difficult. The examination of items in the laboratory as they would have been worn, three-dimensionally is recommended and for court presentation purposes, the use of a mannequin can help to produce visual aids that assist the jury in understanding the bloodstain patterns that have been found (Figure 4.7).

Although assaults in which strikes are delivered into blood can develop diagnostic patterns, it is difficult to define a pattern that would be indicative of a stabbing. The infliction of a single stab to the body rarely results in blood being transferred to the assailant. This is because most of the bleeding occurs internally or is retained by the victim's clothing. If multiple stab injuries are sustained in or near the same area and there is build up of blood on the surface of the skin for example, the potential for spatter to be generated increases. In these instances, the stabbing activity mimics that of a punching assault and the pattern described previously can be observed.

Impact patterns generated as a result of gunshot injuries are again affected by a number of factors. These include the nature of the injury site, whether it is covered or exposed for example, the type and calibre of the weapon, the distance and angle between the muzzle and the skin, the number of shots and the presence or otherwise of an exit wound. A contact or near-contact injury to the head from a shotgun

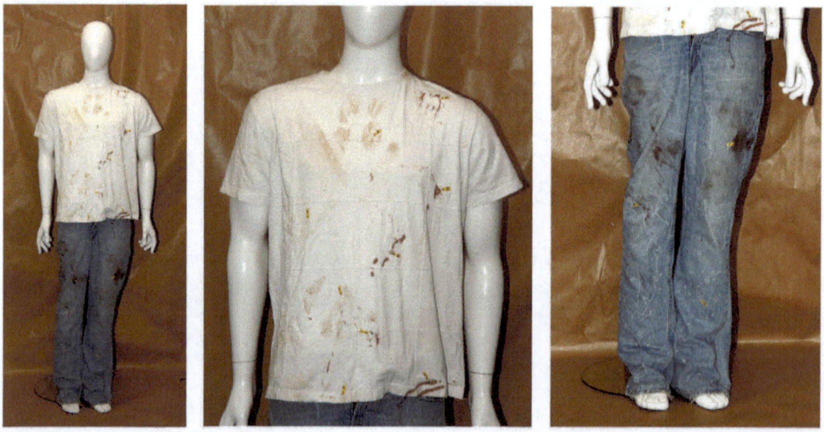

Figure 4.7 3D representation of bloodstain pattern findings on clothing using a mannequin.

can generate extensive spattering of blood, body tissue and bone fragments. The majority of the spatter is projected from the exit wound, away from the victim in the general direction of the initial shot. This spatter creates forward spatter patterns that can be used to identify the position of the victim at the moment that they were shot. Because this spatter travels in the same direction as the impacting force it is directed away from the shooter. At the entry site, an aerosol of blood droplets is created that travel in the opposite direction to the force that has been applied. The significance of this backspatter pattern is that it is projected back towards the weapon and the person firing the weapon. The mist-like blood droplets which produce backspatter patterns, which in themselves express a wide range of stain sizes albeit microscopic, are unlikely to travel more than 1 m from the injury site and in reality the majority travel considerably less than this. These micro-droplets are easily overcome by air resistance and gravity and their flight path can be significantly influenced by air currents caused as a result of hot gases ejected from the weapon. Backspatter is not easily seen on clothing, especially if the background material is dark in colour, but its presence on clothing and a weapon can be extremely probative. It is critical that examination strategies that use adhesive surfaces to speculatively recover DNA, fibres and/or gunshot residue particles, as examples, consider the impact that these techniques can have on BPA.

4.2.2.3 Determining the Origin of Impact Patterns

Impact patterns can play an important part in crime scene reconstruction as their features can be used to relocate the blood source when impacts occurred. This can be particularly relevant if the position of the victim is unknown. In some cases, it may be sufficient to estimate the area of origin by eye, but in other scenarios a more accurate determination will be necessary. A number of reconstruction methodologies are available to the analyst, most recently digital models. Impact pattern reconstruction for many years was based on a technique which used string to represent the trajectory of blood droplets within the spatter pattern. This method, known as 'stringing', was based on trigonometric principles and relied heavily on manual selection and measurement of representative stains. Although stringing is still used today, it has been largely superseded by bloodstain pattern analysis software, such as HemoSpat® and 3D modelling. Digital approaches facilitate a robust and timely approach to the directional analysis of impact patterns and generate data that can be archived and easily re-interrogated.

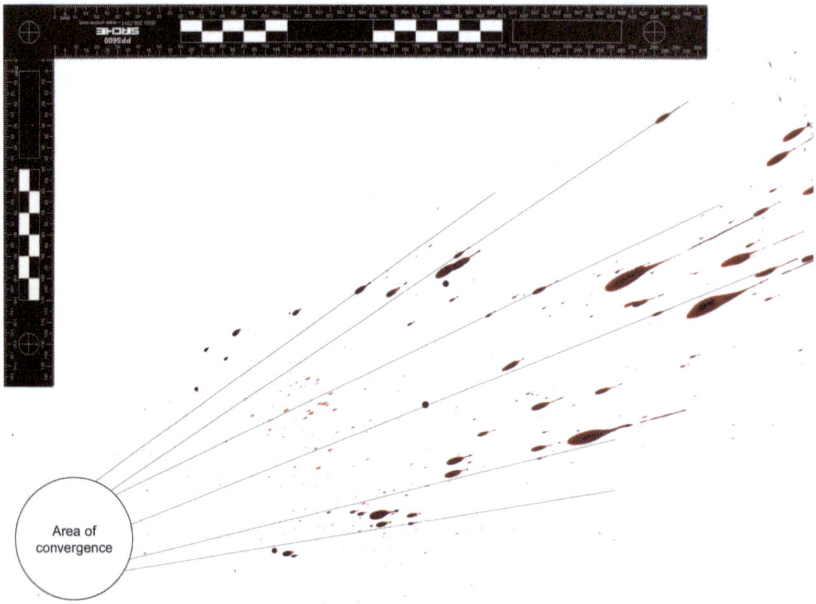

Figure 4.8 Determining the area of convergence of an impact pattern: top view showing the convergence of lines through the long axes of representative bloodstains.

Whether the analysis is conducted manually or using a computer, the underlying methodology follows a number of steps. If an impact pattern on a horizontal surface is considered, a straight line is first drawn through the long axes of a representative number of individual bloodstains which radiate with respect to each other. The area where the lines intersect is called the area of convergence (Figure 4.8) and indicates, in two dimensions, the original location of the blood source.

The three-dimensional location of the impact site, which will lie perpendicularly above the area of convergence, is then calculated by first measuring the maximum width and elliptical length of the selected stains. The ratio of these parameters approximates to the sine of the angle of impact and by applying this data to the equation (Figure 4.2) the angle at which the droplet of blood landed on the surface can be calculated. Each string, or line, is then anchored at the leading edge of the stain, lifted above the stain to the angle that has been calculated and fixed in position. The location where the lines intersect is known as the area of origin (Figure 4.9).

Although these methods provide a good estimation of the original location of the blood source at impact, there are a number of limitations. This includes the assumption that blood droplets travel in a straight

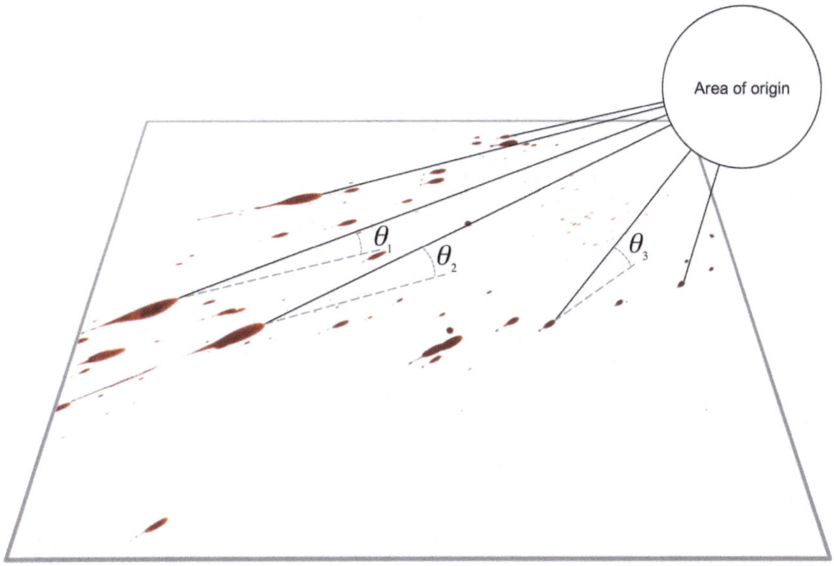

Figure 4.9 Determining the area of origin of an impact pattern: simulated view showing the 3D position of the impact site.

line, but in reality, the flight path is markedly affected by gravity and the trajectories are parabolic. The consequence of this is that the height of the actual impact site is over-estimated.

Despite this, origin of impact estimations can be used very effectively in excluding scenarios that might be presented, for example 'they were running at me when I shot them' *versus*, 'they were shot whilst they were on the ground'.

4.2.3 Cast-off Patterns

As the term implies, cast-off patterns are generated when blood is cast-off or released from the surface of a moving object, either by the action of centrifugal force (swing cast-off), gravitational force (drip cast-off/drip trails) or as a result of the bloodstained object coming to an abrupt stop (cessation cast-off). The resulting patterns can be complex in nature but typically exhibit features that are indicative. Drip trails have been considered previously and so only swing cast-off and cessation dynamics will be discussed here.

4.2.3.1 Swing Cast-off

Swing cast-off is commonly associated with a swinging object such as an arcing weapon or a moving hand that is wet with blood. Wet blood first runs towards the end of a weapon under the influence of centrifugal forces. Depending on the shape of the object, the blood may collect at one or more sites until the momentum applied is sufficient to overcome the surface tension of the blood and it is cast from the surface over a range of time and space. The blood rapidly forms spherical droplets in flight and the resulting pattern comprises a linear arrangement of similarly sized stains which do not have a common point of origin. Distributions of this nature are often referred to as 'in-line staining' (Figure 4.10A). The trajectory of drops projected from a swinging object can be determined using vector analysis of the competing forces. Depending on the velocity of the swinging object and the amount of blood available, drops can leave the surface at any angle that is radial or tangential to the arc of the swing. This means that all of the droplets are projected away from the arc of the weapon's travel (Figure 4.10B).

The number and size of blood droplets flung from a swinging weapon, for example, will depend on the shape and length of the object, the nature of its surface (whether rough or smooth), the velocity and dynamics of the swing and the amount of blood present.

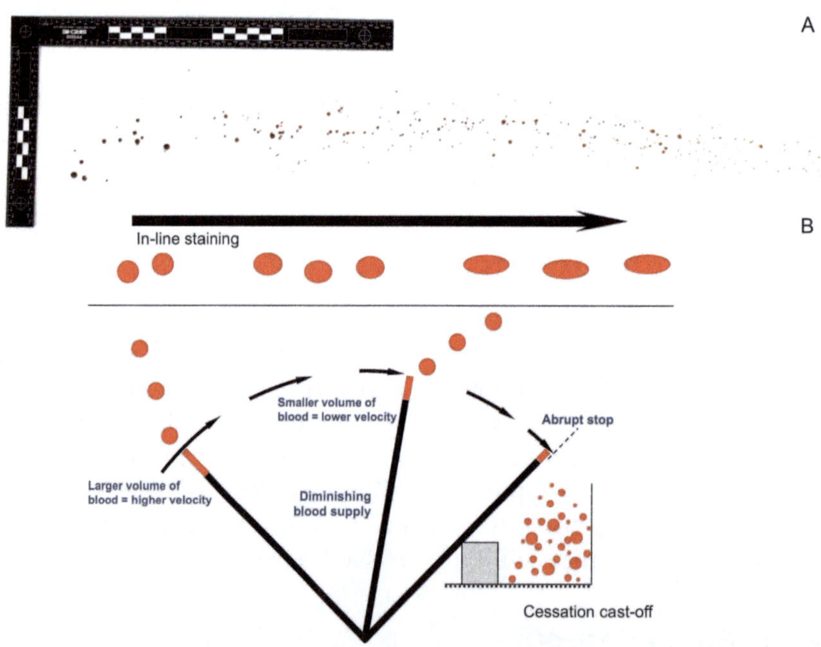

Figure 4.10 (A) Cast-off pattern/swing cast-off deposited onto a smooth white surface by a bloodstained object travelling from left to right. (B) Schematic representation of cast-off blood dynamics.

As the swing develops, unless the blood source is replenished, the resulting stain size will decrease as a factor of the reducing blood volume.

Pronounced cast-off staining can be achieved when a long, relatively light, weapon such as a baseball bat is used since these can be swung with ease. Under optimum conditions cast-off staining can be projected some distance from the site of the attack including onto the ceiling if present. This is due to the forces associated with rapid acceleration and deceleration which are particularly evident at the extremes of the swing. The ability to swing short, heavy weapons tends to be limited and the production of cast-off patterns from these items is less likely. Taking into account anatomy and assault dynamics, the forward swing is often more forceful than the back swing. This increase in force promotes the generation of cast-off droplets that are typically smaller in size because the blood can more readily overcome its inherent surface tension. On average, forward cast-off patterns comprise stains that are smaller than those caused as a result of the back swing, although the replenishing of the weapon before the back swing can blur the trend. In practice cast-off patterns due to back swing comprise stains that are comparatively larger simply because the volume of blood is greater and is depleted by the time that forward motion occurs.

Although diagnostic cast-off patterns can be observed in the crime scene, it is rare to see cast-off patterns on the offender simply because they are usually located within the arc of motion of the weapon. An exception to this can include if the individual is kneeling and wielding a weapon of limited length. This can lead to blood being projected onto their back as the weapon changes direction. There is also the potential for blood droplets in flight to be overcome by gravity and to rain down onto the perpetrator. This may lead to the deposition of a small number of discrete bloodstains that may or may not show sufficient features to be confidently assigned to cast-off mechanisms. This task may be particularly difficult if other types of staining are present. In fact, other than 'in-line' patterns, which are often distinctive, cast-off bloodstains can be indistinguishable from other spatter stains within a blood scene. Even when located the mechanism that caused cast-off patterns can rarely be identified.

4.2.3.2 Cessation Cast-off Patterns

Cessation cast-off patterns are caused when a moving object, wet with blood, comes to an abrupt stop (Figure 4.10B). During rapid deceleration extreme forces, which include inertia, are transmitted through the object to the blood causing it to be cast-off. In this situation the forces acting on the blood are similar to those of an impact and the resulting stain pattern, which may appear as a confined area of small circular spots of blood, can exhibit characteristics that are more in keeping with an impact pattern than cast-off.

The production of cessation cast-off, which is typically projected into the scene, can be associated with the development of characteristic bloodstaining on the weapon itself. As the weapon is swung, blood on the weapon can be forced to run towards the distal end producing stains that are sometimes referred to as centripetal or centrifugal runs. At the moment that the swinging weapon comes to an abrupt stop, residual wet blood, either within the run or in other staining on the weapon, can be forced forward in the direction that the weapon was travelling when it came to that stop. This can create staining that has a distinct feathered appearance along one edge. This characteristic feathering, which consists of fine parallel spines of blood, is termed percussive staining and an example is illustrated in Figure 4.11. Percussive staining can be particularly probative if it is seen on footwear as it demonstrates that the object has experienced forces necessary to disrupt the blood, such as might be experienced

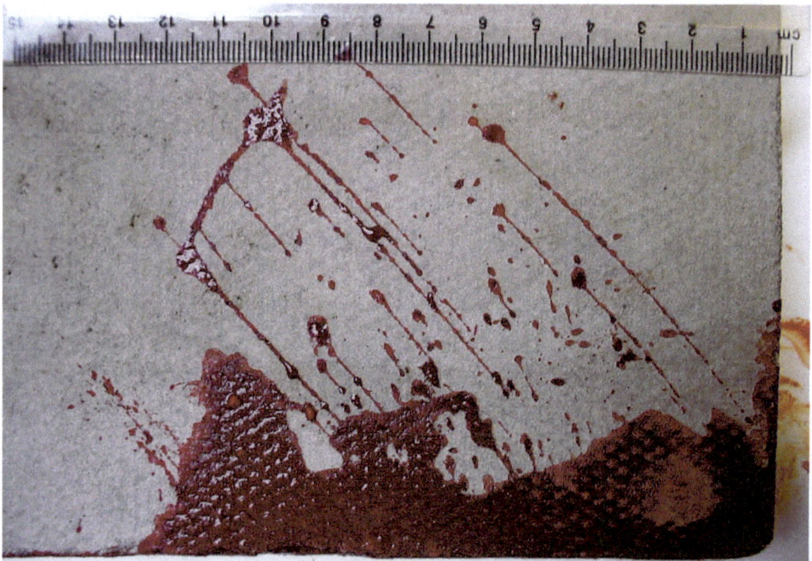

Figure 4.11 Percussive staining on a cement block produced when the block stopped abruptly causing the blood to be projected downwards in the direction of the swing.

in assault-related activities. Furthermore, the directionality of the feathering indicates the direction that the shoe was travelling at the moment that it came to a stop. If this points towards the toe then it can be indicative of the shoe having been involved in a kicking type action.

4.2.4 Projected Bloodstain Patterns

Projected patterns are caused as a result of the ejection of blood under pressure and typically blood that has been projected from damaged arteries under circulatory pressure. The resulting patterns are often referred to as 'arterial damage staining'.

Projected bloodstain patterns are unlike other spatter types because the resultant stains are not directly dependent on the action which caused the damage. In fact, once a blood vessel has been damaged the pattern will develop autonomously under the influence of internal forces alone.

In general terms the human body has evolved to protect the major blood vessels as they are critical to life and severe injury such as penetration with a sharp bladed instrument is often required in order to damage them.

Once an artery has been damaged, blood leaves as a column of blood which breaks up into drops of relatively uniform size. If the breach to the artery is small, a thin column of blood is produced which breaks up into small droplets. If the damage is large, the size of the column increases proportionately, and larger stains develop.

A number of factors affect the development and appearance of a projected pattern:

(1) Size of the damage to the blood vessel.
(2) Nature of the injury site. For example, if it is covered by clothing or if the projection of blood has been affected by the elasticity of the skin.
(3) Nature of the target surface.
(4) Direction of spurting, distance travelled by the blood droplets and movement of the bleeding source relative to the target. In other words, whether the injured party is stationary or moving.

In flight the droplets of blood are markedly affected by gravity, but once the droplets of blood land on a target surface the projected pattern will start to develop. On a vertical surface the resultant bloodstains are markedly parallel in alignment, with respect to each other, and runs can be produced when stains begin to coalesce or when large volumes are at play. On a horizontal surface the stains are distributed uniformly.

The resulting projected patterns can be sub-classified as follows:

Arterial Spurt: this is usually associated with movement of the injured artery/person. As the initial column of blood breaks up into separated drops, a series of uniform sized stains can result. Sometimes 'wave-' (V- or W-) shaped patterns can develop which reflect the rise and fall of circulatory pressure (Figure 4.12A).

Arterial Gush: this occurs when the injured artery projects blood in or near one position creating a pattern that is often dense and circular. Gushes can be associated with exaggerated satellite stains (secondary spatter).

Arterial Rain: this is typically seen when droplets of blood fall or 'rain' down under gravity resulting in the random deposition of numerous similarly sized roughly circular stains on a horizontal surface (Figure 4.12B).

It is important to recognise that the venous system is also under pressure (back pressure) and although lower than arterial pressure, bloodstains with features very similar to those described above can develop as a result of injuries to veins. Venous spurting may be

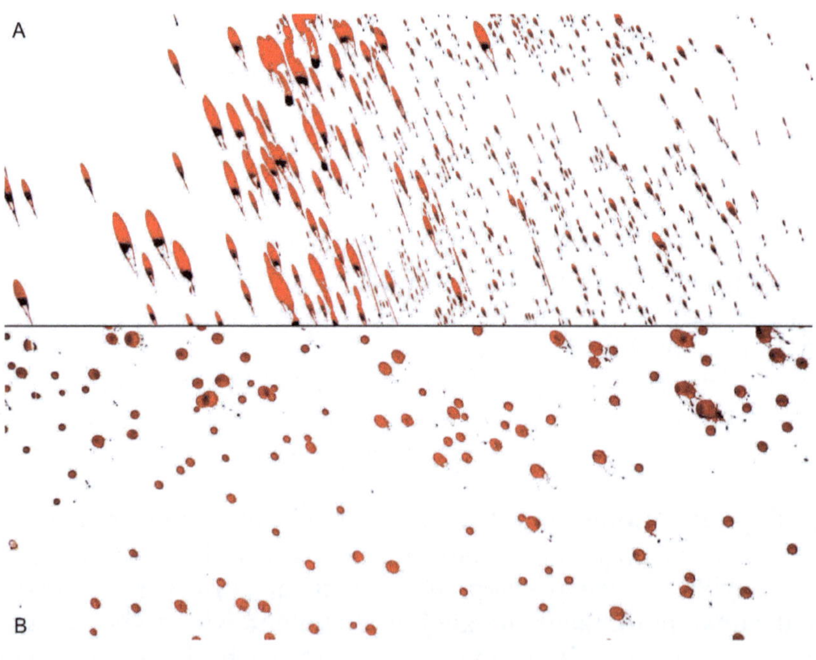

Figure 4.12 (A) Projected patterns: arterial spurt. (B) Projected patterns: arterial rain.

present if a victim suffers from varicose veins, as these tend to be close to the surface of the skin and prone to damage/injury.

4.2.4.1 Simulated Projected Bloodstain Patterns/Arterial Bleeding

The experimental generation of projected patterns allows analysts to examine patterns generated on clothing and surfaces that are commonly encountered at scenes of crime in a training environment. In order to simulate arterial bleeding, the Arterial Pump was conceived by Anita Wonder in the USA, modified by Adrian Emes and Chris Price in London and is now used routinely in our training programmes. The pump is made up of several components aimed to simulate the human circulatory system: a blood reservoir (blood volume), peristaltic pump and pulsing arm (the heart), silicone tubing (blood vessels), manifold and pre-damaged tubing which mimic wounds of a particular size (the injured party). Blood which is ejected from the tubing can be directed towards various surfaces at various distances and angles in order to demonstrate the wide range of characteristics and patterns that can be produced by arterial bleeding (Figure 4.13).

Figure 4.13 The 'Arterial Pump' being demonstrated by the author during the delivery of bloodstain pattern training.

4.2.5 Pools of Blood

Larger volume bloodstains, or blood pools, are typically caused when there is prolonged bleeding, or a large quantity of blood is released suddenly. This type of staining can quickly develop if an individual is critically injured, such as if an artery has been severed, but can also occur if a build up of blood internally is lost into the scene from the mouth or wound for example. Because of their potential size, and because large volumes of blood can remain wet for some time depending on the environment, blood pools can present a source of blood from which other patterns can develop as a result of post-incident activities.

The examination of large volumes of blood at scenes where the victim has been removed, in conjunction with the opinion of a medical practitioner, can be useful in providing an indication as to whether or not the injured party could have survived. Due to a number of variables, such as the surface on which blood is present and its porosity and whether the blood is wet or dry at the time, the precise calculation of blood volume by measuring the area and depth (or by weighing dried samples) can be difficult, but estimations of volume may be probative.

4.2.6 Altered Bloodstains

Altered bloodstains, sometimes referred to as physiologically altered bloodstains (PABS), are formed when blood undergoes a physiological, non-physiological or biochemical change before or shortly after it has been shed. This includes clotting, blood being mixed with other substances for example water or body fluids such as saliva. One of the defining features of altered blood is colour change and any significant discrepancy in the colour of blood, from the default red/brown, is usually an indication that blood has been changed in some way.

4.2.6.1 Clotting

The clotting process involves a cascade of complex chemical changes within blood which begin seconds after the injury and are accelerated if there is damage to the surrounding tissues. Although the initial stages of clotting are not visible to the naked eye, at some time, which could be several minutes later, the blood becomes visibly thicker, known as 'clot firming', and the products of clotting are deposited around the injury. In some circumstances clot retraction is seen where the solid clot separates from the liquid serum. If clotted blood is distributed at a scene, because subsequent blows have been delivered for example, it can indicate that some time has elapsed since the original injury. The precise calculation of time since injury from an examination of clotted blood is impossible due to variables that affect the rate of clotting (such as environmental conditions of temperature and humidity), medical factors and blood composition, which can be altered by drugs and alcohol. Spattered clotted blood can exhibit distinguishing features such as the presence of clot material within irregularly shaped stains or pale-coloured stains. However, it is critical that these apparently diagnostic features are not confused with stains that look similar but have been generated by other mechanisms *e.g.* diluted blood, because body tissue is present within the stain, or because the substrate has affected the usual appearance of blood.

4.2.6.2 Expiration Bloodstain Patterns

Expirated or exhaled blood is formed when blood is forced by airflow out of the nose, mouth or injury site. If a wound has been sustained to the mouth or throat, blood can mix with saliva and if this is present in sufficient quantities the resulting stains can appear diluted and/or become associated with air bubbles. Frothing can be particularly pronounced in blood that has been produced from penetrating chest injuries and can be mixed with lung surfactant.

If the viscosity of the blood is changed, particularly if the blood is mixed with sticky mucous, the resulting expiration pattern can be associated with a distinctive bead-like or stringing features which demonstrate that the blood is no longer able to break up into droplets (Figure 4.14). Blood which is projected as a consequence of coughing, sneezing, snorting, breathing or by medical intervention such as chest compression, can produce indicative patterns that resemble those caused as a result of impact. Although the absence of stringing or air bubbles can mean that unequivocal identification is not possible, the presence of these features can be diagnostic.

4.2.6.3 Insect Stains

The post-incident activity of insects once blood has been shed can result in the production of artefact stains that have a distinctive appearance. Insect stains, or 'fly-speck', is frequently found at crime

Figure 4.14 Exhaled blood: blood mixed with saliva showing pronounced dilution and 'stringing'/bead-like features.

Figure 4.15 Insect staining: regurgitation of blood products by fly to create fly-speck or fly artefact stains. Courtesy of A. Nambiar.

scenes where the corpse has remained *in situ* a few days after death and flies, or other insects, have colonised. After feeding, blood-based products are deposited as a result of regurgitation (Figure 4.15) or defecation and small discrete sometimes comma-shaped stains or stains that have a diluted appearance can be produced. In isolation these stains may be visually similar to blood spatter, although in reality they are typically found in areas that do not easily correlate with the rest of the scene, such as hot spots around a window frame or where lights are located.

4.2.6.4 Vomited Blood and Products of Decay

Blood that is mixed with contents of the alimentary canal can produce distinctive staining, not least blood which has become mixed with the contents of the stomach. If blood is ingested or if the abdomen is injured, the acidification of blood in the stomach causes the red blood cells to clump together. If vomited, this blood can generate staining which has a brown granular appearance equivalent to coffee grounds.

The biochemical products of decay can also lead to the production of stains that have a distinctive appearance. These are often dirty brown and not immediately blood-like in appearance.

4.2.6.5 Non-physiological Actions

The distribution of blood in a crime scene can generate significant evidence and it is not uncommon for attempts to be made to clean up blood or disguise its presence. Although washing or wiping will remove a substantial proportion of blood, in nearly all instances a trace of blood will persist. Non-visible residues, which have often been pushed into cracks and crevices or permeated absorbent surfaces can be detected using specialised light sources or chemical reagents such as luminol, and in some cases these methods have had success in detecting blood that has been simply over-painted or has been concealed by fire.

Whilst blood is wet it can be easily disturbed as a result of contact with another surface. If the original stain has dried to some extent the action of wiping may remove only a portion of the blood and leave an outline or perimeter stain behind. If the remaining halo continues to exhibit features of the original stain, it is possible to use these successfully to assess the original pattern. A wipe pattern relates to a bloodstain that has been altered as a result of an object moving through pre-existing wet blood. In contrast a swipe pattern (Figure 4.16A) is caused when blood is transferred from a blood-bearing surface onto another unstained surface. The resulting smear of blood can show features that are directly related to the movement, such as striations. In broad terms these stain types all fall into the category of transfer stains. The deposition of blood onto a wet surface can produce a diffuse stain

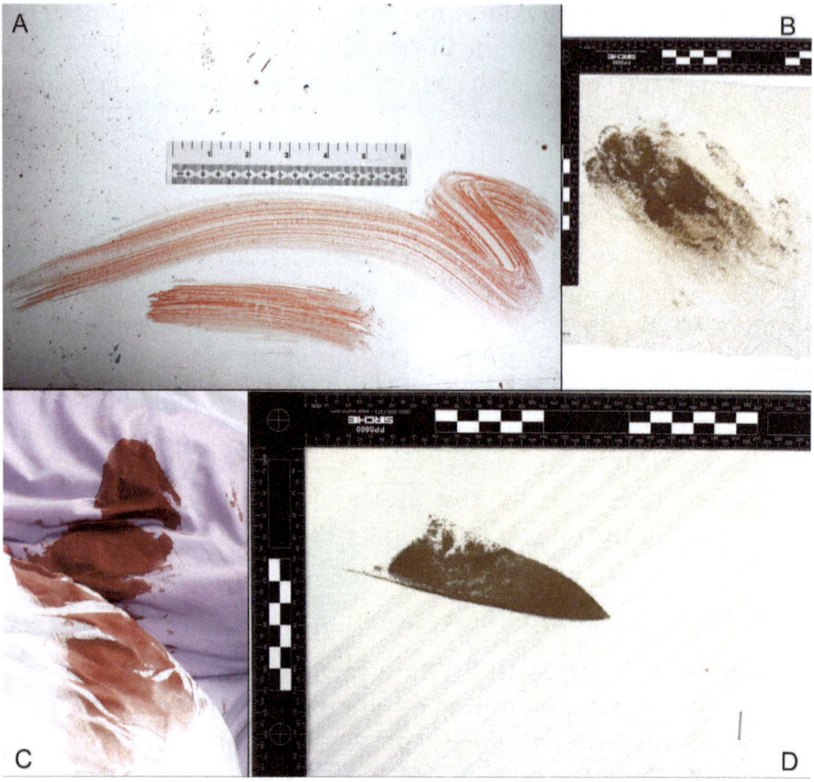

Figure 4.16 (A) Transfer stains: swipe. (B) Contact on fabric. (C) Voids in bloodstaining due to folds and creases in fabric when blood deposited. (D) Impression in blood of knife blade.

that has an indistinct perimeter. In contrast blood that has been pre-diluted has a tendency to form light-coloured homogeneous stains.

4.2.7 Transfer Stains (Contact Stains)

Transfer stains are caused as a result of contact having occurred between a blood-bearing surface and another surface. They are often abundant at crime scenes and on clothing or items examined in the laboratory where there has been continued activity after bloodshed. Swipe and wipe patterns are examples of transfer stains and in the majority of cases the resulting stain has no clearly defined shape (Figure 4.16B). However, if the stains are associated with other features their interpretation can be progressed. The presence of voids within a transfer stain for example can indicate that the surface was creased or folded when the blood was deposited (Figure 4.16C).

Transfer stains are particularly probative if they are associated with pattern features that allow the blood-bearing surface to be identified *e.g.* ridge detail associated with finger marks, footwear marks, fabric or weapon impressions (Figure 4.16D). In these cases the coordinated examination by experts from other disciplines is paramount.

4.2.8 Composite Stain Patterns

This category largely deals with patterns that exhibit elements caused by different or complex actions (Figure 4.17). For example, blood that has been projected from arteries combined with blood that has been cast-off as a result of the injured party having moved through the scene. This can result in patterns that are difficult to resolve into their individual components and interpret. In these situations, the systematic assessment and analysis of key features is paramount and as with all bloodstain patterns, the application of objective criteria can support robust conclusions about how the pattern may have been caused even when the mechanism is not readily apparent. The 'objective criteria' for identifying the main spatter groups are summarised in Figure 4.18.

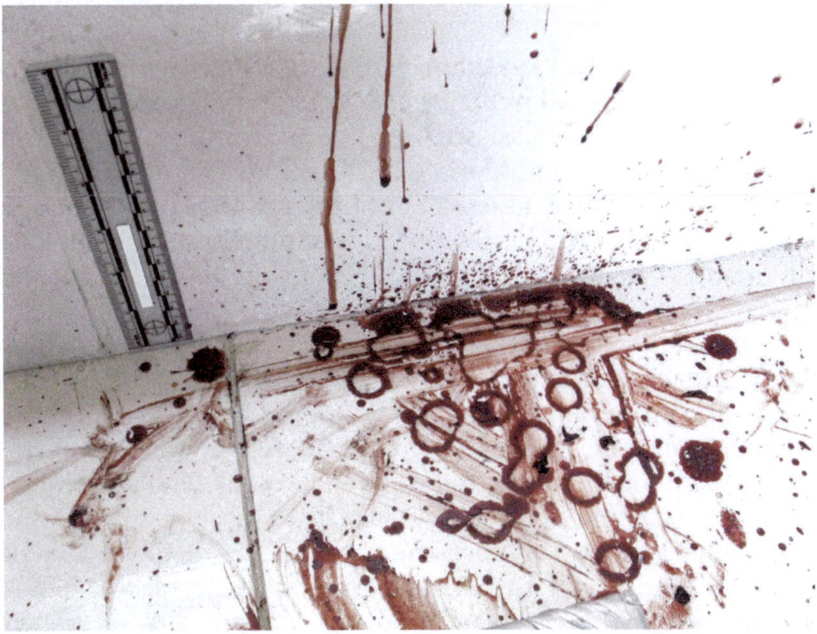

Figure 4.17 Composite patterns: multiple pattern elements including drip stains, transfer stains (wipes/swipes) and satellite stains.

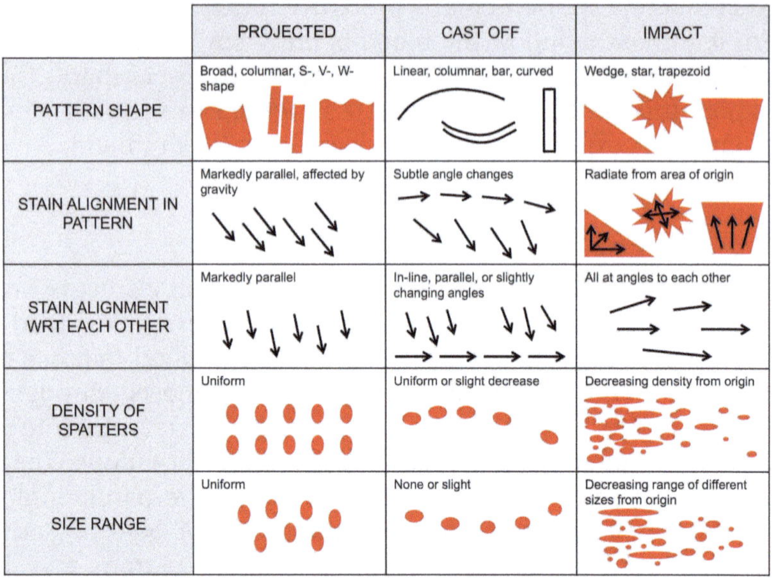

Figure 4.18 Objective criteria for the main spatter groups.

4.3 Evaluation of Bloodstain Pattern Evidence

A robust approach to bloodstain pattern evidence is multi-faceted. It starts with confirmation that the substance under examination is in fact blood, which is assisted with the use of presumptive tests. It is then followed by appropriate documentation and recording of the overall pattern type, assessment of the individual stain features and stain classification, representative sampling, interpretation and then finally evaluation of the findings in context with the case circumstances.

Records relating to bloodstains should include a detailed description of their physical characteristics including size and shape, location and overall distribution and should be supported with photographic documentation. The process of 'road-mapping' combines overview and close-up photography with the use of clear labelling and scales to support the systemic documentation of pattern elements and individual stain features. The systematic referencing of bloodstain patterns allows the viewer to maintain context and critically allows a third party to review the blood findings without necessarily being present at the scene. This is particularly useful for peer review and when demonstrating findings to the court. Contemporary methods utilise laser scanning technology, but these do not replace the requirement to inspect the individual stain features by eye and make careful observations.

The classification of bloodstain patterns at a crime scene or on clothing worn by a suspect is obviously important and it can provide information at key stages of an investigation.

At the crime scene an analyst may be able to provide information that allows possible scenarios by which the patterns may have been produced to be discussed. In this context the interpretation may be incomplete, but it can provide intelligence which can help the police form their initial investigative strategy. Because the potential exists for information at this stage to lead or mislead it is important that caveats are in place that reflect the strengths and limitations of any early opinions and that these are re-evaluated as the investigation progresses. An early assessment of bloodstaining will invariably include the recovery of representative samples in order to establish from whom the blood may have originated, and this is particularly important if the victim is missing, or if an individual is arrested away from the scene, in order to establish if the two are linked.

The propensity for blood to transfer in assault situations has been a driving force in the development of forensic techniques that can individualise bloodstains. Formative methods included antigenic and biochemical blood grouping systems, but these have been largely superseded by the advanced DNA profiling technologies that are employed today. DNA methods in particular provide forensic practitioners with the ability to discriminate between the blood (and body fluids) of different individuals such that if the DNA profile from a bloodstain matches the reference DNA profile of a particular individual there is a high degree of certainty that the blood came from that person. If the source of the blood is ultimately accepted then an evaluation as to how that blood may have been transferred must be considered. In most cases alternative scenarios in respect of that transfer will be provided, one on behalf of the prosecution and the other on behalf of the defence. In 'evaluative mode', the bloodstain pattern analyst considers the appearance and distribution of the staining in context with other relevant information in the case, often referred to as the 'framework of circumstances' and assesses the probability of the evidence if the prosecution proposition were true *versus* the probability of the evidence if the competing defence proposition were true. The ratio of these probabilities, the likelihood ratio, is then presented and communicated to the court in the form of an evidential strength of support. Guidance on the development of evaluative opinion has been published by the UK Forensic Science Regulator.

In order to be able to outline what types of bloodstaining might be seen in light of the alleged activity, it is important for the analyst to have a referable database of expected outcomes for various assault-related scenarios. If the allegation of repeated punching is considered for example, and blood matching the victim has been detected on the cuff of the defendant's sweatshirt, then it is important to be able to assess if the bloodstaining is consistent with that or not. Equally it is important to be able to define the potential outcomes given the defence alternative. Such as if it is proposed by the defendant that the blood transferred indirectly from the victim to his cuff by a third party.

In order for a bloodstain pattern analyst to be able to robustly outline their expectations of blood transfer in any given scenario, it is critical that they have practical experience of BPA and this includes having been involved in experiments where they have had an opportunity to simulate assault scenarios: punching, kicking, beating with a weapon. Formal training courses are available for this purpose and allow participants to wear different types of clothing and footwear and simulate different violent attacks by striking a bloody surface, typically a meat substrate which is known to mimic human tissue and using a blood proxy which is usually animal-derived. In doing so, the characteristic features that might be seen can be examined and recorded and used to build up a database of expectations. In the case of repeated punching, transfer/contact staining at the cuff of a garment with associated directional spatter travelling across the sleeve and front of the chest depending on the relative positioning of the assailant and victim. Therefore, if this type of staining was observed on the sweatshirt of the defendant, in the case scenario introduced previously, its presence tends to provide support for the prosecution allegation rather than the defence.

If the blood had transferred to the sweatshirt indirectly by way of a third party, then the presence of transfer staining would not be unexpected, but in conjunction with blood spatter the finding would be extremely uncommon, especially if that spatter was concentrated at the cuff.

If the transfer bloodstaining was associated with voids but there was no spatter then this would be in keeping with the blood having been deposited by a third party as proposed by the defendant.

This approach to the evaluation of scientific evidence relies on the Bayes theorem and it is widely considered to offer a robust framework against which scientific findings, such as blood patterns, can be assessed.

To be applied most effectively the development of reliable databases, which demonstrate bloodstain features that are expected in different scenarios, is paramount. This is an area of BPA that continues to develop but requires additional investment and the scope of a project such as this cannot be underestimated. To be reliable, the development of a database requires the systematic recording of bloodstain pattern features that can develop in all possible scenarios so that in any given framework, the frequency of an observation can be defined. Notwithstanding the practical difficulties of establishing a database such as this, and coordinating the work that has been done into an open-source resource, the wider pressures on evaluation must be openly recognised. This includes contextual bias. Contextual or cognitive bias applies to BPA as much as it does to all other forensic disciplines. Arguably it is more applicable to BPA because it is at least partially subjective and relies heavily on the experience of the analyst who during the course of an investigation may be exposed to a great deal of contextual information. It is a fact that our past experiences, the information that we are provided with, and our motivation and beliefs all have the potential to impact on our decision making. Our ability to make decisions is broadly influenced by top-down and bottom-up processes. The latter are based on data that is relevant to the situation. In the context of BPA this means the bloodstains that have been observed. Top-down processes incorporate additional, sometimes extraneous, information. In situations in which the data is complex or ambiguous we unconsciously incorporate more external drivers in order to make decisions. It is in this mode that the impact of contextual bias on BPA is considered. First of all bloodstain patterns are frequently ambiguous. Seemingly similar actions can give rise to patterns that show variable characteristics and because of this, interpretations are rarely categorical.

It is recommended that the likelihood of the findings is evaluated against two competing propositions. This approach necessitates the provision of relevant information but often extraneous information is assimilated. This can include information that the analyst unavoidably collects, consciously or unconsciously from the wider scene, if they attend, and/or from witness statements and investigative data that they are exposed to simply as a result of being part of the investigative team. The influence of the latter may be particularly relevant if the analyst is employed by the police.

Lastly, in the absence of a standardised approach to BPA, subjectivity is unavoidable simply because the majority of conclusions draw on the experience and expertise of the analyst. Studies have demonstrated that the interpretation of patterns by experienced analysts is influenced, both positively and negatively, by the contextual information that is provided. Work commissioned by the National Institute of Justice showed that the error rate in pattern classification increased when incorrect information suggesting how the pattern had been generated was provided (negative biasing) and decreased when the contextual information deliberately directed analysts to an accurate classification. In other words, research has unequivocally demonstrated that bloodstain pattern classification is vulnerable to contextual influence. Although the potential effect of contextual bias on bloodstain pattern interpretation is significant, steps must be taken to minimise its influence. In broader terms this requires investment in research and training, but in practical terms its impact on decision-making can be addressed by ensuring that the following controls are in place:

(1) That only information that is relevant to BPA is considered by the analyst.
(2) By ensuring that BPA interpretations are subject to independent peer review.
(3) By the scientific community adopting a standard approach to BPA which is based on objective classification.

In this framework, BPA can provide an effective and objective investigative tool that can robustly assist the court.

4.4 Conclusion

The examination of the distribution and patterns of bloodstaining at scenes and on weapons and clothing has the potential to provide a unique understanding of the activities that took place once bleeding injuries were sustained. When those bloodstains can be evaluated in the context of the case circumstances and accounts provided, they have the potential to be a powerful tool in the investigation of crime.

Further Reading

B. Karger, R. Nüsse, G. Schroeder, S. Wüstenbecker and B. Brinkman, *Int. J. Leg. Med.* 1996, **109**, 66–74.

B. Karger, R. Nüsse, H. D. Tröger and B. Brinkman, *Int. J. Leg. Med.* 1997, **110**, 27–30.

Scientific and Legal Applications of Bloodstain Pattern Interpretation, ed. S. H. James, CRC Press, Boca Raton, 1999.
A. Y. Wonder, *Blood Dynamics*, Academic Press, San Diego, California, 2001.
A. Y. Wonder, *Bloodstain Pattern Evidence: Objective Approaches and Case Applications*, Academic Press, San Diego, California, 2007.
T. Bevel and R. M. Gardner, *Bloodstain Pattern Analysis With an Introduction to Crime Scene Reconstruction*, CRC Press, Boca Raton, 2008.
T. L. Laber, P. E. Kish, M. C. Taylor, G. Owens, N. K. Osborne and J. Curran, *National Institute of Justice*, Grant #2010-DN-BX-K213, 2014.

5 Forensic Collision Investigation

Mark Crouch

Forensic Collision Investigation and Reconstruction Ltd, Suite 134, Capital Business Centre, Sanderstead, Surrey, CR2 0BS, UK
Email: Office@FCIR.co.uk

5.1 Introduction

On the 17th of August 1896 in Crystal Palace, London, Bridget Driscoll gained the undesirable accolade of being the first recorded fatality when she was struck by a car that was described by witnesses as travelling at "*a reckless pace*", a speed that was evidenced at Coroners court as 4 mph. The Coroner, Percy Morrison, is recorded as noting that he hoped "*such a thing would never happen again*".

Sadly, in the UK, over 1500 people are killed every year on the roads with over 10 000 more seriously injured. There are approximately 550 homicides in the UK a year, meaning that the number of people killed in road traffic collisions in the UK is broadly triple that of the number of deaths recorded as homicide victims – therefore significant forensic resource is used in considering the events of a car crash. The recent legislative and sentencing changes that came into force in June 2022 mean that killing someone by dangerous driving carries a maximum sentence of life in prison, thus the consequences of crashing a car are the same as murder.

Crime Scene to Court: The Essentials of Forensic Science, 5th Edition
Edited by Niamh NicDaeid and Peter C. White

Published by the Royal Society of Chemistry, www.rsc.org

The attendance at the scene of a fatal road traffic collision is almost exclusively the remit of the police. Much in the same way as a murder scene, specialist teams of forensic investigators attend the crash and gather the evidence from the scene. These teams are typically not the same forensic teams that would attend murders, more often this function would be performed by highly specialist, forensically trained traffic police officers. There are a number of UK police forces that also employ civilian investigators in the same role, but the principle is the same, these individuals will only typically be deployed to crash scenes involving motor vehicles.

The events of a car crash happen in seconds, indeed the deformation phase (the time between the vehicles first making contact, crushing to their fullest extent, and then starting to move away from one another) takes in the order of 80 ms to 200 ms. The forensic discipline of collision investigation is therefore the detailed piecing together of the events of a crash that took only a few seconds.

5.2 Human Factors

It is important to start any topic on collision investigation by understanding that cars do not crash, drivers do. For now, at least, a driver is required to make the vehicle move and therefore, with the exception of the rarest of collisions, a crash is caused by driver action or indeed inaction. That is why the language surrounding car crashes shifted, over a decade ago, from the term accident to collision. The main thinking behind such a step was to remove the implication that an accident was just one of those things, whereas the term collision more readily reflects the fact that there is often one or more parties at fault in a collision.

The task of driving is a complex one. A huge amount of information is received by a driver, processed, and then actions and decisions chosen. The human brain, however, for the most part is a sequential engine, meaning that tasks and actions are processed one after another – multitasking is not really something that exists. The more complex the situation faced by a driver, the longer it will take them to respond. The same is true if the hazard facing them is not easy to discern or identify. A prompt reaction to a hazard is further eroded by the introductions of additional tasks that draw some of the driver's attention away from the driving task, such as making a telephone call or other distraction. A further slowing of a normal response can be caused by the introduction of drugs, both illegal such as Δ^{-9}-tetrahydrocannabinol, (Δ^{-9}-THC), the active ingredient in cannabis, or legal such as alcohol (albeit with a limit), drugs *via* prescription or even uncontrolled substances (including caffeine).

The time in which a driver takes to respond to a hazard is a critical element in understanding the events of a collision and the conditions that would have made the collision avoidable. It is perhaps well understood that a vehicle's overall stopping distance has two components – firstly a driver reaction phase, and secondly a vehicle braking phase. What is less understood is that the commonly quoted overall stopping distances quoted within The Highway Code are more often than not an underestimate of the true stopping distance of a vehicle. These distances originated loosely from allowing 1 foot per 1 mile per hour of travel, which whilst readily understandable, does not have a scientific basis. This comes as a surprise to many, because the commonly held belief is that modern day vehicles can stop much quicker than the distance quoted within the guidance. It is true that modern day vehicles have a braking phase that is a little shorter than quoted, however this modest improvement is dwarfed by the significant underestimation of the driver's reaction time.

The reaction time, is more properly known as a Perception Response Time (PRT) which is the time required for a driver to detect, identify, make a decision and then commence that action. Krauss found that "*given a readily identifiable hazard that first appears nearly directly in front of the driver… a fairly straightforward situation and minimal time available for a response, most drivers i.e., about 85 to 95%, will respond by about 1.5 seconds after the first appearance of the object or condition of concern. The evidence also indicates that the minimum perception reaction time for the situation described is probably about 0.75 second*". The mean value was found to be approximately 1.1 s. By contrast, The Highway Code employs a value of 0.66 s which is, at best, about twice as fast as a driver's response time in real world cases. This in turn means that real world thinking distances are in the order of twice as long as those quoted in The Highway Code and therefore, real world overall stopping distances are about 10% longer for vehicles travelling at 30 mph. This is of course an oversimplification. For example, if a driver were to be primed to a hazard, say when passing a school in mid-afternoon, there is potential to respond more promptly, or if driving along a poorly lit road at night with little prior warning, a driver may respond more slowly. Therefore, it is important not to take a 'one size fits all' approach when it comes to PRTs and human factors and the circumstances of the collision needs to be considered on an individual basis.

With this is mind, human factors play a significant role in understanding a collision and whether or not a driver performed well, or poorly, when faced with the incident in which there was a collision.

It is also why studying the collision scene is of paramount importance, and why the Collision Investigator is also interested with the areas of the scene prior to any marks being left. Neither the Point of Impact (POI), nor where the vehicles have come to rest are areas where the events that lead to the collision occurred. Indeed, the collision began when the drivers commenced their PRT or on occasion, even earlier.

5.3 Collision Types

There are a number of different types of collisions that are routinely investigated. Below is a list of those commonly encountered on the roads within the UK:

- Single car collisions
- Multiple vehicle collisions
- Pedestrian collisions
- Lorry collisions
- Motorcycle collisions

This is not intended to be an exhaustive list but assists with discussions surrounding the different evidence that is likely to be recovered and/or the key elements that an investigator will be considering when analysing a collision.

5.3.1 Single Vehicle Collisions

A single car collision almost always involves a car leaving the carriageway, to a lesser or greater extent, and striking some roadside furniture causing death or serious injury to one or more of the occupants of the vehicle. In the absence of any mechanical defect, or significantly compromised road surface, the causes of such collisions are down to driver error. Also, a common theme in many of these collisions is excess speed for the vehicle. The questions for the investigator then become centred around identifying or excluding any road/mechanical defects, establishing the speed of the vehicle, and considering what a safe speed would have been.

Collisions such as these commonly occur on bends, whereby the car is driven at such a speed that it is unable to navigate the curve.

The tyres are the only thing in contact with the road surface when a vehicle is driving along the road, and as these surfaces are in contact, friction exists between them. This friction is exploited when accelerating, slowing, or turning the vehicle, causing the vehicle tyres to slightly deform, and generating the force that ultimately changes the speed or direction of the vehicle.

There is, however, a finite limit of friction that exists between the tyres and the road surface. The demands placed on the tyre and the road surface are a function of both the speed of the vehicle and the radius of the driven path, as circular motion dictates that:

$$F = \frac{mv^2}{r} \tag{5.1}$$

Therefore, the force (F in newtons) experienced by the vehicle with mass (m in kilograms) is proportional to the square of the velocity (v in metres per second) and inversely proportional to the driven radius (r in metres). So, in essence the tighter the bend and/or the faster the vehicle is driven around a bend, the more and more frictional force is required at the tyre/road interface. When the vehicle is cornered at an excessive speed for the intended curved path, the level of friction required to maintain the path exceeds the maximum limit available. In this case, the car begins to drift wide (typically understeering), moving to the outside of the bend.

If a vehicle is cornered in an even more extreme way it will begin to 'yaw', a rotation about the vehicle's centre of mass (oversteer) and it is possible for a curved tyre mark to be left on the road surface. These tyre marks are commonly referred to as critical speed marks or 'critical curves' and their examination at a scene is discussed more fully in Section 5.4.1.

Providing the vehicle has not experienced an excessive rotation (or 'yaw'), chord and mid-ordinate measurements can be taken of the front offside tyre mark to obtain the geometric path of the front offside tyre. Testing by Lambourn has shown that by applying a value for sliding friction to the curvature of the front offside tyre mark, it was able to approximate the speed of the vehicle to a tolerance of $\pm$10%.

The introduction of Electronic Stability Control (ESC) has however been mandatory on all vehicles sold in the EU since the latter part of 2014. As the name suggests this is an electronic function of the vehicle designed to supplement any driver input to maintain or regain stability of a vehicle in the early stages of a loss of control event. When activated, in broad terms, the ESC is designed to reduce engine power to the driven wheels and apply a degree of braking to the individual road wheels in an attempt to correct any yaw effects. This in turn reduces the likelihood of a loss of control, although does not preclude the likelihood of the front outer tyre mark being visible in a critical speed event. However, as the principles of circular motion are still the same, if the centre of mass of the vehicle can be plotted from any of the physical evidence, Cash and Crouch found that the speed can be calculated more accurately and also identify the level of braking applied to the vehicle.

5.3.2 Multiple Vehicle Collisions

For two (or more) vehicles to be involved in a collision, at least one of the parties needs to have moved into the path of the other, be that by straying across the centre line of a road on a bend, or for example, by turning out of a junction into the path of an oncoming vehicle.

When there is a reasonable degree of speed involved in a collision, such as a car striking another car, then within the deformation phase, damaged components can be forced downwards and onto the road resulting in scratching and gouging of the road surface. These marks on both vehicle(s) and road assist in assessing the point of impact. This may be a critical piece of evidence, particularly when identifying which of the parties strayed across the centre line in the moments leading to a collision.

There is a convention with respect to describing the nature of the initial interaction between the parties, specifically the degree to which they engage, known as an overlap. Considering a collision, where the fronts of both of the vehicles collide, the degree to which they overlap is measured as a percentage, where 100% would be a full overlap/full head on collision and 0% would be when the vehicles just miss each other.

In the case of a 100% overlap collision, or where the centre of masses of the vehicles are aligned it is possible to calculate the speed of the vehicles at impact by way of a momentum exchange calculation. Within a road traffic collision, momentum (the product of a vehicle's mass and its velocity) is conserved – meaning that the sum of the momentum immediately after the collision is the same as the total of the momentum at impact. In short, if the post impact velocities of the vehicles along with their masses are known, then their pre-impact velocities can be calculated using the following equation:

$$m_1\tilde{u}_1 + m_2\tilde{u}_2 = m_1\tilde{v}_1 + m_2\tilde{v}_2 \tag{5.2}$$

where, m = vehicle masses, kg, u = pre-impact velocities, m s^{-1}, v = post-impact velocities, m s^{-1}, and the subscripts 1 and 2 relate to the vehicles involved. However, velocity is a vector quantity, meaning that it possesses both a speed and a direction, therefore the angle of the impact needs to be catered for by modifying eqn (5.2) in the following way:

$$m_1\tilde{u}_1 \cos \theta_1 + m_2\tilde{u}_2 \cos \theta_2 = m_1\tilde{v}_1 \cos \varphi_1 + m_2\tilde{v}_1 \cos \varphi_2 \quad (5.3)$$

where, θ = pre-impact angles, φ = post-impact angles of the vehicles.

In the case of a fully, or largely head-on collision if both of the parties had equal (but opposite) momentum in the moments prior to the collision, both of the vehicles would come to a complete stop at the point of impact, with neither vehicle being displaced rearwards as a result of the impact. If, however, one of the vehicles is pushed rearwards, it means that this vehicle had less momentum than the other at the moment of impact. Therefore, if both of the vehicles had broadly similar masses at the time of the collision, the vehicle that was pushed rearwards was travelling slower. The degree of the speed imbalance can be calculated directly from the distance it was displaced in the collision.

When the centre of masses are not aligned in a collision, say in the case of a right-angled collision, but one of the vehicles is struck towards the rear, a rotation is introduced, which results in the rear of the one car rotating away from the impacting force of the other car. It is important therefore, for a Collision Investigator to identify the post impact movement of both the vehicles to see how the vehicles reacted when struck *i.e.*, pushed rearwards and/or rotated as well as an assessment of the nature of the damage itself – all of which assists in understanding the magnitude of the impact.

5.3.3 Pedestrian Collisions

Pedestrian collisions often give rise to some of the more subtle collision scene evidence and a Collision Investigator needs to have a very keen eye in identifying the pertinent collision scene information. Along with the identification of the point of impact, it will also be essential to establish the pedestrian's crossing direction, the time they spent in the road prior to impact and the impact speed itself. Furthermore, particularly in collisions that occur during the hours of darkness, the conspicuity of the pedestrian is also a consideration.

In general terms, the longer the pedestrian was detectable in the road, the greater the opportunity a driver would have had to take avoiding action. Therefore, it is not only the identification of the point of impact that is of importance, but also the point at which the pedestrian left the footway. In many cases the direction of the pedestrian crossing and whether they were walking or running is also not clear cut, and the physical evidence present on the vehicle will also be important. Of course, running into the road from the kerb edge closest to the vehicle will provide a much smaller time in which a driver can respond than if the pedestrian was walking at a slow pace from the footway furthest away.

The principles of conservation of momentum discussed regarding vehicle-to-vehicle collisions apply equally to pedestrian collisions. In a right-angled collision a pedestrian will move laterally across the front of the vehicle as they are engaging with it and the greater the lateral movement the more pronounced this displacement – this provides a strong indication of the pedestrian's crossing direction and a suggestion as to their crossing speed. This movement can sometimes be prominent with denting or scratching or in other instances, may be more subtle with cleaning marks present on the edges and bonnet where a thin film of dirt is often present on most vehicles.

So, what is meant by a pedestrian engagement with the front of a vehicle? In a head on collision with a pedestrian, their legs are typically struck by the bumper of the car. As they are struck below their centre of mass (the hip area), their upper body rotates towards the vehicle. Depending on the speed of the collision, there will often be hip contact with the bonnet and a head impact on the windscreen. If the impact speed is sufficient, the windscreen will smash with cracking extending radially outwards from the point of contact – this is colloquially referred to as a bullseye.

After striking the vehicle, particularly when the vehicle is undergoing emergency braking at the time, the car will begin to slow more rapidly than the pedestrian. This causes the pedestrian to begin to separate from the vehicle and they are projected forwards, a period known as the flight (or throw) phase, before landing on the road surface and sliding to rest.

There is a relationship between the distance that the pedestrian is projected forwards in the impact and the speed of the vehicle. In very broad terms, Wood *et al.* have shown that the distance a pedestrian is thrown is proportional to the square of the speed of the vehicle, meaning that a pedestrian would be thrown approximately 4 times further when struck by a vehicle travelling at 40 mph than if they had been struck at 20 mph. This is somewhat of a simplification, given that the engagement the pedestrian has with the vehicle, the height at which they are struck and their interaction with the road surface all play a key role. It does however provide a Collision Investigator attending a scene with a rough indication as to the nature of the collision they will have to investigate.

Another indicator is the level of damage caused to the vehicle itself, and specifically the damage present on the windscreen. An impact at 30 mph may typically result in a bullseye towards the centre of the windscreen with impacts at closer to 40 mph producing a bullseye closer to the top of the windscreen and speeds above this may result in a pedestrian striking the roof lining, or in extreme cases moving over the top of the vehicle completely. This is however only a loose guide, given both the height and the weight of the pedestrian can dramatically influence the level of damage present. As with all elements of collision investigation, the evidence needs to be considered as a whole.

With an approach speed established, and an indication of crossing speed and direction, a Collision Investigator may be in a position to consider how a driver responded. However, as discussed previously, a driver's response time is typically measured from a position when the hazard they are faced with becomes conspicuous. This may not be straightforward in pedestrian collisions, for example, consider the situation where a pedestrian all dressed in black is walking along an unlit country road at night. A driver is only likely to discern the presence of the pedestrian when there is enough light from the headlights of their vehicle falling on the person – they may have been in the road for a very long period of time prior to this, but a driver would not have been in a position to commence their PRT phase. Conversely, in an ultra-urban environment like a well-lit city centre, a pedestrian may be conspicuous from 100 m – therefore aspects such as lighting and contrast all become relevant to the Collision Investigator and should be measured using lux and luminance meters where appropriate.

5.3.4 Lorry Collisions

Lorries and buses, by virtue of their size, have restricted vision not only to the front but also to the sides. In collisions involving vulnerable road users such as pedestrians and pedal cyclists, one of the paramount considerations is to the available view a driver is likely to have had in the moments prior to a collision. These areas of available vision are known as sightlines and can be analysed in a number of different ways.

Vehicles can be repositioned at collision scenes, with parties positioned at known intervals, and photographs taken from the driver's seat. In addition, or as an alternative to this, 3D measuring techniques can be utilised where an investigator (or even the actual driver) sits in the seat of the lorry and an assistant can be directed to position markers at ground level relating to what can be seen by the driver in direct view (out of the windscreen) and in indirect view (*via* the mirrors). Note, most HGVs are now fitted with additional mirrors including a front proximity mirror. (This mirror is attached on the outside of the cab above the windscreen and provides a driver with a downward view of the road immediately ahead of the lorry.) The positions of the lorry and the markers can then be 3D laser scanned and by using specialist software, construct a sightline model which identifies the possible views (sightlines) a driver has from their cab as illustrated in Figure 5.1 and described in the accompanying legend.

Figure 5.1 An example of a sightline model to show the extent of a lorry driver's view. The boundary of the green area shows the extent of view seen through the front proximity mirror. The boundary of the blue area establishes beyond which the driver has a view of the road.

As shown in Figure 5.2 this sightline model can then be placed within the locus of a collision to provide a 3D graphical representation of the view available to the driver. In this example the driver would not have any view of the area between the green and blue sightlines.

Utilising specialist software, it is possible to simulate the movement of the lorry (or any vehicle) through the 3D environment, so in 'real time' the court can visualise how areas of vision, or blind spots may have moved through a collision scene – or to replicate the event of a collision from a driver's eye view as seen in Figure 5.3.

Figure 5.2 The sightline model from Figure 5.1 when positioned within the 3D view of the scene of a collision showing that the driver would not have any view of the scene between the green and blue sightlines.

Figure 5.3 A 3D view from the driver's seat of the lorry as it moves through the scene of a collision. Below the blue line the driver would have no view of the road ahead. The white dots are from the body of the lorry and the black dots are from items such as the dashboard, steering wheel, *etc.*

There is however one important consideration between a simulation and animation with which the investigator needs to take great care. Animation software essentially allows vehicles to be moved into any position considered by the animator, whereas simulation software constrains the movements of the vehicles to the laws of physics.

Whilst this specialist software is a powerful tool, it is also important for a Collision Investigator to consider if and when the driver may be legitimately looking in another direction for example, in one of the nearside mirrors if approaching a left hand. Therefore, care must always be taken when presenting any type of 3D modelling, to ensure that it does not inadvertently mislead.

5.3.5 Motorcycle Collisions

In many ways collisions involving motorcycles are similar to other multiple car collisions. The differences lay more with the geometry and handling characteristics of the motorcycle than any other property. Motorcycles, unlike cars, are inherently unstable, they balance on two wheels and when not moving fall onto their side. A motorcycle generates stability in part by the gyroscopic and trail effects of the wheels when it is moving – the slower the speed, the lower this stabilising effect. Therefore, if knocked at a low speed the machine has a tendency to fall to its side.

With the exception of a few models, the front and rear brakes of motorcycles are operated independently of one another, and on two wheeled vehicles there are advantages to being able to alter the degree to which the rear wheel is braked compared to the front (such as in poor weather conditions). However, braking a motorcycle can dramatically affect its ability to corner. Unlike a car, the cornering forces required to turn a motorcycle are generated almost exclusively by leaning the machine to one side, rather than by physically turning the handlebars. Applying the brakes to a motorcycle that is leaning, causes the machine to begin returning to an upright position (and therefore dramatically reducing the rider's cornering ability). So, imagine the situation where a motorcycle enters a bend at a speed it can navigate the curve, but the rider instinctively applies the brakes – the effect is to straighten the motorcycle and push it out wider on the bend.

Unlike a car, applying the brakes sufficiently hard to lock the front wheel (motorcycles manufactured prior to 1st January 2016, and those under 125cc are not required to be fitted with an Antilock Braking System (ABS)) results in the motorcycles falling to the ground shortly after this application (or on occasion when a rider performs a stoppie). This is the reason why, when faced with a collision, riders often instinctively brake as hard as they can, locking the front wheel and end up falling to one side and sliding into a collision. The marks present at a collision scene where this has occurred give a good indication of when and where events began to occur, and when a rider took evasive action, be that a conscious decision or an unconscious reaction.

5.4 Examination of the Scene

5.4.1 Identification of Scene Evidence

It is said that no two collisions are the same. This is certainly true; however, the types of evidence that is likely to be found at collisions of differing types can be, at times, similar. It is the role of the Collision Investigator to identify the scene evidence and understand its importance to the overall scientific analysis to be performed. They are likely to be required to identify elements such as, but not limited to, the point of impact (POI), vehicle speed, positioning on the road, possible sightlines, and alternative outcomes of the collision if a party had behaved differently. Many of the answers to these questions begin with a detailed analysis of the collision scene.

In its purest form, a collision is simply a transfer of energy, almost exclusively kinetic energy. Parties, save for those that are stationary at impact, possess kinetic energy. Within a collision, energy is transferred between the parties either partially or fully, resulting in post impact movement and damage. Collision Investigators arrive at the scene of a collision faced with the resultant product of this energy exchange and dissipation. It is often quite a dramatic and sometime traumatic scene, with severely damaged cars, injured victims and debris fields that can extend over dozens of metres. It is the identification of the marks, or more specifically the areas on the road upon which the parties struck one another, and their post impact movement, which allows the Collision Investigator to look through the wreckage and unlock the secrets as to how the key moments of the incident unfolded.

Given the energy transfer involved in a road traffic collision, investigators have the luxury of marks involved in a collision often being more readily apparent than in other forensic disciplines. Although subtle, almost all of the marks involved in the analysis of a road traffic collision scene are identifiable with the naked eye, without the need for any chemical enhancement.

When people think of collision investigation, one of the first things that often spring to peoples' minds are skid marks. But what are they, and with vehicle developments such as ABS are they still relevant? In order to understand this question, it is important first of all to understand how such marks are generated in the first place.

The generation of a tyre mark (other than a rolling tyre print on something like a muddy surface) requires one key element, heat. When a wheel is braked so firmly that it locks, the tyre stops rotating and slides along the road surface. The force that exists between the locked tyre and road surface is friction, and the frictional force that exists at the tyre/road interface generates heat. This heat is sufficient to melt the section of the tyre in contact with the road, and also the top layer of the bitumen in the road surface. This means that not only are parts of the tyre transferred onto the road, but also parts of the road are transferred onto the tyre. These are colloquially known as burn marks. An example of locked wheel marks, and accompanying burn marks on a vehicle tyre are seen in Figure 5.4(a) and (b) respectively.

It is possible to identify the direction in which the vehicle was travelling when the tyres locked. Towards the top of the burn mark on

Figure 5.4 An example of locked wheel marks, and accompanying burn marks on a vehicle tyre are seen in (a) and (b) respectively. Skid marks from a 'conventional' locked wheel skid can be seen developing in the lower portion of the image in (a). Burn marks and small stone chippings left on the surface of a tyre that has recently experienced locked wheel braking are shown in (b).

the tyre shown in Figure 5.4(b), the light-coloured road stones and grit can be seen. This collection of grit defines the leading edge of the tyre, as it slides along the road it is, in essence, sweeping the road with the fine road debris collecting at the leading edge of the tyre as it is unable to pass under it.

Directional analysis is possible from the marks left on the road surface. As discussed previously heat is required to generate the marks and in the early stages of a tyre skidding the generated heat is building at the road/tyre interface. This in turn means that there is quite often a fading in at the commencement of the marks, which can be observed in Figure 5.4(a). Conversely, a more abrupt end to the marks is expected as a vehicle comes to rest, or when the vehicle brakes are suddenly released thus ceasing the locked nature of the wheel. In both instances a defined termination of the mark should be observed.

Returning to antilock braking system (ABS) marks; ABS is designed to prevent the road wheels from locking, allowing a driver the ability to maintain some steering control of the vehicle. In its simplest form the brakes are applied and partially released very quickly, with some modern systems able to apply and release dozens of times every second. This partial releasing of the brakes means that the tyres are not fully sliding on the road surface and are only slightly deforming (known as slipping). Without full lock-up, less heat is generated between the tyre and road surface, and therefore there is less opportunity to see the dark tyre marks more commonly associated with a locked wheel. However, the tyre is still generating heat at the road surface interface, and as such marks are still possible. These can vary from quite pronounced marks, simply subtle stone scratches, or a hybrid of the two with more of an intermittent tyre mark, as identified and marked out on the road surface in Figure 5.5.

As touched on earlier, ABS provides the driver with an opportunity to steer the vehicle. This therefore highlights the fact that in a non-ABS vehicle, when the wheels are locked, a vehicle cannot be turned by a steering application. Only minor deviations on the vehicle path are observed due to the natural slope (crossfall or gradient) of the roadway, or by way of a pronounced change in path due to an impact. However, it is possible to observe tyre marks that describe a smooth curved path and as described below, there are two main ways in which curved marks can be generated on the road surface.

The first of these are acceleration scuffs or marks and a good example of such marks can be seen in Figure 5.6. These are produced when a vehicle is being accelerated so firmly that the powered wheels

Figure 5.5 The faint, intermittent marks left by a vehicle when ABS braking. Yellow road marking crayon is added to assist in the identification of the path of the marks.

exceed the tyre/road grip and rapidly rotate on the road surface, but enough grip is retained by the wheels to steer the vehicle. In all likelihood these marks are caused by a rear-wheel vehicle, as the acceleration force is being applied to the road surface by a different set of wheels that are responsible for the steering.

Figure 5.6 Curved tyre marks, demonstrating the characteristics of acceleration scuff marks as the vehicle turned right at the junction.

Electronic Stability Control (ESC[†]) systems have now been designed and developed to either largely or fully mitigate such a scenario, and therefore when such marks are identified on the road surface, the attention of the Collision Investigator must be drawn to the vehicle itself, to consider whether the vehicle was fitted with such a function in the first place, if it was defective, or was deliberately deactivated.

Location is also a consideration when examining marks of this type as they are most likely to be found where there is a need for drivers to apply a sudden and vast throttle application, especially where a steering input is also required. Hence, areas such as junctions are one of the most common places where these types of marks are found. That is, of course, leaving aside large open spaces such as car parks where, of an evening or early hours of the morning, some spirited driving activities may take place!

The second way that curved tyre marks are produced at the scene of a collision is when a vehicle has been driven around a curved path at such a speed where the lateral forces on the vehicle (and tyres) reach and exceed the limit of the road/tyre friction as commented upon earlier (Section 5.3.1). In these circumstances the tyre laterally

[†] Some manufacturers have slightly different names such as Dynamic Stability Programme (DSP) and have subtle operating differences but in the main perform the same core functions – maintaining the vehicle stability in the early stages of a loss of control.

deforms to such an extent that the sidewall of the tyre begins to come into contact with the road surface. As the sidewall elastically deforms and oscillates, a distinctive striped, or striated mark is generated on the road surface, often with fine stone scratches also contained within the mark. An example of such a mark can be seen in Figure 5.7. This type of mark is known as a critical speed mark or also colloquially, given their curved nature, called a critical curve.

It is not only identifying the types of marks which is important, but also their nature or any changes in the marks. A change in the nature of the marks help a Collision Investigator, at first blush, to identify an area 'where something happened'. Newton's First Law states that "*An object at rest will stay at rest or remain in constant motion unless acted on by an external force*". Therefore, if there is a sudden change in direction of a vehicle, it must be the case that it was acted on by an external force *i.e.*, an impact.

An example of such an impact can be seen in Figure 5.8 where a vehicle has been in a collision with a pedestrian. Locked wheel marks can be seen to extend from the bottom of the image. At a point approximately centrally within the crossing there is a short gap in the marks followed by a sharp deviation to the left. The gap is caused by a sudden loading of the tyres overcoming the brake force (as the pedestrian transferred weight onto the front of the car). The change in direction is caused by a lateral movement of the vehicle *i.e.*, compelled to move sideways (both by the movement of the pedestrian and also the applied steering now taking effect in the sudden momentary release of the brakes and rotation of the wheel). In essence, the gap in the mark defines the position of the tyres when the compelling force as described by Newton was applied – this is the point of impact, an essential element for a Collision Investigator to identify in nearly every collision they are instructed to consider. Additionally, the change in the profile of a tyre mark is also an important element of any investigation.

When driving a vehicle, it is only the tyres that are in contact with the road surface. However, in a collision that is often not the case. Metallic components, such as substructures of the car, can become deformed and come in contact with the road surface. In the case of motorcycle collisions, the machine very often falls to its side, bringing some of the metallic components such as foot pegs or hand levers into contact with the road surface. These metallic objects in contact with the road surface can leave scratch marks, or even deep gouges to the surface of the road, as illustrated in Figure 5.9.

These marks are useful in that they describe both the path and the position of the vehicle that left them. It is often possible to

Figure 5.7 A view of a section of a curved critical speed mark showing characteristic striations.

identify the exact component that generated the mark, either at the scene or at a subsequent vehicle examination, thus being able to position the vehicle with a high degree of accuracy. With accurate measurement of the marks, it is also possible to establish the

Figure 5.8 A set of locked wheel marks, with a deviation towards the centre of the pedestrian crossing which helps to identify the point at which an external force *i.e.*, collision with a pedestrian, has acted on the vehicle.

Figure 5.9 Scratches and gouge marks present on the road surface (amongst other collision debris) depicting the post impact path of a vehicle.

trajectory of the vehicle in the moments between impact and coming to rest.

A similar phenomenon occurs in many collisions where pedestrians are involved. After the initial impact and flight phase, as described earlier in Section 5.3.3, they then engage and slide across the road surface. During this slide phase, heat is generated much in the same way as a locked tyre (although to a much lesser degree), and scuff marks both from skin and also from clothing can be transferred onto the road surface by a wearing away and transfer of the two objects in contact. An example of this type of transfer is shown in Figure 5.10 where the blue scuff mark was caused by the pedestrian's jeans coming into contact with a wet road surface.

As a slight aside, the generation of heat by way of friction is another way that it is possible to identify whether a seatbelt was being worn by a vehicle occupant. In a collision, pretensioners fire a small pyrotechnic charge that rapidly tightens the seatbelt before the mass of an occupant moves forward into the belt stretching it slightly. In this process the webbing of the belt rapidly moves through the anchoring loops which generate heat and often leaves burn marks on the surfaces in contact. The location of these marks identifies if the belt was

Figure 5.10 A blue mark found on wet road surface following a vehicle collision with a pedestrian. This blue mark was found to originate from the pedestrian's jeans. This trace evidence is often difficult to see when garment colours are of similar colour to the road surface and hence, as in this instance, highlighted with yellow road marking crayon.

being worn, as does the position in which it ultimately locks *e.g.*, in a position consistent with being worn, or alternatively tight against the structural pillar (not being worn).

When trace evidence is found, such as a scuff or a gouge, it is important for the origin of this mark to be identified, as it is ultimately both the mark and how it was generated that allows the science to unlock the secrets of the collision. Consider this incident, where a vehicle has lost control and crashed. The front offside tyre was found to be deflated with damage to the wheel rim and tyre as seen in Figure 5.11(a). The vehicle in question in this collision was travelling on the wrong side of the road and collided with another vehicle. The questions for the Collision Investigator are when, why, and how this happened? Was it in the impact with another vehicle? Was it in the initial stages of a loss of control, prior to striking the other vehicle? Was it in existence prior to the incident? The cause of the deflation (or indeed any other plausible explanation) will have a key bearing in the understanding of the events of the incident.

With careful analysis of the scene, a Collision Investigator can identify the cause of the deflated tyre. In this example, by examining the kerb stones in the location (sometimes for tens, if not hundreds of metres on the approach), fresh kerb damage was found as shown in Figure 5.11(b). Of course, more analytical work such as speed, time and distance calculations are required to conclusively tie the wheel and kerb evidence together, and also establish the precise position of the initial loss of control – but it is this contact trace work that is the key to a thorough analysis.

Figure 5.11 Shows damage to (a) the wheel and tyre of the vehicle and (b) after its striking of the edge of a kerb stone – marked with yellow crayon at time of the investigation and highlighted with an arrow.

In this case, as alluded to earlier, it was the front offside wheel that was damaged. Therefore, by identifying which wheel struck which kerb this provides proof of the direction of travel of the vehicle.

The final consideration within this section is the debris field itself. The greater the energy involved in a collision the greater the degree to which vehicles are damaged and deformed. With a greater degree of damage, the larger the field of debris is likely to be present on the road surface.

Perhaps an obvious, yet easily overlooked principle, is that the debris field is made up of components that were originally travelling at the velocity of the vehicle. Therefore, when they break away from the vehicle, they are projected forwards, and without any secondary contact, travel generally in the initial direction of travel of the vehicle. The spread of debris can often take a fan-like form, where the various components all appear to converge on a single point. This view of the scene, with a consideration of any forward projection prior to the components falling to the road surface, is a good indication of the point of impact. That is why a Collision Investigator will first look at the vehicle to identify what is missing from it and then return to the road surface to see where it can be found. Sometimes this can be quite obvious to find, such as a door mirror casing that has broken from a vehicle and fallen to the road. Other times it can be more challenging, for example, tiny glass fragments from a broken windscreen or paint flakes that have broken free of the car as one of the body panels has deformed.

Of course, this is not intended to be an exhaustive description of the physical evidence at a scene, as elements such as road layout and environmental factors, impact marks on kerb stones and the nature of interactions with roadside furniture will all form part of a Collision Investigator's scene analysis. The key point here is perhaps returning to the earlier statement that no two collisions are alike, and the Collision Investigator needs to be alive to the fact that a critical piece of evidence in the previous collision they attended may be a completely different one in their next collision investigation.

5.4.2 Scene Management

Measurements are of significant importance to a Collision Investigator, as many of the questions needed to be answered rely on the position of objects relative to one another. For example, calculating a vehicle speed at impact often requires its rest position to be known along with its post impact displacement. As discussed previously, the distance in which a pedestrian crossed into the road before they were struck is a critical piece of information required. For these reasons great care must be taken when measuring the scene of a collision. Traditionally, such measurements would be taken with tape measures or measuring wheels, which are still valid techniques to this day. However, technological advances have allowed for faster and better data capture at the scene of a collision. Total stations, like those used on building sites, where a surveyor walks through a collision scene with a pole mounted prism, followed by a robot head taking a series of individual laser measurements, allows for the quick production of electronic two-dimensional locus plans. The main downsides to either of the methods above are that the investigator is only taking measurements of the things that they have identified as potentially being important at the scene.

One of the most recent developments in scene measurement is the use of three-dimensional laser scanners, which take a series of individual laser measurements. These measurements can be taken in a series of positions and stitched together in a process known as registration. This process provides a 3D model of the collision scene, capturing those positions as identified as important at the time, and also many more for later consideration. Figure 5.12 shows an example of a laser scanned 3D view of a collision scene.

These 3D models are not computer 'mock ups' of a scene, they are tens of millions of individual measured points, which explains the origin of their colloquial name, 'point cloud'. These point clouds allow for practically anything to be measured within a scene, and also to allow for advanced techniques of video analysis and simulation, which are both discussed in more detail in Section 5.6.

Figure 5.12 An example of a 3D laser scanned view of a collision scene which contains over 100 million individually measured points.

5.4.3 Testing at the Scene

There have been previous discussions about the road/tyre coefficient of friction, μ, which is the key component in understanding how a braked vehicle will decelerate on a road. This level of deceleration is dependent on both the tyres and the road surface, and to some degree the suspension geometry of the vehicle being braked. Whilst there are some values that are broadly accepted, such as locked wheels on a dry road surface reaching a μ value in the order of 0.7, and on a wet road in the order of 0.6, with ABS braking achieving 0.8 to 1.0 – these are a 'best guess'. It is a significantly better approach to measure this value using the actual vehicle and the actual road surface at the collision scene and calculate the value of μ using eqn (5.4).

$$v = \sqrt{2\mu g s} \tag{5.4}$$

where, v = velocity (m s^{-1}) at the commencement of the marks, μ = road/tyre coefficient of friction (dimensionless), g = acceleration due to gravity (m s^{-2}) and s = the length of the marks (m).

In order to obtain a tested value for μ, accelerometers are placed in the incident/test vehicle and then from a speed in excess of 30 mph the vehicle is emergency braked to a halt through the area of interest at the collision locus. This is repeated, typically three times, and averaged. This measured value is then taken forwards into the calculations.

The testing above works well for establishing the road/tyre coefficient of friction, but a different approach is adopted to calculate a road/motorcycle coefficient (after all, riding down the road with accelerometers fitted, and throwing the motorcycle to the ground is a test even the most hardened of Collision Investigators is unlikely to volunteer for). Instead, a load cell is attached to a towing arrangement affixed to the motorcycle in question and the whole arrangement is coupled up to a towing vehicle. The vehicle is then driven through the area in question (at a slow speed for safety reasons) and the profile of the force recorded on the load cell is averaged – an experiment known as a 'drag test'.

The motorcycle is weighed, and the coefficient of friction (μ) is then obtained from eqn (5.5):

$$\mu = \frac{F}{W} \tag{5.5}$$

where, F = lateral force (from the load cell) (N), W = weight of the motorcycle (N).

5.5 Evidence from the Vehicle

5.5.1 Damage Profiles

Having previously discussed the scene evidence, it would be remiss not to consider the information that the vehicle itself holds with respect to the analysis of a collision. The damage profile can give a strong indication as to how the parties initially interacted and the orientation of the forces that initially acted on the vehicle, known as the Principal Direction of Force (PDOF).

The degree to which a vehicle is damaged provides an indication of speed. As stated at the very commencement of this chapter, a collision is a transfer of energy between parties. The greater the energy absorbed by a vehicle, the greater the degree of damage present. Leaving aside low speed collisions, where components elastically deform (where they return to their original form after the collision forces are removed), the degree of permanent, or plastic, deformation provides an indication of the severity of the impact. Generally speaking, the greater the deformation, the greater the energy absorbed by the vehicle and there are a series of algorithms that allow a Collision Investigator to ultimately estimate vehicle speeds from the level of damage present. However, care must be taken here, as some areas of a vehicle are designed to deform – colloquially known as crumple zones. The principle being that by deforming, the crash pulse (the time in which the forces are exerted on the vehicle) is extended, reducing the forces experienced by the vehicle occupants, in order to reduce injury. Therefore, attention needs to be paid to which parts of the vehicle have been struck and their structural properties.

5.5.2 Lightbulbs

It may be a requirement in a day or night time collision to determine if any of the lightbulbs of a vehicle were illuminated at the time of the collision. Examination of tungsten filament lightbulbs may provide evidence as to whether a vehicle's head, brake or indicator lights were in operation at the time of a collision. Therefore, an examination of the lightbulbs may yield compelling evidence in this regard. Unfortunately, there is currently no way of discerning the operation of a Light Emitting Diode (LED) or High Intensity Discharge (HID) lamp – both of which are becoming more and more prevalent in modern vehicles. There is, however, some useful analyses that can be performed on a conventional filament lightbulb.

In a very simplistic sense, a filament bulb is a coil of thin tungsten wire that is encased in a glass envelope containing an inert gas (such as xenon). In operation, a current is passed through the coil so that it glows white hot – producing white light (some with an amber tint, giving a warmer white light, and some with a hint of blue giving the appearance of a brighter white). When the tungsten is glowing white hot, it is malleable, meaning that it can stretch and deform when a force is applied, as in a collision, whereas a cold filament could simply fracture prior to any permanent deformation of the nature of the coils.

The Collision Investigator may, following a collision, collect or receive tungsten filament lightbulbs where the glass envelopes are either intact or in a shattered state. In the case of intact envelopes, when a deformed or distended filament is observed, as shown in Figure 5.13, it is possible to state that the coil was hot when it experienced the force of the collision and hence, the bulb was illuminated at the time of the collision.

A coil that has not deformed does not necessarily mean that the lightbulb was illuminated. In this instance and also where a filament has fractured, a detailed understanding of the collision dynamics and perhaps further examination of the fracture face by Scanning Electron Microscopy (SEM) may be required. This technique offers an extremely high power of magnification and resolution.

Figure 5.13 A light bulb with a distended filament (highlighted with an arrow), indicating that this bulb was illuminated at the time of the collision.

When the glass envelope has been shattered, the filament, dependent upon the collision dynamics, could be intact, deformed, oxidised or broken. Oxidation occurs due to the sudden influx of air which hits a hot filament when the glass envelope is shattered. This might indicate that the lightbulb was illuminated at the time of the collision. However, in all instances where the envelope has been shattered further examination by SEM would generally be required.

5.5.3 Frozen Speedometers

On occasion a Collision Investigator may come across a collision vehicle that had a speedometer showing a speed even though the vehicle was at rest. This is typically referred to as a 'frozen speedo.' To understand how this occurs, a brief consideration of how speedometers display their speed is required. The car 'knows' how fast it is travelling by measuring the rate at which the road wheels are turning. This is communicated *via* the Canbus system (discussed in more detail in Section 5.5.4), to the control unit that is responsible for displaying the speed to the driver. This control unit powers up and down a series of small electromagnets in a stepper motor, to move the needle both up and down as vehicle speed increases and decreases, respectively. The important point to note here is that the speedometer needle is also moved down by electromagnets which require power to energise their coils. Hence, if there is no power being supplied to these electromagnets, the magnetic field cannot be generated, and the needle is not moved down.

If, in a collision, there is a sudden power interruption, such as the electrical loom being severed, then there is no way for the electromagnets to be powered and therefore the needle is held in the position it was in immediately prior to impact. An example of a frozen speedometer found after a collision is shown in Figure 5.14.

The question for the Collision Investigator therefore becomes, when did the power interruption occur *i.e.*, was it at a primary impact or some secondary contact and what were the wheels doing in the moments prior to the impact. One further important consideration is how robustly will the considered speedometer hold its needle position – its static torque. For example, under a gentle vibration will the needle move, or can the speedometer be subjected to significant collision forces without the needle moving? This of course determines the reliability of such a displayed reading. To understand this, the speedometer unit needs to be removed and the motor components be

Figure 5.14 The dashboard dials showing the speedometer from a crashed vehicle with the needle 'frozen', and displaying a speed of just less than 40 mph. The photograph also confirms that the airbag had been deployed.

examined and compared with the research in this matter, such as described in the work of Goddard *et al.*

5.5.4 Airbag Control Modules and Infotainment Systems

As mentioned previously, there is a Canbus system within nearly every vehicle commonly encountered on the roads today. The Canbus can be considered, for simplicity, a messaging system whereby all sensors communicate their signals, and all actuation units receive their instruction – analogous to the central nervous system of a human.

One of the actuation units contained within a vehicle are those designed to activate the airbags and supplementary restraint systems of the vehicle (such as the seatbelt pretensioners) when a crash is detected. This is known as the Airbag Control Module (ACM) which is an Event Data Recorder (EDR) – and the two terms are often used interchangeably (though the EDR function may be performed by another module). In an impact this ACM decides which airbags and restraints to deploy and when, and complex and commercially sensitive algorithms surround this process. However, this unit stores data, both prior to and immediately after the collision, which is becoming more readily accessible to an investigator. When analysed, an investigator can get such information as vehicle speed on the approach to the collision, throttle position, brake application and steering input. Through the impact itself detailed acceleration profiles can be stored at tens or hundredths of second intervals, which can assist greatly in momentum exchange calculations. Therefore, it is sometimes possible to calculate the speed of both vehicles involved in a collision from the data recorded by one EDR.

This is not the only digital data that can be harvested from a vehicle. The infotainment system, an amalgamation of information and entertainment; the stereo/GPS/telephone console, contains a lot of information. Similar to the ACM, but also sometimes additional information, such as GPS breadcrumbs of data of where the vehicle has been, various times and routes taken can be determined. It may also be possible to ascertain which mobile phones were paired to the vehicle and which keys were used to start and stop the vehicle. These units can therefore contain significant amounts of information, not simply for collision investigation, but also for the investigation of other crimes.

5.6 Video Evidence

Undoubtedly, in recent years the number of collisions or their immediate aftermath that are captured on video has increased dramatically. The video may be captured on CCTV systems, dash cameras or even mobile phones. Indeed, as the cost and power requirements of these devices decrease both motorcycle and pedal cyclists appear to be more readily fitting video cameras to their helmets or bikes. For this reason, a Collision Investigator's skill set is being called upon to diversify, and enter the world of video forensics, which is somewhat different to the historical skill set required to perform this role.

Whereas conventional video forensics may be more concerned with the processing of image data for the identification of persons, clothing or movements, collision investigation is more concerned with positioning of vehicles and time intervals, from which speeds can be calculated. Such calculations typically require careful distance measurements and a very accurate assessment of the time interval elapsed between the recordings of images – often down to the hundredth of a second.

In addition to the assessment of the metadata of a video file (the digital fingerprint that contains details of how the images were encoded) specialist equipment such as a frame interval timer, colloquially known as a lightboard, can be deployed which, with careful analysis, can provide a detailed profile of the timing pattern of a CCTV system. Positioning techniques based upon both conventional photogrammetry and also modern-day software advancements, as outlined in the work of Crouch and Cash, allow for positions to be obtained to a high degree of accuracy, and not only average vehicle speed, but also vehicle speed trends *i.e.*, braking or accelerating to be established.

5.7 Conclusions

As highlighted in this chapter the Collision Investigator needs to be able to draw upon their knowledge of several scientific disciplines. Furthermore, the investigator requires the knowledge and skills of a range of forensic science disciplines including; crime scene investigation, marks and impressions, trace and contact evidence and digital forensics and be able to present complex findings in courts. Additionally, with a significant shift towards the realism of digital data, be that from video analysis or data stored in vehicle control units, the investigator is required to undertake careful consideration and analysis of such data.

However, one of the most dramatic and exciting changes within this field, without question, has to be in the introduction of autonomous vehicles. It will simply not only be sufficient any longer to consider the actions and inputs of the driver, but also a detailed analyses of the actions and responses that the vehicle itself chose to perform will be under great scrutiny – evidence for the Collision Investigator to find!

Websites

Road Casualties Year Ending June 2020, https://assets.publishing.service.gov.uk/government/uploads/system/uploads/attachment_data/file/956524/road-casualties-year-ending-june-2020.pdf.

Homicide in England and Wales Year Ending March 2021, https://www.ons.gov.uk/peoplepopulationandcommunity/crimeandjustice/articles/homicideinenglandandwales/yearendingmarch2021.

Further Reading

D. Krauss, *Forensic Aspects of Driver Perception and Response*, Lawyers & Judges Publishing Company, Tucson, AZ United States of America, 4th edn, 2015.

R. F. Lambourn, *J. Forensic Sci. Soc.*, 1989, **29**, 371.

S. Cash and M. J. Crouch, presented in part at ITAI Conference, Birmingham, June, 2014.

D. P. Wood, C. K. Simms, and D. G. Walsh, *Proc. Inst. Mech. Eng., Part D*, 2005, **219**, 1.

C. H. Goddard and D. Price, *Speedometers and Collision Reconstruction*, SAE Technical Paper 2017-01-1412, 2017, DOI: 10.4271/2017-01-1412.

C. H. Goddard, Motorcycle Instrument Clusters for Collision Reconstruction, *SAE Int. J. Adv. Curr. Pract. Mobility*, 2020, 2(3), 1614–1622.

C. H. Goddard and S. Anderson, *Validation of Frozen Speedometer Readings Against EDR Data*, SAE Technical Paper 2022-01-0821, 2022, DOI: 10.4271/2022-01-0821.

M. J. Crouch and S. Cash, *Video Analysis in Collision Reconstruction*, Independent Publishing Network, London, 2017.

A. Yip, A. P. Sutton, P. Jennings, R. Lambourn, M. Langham and J. Neades, Forensic Collision Investigation, a Primer for the Courts, The Royal Society and Royal Society of Edinburgh, 2022, https://royalsociety.org/-/media/about-us/programmes/science-and-law/DES7309Science-and-the-Law-PrimerCollision63101.pdf.

6 Forensic Archaeology and Anthropology

Karl Harrison*[a] and Lucina Hackman*[b]

[a] Archaeology and History Department, University of Exeter, Amory Building, Rennes Drive, Exeter, EX4 4RJ, UK; [b] Centre for Anatomy and Human Identification, University of Dundee, Dundee, UK
*Emails: k.harrison2@exeter.ac.uk; l.hackman@dundee.ac.uk

6.1 Forensic Archaeology – Introduction

As the breadth of this book shows, forensic science, like criminology, is a "rendezvous subject"; practitioners from a spectrum of sciences apply their specialisms to fulfil the needs of the court. Forensic archaeology has been defined recently as a discipline of forensic science using archaeological theories, methods and techniques in a legal framework, and the integration and application of archaeological methods within criminalistics is the key to differentiating forensic archaeology from traditional archaeology. As this section outlines in some detail, forensic archaeology is most closely associated with the excavation and recovery of human remains from concealed burials (referred to as clandestine graves in much of the literature), but offers valuable scope for use in other stratified scenes, such as structural fires and collapse scenes.

6.1.1 History

The earliest publications dealing with the application of archaeological methods for use within police major investigations derive

Crime Scene to Court: The Essentials of Forensic Science, 5th Edition
Edited by Niamh NicDaeid and Peter C. White

Published by the Royal Society of Chemistry, www.rsc.org

from the United States. The four field approach of anthropology in North American universities that encompasses both cultural and physical anthropology in addition to archaeology and linguistics facilitated the leap from forensic anthropology, an older discipline with deeper academic roots, to forensic archaeology. As a consequence, in the USA and Canada techniques of excavation and body recovery are regarded as part of the toolkit of the forensic anthropologist, although this position is not without recent challenge.

Prior to this, the importance of excavation and the recording of grave contents in relation to atrocity investigation was noted in the work of Archibald Reiss during and after the First World War, while early uses of archaeology in mass killing contexts were associated with the Second World War. The most influential of these early works in the British literature was the work of Keith Mant, who performed numerous exhumations as part of his investigation at the Ravensbrück Concentration Camp in the years immediately following the end of the Second World War. At the time, his work formed part of the case for the prosecution presented at Nuremberg, but much later a summary of his work was published in the book 'Death, Decay and Reconstruction'. This book reflected a growing interest in palaeopathology across British archaeology and identified a specific link between forensic and archaeological sciences. This link was in part inspired by the high profile discoveries of fleshed archaeological bodies around the time, such as the medieval lead coffin interment found at St. Bees in 1981, whose remains were examined by the forensic pathologist Dr Edmund Tapp, or the Iron Age remains of 'Lindow Man', preserved in a Cheshire bog and when discovered in 1984 was so well preserved it was initially subject to a police investigation.

At the same time, across the same landscape and reported by the same media, high profile police searches for what would later be conceptualised as 'no-body murder' victims were being widely aired. Most especially the press interest around the 1985 reinvestigation of the Moors Murders culminated in the recovery of the remains of Pauline Reade in 1987 from a shallow grave on Saddleworth Moor – it is interesting to note here that when considering the potential for the survival of Pauline Reade's remains prior to the 1985 reinvestigation of the Moors Murders, the Senior Investigating Officer Peter Topping cited the case of the preservation of Lindow Man as giving confidence to the enquiry team.

Given this background to development, and the general growth in the sophistication and importance of forensic science from the late 1980s onwards, it is perhaps no surprise that the first recognised formal use of forensic archaeology is recorded soon after. In 1988 John Hunter was asked to assist West Yorkshire Police in the recovery of the remains of Stephen Jennings from beneath a collapsed stone wall.

John Hunter's role in the foundation and formalisation of forensic archaeology in the UK is hard to underestimate. The efforts of academics based principally at Bradford University, Hunter among them, were pivotal in the early adoption of forensic archaeology by British police forces. Hunter's writing in particular has defined the role of the forensic archaeologist in the UK, and prior to his retirement he coordinated the organisation of an Expert Panel of forensic archaeologists, under the auspices of the Chartered Institute for Archaeologists. A student seeking to understand the development of British forensic archaeology and its context of use within major crime investigation from the 1990s into the 21st Century will be well-served by close reading of his three seminal texts given in the Further Reading section at the end of this chapter.

While anthropologists have a long history of working internationally to assist with the identification of victims of mass killings, UK-based forensic archaeologists began their involvement in such large-scale investigations as part of the International Tribunal for the Former Yugoslavia (ICTY) in the mid-1990s. Although the numbers of individuals actively involved in mass grave excavation peaked soon afterwards, they have gone on to play an important part in the work of the International Commission for Missing Persons (ICMP). Initially, the small number of UK forensic archaeology practitioners worked both for British police forces on domestic major crime investigations, as well as international organisations concerned with mass grave sites abroad. As both types of work have matured and become more nuanced and complex, this movement from one to the other has become less commonplace, with individuals now usually specialising in one of the two types of investigation.

6.1.2 Organisation and Deployment

The forensic archaeologists registered as members of the Expert Panel under the Chartered Institute for Archaeologists (CIfA) are generally employed by universities, commercial archaeological units and forensic service providers. Police forces requiring their assistance would usually commission them independently, but as with other areas of specialist or 'niche' forensic science, contract-based availability is becoming more commonplace. Full membership of CIfA requires the demonstration of both skill and experience, but aspiring junior members can gain exposure to forensic archaeology via an open special interest group. In addition, the external advisers database maintained by the National Crime Agency (NCA) retains contact details for some experienced forensic archaeologists (amongst other specialists), to be provided to police forces on request.

The balance and nature of work in forensic archaeology is unusual when compared with most other forensic sciences. Casework is almost entirely focused on and around the crime scene, with work at a laboratory bench being restricted primarily to the sieving and searching of retained soil. As a consequence, forensic archaeologists frequently work closely with Police Search Advisers (PolSAs) and Crime Scene Managers (CSMs) in developing search and location strategies for clandestine burials, and with forensic pathologists and anthropologists when communicating the position and nature of remains in the ground on their recovery. Furthermore, the organisational structure of forensic laboratories has grouped forensic archaeology with the wider disciplines of forensic ecology, and forensic archaeologists work closely with forensic palynologists, botanists and soil scientists in search efforts, particularly where they relate to 'no-body murders'.

No-body murders are recognised by UK investigators as posing specific and significant challenges to successful prosecution; proving a victim's life is extinct, developing forensic lines of enquiry when no murder scene can be located, and facing the prospect of building a case for the prosecution without ever locating the victim's remains are all complicating factors that characterise these investigations. No-body murders frequently follow a two-pronged strategy; firstly the identification of material related to the commission of a crime, and secondly to the location and recovery of the victim's remains. It is in the fulfilment of this latter priority that forensic archaeology is most often employed, through a process encompassing search, location, excavation and recovery.

6.2 Search

Techniques of search look at broad areas rather than specific locations that might be identified as crime scenes. As such it is frequently undertaken initially from the desk, utilising maps, aerial photographs and physical reconnaissance over the ground. If specific intelligence exists to indicate the general site of a clandestine burial, an investigation might happily bypass the considerations of search and begin with location techniques.

Because the search phase of a no-body murder is frequently forced to consider expansive and varying landscapes, it is often coordinated around the formulation and testing of scenarios derived from intelligence built around the known predilections and resources of suspects, the location at which the victim was last seen alive (the 'Last Known Sighting' – LKS), movement indicated by mobile devices and intelligence on the nature of the offence that might be gathered from the murder site if it has been identified. The Senior Investigating Officer (SIO) in charge of the no-body murder would normally delegate responsibility for search to a Police Search Adviser (PolSA), who will develop a search strategy. A wealth of expertise and resources exist to assist the PolSA in the search phase including satellite and aerial imagery, Automated Numberplate Recognition (ANPR) systems to track vehicles on major roads, the interrogation of digital devices, police search trained officers, underwater search teams and civilian Search and Rescue (SAR) organisations.

The risks incumbent in the search phase are the lack of physical parameters to the search area, and the reliance on intelligence-based scenarios of body deposition. Poor recording of searched areas, ill-discipline or poor choice of search assets all risk areas having to be searched a second time (note though that this does not mean that all areas searched a second time have been subject to a poor initial search; sometimes the intelligence in an investigation changes so radically that the object of the search is entirely changed). Of fundamental importance is the recognition of search policy ownership by the PolSA. The forensic archaeologist may add value but should not be the coordinators of widescale search efforts.

The role of the archaeologist in the search phase will differ depending on the circumstances of the case, but they might be asked to consider the likely nature and appearance of the remains, or what the ground around or above the remains might look like given the durations since the victim was last seen. They might be asked to assist the PolSA in interpreting the various overlays of data generated by mapping and search assets. The archaeologist's ability to interpret the signatures of past land use can be useful here, especially when combined with the interrogation of historic maps and aerial imagery.

The purpose of the search phase is to identify locations that are more likely to harbour the disposed remains, based on a given scenario of deposition. Once identified and prioritised, such areas can then be subject to appropriate techniques of location. Whereas search utilises techniques over broad geographical areas, location uses a related set of tools to provide an optimal examination of a defined area that is suggested to investigators either directly by intelligence or as a result of prior search work. More so than search, location requires the direct deployment of staff and resources over the identified area of interest, and as such it relates directly to the forensic strategy established by the Crime Scene Coordinator (CSC) and the Crime Scene Manager (CSM). The establishment of scene cordons, the adoption of suitable Personal Protective Equipment (PPE) and the creation of a body recovery strategy are all examples of planning that form part of the forensic strategy. The forensic archaeologist is frequently one of the experts best placed to advise on this shift in emphasis from search to examination.

6.3 Location

The purpose of location is to identify, pinpoint and characterise as closely as possible any subsurface features that might represent anomalous characteristics that represent a buried body. Such features are generally referred to as 'anomalies' or 'targets', as they are located by the differences in composition they feature when compared with the undisturbed ground surrounding them (generally referred to by archaeologists as the 'natural', whether its formation was truly natural or not). A range of location techniques that can be used are listed below.

- Geophysical search techniques – *e.g.*, Ground Penetrating Radar (GPR), electrical resistivity, electrical conductivity, magnetometry and magnetic susceptibility.
- Victim Recovery Dogs (VRDs).
- Landform awareness – *e.g.*, ground disturbance, topography, botanical indicators.
- Probing techniques – combined with use of VRDs and depth penetration assessment.
- Intelligence – *e.g.*, witnesses, informants and aerial imagery.

This list is by no means exhaustive, but it captures the majority of resources commonly at the disposal of the forensic archaeologist.

Geophysical, landform and probing techniques will identify underground anomalies, but they will not definitively detect bodies. However, they will find changes in soil compaction associated with the digging and subsequent filling of a grave, solid buried objects such as past construction activity or geological features, and in the case of magnetic geophysical techniques, significant heating of soils. Any of these characteristics might be indicative of a clandestine burial, but they might also identify natural or 'traditional' archaeological features, hence the forensic archaeologist requires a broad understanding of field archaeology beyond their specialist area.

In contrast with the above techniques, Victim Recovery Dogs (VRDs) will indicate to their handlers when they smell the gaseous products of decomposition. Research is ongoing in this area, as it currently remains unclear what constituents of the complex and dynamic mix of decomposition gases are detected by VRDs, and they will on occasion indicate on other related scents (decomposing animals and vegetable matter, sewage or other waste material). The main role of the forensic archaeologist during this location phase is one of coordination. They are responsible for recording the crime scene in a manner that allows the representation of the various location techniques employed (usually either a scale plan or an optical survey). They will then advise on which techniques might be most appropriately applied, where on the scene they might be used, and in what order they might be best deployed. The forensic archaeologist then synthesises the various outputs of these location techniques, recording all anomalous targets on their plan, noting the character and appearance of each target and advising the CSM on how each should be examined.

A successful end to the location phase sees the location and extent of each anomalous target clearly mapped on the archaeologist's plan, and their character understood within the context of what is known about the broader archaeology of the site.

6.4 Excavation

While location techniques can suggest the nature and characteristics of buried targets, only archaeological excavation can definitively demonstrate what the target actually is. Anyone can dig a hole; they may do so more quickly than an archaeologist can excavate and record to the same depth, and if they make good use of location techniques, they may find the buried body they seek. However, the non-archaeological hole-digger will destroy much more of this vital crime scene than they will recover. Items found within the dug area will be recovered without any spatial context, so a cigarette butt that might have been significantly located beside a buried body and one insignificantly lying on the ground surface will appear to have the same evidential value. More ephemeral evidence types, such as tool marks left by the offender in the grave sides or footwear marks in the base would be destroyed unrecorded. As well as this, no attention would be paid to the grave outline or the material filling it. Perhaps worst of all, the archaeological excavation is an 'unrepeatable experiment'; once completed, it cannot be returned to. To assist with this explanation a generic process of forensic excavation would be expected to be carried out as follows;

- **Identification of the anomaly and preparation for excavation.** Location techniques suggest anomalous target to be excavated. A section line across the target and tape lines around the feature are established. As the outline of the grave becomes established it is recorded on a scale-plan (see Figure 6.1).
- **Begin excavation of a half section of the identified grave.** Excavation begins on one side of the section line. Material is removed in horizontal slices (contexts 001–004), exposing the grave cut 005 made by the offender. With removal of context 004 the lower half of the body is shown (see Figure 6.2).
- **Complete excavation of half section.** Presence of human remains confirmed and section across the grave beneath the section line is recorded. Repeat of excavation process for second half of the grave. Produce a full plan of the grave when completely uncovered (see Figure 6.3).
- **Complete excavation with body ready for removal.** All visible detail on exposed remains is recorded. Plastic bags are loosely secured over the head and hands to protect any *in situ* trace evidence. The graveside is extended and body rolled onto body sheet see (Figure 6.4).

Every opportunity to interpret the act of clandestine burial; to find the small, broken, deformed or otherwise obscured fragments of human remains or associated items must be taken, not only to ensure that as much material as possible is available for pathological and anthropological examination, but also for the duty to the victim's family and loved ones. In addition to this, archaeological excavation of targets allows the understanding of the effort and preparedness of the offender in creating the grave and concealing it. Locating and recording marks that might match the offender's digging tools and footwear provide a forensic link between them and the act of excavation, and being able to locate recovered evidence within and around the grave to a specific place allows them to be appropriately

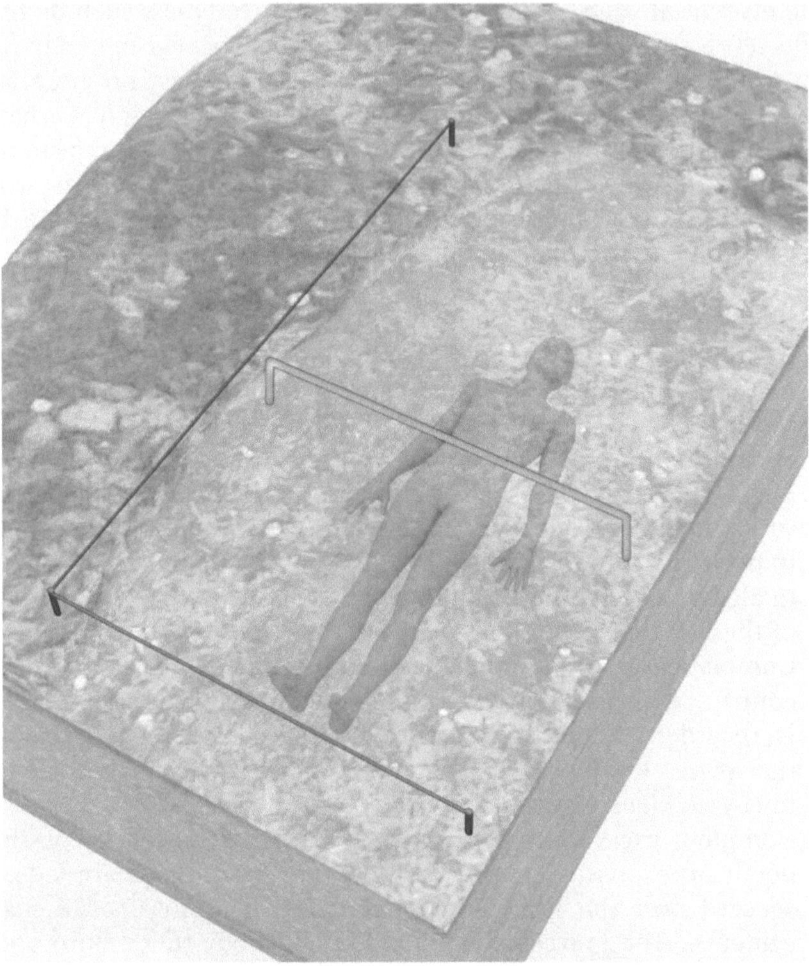

Figure 6.1 Identification of the anomaly and preparation for excavation.

prioritised by the CSM. Above all, the lack of any form of recording of the grave means that nobody qualified is able to report to a jury the nature of its examination, or indeed to mount a competent defence from contemporary records. All of these opportunities are not only lost by the hole-digging fool; they risk being utterly destroyed beyond the ability of any archaeologist to interpret from photographs after the fact.

Across the world, traditional (*i.e.* non-forensic) archaeologists have devised numerous systems and subtle variations of forms of excavation and recording in order to control and reconstruct (on paper or computer) the act of excavation and the archaeology revealed. These

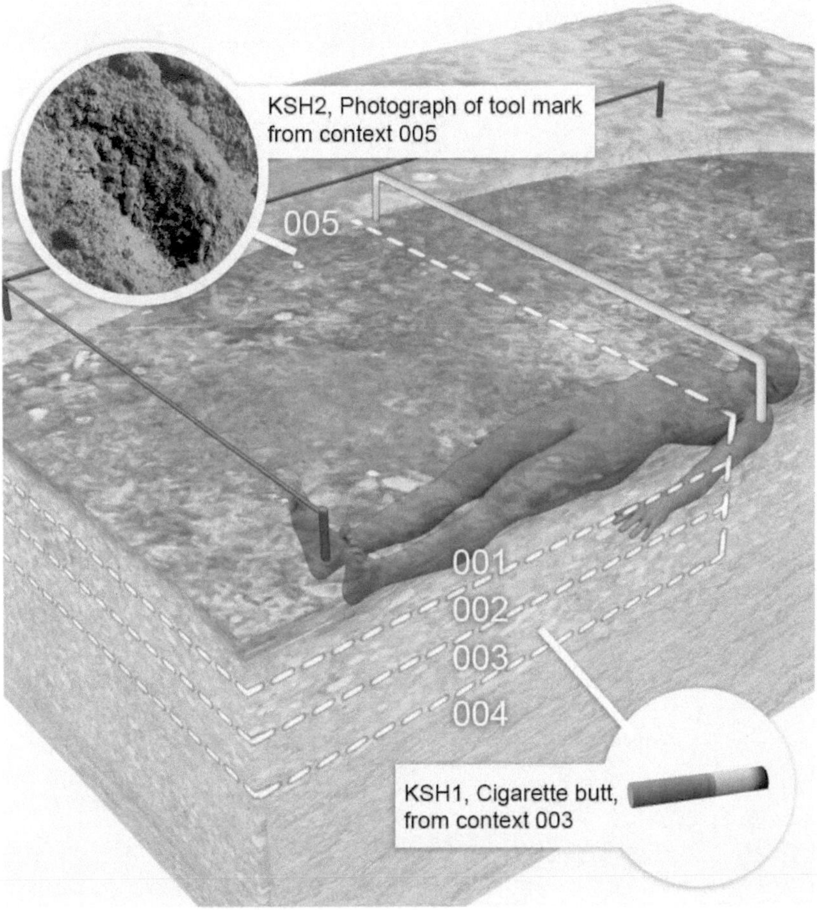

Figure 6.2 Beginning of excavation of a half section of the identified grave.

systems frequently reflect the archaeology they seek to record, perhaps prioritising horizontal (*i.e.* spatial) relationships in parts of the world with a history of limited settlement, and vertical (*i.e.* chronological) relationships where there is a long tradition of intensive settlement on sites. The single context recording system that has dominated UK archaeology since the 1970s derived from the latter requirement, being developed on deeply stratified urban excavations at the time.

The concept of 'context' is central to UK-based archaeological excavation, where it is used as a technical term distinct from, but related to its more general meaning. A context is a discrete unit of

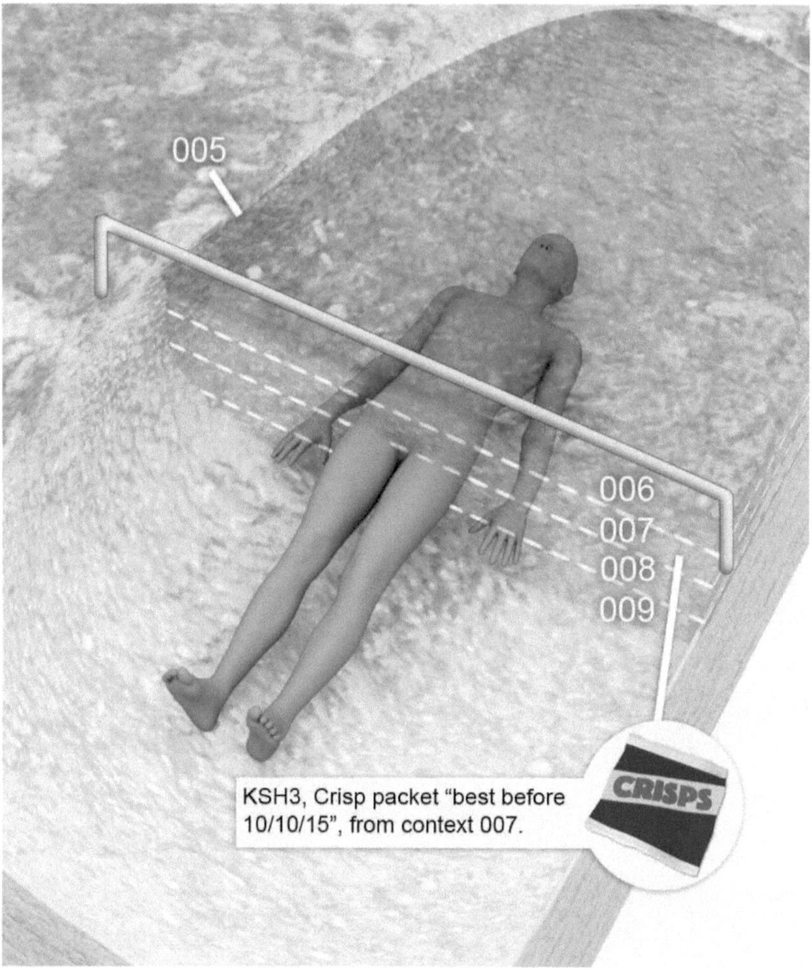

Figure 6.3 Half section excavation completed.

archaeological information, whether it be the fill of a wall foundation, the surface of a packed earth floor, or the cut line of a grave's edge. It is recognised and excavated separately from other contexts that surround it, it is given its own unique context number, and items found associated with it, whether finds in traditional archaeology or exhibits in forensic archaeology are given a location and related back to this context.

The single context recording system is a powerful tool for breaking down complex clusters of past activity, retained in the ground as superimposed cuts, fills, deposits and layers, but at times it struggles to accommodate the needs of the forensic archaeologist. Filling in a

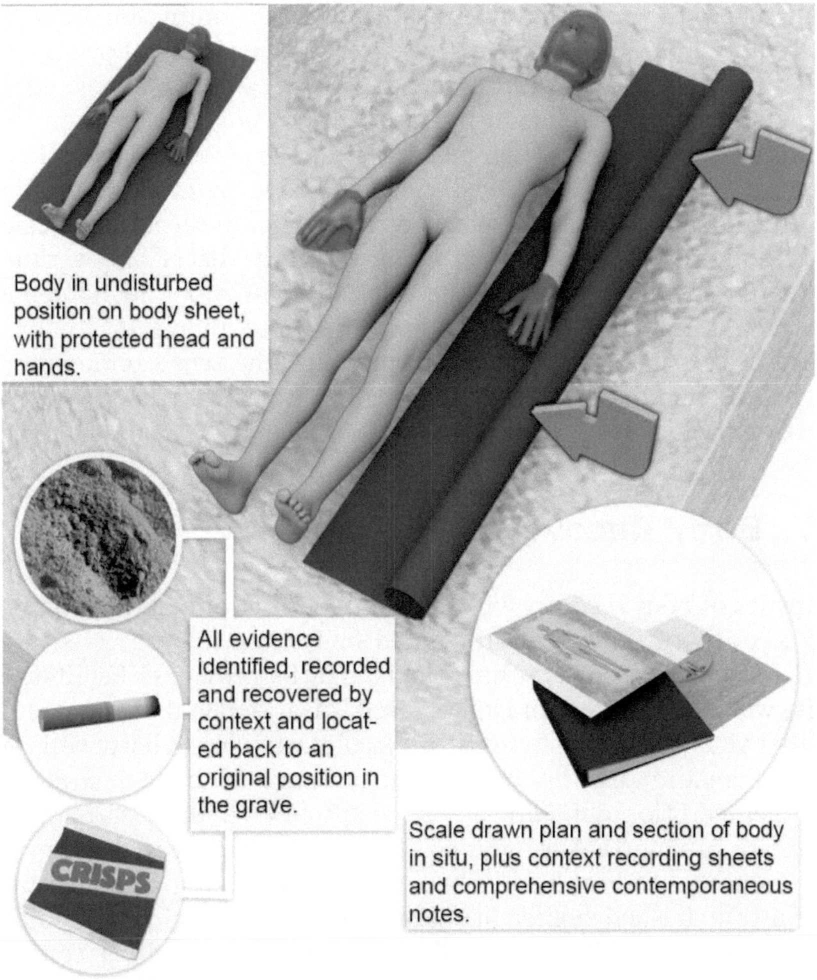

Figure 6.4 Completion of excavation with body ready for removal.

grave is routinely a single activity, the same as backfilling any other dug hole. As such, it might be expected to be recorded by an archaeologist as only one context. This fails to give the level of detail regarding depth that any buried exhibits might be found at, so the forensic archaeologist routinely subdivides the excavation of grave fill into numerous horizontal slices; a tool adapted from the planum technique of excavation.

A successful excavation phase sees a controlled removal of grave fill, faithfully revealing the cut edge as it was originally dug by the offender; neither undercutting to leave fill adhering to the sides, not overcutting and destroying all of the characteristics of the grave's edge and mixing in unrelated material from the undisturbed surrounding soil. Contexts are removed and recorded in a systematic manner, allowing the undisturbed exposure of the remains as they were deposited in the grave by the offender. At the culmination of the excavation, the remains are revealed and can be accessed for recovery. All associated evidence is recorded, packaged, and identified by context number, and the archaeologist has a set of notes that include accurately scaled plans and sections, completed context record sheets and a timed contemporaneous account of the excavation in a bound notebook. The excavation having been completed, these represent the only record of the grave construction from which the scene photography can be understood and the archaeologist's expert witness report can be written.

6.5 Body Recovery

Strategies of body recovery will depend upon the state of the remains. The recovery of a fleshed, articulated body might be observed by a pathologist, whereas an anthropologist may be present to facilitate or assist with the recovery of skeletonised, disarticulated remains. It is another vital part of the forensic archaeologist's role to liaise with the CSM to ensure that an appropriate strategy of body recovery is documented prior to making any attempt to lift a body. Disarticulated bones are light and easily handled, if sometimes fragile in damp grave environments, and are relatively easy to recover as individual bones. In contrast, fleshed bodies are heavy, frequently fragile and notoriously difficult to lift and handle, particularly when in a state of advanced decomposition. A poorly thought-out or rushed recovery risks compromising trace evidence potentially associated with the body, and even damaging the body itself. The process listed below reflects a good generic strategy when directed by the forensic archaeologist and with the assistance of 3–4 other scene attendants.

(1) The victim's body is exposed in grave; detailed photography conducted.
(2) The victim's head and hands are protected with loosely secured plastic bags.
(3) The body is then photographed again.
(4) The victim's body is then protected with board and sheeting while one long side of the grave is excavated down to the level of the grave base to allow access from the side.

(5) Board and sheeting is removed; the body is photographed again, to show its undisturbed position in the now-extended grave.
(6) The victim's body is turned sideways through 45° to allow a body sheet to be slid beneath them.

6.6 Reporting and Court Attendance

Many different people attending a body deposition site will have notes and records pertaining to the burial and excavation; the CSM's strategy, the CSI's exhibits and photographs, the detective's policy book and the anthropologist's recording sheets. None of these people can give an expert interpretation on the nature of the burial or the location of exhibits, so frequently it is the written report and oral court testimony of the forensic archaeologist that provides the key narrative to this examination in a trial. As a consequence, although buried bodies are a relatively rare component of UK homicides, the forensic archaeologist can expect no shortage of court appearances in their careers. As a guide, in the past ten years, one of the authors (KH) has undertaken 80 cases as a forensic archaeologist, which has required the examination of 110 scenes and 30 body recoveries. As a consequence, he has been required to give evidence in court on 14 occasions.

The forensic archaeologist must be fully aware of the rules governing disclosure, reporting and expert conduct, and in addition must be mindful of the need to separate their observations from any consequent interpretations. Unlike many forensic sciences, there is rarely any opportunity for a defence expert to retest any of the prosecution archaeologist's findings, thus their credibility rests on the comprehensive nature of their notes and their testimony. To produce a partial record that fails to afford a full defence to the accused is to risk censure from the court.

6.7 Beyond Body Recovery

By necessity, this section has focused on the role of the forensic archaeologist in the no-body murder where the victim has been subject to clandestine burial. As has been touched on, the operational and academic role of the forensic archaeologist has been broadened by its practitioners to include international mass graves, scenes of past atrocities, such as those associated with the Holocaust, the application of excavation techniques within forensic fire scenes, and the application of fire investigation techniques on site of archaeological fires. As an academic discipline led by casework practitioners, forensic archaeology is a young discipline in only its second generation in the UK. It still has significant untapped potential to offer both forensic science and archaeology.

6.8 Forensic Anthropology – Introduction

Forensic anthropology was once defined simply as the application of physical (or biological) anthropology techniques of skeletal analysis to the law. This meant not only that bones needed to be determined to be human (not animal), and that estimates of the population affinity, sex, age, stature and signs of injuries or diseases were noted, but that this had to be done in a legal context. Thus, although the methods used by a forensic anthropologist are similar to those used in the analysis of human remains from historic and ancient archaeological sites, the context of the discovery of the remains assumed greater importance. The rules governing any evidence associated with a crime scene are applied not only to the recovery and analysis of the remains themselves, but also to the handling and documentation of that evidence in the laboratory. Over the past few decades, even this definition and relatively narrow practice of working with skeletal remains in the legal context has expanded considerably.

A forensic anthropologist engaged in case work today must be trained and experienced in a variety of situations as the work incorporates: the identification of fragmented, decomposed and burned individuals; the recovery of human remains, especially where fragmentation or scatter has occurred, the estimation of time since death based on stages of decomposition, insect exposure and the accumulated environmental temperatures to which the remains have been subjected; analysis of peri-mortem trauma (that which occurred around the time of death and may have contributed to the cause of death) – this analysis includes estimating the type, number and direction of blunt force, sharp force or projectile trauma; human identification in mass fatality situations, whether they be natural disasters (*e.g.* the 2006 Asian tsunami, Hurricane Katrina, *etc.*), transport accidents (Marchioness and Bahrain pleasure boats, Paddington station rail crash, *etc.*), terrorist incidents (the London bombings, 11 September 2001, *etc.*), post-conflict investigations (*e.g.* Guatemala, Rwanda, the

Balkans, Iraq, *etc.*) or investigations into political disappearances (*e.g.* Argentina). In addition, forensic anthropologists (and odontologists) are being consulted more frequently to estimate the age of living individuals in cases concerning immigrants, asylum seekers, accusations of benefits fraud, age of consent, age of criminal responsibility and age at which abuse or human trafficking occurred. The technology involved in these types of age estimations is of medical origin, *e.g.* radiographs and CT scans, and the forensic anthropologist must also now be familiar with cross sectional anatomy and various software programs for reconstructing 3-dimensional skeletons from these digital images.

To become a forensic anthropologist, training in several areas is required. Certainly, a first degree in human anatomy, one of the "hard" sciences (*e.g.* biology, chemistry) or a closely related field is desirable and a masters and/or doctoral degree in forensic anthropology should follow. Training in both skeletal and soft tissue anatomy as well as an understanding of growth and development is critical to understanding variation in sex, age and stature. Comparative anatomy is useful for determining whether remains are human or animal. Evolutionary biology and genetics aid in the understanding of the concepts of population affinity and their application in forensic anthropology. Familiarity with human skeletal population variation is essential in order to evaluate how the individual under examination relates to the increasingly racially, ethnically, and geographically diverse population of the UK.

The Royal Anthropological Institute (RAI) now oversees the accreditation of forensic anthropology practitioners and allows them to achieve Chartered status. Certification is possible at three levels: at the FA-III level, there are currently 23 qualified individuals; at FA-II there are 6 practitioners; and at the Chartered level, which was the previous FA-I there are 9 practitioners qualified at the highest level. Certification at FA-III is based on a review of an application and portfolio documenting one's involvement in the field; FA-II requires evidence of active practice and mentorship and approved applicants must pass a written and practical examination; and an applicant for Chartered status must show evidence of independent casework, court testimony, continuing professional development, *etc.* and, if approved, sit a *viva voce* examination.

The Forensic Anthropology Society of Europe (FASE), a section under the International Academy of Legal Medicine, offers a two-tier certification scheme for forensic anthropologists operating in Europe; application to this scheme requires demonstration of casework, a master degree at level II and a PhD or MD at level I; examination at these levels mirrors the RAI scheme.

Both the oldest certification process and that with the greatest number of practitioners is in North America, which has a different, single-tier certification process. The American Board of Forensic Anthropology (ABFA) is the body which certifies the competency of forensic anthropologists, insures them and provides guidance. In order to become a Diplomat of the ABFA, one must hold a PhD, show proof of casework and pass a written and a practical laboratory-based examination. Since ABFA's inception in 1977, there have been 111 individuals who have achieved Diplomat status, but not all of these are still actively practising (34 have either died, retired, or been decertified due to lack of professional activity). One of the largest concentrations of forensic anthropologists today is in Latin America, where most are situated in national teams (*e.g.* EAAF in Argentina, FAFG and CAFCA in Guatemala, EQUITAS in Colombia) that function primarily as non-government organisations (NGOs). The Latin American Forensic Anthropology Association (ALAF) was founded in 2003 and serves those practising forensic anthropology in Central and South America. The ALAF plans to establish both standard professional requirements for forensic anthropologists and an accreditation board, which will implement these requirements.

In the UK, a forensic anthropologist may either be called to examine and/or recover human remains at the scene of the crime or in the mortuary on the recommendation of the investigating police and/or Home Office pathologist. The examination may be conducted at the mortuary itself or in a secure laboratory setting, but in either case, all the protocols of logging receipt of evidence, documenting what has been done (cleaning, testing, *etc.*) to the evidence and sealing of evidence apply, just as they do in all other aspects of forensic science. Ultimately, the forensic anthropologist will submit a report to the coroner or procurator fiscal in Scotland (copied to the lead investigator and the Home Office pathologist) detailing the process of the examination, the results of the examination and an analysis of the findings in the case. As with any interpretation of results, the practitioner must take care to stay with the boundaries of defensible knowledge, citing published and accepted research to support his/her conclusions and stating with what probability these conclusions are correct in the given context. A forensic anthropologist may also have to testify in court in criminal or civil proceedings or at an inquest at a coroner's court. Forensic anthropologists may also be consulted in cold case reviews where there are unidentified remains to see if new techniques might resolve the victim's identity. Without first knowing who is dead, the likelihood of finding a perpetrator of the crime is greatly diminished.

Forensic anthropologists are not medical doctors and thus may not comment on the actual cause of death (*i.e.* the physical reason for death, such as a myocardial infarction, or heart attack); yet, they do evaluate peri-mortem injuries to the skeleton and may offer interpretations of events occurring around the time of death that are relevant to the manner of death (homicide, suicide, accident, natural, or unknown). Like forensic pathologists, in cases of multiple deaths, they will look for and try to identify patterns of similarity linking one death to another. This might be useful in identifying multiple victims of a serial killer or supporting a charge of torture when a mass grave of the disappeared or prisoners of war is exhumed.

The role of the forensic anthropologist in the UK is still a relatively new one, although the use of the forensic anthropologist is gaining in popularity as more investigators become familiar with what a forensic anthropologist can contribute to an investigation. For example, many investigators immediately turn to DNA profiling when unidentified skeletal remains are found. This can be effective but relies on the fact that the missing person has been reported as missing and their DNA profile uploaded to the missing persons DNA databases. If this has not occurred a forensic anthropologist can provide a biological profile and individuating features that might narrow the field considerably and can be used by the police in media appeals as well as searches for the potential identity of the victim. Likewise, dental radiographs and fingerprints proved more powerful and expedient identifiers of badly decomposed remains in the Asian Tsunami than DNA.

6.9 Are the Remains Human?

Most forensic cases begin with the forensic anthropologist being called to the morgue or scene to examine human skeletal remains. Scene types vary enormously from single fatality scenes that are crime scenes, to fatal fires and deaths which involve accidental deaths or suicides where the bodies have not been found for a long period resulting in extensive decomposition, skeletonization and often resultant scattering of remains. They are also involved in the identification and recovery of bodies after mass fatality events. Equally likely, however, bones are delivered to the forensic anthropology laboratory by a police officer seeking a consultation concerning the most basic question: are the remains human? With the exception of an entire skull, most lay persons, many of the police and even some veterinarians and medical doctors cannot tell with certainty if skeletal remains are of human or animal origin. A forensic anthropologist, however, is familiar with the details of the human skeleton, both adult and juvenile, and he/she can identify with certainty that remains are human (or not) and, if they are human, can identify with precision the name of the bone, the exact portion of that bone (and the names of all the structures comprising it), and the side of the body the bone came from, even with fragments the size of a 50 pence piece. Potentially, a forensic anthropologist can provide even more information about that small piece such as, depending on which bone, whether it is from an adult or child, male or female. Many forensic anthropologists are also trained in comparative skeletal anatomy or archaeozoology (animal bones from archaeological sites) and are thus also adept at identifying animal bones to species.

6.10 How Many Individuals?

After identifying the remains as human, when a case involves more than a single bone find, a forensic anthropologist will then determine how many individuals are present. In routine inquiries, this is not usually a problem, as most cases requiring a forensic anthropologist consist of single individuals found in indoor or outdoor settings. When an individual dies and decomposes in a remote location, they may not be found for years and then only by chance (*e.g.* a dog walker). Bones may have become buried or scattered, and those recovered in the field may only represent a percentage of the individual's skeleton; and, if more than one individual died there, their bones may have become commingled (mixed together).

Determining the Minimum Number of Individuals (MNI) is based on knowing the number of each bone in the body; some bones such as the sternum (breastbone) occur only once, some occur in pairs (a right and left femur, or thighbone), whereas some occur in higher numbers (*e.g.* 12 right ribs, 12 left ribs, 7 cervical and 12 thoracic vertebrae, *etc.*). If there are two right femora, there must be at least two people represented in the assemblage of skeletal material. Sex and age can also inform the MNI calculation in a sample if there is, for example, one large right femur from an adult male, one right innominate (pelvic bone) from an adult female and one right humerus (upper arm bone) from a child or teenager; even though all the skeletal elements are different, each can be attributed to a different age or sex and therefore do not belong to the same individual. If more than one individual is present, a separate biological profile must be constructed for each person, even if the individual is represented by as little as a single bone.

Estimating the MNI becomes more critical when dealing with events that result in extreme fragmentation or commingling of remains. This may occur in post-conflict excavations of primary and secondary clandestine mass graves where hundreds of remains were hastily disposed of in an attempt to conceal crimes of war, crimes against humanity or genocide. The excavation of mass graves requires archaeological skill in combination with forensic osteological knowledge so that the integrity of whole bodies is maintained and remains are not separated due to unfamiliarity with what constitutes a complete human skeleton. Remains may become both fragmented and commingled in bombings and high speed mass transportation crashes. In both cases, the remains of a single individual may become widely scattered and may even become embedded in the remains of other individuals, for example in the World Trade Centre terrorist attack.

6.11 The Biological Profile

When dealing with remains that are human and unidentified, after determining the MNI, the next task for a forensic anthropologist is determining the biological profile of the individual. This involves estimating the population affinity, sex, age and stature from information gleaned from the person's bones. The anthropologist will then examine the remains for any indications of ante-mortem trauma (injuries to the skeleton that occurred while the person was still living) and the presence of pathological conditions that might indicate diseases suffered by the individual during his/her lifetime.

6.11.1 Sex

Like many mammals, humans are a sexually dimorphic species. This means that males and females differ according to both the size and the shape of their bones. Males are, on average, larger and more robust than females. The primary cause for these differences relates to the release of male hormones, such as testosterone, during the adolescent growth spurt, causing a longer growing period and an increase in stature, ribcage dimensions, muscle mass and its associated skeletal robusticity (surface area of joints, muscle insertion markers, *etc.*). When examining unknown remains, it is very important to consider the population origin of the skeletal individual – sexual dimorphism is universal, but population specific. For example, a Japanese male may look very gracile indeed when compared to a European female, but will still appear more robust than a Japanese female. It follows then that this is true for all features of sex estimation, and one must be cautious when working with new populations or in new geographic areas, as what once looked female may now represent a male. As many sexually dimorphic features of a skeleton should be examined as possible, in order to form a complete and more reliable estimate. Please note that sex is a genetically determined biological trait that is reflected in the development of an individual's body into a sexually mature human being; gender is a social construct based on an individual's decisions and self-perception and how that person portrays himself or herself in society. A forensic anthropologist will estimate sex, not gender. Gender might be inferred by clothing found on a body, but never from the remains themselves.

The most reliable means of estimating the sex of an individual from skeletal remains is a visual observation of the innominates and sacrum, the bones that form the pelvis. A male skeleton differs from a female skeleton primarily in the pelvic region, because the pelvic bones of females have undergone an evolutionary adaptation for childbirth. In general, the pelvis of a female tends to be shorter in the vertical dimension, broader in the horizontal direction and has a larger pelvic inlet in order to allow for passage of the infant's cranium during birth.

The pelvis consists of three bones, a right and left innominate and the sacrum, all of which are sexually dimorphic. In females, the sacrum is shorter and broader with less curvature than that of males. The innominates themselves are composed of three separate bones (the ilium, the ischium and the pubis) that fuse together during early adolescence and possess several features that distinguish males from females. The greater sciatic notch is formed by the joining of the ischium and ilium and can also be a good indicator of sex. In general, males have a deep and narrow greater sciatic notch, whereas this feature is more open and shallow in females. There is a tremendous amount of both inter- and intra-population variation in this anatomical region, however, and caution is warranted when no other supporting features indicative of sex are present.

The best indicator of sex is found at the anterior joint between the two innominates, the pubic symphysis. In females, the subpubic angle formed by this joint is more obtuse and the edge of the descending pubic ramus has a tendency to "fold" back on itself anteriorly, forming what is known as a ventral arc. This combination of features appears to manifest as universally female, no matter the population. Although some males have a ventral arc, it is a relatively rare feature and most often not associated with the more reliably "female" presence of a wide subpubic angle. The accuracy of determining sex from the subpubic angle alone is around 95%, which is virtually the same as taking into account all the other features of the pelvis.

Estimating sex from the skull, consisting of the cranium and mandible, is the next most reliable, yielding about 85–90% accuracy. In the skull, the robusticity of males is evident in such features as pronounced supraorbital ridges (brow ridges), rounded rather than sharp supraorbital rims, enlarged mastoid processes and nuchal lines, as well as more square, rather than pointed, chins.

These typically male features begin to form during puberty with the release of more male hormones as the individual becomes sexually mature. Females remain more gracile, although due to certain activities (such as basket carrying on the head) they may develop strong neck musculature (related to the nuchal lines and mastoid processes) in some populations. Male crania and mandibles are generally rougher in texture than those of females, again due to presence of larger muscle masses and their attachment sites. Females, particularly in African populations, are more likely to exhibit vein impressions on the external surface of the cranium, usually on the lateral part so the frontal bone. In some populations more gracile than Europeans, *e.g.* southeast Asians, male skulls may be characterised by an extension of the root of the zygomatic process posteriorly onto the occipital bone for someone less familiar with Asian sexual dimorphism in cranial morphology, the presence or absence of this feature may be one of the only ways to determine sex from a cranium alone.

Other bones of the skeleton are also sexually dimorphic, albeit to a considerably lesser degree, with the best yielding accuracies of only 70–85% when measuring specific dimensions (such as the vertical diameter of the femoral head). The vast majority of these features relate to occupational (or usage) stress on bones from the musculature. The more a muscle is used, the more it pulls at its insertion into the periosteum, the membrane surrounding all bones external surfaces. This repetitive action, in turn, causes the activation of osteoblasts (bone forming cells) underlying this membrane to lay down new bone along the ridge of the muscle insertion. Other types of stress on bone during growth can cause joint surfaces to enlarge. Certain well-defined measurements of the long bones can be taken with osteometric callipers and an unknown individual can therefore be classified as male or female in relation to known standards for a population. If, however, North American standards are applied to, *e.g.* Guatemalans, males from Guatemala might appear to classify on the female side of the scale, as they are smaller and more gracile people.

Sex cannot be determined for juvenile skeletal remains as the features associated with sexual dimorphism do not appear until puberty is advancing. Sixteen years of age is a somewhat arbitrary but reasoned cut-off point as by this age most females have reached their adult height and are sexually mature, whereas males may continue to grow until their early 20s. Therefore, a strictly speaking sub-adult (*e.g.* 15–17 years) skeleton that possesses markedly female pelvic characteristics discussed above is likely to actually be female. For the purposes of the discussion here, female puberty can be said to begin with the onset of menstruation, which occurs in most European populations between the ages of 11–14 years of age; however, onset of menstruation can be as early as 9 years of age or as late as 18 years of age in some populations around the world. Since the onset and completion of male puberty is delayed by 2–3 years beyond that of female puberty, all skeletons appear generically "male" prior to *ca.* 16 years of age. Generally, it is not considered possible to determine the sex of a child under the age of 16 years from skeletal remains alone.

6.11.2 Age

If determining the sex of skeletal material is relatively straightforward, age estimation presents greater challenges. Estimating age accurately relies on assessing multiple regions of the skeleton and even more on the forensic anthropologist's experience and the number and variety of skeletons examined over the years. Many practitioners in the field have described the age estimation process as more of an art than a science. Age estimation is always provided as range (*e.g.* from 17–21 years), rather than an exact figure because the indicators of age are influenced by both genetics and life experiences and chances. Providing an age range allows for these various influences to be taken into account. The aim is to bracket the age of the unknown individual within that range, it is unimportant where within the range the individual's true chronological age may be as long as it is within the range. Many techniques have been published and visual aids, such as casts depicting the appearance of bony features in various phases of age progression, are available. Yet, providing an accurate age range for an unknown individual requires knowledge of developmental and degenerative anatomy as well as a need to discern something about the lifestyle of the individual. This is particularly true with regard to alcohol and drug abuse, as both of which can produce premature ageing in the skeleton.

The first step in age estimation is determining whether the skeletal remains are those of a sub-adult or adult. The techniques for age estimation during childhood are based on the genetically programmed timing of 'developmental event sequences' of bones and teeth, which, when examined as a whole, can provide a fairly narrow age range estimate (probably within 2–3 years at most). These types of changes may be recognisable even to the lay person although their application and significance to age estimation may not. For example, everyone knows that babies have a "soft spot" (the anterior fontanelle) near the top of their heads; the skeletal closure of the anterior fontanelle by age two years is a developmental landmark. In another example, lay people understand that babies have all their teeth by around age two years as well; but they may not know the stages of tooth development or the exact timing and sequence of tooth eruption. A child actually forms more bones after birth, but these are primarily the "growing ends" (or epiphyses) of bones already present. These epiphyses (as well as the carpal and tarsal bones – those of the "wrist" and "ankle" – and the patella or "kneecap") all form according to a genetically predetermined sequence at genetically predetermined ages. Both the sequence and the timing are known, and even though they may vary slightly from one individual to the next (and girls go through the sequence earlier than boys), excellent data exist from which one can estimate age accurately from both bones and teeth. Eventually, when the individual completes maturation (puberty) and ceases growth, the epiphyses will fuse to the other parts of the bones to make whole adult skeletal elements. The lengths of the long limb bones in children (as well as dimensions of other bones in the foetus) can be measured, and these dimensions correlated with an age range. The clavicle (collarbone) is the first bone to form in the developing foetus (at *ca.* 6–7 week *in utero*) and has the last epiphysis in the body to close (between *ca.* 17–29 years of age). Working with sub-adult skeletal remains involves recognising many more pieces of bone and their placement with the skeleton than in the adult.

Almost as soon as individuals stop growing, their bodies begin to deteriorate, although these changes may be quite subtle. When examining adult remains, the methodologies employed are those dependent on 'degenerative changes' in various skeletal regions, which are neither as well understood nor predictable as those of childhood developmental stages. As with sex estimation, the pelvis provides two of the best bony landmarks used as age indicators namely, the pubic symphysis (the single anterior joint of the innominates) and the auricular surface (the posterior joints between the ilium and the sacrum on both sides). Both of these regions go through a predictable series of changes in their morphology from early adulthood (*ca.* 16–20 years of age in males and females) to older age.

The pubic symphysis phase system is easier for most practitioners to apply, because there is a series of 12 reference casts exemplifying the most typical configurations of each of the six phases, but the resulting age ranges are very broad and increase with age. Another problem with the pubic symphysis is that males pass through a more regular series of morphological changes than females; this is related to the fact that the joints of the female pelvis may be stressed through childbirth and develop in different ways thereafter. There are therefore different sets of casts depicting changes associated with age for males and females and these are equated with different age ranges. Thus, when using the pubic symphysis for a young adult male it should be possible to provide an age range of 5 years, but for a postpartum older female the pubic symphysis might provide an appropriate age range of 15 years.

The auricular surface is less commonly used for age estimation for several reasons – many practitioners have not been taught to use it well and thus find it difficult to apply, there are as yet no reference casts of the stages and lastly, there are at least three different sets of standards for the age ranges currently published. However, for those who are comfortable examining the traits of the auricular surface, it, too, provides an accurate estimation of age. The other advantage of the auricular surface is that the age-related changes in morphology are the same for males as for females.

The thoracic (or chest) region is also commonly used in age estimation. Another morphological change phase-based system exists for the sternal (anterior chest) end of the right fourth rib. As in the pubic symphysis, there are eight phases of changes which are different for each sex. Casts depicting the classical appearance of each phase are available, but the age ranges provided for them are somewhat unrealistic as there were problems with the sample size and type of statistics used to derive the age ranges. Radiographically, the joints between the anterior ribs and the sternum also show age-related changes that are sex specific. Males have a tendency to develop claw-like extensions on their sternal ribs over the lengthy cartilage between the end of the rib bone and the sternum; females, on the other hand, develop granular bits of bone within the cartilage itself.

In the skull, the phased closure of the cranial sutures (the non-moveable joints between the bones of the cranium) may also be useful in age estimation, but these also provide wide age ranges and are often not relied upon as anything other than a measure of last resort. A forensic odontologist (dentist) may also be adept at estimating the age of an adult from the wear patterns on the occlusal (biting) surfaces of the teeth as well as other features of the teeth and jaws. Whilst published standards exist for osteologists performing age estimation on archaeological skeletons, the differences in diet between those populations and modern ones preclude the use of these for forensic cases.

Many of the techniques discussed in this section may also be performed using 3-dimensional reconstructions of the skeleton achieved by using special software applied to CT scans. Like any forensic technique, the application of these traditionally "dry bone" methods to this medium is currently undergoing validation studies to determine its applicability and accuracy, with good results. Not all forensic anthropologists are accustomed to working with these types of images and may require specialized advanced training to interpret them, although the advent and recently reduced price of 3D printing allows for reconstruction of the images to a more, traditional hands-on approach if needed.

In summary, estimating age in the adult skeleton is challenging, but when done well can assist in the identification of a missing person by narrowing the possibilities that should be explored.

6.11.3 Population Affinity

Any discussion of the estimation of population affinity from skeletal remains must be prefaced by a series of definitions and explanations. There are a number of terms that have been used in relation to the determination of population affinity and which can still be seen in older literature. More recently the term ancestry came into use to try and avoid the negative connotations associated with terms such as race, however the most correct term of use is population affinity. Even this term can become confused with the concept of ethnicity.

In the past the term race was used however, this was then replaced with the term ancestry which refers to the physical and genetically determined traits of the body's external appearance (phenotype) that an individual inherits from his/her parents and whose recognition can be culturally specific, although tied in with the biological and geographic origin of the 'populations' from which one's parents are derived. In other words, in the USA today, people with darkly pigmented skin, tightly curled head hair, broad and flatter noses, more projecting upper teeth, *etc.* are considered to be "black" – and probably refer to themselves as "black" or African-American. But for example, in New Orleans, Louisiana, in the early 1900s there were subdivisions of individuals with the same darkly pigmented phenotype – *e.g.* octaroons, quadroons, *etc.* – which reflected their percentage of "black blood" or ancestry suggested by the degree of darkness of their skin tone. In present-day Brazil, people with these same phenotypes might be culturally divided into many more classifications and might be categorised (named) differently as well. Today the term population affinity is used which refers to skeletal attributes which are associated with socially bound groups of people. This is a more inclusive term that takes into account the diversities that are present in everyday society.

Ethnicity is more often based on self-identification, or how one perceives, categorises and presents oneself in one's contemporary socio-cultural setting. It may often contain some element of ancestry categorisation, but is more likely reflective of religions, cultural and ancestral nationality, *e.g.* someone can be white or black or brown skinned and also be ethnically Muslim (or Christian), or "Asian" (or "British"), *etc.* Ethnicity does not necessarily reflect ancestry (skeletally speaking), and thus ethnicity is not addressed by a forensic anthropology examination. Again, ethnicity might be inferred from the mode of dress, jewellery, tattoos, *etc.* on human remains, but not from the skeleton itself.

Most categorisation systems for population affinity are based on skull shape. The skull is very 'plastic' evolutionarily, which in this context means that it subtly changes its form in relation to the genetic dynamics of population mobility (predominantly seen in the form of changing marriage and reproductive patterns) over time. There is very often no adaptive reason for skull shape, it simply takes its form after decades and centuries of continued sexual reproduction within a relatively restricted gene pool (the evolutionary mechanism of random genetic drift) and thus varies according to population and geography. If one examines morphology, measures enough skulls and compares one population to another, differences that distinguish these populations (though subtle) become clear. Secondly, most character lists and measurement databases (computerised programs) of population affinity for skulls are based on socio-cultural ancestry categories – in other words, when an identified person died and an anthropologist set out to observe and measure his/her skull, the anthropologist knew the ancestry group of the individual from a photograph, or records, or remaining skin/hair on the remains provided a phenotype. Therefore, all the standards for determining ancestry based on the skull are indeed based on the perception of 'biological ancestry' as it is culturally known today – and not, as some researchers have suggested, on the 'sociological ancestry' of these individuals. Ancestry is hard to perceive from a photograph and very few (if any) public records identify the actual ancestry (*e.g.* half Norwegian, a quarter French and a quarter Native American Indian) of an individual. Most, however, will identify the ancestry grouping. Thirdly, the assessment of ancestry is expected by those searching to identify missing persons, as it, too, narrows down the possibilities. When taken in conjunction with an estimate of sex and age, as well as the context where the remains were found, the list may be narrowed enough to suggest a specific individual. Population affinity is also not usually estimated for sub-adult individuals, although many forensic anthropologists would argue that it is possible to do so.

6.11.4 Stature

In contrast to estimating age and population affinity, estimating the stature (height) of an unidentified skeleton may be accomplished with relative ease, particularly if the long limb bones of the skeleton are intact. While the process is simple – measure the indicated long bones, input the measurement into a regression formula and produce a stature estimate plus/minus the standard deviation – there are some theoretical considerations underlying the practice. Stature estimation is based on the principle that long limb bones (the femur, the tibia, the humerus, *etc.*) comprise a proportion of an individual's total height. Although it is recognised that if two people are both 1.75 metres tall, one may have very long legs and a shorter torso while the other has a long torso and shorter legs (*e.g.* men's trousers may be purchased in many different standard lengths), the formulae are based on averages of many people and most formulae will produce a stature range that will describe the missing person. Using the right formula, based on sex and race is crucial (as statures and body proportions differ amongst populations), so sex and race are determined before stature. Stature cannot be determined for sub-adults whose epiphyses are un-fused, as all formulae are dependent upon complete long bones. Likewise, it is not recommended to use stature formulae on one population that were developed on another. This holds true not only for modern geographically distinct populations, but also for chronologically distinct populations. Some of the best known USA stature formulae were developed from skeletal collections of people who died between the late 1800s and the 1940s. Secular trends (long term population level non-directional fluctuations in measurements), however, can affect populations over time, and these formulae are no longer applicable to modern forensic skeletons. Nutrition, socioeconomic status, altitude and genetics are all known to affect stature in populations.

Whatever formula is used, the practitioner must be familiar with the measurement definition for the value input into the equation. For example, in two stature formulae developed contemporaneously, one determined stature from the maximum length of the femur and the other from the bicondylar length of the femur. This may not sound important, but if the maximum length was 425 mm and the bicondylar length was 419 mm, these could produce different final stature estimates where the stature range did not accurately reflect the individual's height. That person's skeleton might remain unidentified. Thorough reading and correct application of the primary source's instructions is vital. In addition, measurements should always be taken three times before being input into the formula to ensure accuracy and consistency.

There are two types of height estimates that can be produced: measured stature and forensic stature. The former refers to the height of an individual that is measured by, for example, a health professional, and the latter is the height that is self-reported by an individual for other purposes (driver's licence, telling a friend, *etc.*). While measured stature is certainly more accurate, it is difficult to locate these records for most people. Forensic statures, although somewhat inaccurate records of true height, are more readily available – and their inaccuracies tend to be directional (*e.g.* young males and shorter people over-report stature) and have been well-documented. Most forensic anthropologists will use a formula to estimate forensic stature, as these equations are generated from modern forensic skeletal collections.

Stature estimation in mass disasters and post-conflict situations is less often critical to positive identification. This is because many people (particularly in post-conflict situations) may have similar characteristics as they were representative of a specifically targeted segment of the population, *e.g.* military age males. With the estimation of stature, the basic biological profile of sex, age, race and stature is complete. The further individuation of remains, leading to positive identification, occurs next and begins with an assessment of conditions that may have affected the individual's skeleton during the time the person was alive – ante-mortem skeletal trauma and pathology.

6.11.5 Ante-mortem Trauma and Pathological Conditions

The forensic anthropologist will examine skeletal remains for indications of healed injuries or disease lesions and also for signs of "reactive bone" indicating active disease or injuries in the process of healing at the time the individual died. Bone reacts to injury or disease insult by producing lesions that are either lytic (destructive of bone) or proliferative (creating new bone). These are both facets of the skeleton's ability to heal itself by removing older bone and regenerating new bone (remodelling), and this process requires the resorption of damaged (essentially dead) bone as a first step before new bone is laid down. Wolff's law states that bone is resorbed where not needed and put down where required – this process relates to all aspects of changes within the skeleton, from growth in length and changes of shape with age, to epiphyseal union, fracture healing, and to reactions to the progression of bone diseases.

Evidence of healed fractures (as well as dislocations, muscle injuries, *etc.*) are fairly easy to detect on skeletal remains both through observation and the sense of touch. A healed fracture presents as a callus, a proliferation of bone that goes through several stages in the healing process. Initially, union of fractured ends is accomplished with a large fracture callus surrounding the fracture site (Figure 6.5). This callus is made up of what is known as "woven bone", which has a very fibrous and porous appearance and is somewhat like a dry, rough and hard sponge in texture. As the remodelling process continues, this callus transforms into cortical (smooth) bone and reduces in size as the fractured ends begin to unite internally within its protective circumference. Eventually, the well-healed fracture will potentially be visible as a slight misalignment of the bone (in long bones, hand bones or the clavicle, for example) with some smooth, extra bone surrounding or filing in areas, or, as a slight lump, almost like a swelling, on a bone of reasonably normal shape (such as a rib).

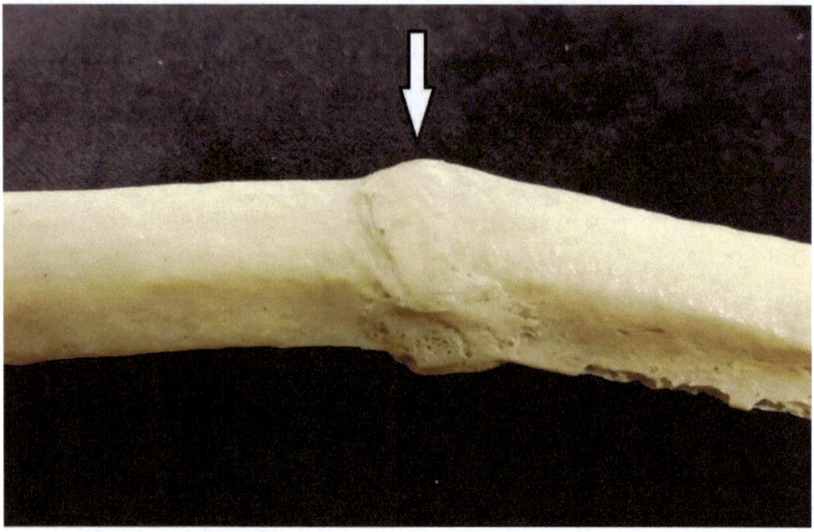

Figure 6.5 Remodelled bone evident at the site of a previous fracture (indicated by arrow).

In a forensic examination of skeletal remains, rib fractures are often more easily detectable by touch than sight. Other musculo-skeletal injuries like dislocations, severe sprains or tears at a muscle insertion site will be visible as exostoses, or spicules of bone growth at the site of the injury.

Whilst healed fractures are fairly commonly observed, it is important to note that most common diseases of modern times do not leave markers on the skeleton. The primary exceptions to this, though still relatively rare in living populations indeed – and by extension even more rarely found in a forensic case – include:

(1) Certain types of tumours, whether malignant or benign.
(2) Various infectious diseases, including tuberculosis.
(3) Blood disorders (*e.g.* severe anaemia).
(4) Metabolic conditions (*e.g.* osteoporosis).
(5) Congenital abnormalities (*e.g.* spina bifida, club foot, extra ribs or digits).
(6) Skeletal dysplasias (*e.g.* achondroplastic dwarfism).

If present, any of these may be very helpful in individuation if there is extant and locatable medical documentation. As with all aspects of the biological profile, identification can only be achieved if there are ante-mortem records with which to compare the information gleaned from the skeleton.

6.12 Time Since Death

One of the most important aspects in identifying decomposed or skeletal remains often revolves around being able to state when the individual died. This allows the police to compare the post-mortem interval estimate with records of people reported missing during that interval, thus narrowing down the possible matches. The scientific basis of estimating time since death is a relatively new aspect of the forensic anthropologist's repertoire and one with which many are not yet proficient. The science of 'forensic taphonomy' is concerned with everything that happens to an individual from the moment of death until the remains are recovered. As such, it is primarily concerned with the processes of decomposition and those factors that may influence its rate and pattern.

Much of the early research in this area involved observations from the University of Tennessee, Knoxville's Anthropological Research Facility in the USA, where donated human cadavers are allowed to decompose naturally. Since then, numerous institutions in the USA, UK, Canada and Australia have developed taphonomic research facilities using either human or animal models. The advantage to using human cadavers is that no other adjustments are needed to the data; the disadvantage is that it is rare that well-designed experiments are conducted using human cadavers because it is difficult to obtain and store (refrigerate or freeze) such large items until a big enough sample size is achieved to make the decomposition experimental results meaningful. Using animal models (usually pigs, as they are most anatomically similar to humans with respect to skin type, having an omnivore's gut design, *etc.*) is advantageous as experiments can be conducted with enough cohorts, replicates and controls to make the data obtained statistically meaningful, but of course, certain adjustments to the data (namely body and limb size) must be made prior to applying them to humans. Retrospective studies on human cadavers (seen at autopsy) have been employed to great effect in the past 10 years, producing equations for the estimation of time since death based on:

(1) The appearance of the corpse.
(2) The accumulation of temperatures to which it has been exposed.
(3) Whether or not it was exposed to insect activity.

Calculating time since death relies on the assignment of a Total Body Score (TBS) and assigns a cumulative numerical value to the body, based on the appearance of three regions (the head and neck, the torso and the limbs) according to defined characteristics along a continuous scale. The Accumulated Degree Days (ADD) are the average daily temperatures (in °C) to which the body has been exposed for the duration of its time in the depositional environment. So a body left on the surface (or in water, or indoors or buried) for 5 days at 20 °C will have an ADD of 100, whereas a body exposed for 10 days at 10 °C will also have an ADD of 100; the concept is that any body with the same ADD will have a similar appearance (or TBS). The two factors that appear to influence this equation of ADD to TBS are the size of the corpse and the presence/absence of insect exposure. Insects cause the body to decompose at a more rapid rate than bodies protected from insects (whether buried, indoors or submerged in water). In the presence of insects, smaller bodies decompose more quickly than larger ones, but when insects are absent there is no change in rate. The pattern of deposition appears to be altered slightly by perimortem penetrating injuries (gunshots, stabbing, *etc.*), but the rate at which the body decomposes is not altered. Likewise, superficial burning of a body appears to affect the type of insects attracted to the corpse as well as the pattern of decomposition, but burnt and un-burnt corpses still decompose at the same rate (in the same ADD). On the other hand, hanging bodies are subject to different decomposition rates, most likely owing to their greater exposure to drying winds, the inability of certain insects to access them and gravitational forces. Decomposition in bodies where insect access is prevented (*e.g.* remains found indoors, in water or deeply buried) occurs much more slowly.

6.13 Human Identification

The identification of human remains requires the matching of ante-mortem and post-mortem data. The former is collected by interviewing the relatives of the missing person (usually by community liaison officers), and collecting medical, dental and fingerprint records of the missing person. Whilst in the investigation of a single individual's skeletal remains, one must match that single set of post-mortem data to ante-mortem data collected from multiple candidates, in a mass disaster or post-conflict setting, the situation is compounded by there being many missing persons' remains (possibly in fragmentary condition) and many possible candidates for identity. Some mass fatality events are defined as 'closed events', such as an air transport disaster where the identities of the crew and passengers are known from the plane's manifest. There will thus be an equal number of individuals missing as remains of individuals found (see discussion of MNI above). Other mass fatality events are defined as 'open events', where both the number and names of victims are unknown, such as mass graves in post-conflict settings, natural disasters such as earthquakes or transport disasters such as commuter train crashes. Identifications in open events are much more difficult to manage as it may take some time to discover who is really missing in relation to the event and also to gather the relevant ante-mortem information. In all cases, a positive identification must be achieved before remains can be repatriated to families and any legal matters relevant to the deceased's death be resolved.

A 'positive identification' may only be achieved through a high probability match of fingerprints, DNA, or dental radiographs. The number of missing, and hence the potential number of matches, has direct bearing on the degree of probability necessitated to declare a match and make an identification. The more victims involved, the higher the probability of false positive matches for any of these techniques. Alternate and supporting means of identification may consist of a matching biological profile, ante-mortem medical radiographs, serial numbers from medical or dental implants, truly unique items of jewellery or tattoos unique to image content and placement, *etc.* Clothing and other personal effects, however, are only a means of supporting identification and would never suffice in themselves as a positive identification. In many other parts of the world, 'presumptive identifications' (those made based on, *e.g.* the family's recognition of the deceased's facial features, clothing, *etc.*) are still common.

Thus, while a forensic anthropologist contributes to the gathering of evidence used in making a 'positive identification', the techniques and methodology of the forensic anthropologist are not means of positive identification by themselves. In some countries, forensic anthropologists may perform dental radiographic matching and other types of radiographic matching as a means of positive identification, for example frontal sinus patterns, but this is not common practice in the UK. Forensic anthropologists have proved particularly useful in mass fatality events because of their ability to recognise and identify tiny fragments of tissue to sex, side and age. For example, at the World Trade Centre (WTC) anthropologists were responsible for stopping identifications being made when they recognised that the sample taken for a DNA profile did not actually belong to the majority of the remains (a part of one person had been blown into the body of another, and the match related to the small fragment of someone else).

Likewise, at WTC anthropologists were used to sort through the removed rubble and identify human body parts, distinguishing them from the food waste from the numerous restaurants housed in the twin towers.

6.14 Trauma Identification and Interpretation

Identifying the presence of trauma to the skeleton, distinguishing what type of trauma it is (projectile, blunt force or sharp force) and interpreting force trajectory from the characteristics of the wound are some of the most important things a forensic anthropologist does that have legal ramifications. Trauma can be inflicted ante-mortem (while the individual was living), peri-mortem (occurring at or around the time of death) or post-mortem (deliberate damage occurring to the skeleton after death, for example dismemberment). Ante-, peri- and post-mortem injuries exhibit different qualities consistent with the character displayed by the bone itself in a living or deceased state. Bone is a living tissue when the individual is alive and it dies when the individual dies. It primarily contains a mineral component, which renders it strong and hard (hydroxyapatite), and an organic component (collagen) which imbues it with elasticity, resilience and resistance to stress. For a time after the individual dies, bone will still behave like a living tissue, and injuries to it cannot be readily distinguished from those occurring during life, or rather at the time of death because healing will not have had a chance to begin. The crux of the matter is that there is little known concerning when (chronologically) the bone tissue begins to become drier and more brittle and react like dead bone rather than live bone.

Ante-mortem injuries are, as noted above, distinguished by the process of healing with reactive bone at the site of the injury. A peri-mortem injury is an injury that evinces no signs of healing and thus appears as fresh damage.

Characteristics of peri-mortem injuries include adhering fragments of bone depressed into the injury site (in blunt force trauma), well-defined and "clean" cuts with lifted spurs of bone (in sharp force trauma) and well-defined entrance/exit defects from which one can estimate bullet/projectile trajectory (in projectile trauma). Some trauma may be a combination of two of these types, such as a machete or axe injury, where the implement inflicting the trauma is both sharp and heavy; this may result in characteristics of both type of trauma appearing on the bone, for example a sharp cut that has a depressed, crushed area continuous with it. Post-mortem trauma, usually caused by animal scavenging, trampling or excavation damage, is defined by differential colouration of the cut or broken surface from that of the rest of the bone. Bones damaged post-mortem have a tendency to break transversely, rather than in a more spiral manner as they do in living tissue. The location of fracture and sharp force trauma on the bone is also a clue to its chronology – post-mortem breaks and cuts may occur anywhere on the bone, whereas breaks occurring peri-mortem are focused in injury sites known to occur in attacks or accidents (ankle, collarbone and wrist for injuries occurring from falls; mid-shaft of the ulna for defensive injuries from direct blows; palmar surfaces of hand and finger bones from sharp force defensive injuries, *etc.*). Other patterns of injuries are common to hit and run motor vehicle *versus* pedestrian accidents or motorcycle accidents.

It is important for the forensic anthropologist to be conservative in his/her interpretation of peri-mortem trauma, particularly when these occur in multiples. A single stab wound to the ribs can easily damage the upper border of one rib and the lower border of the rib above it. The anthropologist has seen two sharp force trauma injuries, but they represent a single stab wound. It is very prejudicial in a court case to overstate the number of stab or blunt force injuries. This is true as well for blunt force trauma, and is particularly relevant for blows to the cranium where one blow may overlap with and obscure a previous one. It is still professionally more ethical in expert witness statements and in court to err on the side of caution and provide a minimum number: 'Yes, it could certainly be more, but it was no fewer than two.' After all, both the verdict and the sentencing may be strongly influenced by this testimony.

6.15 Conclusion

Although forensic anthropologists are still fairly few in the UK and remain rare elsewhere in the world as well, competent practitioners of the discipline have contributed significantly to the analysis of human remains from high profile murder and serial murder cases as well as mass fatality events and post-conflict investigations. Broadly trained forensic anthropologists possess a multi-disciplinary nature that is rare among specialists and their cooperation as team members in criminal investigations, civil suits and identifications should not be overlooked. Population based studies relating to the establishment of the biological profile remain important, while trauma analysis and time since death estimation represent the cutting edge of forensic anthropology as an emerging experimental science.

Further Reading

Age Estimation in the Living: the Practitioner's Guide, ed. S. Black, A. Aggrawal and J. Payne-Hoames, Wiley-Blackwell, Chichester, 2010.
Handbook of Forensic Archaeology and Anthropology, ed. S. Blau and D. Ubelaker, World Archaeological Congress Research Handbooks in Archaeology, Left Coast Press, Walnut Creek, 2009.
Death, Decay and Reconstruction, ed. A. Boddington, A. N. Garland, and R. C. Janaway, University Press, Manchester, 1987.
Standards for Data Collection from Human Skeletal Remains, ed. J. Buikstra and D. Ubelaker, Arkansas Archeaolgoical Survey Research Series No. 44, Arkansas Archaeological Survey, Fayetteville, 1994.
Forensic Archaeology: A global perspective, ed. W. J. M. Groen, N. Márquez-Grant, and R. C. Janaway, Wiley Blackwell, London, 2015.
Studies in Crime: An introduction to forensic archaeology, ed. J. Hunter, C. Roberts, and A. Martin, Routledge, London, 1995.
Forensic Archaeology: Advances in theory and practice, ed. J. Hunter and M. Cox, Routledge, London, 2005.

7 Fire Investigation

Iain Peck

Prometheus Forensic Services Ltd., Bulrushes Business Park, Coombe Hill Road, East Grinstead RH19 4LZ, UK
Email: iain@prometheus-fs.co.uk

7.1 Introduction

Fires are part of our everyday experience. Without fire, the things we take for granted today would not exist and our lives would be very different. Fire also exerts a fascination over people, and its destructive potential is well known, so it is not surprising that fire is used to attack both people and property. In English and Welsh Law a deliberate act that damages property without the permission of the owner is an offence against the Criminal Damage Act 1971, and if the destruction is by fire then the offence is classed as arson. In Scottish Law when someone deliberately sets fire to someone else's property to damage it or to injure people it is classed as wilful fire raising.

Arson is one of the few crimes that by its very nature destroys forensic evidence; not simply the evidence that might be used to determine who was responsible, but also the evidence of how the crime was committed. Since many fires are the results of accidents, the investigation of a potential arson is unlike many other areas of forensic science because the fire investigator will be asked to prove that a crime has been committed before the question of who was responsible can be addressed. It is this double burden of proof that

Crime Scene to Court: The Essentials of Forensic Science, 5th Edition
Edited by Niamh NicDaeid and Peter C. White

Published by the Royal Society of Chemistry, www.rsc.org

makes prosecution in cases of criminal damage by fire one of the most difficult tasks the courts have to deal with, and it places considerable demands upon the knowledge and experience of the scientist giving evidence on the fire's cause.

Fires are not only investigated for forensic reasons: there may be issues of public safety or regulatory compliance that require an in-depth study of ignition or propagation mechanisms at some incidents. Fire investigation is not, therefore, the sole province of the public and private sector forensic scientists but may be carried out by fire service personnel, who have a statutory duty to fulfil in reporting the cause of fires, and fire engineers and specialists from other organisations. In the last 10–15 years the emergency services have developed a partnership approach to fire investigation, with teams comprising Scenes of Crime Officers (SOCO) or Crime Scene Investigators (CSI) and Fire Service officers taking responsibility for the majority of incidents that are reported to the police as suspicious. In technically complex cases, however, or those that rely on a combination of evidence drawn from work both at the scene and in the laboratory such as fatal fires and fire murder cases, it may be desirable to call upon the dual skills of a trained and experienced forensic scientist.

Forensic scientists who specialise in fire investigation must possess a basic knowledge of chemical and instrumental analysis, combustion chemistry, fire dynamics and fire engineering to augment the vocational competence that they acquire during training at scenes and in the laboratory. Interpersonal and communication skills are also important given that they will meet and manage people from all walks of life at a fire scene and they may be required to present their findings in court.

The current hot topic affecting the forensic science sector in the UK is the expansion of quality standards or the accreditation of examinations and testing activities within forensic disciplines in accordance with the Forensic Science Regulators Codes of Practice and Conduct. Fire investigation is currently not subject to accreditation although it is expected to form part of the Forensic Science Regulator Codes of Practice in 2026 under ISO/IEC 17020: Requirements for the operation of various types of bodies performing inspection.

7.1.1 What Is a Fire?

Real world fires are extremely complex chemical processes, but in simple terms a fire is an exothermic oxidation involving a fuel and an oxygen supply, usually air. Air is all around us but even highly flammable materials do not burst into flames at random because the oxidation must be initiated by the application of a heat source. All fires depend upon the presence of these three interdependent components – air, fuel, and heat – for their existence. If any one of these is missing then the fire will not start, and if you take away one or more of them while the fire is burning it will go out.

Air, fuel, and heat are the three sides of what is commonly referred to as the fire triangle, although combustion scientists will by preference use the term fire tetrahedron because a self-sustaining fire will require, as a fourth side a propagation mechanism if it is to keep burning. Considering the simpler fire triangle, in real life apart from the air around us the other two sides are formed by the combustible items we put in our homes or offices and heat from sources such as flames, smouldering cigarettes and other incandescent objects. The mechanism that produces the heat that starts the fire – the ignition source – is the aspect of the fire that is of most interest to forensic fire investigators, and a large part of the work that they carry out at the scene and in the laboratory is focussed upon answering the question "How did the fire start?".

7.1.2 Ignition of Gases, Liquids and Solids

Ignitable gases and vapours will only ignite if they are mixed with air in the correct proportions. Above a certain concentration the mixture will be too rich and below a certain concentration it is too lean; in either case the application of a flame or spark will not cause the mixture to catch fire. These concentrations, the upper and lower flammability limits, are different for each type of fuel.

When a fuel–air mixture is within its flammability range a small volume of it must be heated to a certain minimum temperature, called the autoignition temperature of the gas or vapour, before it can catch fire. In most cases a flame burning at 900–1200 °C or an electric spark at a temperature in excess of 3000 °C will be sufficient. Studies of ignition by static electricity, however, have revealed that temperature is not the sole criterion: for fuel gases such as natural gas or propane the ignition source must also have a minimum energy of approximately 0.3 millijoules.

Combustion can be a simple process: in an everyday context a lit Bunsen burner will burn natural gas (methane) to produce carbon dioxide, water vapour and heat:

$$CH_4 + 2O_2(\text{air}) \rightarrow CO_2 + 2H_2O + \Delta H_c \qquad (7.1)$$

where ΔH_c is the heat released by combustion. The stoichiometric nature of this reaction is only true for a premixed flame, where the correct amount of oxygen is mixed with the fuel before it burns. If the amount of available oxygen is severely reduced by adjusting the air inlet at the base of the burner then the fuel will burn with a luminous diffusion flame; the fuel–air mixture will be too rich and it will only burn at its interface with the surrounding air. Incomplete combustion can occur in a diffusion flame; its luminosity is due to the presence of unburnt carbonaceous (soot) particles that have been heated to incandescence. Analysis of a sample of the gases in the flame would reveal a more complex reaction taking place:

$$CH_4(\text{excess}) + O_2(\text{air}) \rightarrow CO_2 + 2H_2O + CO + C + \Delta H_c \tag{7.2}$$

The equation has deliberately not been balanced in this instance because the amounts of carbon dioxide, carbon monoxide and carbon produced would depend upon the amount of oxygen available. Real-life fires are generally oxygen deficient, hence their propensity to produce large amounts of soot and unburnt volatiles as well as heat and flames.

At normal room temperatures the vapour pressures of many liquid fuels are too low to produce concentrations of vapour in air that are above the lower flammability limit. These liquids need to be heated to enrich the vapour–air mixture directly above the liquid before the concentration can reach the lower flammability limit. The lowest temperature at which sufficient vapour is present that it can be ignited by a small flame is known as the liquid's flashpoint. There are a number of different methods by which flashpoints are measured, and figures quoted in the literature for any particular liquid will vary depending on the one used. Table 7.1 lists a number of common

Table 7.1 Flammability limits, flashpoints and autoignition temperatures of common fuels and solvents.[a]

	Flammability limits (vol% in air)			
Fuel	Upper	Lower	Flashpoint (°C)	Autoignition temperature (°C)
Methane	5.0	15	−188	640
Propane	2.1	9.5	−104	500
Ethanol	3.3	19	13	365
Petrol	1.3	7.1	−45 to −42	440
White spirit	1.1	6	38 to 60	149 to 232
Diesel fuel	0.6	7.0	>55	>250

[a] Data collated from a number of sources – see Further Reading.

gaseous and liquid fuels with their flammability limits, flashpoints and autoignition temperatures.

Liquids with a flashpoint above room temperature can be ignited by a small flame without the bulk of the liquid being pre-heated, but only if they are soaked into an absorbent material that is capable of acting as a wick. In such circumstances the application of the flame to the wick will quickly heat the small volume of liquid soaked on to it, vaporising it and producing a locally enriched fuel–air mixture that will ignite if it is within the flammability limits. The heat released by the combustion of the fuel–air mixture will continue to vaporise more of the liquid from the wick to sustain the flame, and the fuel that is burnt will be replaced as the wick draws more of it from the reservoir by capillary action.

Common combustible organic solids include natural materials (such as wood, cotton, leather or wool) and synthetic polymers that are primarily hydrocarbon-based. The chemical composition of these materials principally comprises carbon and hydrogen, with varying amounts of oxygen, nitrogen and other elements such as sulphur and chlorine. When heat is applied to an object made from these materials its surface temperature will rise; thermoplastic polymers will melt and other materials will scorch and char. If the temperature is high enough (between 150 and 500 °C for wood and many common polymers) thermal decomposition will take place. The molten thermoplastic, or the charring object, will start to pyrolyse, with the weakest chemical bonds in its molecular structure breaking down to produce smaller molecules called pyrolysis products. Low molecular weight pyrolysis products are released as gases or vapours, many of which are flammable when mixed with air. Once the concentration of the pyrolysis products reaches the lower flammability limit, and assuming that the heat source is sufficient to raise them above their autoignition temperature, they will catch fire. The combustion of the pyrolysis products then generates more heat, which will help to maintain the decomposition process. Provided the heat released by the combustion of a given quantity of pyrolysis products feeds back into the fuel more energy than is required to produce them then the fire will continue to burn for as long as additional fuel and fresh air are available.

7.1.3 Self Heating and Spontaneous (Auto)Ignition

Many organic materials undergo slow surface oxidation reactions even at room temperatures, but normally over a short period of time these will have no perceptible effect on the physical properties of the material (although they may have a distinct effect on the flavour of foodstuffs). If the oxidation takes place on the surface of a solid or liquid then the small amount of heat that is generated will usually be lost to the environment by convection. If the material has a large surface area to volume ratio, however, then heat production is much greater and heat loss will be greatly reduced due to the insulating effect of its bulk. As a result, the internal temperature of the material will rise and, provided the oxygen supply is sufficient, the rate of reaction will increase until there is a balance between heat production and heat loss. If no balance is achieved the reaction will undergo thermal runaway and the material may ignite. The process of self-heating, thermal runaway and ignition is called spontaneous combustion as the material involved can catch fire in the absence of an external ignition source.

In all cases of spontaneous ignition resulting from prolonged self-heating the process starts within the bulk of the stacked or packed material rather than on its surface. Provided that the scene has not been too badly damaged by fire fighting, the investigator will look for signs of this "inside-out" burning when considering spontaneous combustion as a cause of the fire.

A wide variety of materials and products can undergo self-heating, including hay, coal, cotton waste and grains together with unsaturated oils and fats if they are soaked onto fabric or a fibrous substrate with a large surface area. Spontaneous ignition is a problem frequently encountered in the food process industry if heat-dried powders or granular materials are not cooled before bulk storage, and it can also occur in commercial laundries if freshly washed and dried clothing is stacked while still hot and the fabric retains traces of oil or fat on it. The risks of spontaneous combustion are well documented in product and process safety literature, and the fire investigator will draw upon this information when considering whether materials at the scene are capable of self-heating.

7.1.4 Flaming and Smouldering Fires

A fire that is supplied with adequate fuel and air will usually continue to burn until it is extinguished or until the fuel or air supply runs out. The majority of the combustion processes take place in the gas phase, producing the flames we commonly associate with a fire. Smouldering fires, by way of contrast, do not produce flames because the combustion process takes place directly on the surface of the fuel, which glows. Smouldering fires propagate directly through the fuel they are burning on: the combustion front preheats the unburnt material ahead of it until it ignites and starts to glow. Since the minimum ignition temperatures necessary to initiate glowing combustion are lower than the autoignition temperatures of many liquid and gaseous fuels, smouldering fires can be started by heat sources that are incapable of igniting flammable liquids. Thus, a lit manufactured cigarette (rather than a hand rolled cigarette as these tend to self-extinguish it not actively smoked) can easily start a smoulder on many common fabrics and padding materials but it cannot set light to petrol (despite the existence of many apparent instances of this happening in films).

In general, only materials that are capable of forming a rigid char will smoulder. The temperatures necessary to initiate and maintain smouldering would melt and break down thermoplastic polymers. Since a polymer melt will not undergo glowing combustion, fires will only start on thermoplastic material if the temperature of the heat source applied to it is higher than the autoignition temperature of any vapours or pyrolysis products that are generated, and these conditions of course produce flames. Placement of a smouldering ignition source such as a lit cigarette on the surface of a thermoplastic material will therefore only result in a hole being melted in it.

If the air supply sustaining a flaming fire is restricted, then the reduction in available oxygen will lead to a drop in the burning rate and heat output. There may come a point at which the heat fed back into the fuel is not capable of generating sufficient quantities of pyrolysis products to sustain the flames. In such circumstances the fire may die down and go out, or it may smoulder and continue to burn at a rate determined by the air flow.

Fires can also make the transition from smouldering to flaming. As a smoulder grows its total heat output will rise, a greater volume of unburnt material will be pre-heated and increasing quantities of gaseous pyrolysis products will be released. At some point the concentration of these products in the air above the smouldering material may exceed the lower flammability limit and, if the heat output from the smoulder is sufficient to raise them above their autoignition temperature, they will start to flame. The time taken for a smoulder to grow and undergo the transition to a flaming fire is highly sensitive to the nature of the fuel and the environmental conditions. With a lit cigarette, and given the right circumstances, paper tissue can take as little as twelve minutes to start flaming, however it is generally accepted that an induction period of twenty minutes or more is required for foam-padded furniture that has not been treated with fire-retardants. Some smouldering fires never make the transition, and since the outcome for any particular combination of materials and environmental conditions is difficult to predict it is best determined by testing.

As a smouldering fire propagates it almost completely consumes the fuel that it is burning on, leaving an area of severe localised damage that may be surrounded by completely intact materials. In contrast, flaming fires tend to cause relatively superficial and widespread damage as they grow. The appearance of the damage after the fire is one of the clues that the fire investigator will look for when trying to determine how the fire started, as it may indicate that the fire has either flamed or smouldered for much of the time it was alight.

7.1.5 Fire Growth and Propagation

If a fire is to spread then heat needs to be transferred to nearby combustible surfaces that have not yet ignited. This can happen in three ways: by radiation, convection or more rarely conduction. As might be expected, radiative heat transfer from flames has a strong influence on the growth of a flaming fire in its early stages as the unburnt materials around the perimeter of the burning area are heated to ignition temperatures. As the fire grows, however, it will produce increasing quantities of hot gases and smoke. These are less dense than the surrounding air and will rise in a plume above the burning fuel, being replaced by incoming cold air at the base of the fire. Convective flows will elongate the flames, and if the fire is burning against a vertical combustible surface then contact with, and radiation from, the flames and hot gases will pre-heat that surface a greater distance vertically than the fire heats surrounding surfaces horizontally. This will result in the fire propagating upwards faster than it spreads outwards at its base (see Figure 7.1).

Within a compartment the plume of hot gases and smoke from a fire at low level will be confined and deflected by the ceiling, rapidly forming a hot smoky layer across the entire room. Deflected flames can spread across the underside of a ceiling up to ten times further than they would extend vertically if unconfined. As a consequence, combustible materials within the hot gas and smoke layer will rapidly ignite some distance away from the base of the fire, and may fall into areas that the fire has not yet spread to at low level. These pilot ignition sources can start additional small, and apparently separate, fires that may be misinterpreted by inexperienced fire investigators.

(a) (b)

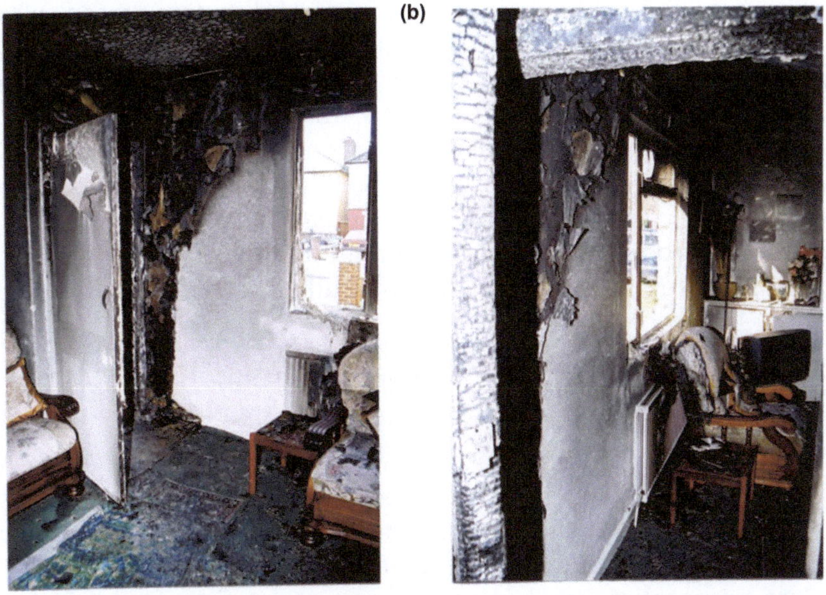

Figure 7.1 (a) and (b) From low level at the bottom of the doorway the fire had propagated upwards faster than it has spread outwards. Host gases spreading under the ceiling have ignited the curtain at the window and it has dropped on to the radiator and armchair spreading the fire.

One of the most commonly encountered examples of pilot ignition involves curtains catching fire and falling to the floor (as is visible on the right of Figure 7.1). If the fire is put out quickly then the patterns of damage may give the appearance of more than one fire having been started in the room.

As the hot gas and smoke layer gets deeper, and its temperature rises, the amount of heat radiation it emits increases. Exposed combustible surfaces throughout the room start to heat up, pyrolysis is initiated and within a short time the gases and vapours that are produced can ignite. This period of extremely rapid fire growth is referred to as flashover when everything combustible in the room has ignited. In a domestic fire a room can go from ignition to flashover in less than five minutes depending upon its contents.

Once the materials in a compartment have become fully involved in a fire the rate at which they burn will depend upon how much air can get into the room through the existing openings. The fire at this stage is said to be ventilation limited. A long period of burning under such conditions may result in greater damage in areas where the air enters and the fire burns more vigorously.

If a fire in a building is not extinguished it can spread a considerable distance from the compartment of origin. Hot gases will flow at high level through openings such as doors and windows and can ignite materials in other rooms. In multi-storey structures, if vertical shafts such as stairwells are accessible to the fire then it will be drawn upwards in a "chimney effect" rather than spreading outwards at the lowest level. This may eventually leave the level on which the fire started much less damaged than the floors above. In extreme cases the entire building may become fully involved in the fire.

7.1.6 Interpreting the Physical Evidence Resulting from Fire Behaviour

The mechanisms by which a fire grows and spreads leave characteristic patterns of damage that can be interpreted by the fire investigator. This has led to the development of some general rules of thumb that investigators find useful when assessing fire damage.

Fire propagation mechanisms result in the fire spreading faster vertically than it does horizontally, and the fire gases themselves flow outwards from the fire at high level. When comparing two fire-damaged compartments connected by a doorway the fire investigator will look for evidence of the direction of this flow – damage at lower level in the originating compartment – to determine where the fire was located. If the room of origin has been identified, and if the damage is not too severe, then the tendency for flaming fires to spread upwards and outwards may allow the investigator to locate the area of origin of the fire at the lowest point of burning.

The rate at which a combustible material will burn is dependent upon both the nature of that material and the fire environment to which it is exposed. In general, however, the longer a fire burns on an object the greater the degree of damage it will suffer, so provided you compare like with like then the worst damage will be found closest to where the fire started.

Many simple fires can be successfully investigated using these two guidelines, however as with all rules of thumb there are a great many exceptions: pilot ignition may cause additional areas of low-level burning, the degree of damage sustained by objects in the scene can be strongly influenced by the way the fire is fought and restricted ventilation may smother the fire at its point of origin while it continues to burn where an air supply is still available. Recognising when the rules of thumb do not apply is an essential skill for the competent fire investigator.

7.2 Fire Scene Investigation

Although every fire scene is different, the investigation of fires benefits from a methodical approach as it ensures that the work is carried out efficiently and that nothing is omitted. However, any general investigative strategy may require modification before or after the work starts if the situation requires it, and the experienced fire investigator will always be prepared to adapt to changing circumstances.

A robust scene investigation process includes most or all of the following components:

- Safety assessment
- Collation of information
- Inspection and strategy decisions
- Clearance and reconstruction of relevant areas
- Recording and retrieval of evidence
- Evaluation of findings

7.2.1 Scene Safety Assessment

Burnt structures probably contain a greater number of hazards than any other type of forensic scene. If the building is to be examined safely then the investigator must ensure that these hazards are mitigated sufficiently to allow work to be carried out.

Structural damage may leave walls, floors and roofs in an unstable condition if the fire has destroyed supporting elements. As a result, parts of the building will be prone to failure if they are disturbed by post-fire settlement or by the weight of a person walking on them. Before entering any part of the premises, the investigator will assess the structure's safety; cracked or leaning walls and wooden floors which have been severely damaged from below will be treated with caution as they could fail without warning. Stonework may be a particular problem as it can crack internally due to thermal stresses generated during the fire and then break up and collapse as it cools and contracts.

Sharp items are almost always present at fire scenes in the form of broken glass, protruding nails, splintered wood and torn metal.

Exposed electrical conductors may pose an electrocution risk if the supply has not been cut off, so the building service inlets will be checked to make sure that gas and electricity supplies have been isolated.

Older buildings may contain asbestos, which was used in domestic properties up until the year 2000 in the UK, that can be released as airborne fibres after a fire and industrial sites may contain specific process hazards. If necessary, the investigator can get the building tested for the presence of asbestos and seek information about the building's history and the nature of the work carried out in it before deciding if it will be safe to work there or what protective measures and equipment are required.

Hazards arising from the nature of the building's occupancy may not simply be confined to industrial premises. Squats and semi-derelict houses are frequently used by vagrants and people who use drugs, and may contain biohazards such as used syringes and needles, faecal material and parasite infestations.

Anyone entering the premises will need appropriate personal protective equipment; at a minimum a face fitted mask, boots with reinforced insoles and ankle protection, sturdy disposable gloves, a helmet and disposable overalls will be worn when examining a fire scene. A face fitted mask protects the investigator from any airborne products of combustion. Fire scenes attended during the pandemic meant that masks were worn all the time rather than just during excavation and reconstruction. Disposable gloves and overalls are worn to protect the investigator from exposure to the products of combustion and should not be washed and reused as the contaminants cannot be fully removed and there is a risk of recontamination. The helmet and boots should be thoroughly cleaned after the fire scene examination, along with any tools used, to remove any products of combustion or contaminants. In cold conditions the protective clothing will include garments chosen for warmth, as prolonged exposure to low temperatures inevitably leads to physical and mental impairment.

Before starting any work inside the scene the fire investigator will take into consideration all the potential hazards before deciding whether it is safe to proceed (see Figure 7.2). Advice may be sought from other better qualified persons if there are specific structural or process safety issues that need to be resolved. Only if all of the potential risks have been reduced to an acceptable level will the site investigation begin.

Safety assessment does not end once the decision is made to start work inside the premises; it is a continuing process in which the risk to persons working in the building will constantly be re-evaluated to ensure that their activities, or changing circumstances, do not expose them to new hazards.

7.2.2 Information Gathering

Prior to the fire investigator's arrival, a considerable amount of information may have been gathered about the incident. Some of this will have an immediate effect on the scientist's overall investigative

Figure 7.2 Scene safety assessment: obvious hazards include unsupported walls and loose masonry as well as potential falls from a height. Given these hazards the scene investigator would not enter the area until safety measures were put into place.

strategy, including professional advice on safety issues or allegations that other serious crimes have taken place.

Other information, such as the state of security of the premises when the fire fighters arrived and the scope of the fire service's subsequent fire-fighting activities, may be relevant both to the surrounding circumstances of the incident and the interpretation of the physical evidence at the scene. Fire fighters often need to break in to locked premises that are on fire, so evidence of forced entry found after a fire is not always a sign of a criminal act. Forensically aware fire crews will take pains to ensure that any of their activities that might impact upon a crime scene examination are made known to investigators.

Many modern alarm systems are computer-based and can store activation and fault logs in their electronic memories. If this information is retrieved by an engineer with the right equipment, it can offer an insight into the time, area of origin and pattern of spread of the fire. Due to the finite life of battery back-up power supplies, and restrictions on memory capacity that may lead to data being overwritten, it is a matter of some importance that alarm systems are interrogated as soon as possible after the fire is put out. Event logs from combined intruder and fire detection systems may offer an additional insight into the cause of the fire if they have recorded evidence of a break-in or, as has sometimes been observed, if a keypad has been used to deactivate sensors to allow someone to enter the premises just before the incident.

Any information about what happened during the incident is useful material for the investigator, and eyewitness accounts can be particularly valuable if they cover the events leading up to the fire. Normally witness information will be taken into consideration before deciding upon a strategy but there is one exception to this: many investigators prefer not to be informed of any admissions that have been made by a suspect until after they have completed their work at the scene. This avoids any suggestion that their findings have been biased towards one particular outcome because of prior knowledge, and it adds weight to the evidence if their findings independently corroborate the suspect's account of events.

7.2.3 Visual Inspection and Strategy

Before anything is disturbed at the scene the investigator will take the time to carry out a visual inspection of the entire premises. During this inspection the general patterns of fire damage will be evaluated to establish the directions in which the fire has spread through the structure. Hot gases and smoke are buoyant; if they vent through an opening from a compartment that is on fire they will predominantly cause damage at a higher level in the area they spread into. A pattern of decreasing damage at high level indicates to the investigator that they are moving away from the fire and conversely, if the damage increases and is found at lower levels, that they may be approaching the area of origin. These simple observations are often sufficient to allow the investigator to place the origin of the fire within a single compartment that can then be prioritised for detailed examination.

As smoke spreads through a property, it leaves sooty or tar-like deposits on exposed surfaces. After a fire the pattern of smoke staining that is left can be a valuable indicator of the positions of doors, window latches and free-standing objects, offering the investigator an insight into the condition of the premises before the fire and indicating whether items have been moved after the incident (Figure 7.3).

During the walk-through the scientist will formulate a forensic strategy, identify any vulnerable evidence that will benefit from early retrieval and decide what records need to be made and what photographs need to be taken before any of the evidence is disturbed. In some instances, the visual inspection will immediately reveal strong

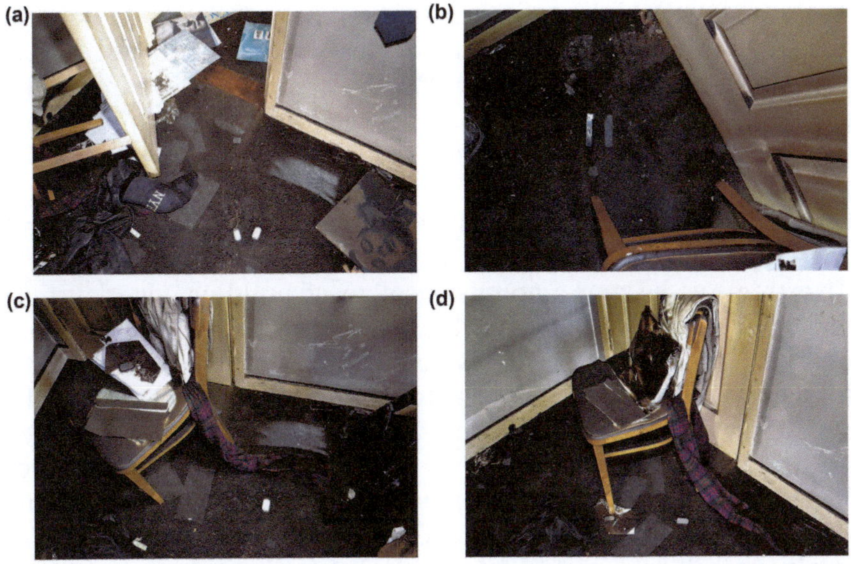

Figure 7.3 (a)–(d) A series of four images showing the entrance door to a bedsitting room where there had been a fatal fire. The images illustrate how smoke patterns aided the reconstruction which showed that the chair had been against the door at the time of the incident. This indicated that there had not been any third party involvement in the cause of the fire.

evidence of a malicious act, such as multiple areas of fire damage that are entirely separate or signs that an ignitable liquid has been used. Early advice to that effect may have a significant influence on the successful outcome of any criminal investigation, as it might result in the timely seizure of items from suspects before trace evidence is lost.

Modern fire-fighting techniques are usually highly effective, and fire crews are increasingly aware of the importance of preserving scenes for forensic examination. If a fire is discovered early and there is a timely response by fire fighters then normally a domestic premises will be left in such a condition after a fire that only one room, or a very small number of rooms, will need to be examined in detail. Although in most cases a small amount of fire debris will still need to be cleared or excavated, the evidence of interest to the fire investigator may survive almost intact.

7.2.4 Detailed Examination, Clearance and Reconstruction

During a fire the contents of the structure will burn and break up; ceiling and wall linings may fail and in extreme cases walls and roofs may collapse. All of this debris must be removed as it may conceal evidence of interest to the fire investigator. Excavation is not simply about shovelling the debris away: the stratigraphy or layering of the materials that have been deposited may well yield useful information on events leading up to, or during, the fire. Thus, evidence that the contents of a room had been strewn across the floor before any fire-related collapse of the linings might indicate that the premises had been searched or vandalised.

In extreme cases large amounts of collapsed building materials may need to be removed mechanically, but once the layers of material from the compartment of interest are uncovered the fire investigator will start to remove the debris using a shovel, a trowel or even by hand. The positions of significant items that are uncovered will be recorded before they are moved, with a view to replacing them later after the debris below has been searched and cleared away. Any evidence of potential ignition sources, such as electrical wiring and appliances, will be retrieved for later examination if necessary.

The layers of debris covering the floor may also act to protect materials of evidential value from the full effects of the fire. Ignitable liquid residues that might otherwise vaporise and burn away will be preserved and after excavation they can be recovered for analysis. In many instances the quantity of ignitable liquid that survives the fire is large enough that it can readily be detected by smell, but there are other aids that can be used at the scene such as portable electronic instruments incorporating Flame Ionisation Detectors (FIDs) or Photo-Ionisation Detectors (PIDs). These devices react to the presence of organic vapours in air; however, they must be used with caution as they are not specifically ignitable liquid detectors but respond to any volatile vapours including pyrolysis products. They are perhaps most useful when guiding the investigator in the selection of suitable samples for laboratory analysis at scenes where large quantities of ignitable liquid have been distributed and olfactory fatigue has set in. Dogs trained to detect ignitable liquid residues are being deployed increasingly at fire scenes, usually by fire services. Properly trained and handled these dogs can be of great assistance to the fire investigator, however any positive indications must be confirmed by laboratory analysis if the investigator is to confirm that an ignitable liquid was present.

Removal of fire debris may also uncover protected areas on surfaces to show where items were standing while the fire was alight. These patterns of protection may have characteristic shapes or dimensions that uniquely identify the object that made them, allowing the replacement of that object as part of a scene reconstruction. In some

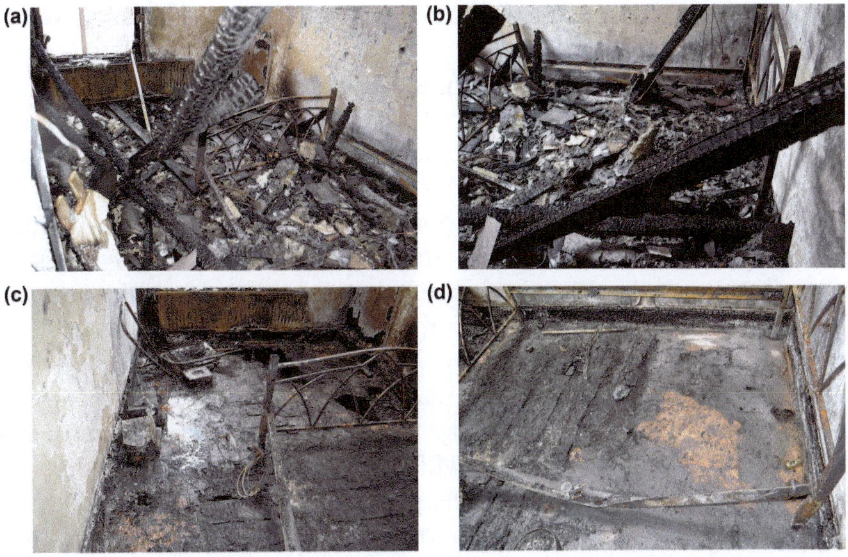

Figure 7.4 Despite the contents of the roof had collapsed into the bedroom the contents of the room can be reconstructed.

incidents the contents of a room could have been entirely cleared during fire fighting and yet floor coverings, furniture and personal effects can all be replaced with certainty in their original positions by the investigator (Figure 7.4).

7.2.5 Recording and Retrieval of Evidence

During the course of work at the scene the investigator may uncover items that need detailed examination in the laboratory. These will be retrieved, appropriately packaged to prevent further damage or cross-contamination and labelled for continuity purposes. Most samples recovered from fire scenes are taken for subsequent analysis for ignitable liquid residues, but items may also be seized for inspection or testing. To package an item for ignitable liquid analysis it should be placed within a nylon bag that is swan neck sealed and secured with a cable tie. Then the item should be placed within a polythene bag, swan neck sealed and secured with a cable tie. A combination of both nylon and polythene bags means that hydrocarbon based ignitable liquids such as petrol and non-hydrocarbon based ignitable liquids such as methylated spirit and specialist solvents will be retained. To ensure that any ignitable liquid residues detected have not originated from contaminated packaging, a control should be prepared, using the same batch of materials as those used when packaging the item, and analysed.

As the results of a fire investigation may have to be presented in court many months, or even years later, the investigator must make a permanent record of the evidence at the scene. This can be done in a variety of ways depending upon the circumstances of the incident and the resources available. Most commonly the investigator will take photographs and draw plans of the relevant areas. They will make written or dictated notes about the construction, internal layout and building services and their interpretation of the patterns of fire damage to identify the point of origin of the fire and its cause. Any handwritten or typed notes and plans that are prepared at the scene will be signed and dated to show that they are a contemporaneous record. Similarly dictated notes will be transcribed and the transcript signed as a true record as soon as possible after the work is done.

Photographs are an essential part of the evidence package and they need to be correctly exposed high-quality images. Given the questionable permanence of digital storage media, and the need to retain some case records almost indefinitely, all forensic practitioners have storage protocols for digital material that are designed to ensure that data is not lost.

The scene record must be as comprehensive as possible because it will be the primary source of information if the investigator is asked to prepare a statement of findings. Before or during court proceedings the scene notes may be scrutinised by other experts under the rules of disclosure, and if any important information is missing then this may result in the investigator's competence being called into question.

It is acknowledged that the arson detection rate is low when compared to most other crimes and therefore to improve the detection rate there needs to be a scene investigation approach that is more focused upon evidence recovery, to link suspect to a scene, as much as the cause and origin of the fire. A senior investigation officer's first question upon being told that there is an arson is more likely to be 'who started the fire' rather than 'how did the fire start'. This makes sense as there is no criminal case or charges if a suspect is not identified however both questions need to be answered.

A domestic fire, at ceiling level, can reach temperatures of 1000 °C however, if you move away from the immediate area of the fire and especially at floor level the temperatures are much lower. DNA starts to denature at around 60 °C but recent research indicates that a full DNA profile could be obtained from blood exposed to 150 °C. Blood can survive in fire situations though the colouration changes and it is not always recognised for what it is. Fingerprints can also be preserved under soot deposits and even be developed by them. It is also possible to obtain forensic evidence preserved under a large item which is present from early in the fire's development that protects the materials underneath it.

7.2.6 Interpretation and Evaluation

Before communicating any findings to interested parties the fire investigator must interpret the evidence that has been recovered during the scene examination. Ideally the evidence will be sufficient to identify the area or areas where the fire started and the ignition sources that are present. If ignitable liquid has been distributed around the scene, or if a number of fires have been started in entirely separate areas, then the investigator's task may be very simple as the presence of a flammable liquid or multiple seats of fire are generally good indicators of arson. In the absence of obvious signs of arson, however, the investigator may have to consider a variety of explanations for the fire. To determine which is the most probable a series of hypotheses will be formulated based upon viable ignition scenarios and these will be tested against the physical evidence at the scene and other reliable information that may be available. Some of the hypotheses will not withstand scrutiny and testing – they will be discarded as unworkable, thus ruling out certain ignition mechanisms as in the following examples:

- A flaming fire is discovered in a room that had been occupied only minutes before. After excavation and reconstruction, the investigator considers the hypothesis that a lit cigarette was the ignition source. The surviving materials found at the seat of the fire cannot support a smouldering fire, so the hypothesis that a cigarette started the fire appears unlikely. Further support for this conclusion comes from the witness evidence. A smoulder will produce copious quantities of acrid smoke as it develops, and this smoke should be readily perceptible to anyone in the room shortly before the transition to flaming. Since the room was occupied just before the flames were seen, the presence of a well-developed smoulder at that time would appear to be highly improbable if the witness evidence is to be trusted.
- Electrical wiring is present within a room where a fire has started. The investigator considers the hypothesis that the fire was the

result of an electrical fault. Examination of the evidence from the reconstruction of the scene reveals however that no electrical wiring or appliances were located within the localised area of damage apart from cables running under the floor. There is a hole in the floor, but the patterns of damage indicate that it has burnt through from above and the insulation on the cables is intact (Figure 7.5). The available evidence is not consistent with an electrical cause of fire and consequently the investigator discards it as a working hypothesis.

In many cases even after hypothesis testing more than one viable explanation for the fire may remain, and the investigator will use their technical knowledge and experience to assign a likelihood to each possible cause based upon the circumstances of the incident.

Throughout the evaluation process the investigator will also test the factual evidence from the scene against information from witnesses to assess the reliability of the latter. Although witnesses can be genuinely mistaken, in some instances there may be such large discrepancies between the evidence at the scene and the account given by a witness that the witness may become a suspect.

Figure 7.5 The fire has burnt a hole in the floor but insulation on the electrical cable in the void below is intact, indicating that the fire did not start in the void as a result of a fault to the cable.

7.3 Fatal Fires

In general, the methodology employed at fatal fires is the same as that used to determine the area of origin and cause at other fires. If the victim is still *in situ*, however, the investigator will need to examine the body if it is still present as it can yield useful evidence.

7.3.1 Causes of Death by Fire

As a fire burns it will produce heat, smoke and toxic gases that can pose a potentially lethal threat to anyone in the immediate vicinity. While some people die of thermal injuries in flash fires, the majority of fire victims do not succumb to flames but are overcome by smoke inhalation. In many accidental fires the body will be found in an area that is smoke stained but unburnt. In sufficient quantities smoke is an asphyxiant, but it is poisoning by carbon monoxide (a product of the incomplete combustion of carbon-containing fuels) that statistically leads to most fatalities. Carbon monoxide binds strongly with haemoglobin, having an affinity almost 200 times greater than oxygen. As a consequence, carbon monoxide will progressively accumulate in the blood of a victim breathing it, inducing a variety of symptoms that culminate in unconsciousness and eventually death. Hydrogen cyanide can also be found in fire gases, but it is rare for a fire victim to die solely of cyanide poisoning because toxic cyanide levels are usually accompanied by potentially fatal carbon monoxide concentrations as well.

7.3.2 Examination of Fire Victims

Any fatal fire is potentially a murder scene, and the fire investigator will take appropriate precautions when dealing with the body to avoid compromising other forensic evidence. If the body is extensively burnt then signs of other non-accidental injuries may well be concealed, however the presence of a large amount of blood under or around a fire victim is grounds for suspicion as thermal injuries alone do not normally lead to blood loss.

After a fatal fire it is generally desirable that the victim is removed as soon as possible to a mortuary where a proper examination can be carried out, but the investigator will record the condition and position of the body before it is moved and supervise the process of removal to ensure that fire-related evidence is not lost. Any personal effects that might aid in the identification of the victim will be retrieved at this time.

When the body is lifted, the presence or absence of fire debris or smoke staining under it may offer some insight into the stage of the fire at which collapse occurred. A clean contact area between the victim and the surface they are resting on may indicate that the victim was prone prior to, or at an early stage of, the fire. Conversely the presence of fire debris or smoke deposits trapped between the contact surfaces indicates that the victim must have been upright and mobile after the fire started. If the fire started as a result of the victim accidentally igniting their clothing, then the garments trapped beneath the body may show signs of burning, and in cases of self-immolation these burnt materials might also retain the only traces of flammable liquid to survive the ensuing fire.

7.3.3 Lifestyle and Fire Fatalities

Due to most fatal fires being accidental, the habits and activities of the victim may be of considerable importance when looking for a cause. A combination of elevated blood alcohol levels and smoking is frequently reported at inquest, and evidence of the victim's drinking and smoking habits may be readily apparent to the investigator at the scene. Advanced age, infirmity, poor living conditions and social or religious customs are also associated with enhanced fire risk. There is a possibility that the fire in which the victim dies is the culmination of a series of similar incidents; evidence of previous accidents with smoking materials, or misuse of candles or heating appliances may be apparent to the investigator when examining the premises. Relatives, friends, neighbours and, in the case of the elderly or disabled, the social services may be valuable sources of information on the victim's lifestyle and activities, however when considering this information the investigator will always bear in mind that the witnesses may not be in possession of the true facts; fire victims are known to have successfully concealed the fact that they smoked even from close relatives. When there is a conflict between the witness information and the facts established at the scene by examination, the investigator will always rely upon the latter when deciding what started the fire.

7.3.4 Post-mortem Evidence

When interpreting the scene findings and reconstructing the events leading up to the fire the investigator will also draw upon information obtained from post-mortem and toxicological examinations. This may provide unambiguous evidence to indicate that the victim died after the fire started. The presence of soot in the airways and elevated levels of carbon monoxide in the blood are indicators that a person has inhaled fire gases and must, therefore, have been breathing while the fire was burning. The converse is not always true as victims of flash fires may not have time to inhale significant quantities of smoke before they die.

7.3.5 "Spontaneous" Human Combustion

In some fires the investigator may find that the body of the victim displays exceptionally severe localised damage while other combustible materials in the immediate vicinity appear to be unaffected. Often the victim's torso is partly or completely consumed and large bones will be reduced to a crumbly grey ash. In the past, many observers of this phenomenon, in trying to reconcile the conditions they know are required to destroy a body with those that must have been present at the scene, explain the damage in terms of a supernatural "fire from within". Spontaneous Human Combustion (SHC) has been used as a popular term for such fires and it has also been used as an explanation for fatal fires where no ignition source has been found.

Alleged instances of SHC involving severe damage to a body arise from a mistaken belief that a process akin to cremation is the only way that a fire can completely burn to destroy flesh and bone. Cremation involves the destruction of a body using a furnace fitted with gas burners operating at high temperatures; equipment designed to make this process as quick and efficient as possible. In contrast, most domestic fires that cause partial or complete destruction of a body are neither quick nor particularly efficient, but even a relatively small flame is capable of causing massive damage if it is allowed enough time to burn. All that is required is that the fire around the body is provided with a source of fuel that will sustain it for the time needed to burn away the flesh and calcine the exposed bone. This fuel is supplied by the body itself in the form of liquefied fat rendered out from the tissues as they are heated. Such fires tend to be more common with victims who are obese, as they can provide a greater quantity of fat to fuel the fire allowing it to burn for longer. As long as there are absorbent materials nearby which are capable of acting as a wick, such as the victim's clothing, then the fat will burn as if in an oil lamp and the process of rendering and tissue destruction will continue, sometimes leaving only the extremities around a hole in the floor. As they involve the burning of fat, these fires tend to leave very greasy smoke deposits within the premises.

Close examination of the scene by the fire investigator will almost always reveal evidence of a viable ignition source in the vicinity of the victim. In some cases, however, the evidence can be complicated by the actions of the victim who may have moved away from or turned off the ignition source before they collapsed.

7.3.6 Multiple Fatalities and Major Disasters

At fire scenes where there has been a major loss of life there may be problems identifying the victims because of the destruction of clothing and personal effects. Severe damage to the body itself may also render it impossible or inadvisable for relatives to provide evidence of identification. DNA profiling is a valuable tool for confirming the identity of fire victims, but it is only effective if there is a sample of known origin available for comparison. At the early stages of the investigation of a major disaster, therefore, preserving any evidence of identity associated with the victims is extremely important. Due to the large-scale nature of such incidents, victim recovery will be a team effort, and interaction with the site teams will be factored into the investigator's strategy. Depending upon the degree to which the investigator is involved with victim recovery, work to determine the origin and cause of the fire may be deferred or will be carried out in such a way as to ensure that any bodies and associated materials remain undisturbed until they can be removed. For example fire investigators can work with DVI (Disaster Victim Identification) teams, anthropologists and archaeologists.

Inevitably at major incident investigations there are a great many interested parties, some of whom have an interest in the cause of the fire. The scene investigation process will take longer under such circumstances even though the work carried out will follow the same principles as with much smaller fires.

7.4 Endangerment of Life

Section 1(2) of the Criminal Damage Act 1971 states in essence that a person who damages any property with intent to, or being reckless as to whether they might, endanger the life of another is guilty of an offence. Although a charge of endangerment could be applied to any type of criminal damage it is almost exclusively used by the criminal justice system in cases of arson. To support such charges fire investigators may be asked to comment on the degree of risk posed by the fire they are investigating.

Endangerment carries with it the suggestion that death or serious injury is a likely outcome of events. The principal life hazards in any fire are exposure to hot gases, soot and toxic combustion products and injuries sustained while attempting to escape. In evaluating life threat the fire investigator will consider whether there is a risk that occupants of the building would be exposed to fire gases for the length of time it takes to produce impairment, or whether there would be sufficient thermal contact to produce burns.

One of the most obvious ways in which endangerment can occur is if potential victims are unable to avoid exposure because their means of escape is blocked. The way the fire has been started, its location and its potential to develop further will all be assessed by the investigator together with the fire precautions that are present and the available means of escape from the premises. Based upon this assessment the investigator will comment on the likelihood that persons in the premises might sustain injuries or die.

7.5 Laboratory Investigations

Within forensic laboratories the majority of all fire-related examinations are directed towards the recovery and identification of ignitable liquid residues. The presence of such residues in a sample recovered from a fire scene provides strong support for the proposition that the fire was deliberately started, although the accidental spillage or leakage of liquids that were legitimately present at the scene must be ruled out before the scientist can say with confidence that the fire was arson.

Other examinations that the fire investigator may be asked to carry out include the inspection of clothing for heat damage, ignition and burning tests on materials recovered from the scene, the identification of improvised and manufactured pyrotechnic or incendiary devices and the inspection and testing of small domestic appliances for evidence of fault or misuse.

7.5.1 Ignitable Liquid Recovery and Analysis

In most instances the ignitable liquids used to start fires are the ones that are readily available to the general public. These tend for the most part to be hydrocarbon liquids such as petrol, white spirit and kerosene-based liquids (paraffin and turpentine substitutes). Of these, petrol is the most frequently used, probably because it is easy to purchase in reasonably large quantities. The laboratory methods for the recovery and analysis of flammable liquids are capable, however, of identifying most hydrocarbon and non-hydrocarbon liquids that can be used to start fires.

Liquid samples recovered from containers need no preparation before analysis, but the majority of samples taken from scenes or suspects comprise absorbent substrates such as debris or clothing. Any liquid residues present on these items need to be recovered before they are analysed. Non-volatile liquids such as mineral and edible oils can be recovered by solvent washing, and hydrocarbon liquids can be extracted by steam distillation, but mostly volatile ignitable liquids are usually recovered using headspace techniques.

Most ignitable liquids are volatile to a greater or lesser degree. In sealed forensic case samples, the vapours that accumulate in the headspace inside the packaging can be recovered and concentrated on a variety of adsorbent media. If necessary, the concentration of these vapours can be increased by heating the item. Passive headspace techniques usually involve placing the adsorbent within the package and incubating it, while dynamic sampling involves the withdrawal of a measured volume of the headspace that is passed over the adsorbent. The adsorbent is then either treated with a solvent, or thermally desorbed, and the resulting sample is introduced into an analytical instrument for identification.

Gas Chromatography–Mass Spectrometry (GC–MS) is the generally accepted method of choice for the forensic analysis of volatile organic compounds such as ignitable liquids. Within the gas chromatograph the complex mixture of chemical compounds that makes up the ignitable liquid is driven by a carrier gas into a long capillary tube coated with an adsorbent. During transit through the capillary these compounds will interact with the coating to an extent that depends upon their physical and chemical properties. This has the effect of slowing the passage of some compounds along the capillary with respect to others, separating them so that they emerge into the detector at different times.

As the chemical compounds enter the mass spectrometer they are ionised and break down into fragmentation patterns that are characteristic of the molecules involved. A mass detector then records these fragmentation patterns. In its simplest form the mass spectrometer's integrated output or Total Ion Count (TIC) can then be reproduced on a chart as a series of peaks. These peaks may have the characteristic pattern of an ignitable liquid (Figure 7.6), but even if

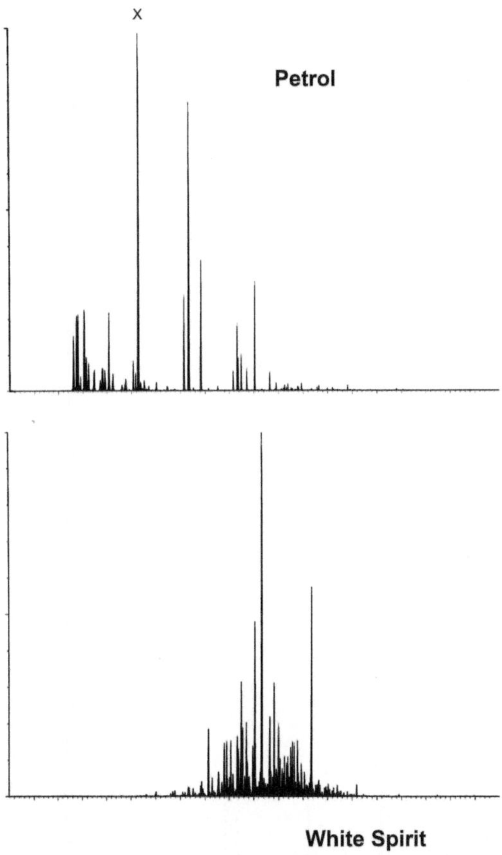

Figure 7.6 GC–MS TIC plots of petrol and white spirit. Note the peak marked X on the petrol plot and refer to Figure 7.7.

they do not the MS data will provide details of the molecular structure of the compound that each peak represents (Figure 7.7), normally allowing the liquid to be characterised. This is an extremely powerful method for identifying ignitable liquids and their residues.

7.5.2 Examination of Clothing for Heat Damage (Flash Burning)

When a volatile liquid such as petrol is spilled or poured it will start to evaporate, and the vapour that is released will mix with air to form a diffuse flammable cloud above the liquid. If the cloud of vapour is ignited then the resulting burst of flame can cause heat damage to vulnerable surfaces nearby, including the clothing of a person applying the ignition source.

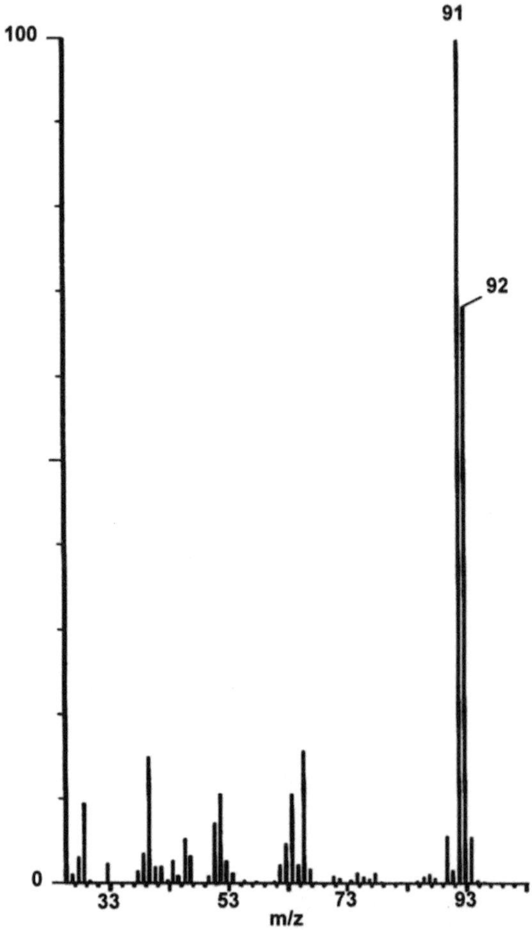

Figure 7.7 Mass spectrum of peak marked X in Figure 7.6. The mass response at *m/z* 92 is the molecular ion of toluene.

The response of garment fabrics to this brief application of intense heat will vary depending upon the nature of the fibres they are made from. Thermoplastic materials will melt and natural fibres will scorch or char. The tips of fibres and threads protruding from the surface of the fabric are most vulnerable to heat, and at high enough temperatures they will either melt back or discolour. This flash burning is not always visible to the naked eye, so microscopic examination of the surface of the garment may be necessary to identify it. The distribution of the damage across various garments may be evidentially significant depending upon the circumstances of the incident – petrol vapour is denser than air so if a pool of petrol is ignited then the burning vapour is likely to cause greatest damage to garments at low level such as shoes and the legs of trousers rather than upper items of clothing.

> Some fabrics and garments are heat-treated or flamed to remove loose threads during manufacture, and there are other ways in which clothing can be thermally damaged by heat after purchase (over-hot tumble drying or ironing are two examples). As a result, the identification and interpretation of flash burning requires considerable expertise on the part of the scientist.

The presence of flash burning on a suspect's clothing is only evidentially significant if it can be shown that a volatile ignitable liquid was used to start the fire. In such circumstances, and provided that alternative damage mechanisms can be ruled out, then the existence of flash burning offers considerable support for the suggestion that the suspect was close to the fire when it was ignited.

7.5.3 Ignition and Burning Tests

The process of hypothesis testing during the evaluation of a scene investigation might indicate that there is more than one possible explanation for the fire. If the investigator has been able to identify the material that was first ignited then tests can be used to establish which candidate ignition sources are viable, and which can be ruled out, as causes of the fire. Equally if there are unanswered questions about certain aspects of the fire's subsequent development, such as the time needed for a transition from smouldering to flaming, then a burning test might provide the answers.

Ignition and burning tests need to be designed carefully if they are to be reproduced as closely as possible to the circumstances at the scene. The more a test differs from the situation at the scene the more likely it is that the results will be challenged in court. Forensic burning tests are not therefore like the national and international standard tests that are used to assess ignitability or fire performance.

If possible, the investigator will use materials recovered from the scene to carry out the tests and will try to replicate as closely as possible the conditions at the time the fire started. This is particularly important if the size of the reconstruction, or the restricted availability of items to burn, means that only one test will be carried out. If a completely intact test object is not available, then a mock-up may be constructed using whatever materials can be salvaged from the original item at the scene.

For small scale burns, replicate testing is desirable as it allows the investigator to explore the effect of changes to the starting conditions. Ignition sources such as smouldering cigarettes are particularly sensitive to small changes in orientation and ventilation to the extent that apparently identical tests can have different outcomes. In such circumstances the investigator may have to carry out a large number of tests to determine the probability of ignition so that they can offer an opinion on the likelihood of it taking place.

For safety reasons, anything other than simple ignition tests are only carried out in purpose-built laboratory facilities fitted with smoke extraction equipment and fire extinguishers.

7.5.4 Examination of Incendiary Devices

Improvised incendiary devices are sometimes used to start fires or to provide a time-delay so that the person responsible can arrange an alibi. Since such devices are often made up from commonly available household materials it is only the combination of items that allows the scientist to say that the recovered item is an incendiary device.

Simple time-delays can be constructed from cigarettes and matches, or employ electrical time switches connected to heating elements or other incandescent sources. Such devices are often deployed with a flammable liquid that will assist the growth of the small fire that is produced, and if petrol is used the delay may allow sufficient flammable vapours to build that there is the risk of an explosion.

Laboratory testing of time-delay devices is normally carried out to determine how long a delay they produce and how reliable they are. As with other burning tests, it is important that the reconstructed device reproduces as closely as possible the one found at the scene if the results are to be relied upon in court.

Incendiary devices that do not incorporate a time-delay are generally intended to start a large fire quickly and efficiently. They will almost always include ignitable liquid, and they tend to be variations on the design of a petrol bomb. While all such devices may have similar effects when used, there are specific legal rulings in case law that apply only to genuine petrol bombs, which are classified as explosive substances. Manufacture or possession of a petrol bomb is therefore subject to the provisions of, and the penalties imposed by, the Explosive Substances Act 1883. For the purposes of the Act a petrol bomb is strictly defined as a glass bottle, containing petrol and with an absorbent wick fitted through its neck. Possession of the components needed to make a petrol bomb is not sufficient; the device must be made up as prescribed if the law is to apply. The forensic scientist will be asked to certify that a particular device conforms strictly to the definition of a petrol bomb if charges are brought against an individual alleging manufacture or possession of an explosive substance.

7.5.5 Appliance Examinations

Electrical and gas appliances can start fires through fault or misuse. When examining such items often the most important consideration is to establish whether the appliance was turned on or not, and in the case of gas appliances whether gas was supplied to them. If the appliance was switched off at the time of the fire then it is unlikely to have been the ignition source, although if the scientist is to say this then the possibility that the controls were interfered with after the fire must be ruled out.

Normally any appliance recovered from the area of origin of a fire will be damaged as a result. Plastic casings and light alloy components will be melted, and metal parts will be oxidised and distorted. The investigator will use non-destructive techniques such as visual inspection and x-radiography to establish the condition of external and internal components before disassembling and testing the appliance. If the fire damage is not too severe then internal fuses and electrical switches can be checked for continuity, and gas valves tested for flow. After the damaged casing is removed inspection may reveal that internal components are intact, indicating that the fire has not originated inside the appliance.

If an appliance is not too severely damaged then it can be repaired and refurbished if tests are required, otherwise the scientist will try and obtain an undamaged example to carry out the work.

7.6 Conclusion

Fire investigation is a fascinating but challenging area of forensic science. The process of investigating a fire is complex and changes, to some degree, at every incident. There is no such thing as a 'normal' fire. The slight changes in the fuel content and orientation as well as airflow, amongst other parameters, affects the way a fire burns. However, each fire investigation should make use of a consistent application of the scientific method to ensure that the outcome is reliable and could be re-examined by another competent person to reach the same conclusion.

The impending inclusion of fire scene investigation in the Forensic Regulators Code with the requirements for accreditation to ISO/IEC:17020 should ensure that methods are appropriate and that they are followed by investigators to achieve the best possible outcome for each case.

A thoroughly investigated fire scene can yield evidence of origin, cause and, when appropriate, the presence of perpetrators.

Further Reading

D. J. Icove and G. A. Haynes, *Kirk's Fire Investigation*, Pearson, 8th edn, 2017.

D. J. Icove, J. D. DeHaan and G. A. Haynes, *Forensic Fire Scene Reconstruction*, Pearson, 3rd edn, 2013.

Fire Investigation, ed. N. NicDaeid, CRC Press, 2004.

E. Stauffer, J. A. Dolan and R. Newman, *Fire Debris Analysis*, Elsevier, 2008.

The SFPE Handbook of Fire Protection Engineering, Springer, New York, 5th edn, 2015.

V. Babrauskas, *Ignition Handbook*, Fire Science Publishers, Issaquah, WA, 2003.

D. Drysdale, *An Introduction to Fire Dynamics*, John Wiley and Sons, 3rd edn, 2011.

N. Carey, N. NicDaeid, M. Svare, C. Holland and K. Robertson, Fire Investigation, A primer for the Courts, The Royal Society and Royal Society of Edinburgh, 2023, https://royalsociety.org/-/media/about-us/programmes/science-and-law/royal-society-fire-investigation-primer-for-courts-digital.pdf.

8 Marks and Impressions

David Baldwin*[a] and Caroline O'Mahoney[b]

[a] Baldwin Forensic Consultancy Ltd, Melbury House, 34 Southborough Road, Bromley, Kent BR1 2EB, UK; [b] Forensic Science Service, 109 Lambeth Road, London SE1 7LP, UK
*Email: davidbaldwin55@googlemail.com

8.1 Introduction

The range of marks covered in this chapter is restricted to the main types routinely encountered in case work and includes footwear marks, tool marks, plastic film and mass-produced manufactured items. There is also some reference to the area of physical fits, but in each case the approach is broadly similar. The examination of any type of mark or impression is based on the same fundamental principles and this chapter will explain and illustrate their use in marks examination.

In simple terms, a mark is left when one object comes into contact with another and leaves a pattern of detail behind. In a forensic science context these are usually fingermarks, footwear, tool, glove and tyre marks, and marks produced during the manufacturing process. In reality, marks examinations have been carried out on all kinds of items which have been in contact with each other, for example, a knife in a leather scabbard, or a piece of hardboard used as a van door panel. These items had been in contact for a long period and when separated, an impression was left behind.

Crime Scene to Court: The Essentials of Forensic Science, 5th Edition
Edited by Niamh NicDaeid and Peter C. White

Published by the Royal Society of Chemistry, www.rsc.org

The most commonly recovered marks are fingermarks. Fingerprint comparisons depend on friction ridge characteristics that do not alter with the passage of time, unless the fingers have become badly burnt or severely scarred. It is this property of unchanging detail that allows individuals to be identified.

On the other hand, footwear and tools do alter with time and acquire their own individual characteristics. Most tools and footwear are mass-produced articles, and when new they are likely to appear the same as thousands, if not hundreds of thousands, of others. But as new footwear is worn, some areas of the soles begin to wear away; particles of grit become trapped, small cuts and gouges appear from walking on rough surfaces or stepping on sharp particles like broken glass. These accidental damage features are acquired randomly and the sole pattern of the footwear becomes individualised. Similarly, as a tool is used, it becomes damaged, the plating chips, it gets thrown in with a load of other tools, gets used on something that is too hard or is dropped; it may pick up rust and corrode. The tool acquires new features that, when taken together can allow a possible conclusive association of a tool with a mark.

When a marks examination is undertaken, it can be divided into two distinct aspects. The first can be considered as the technical aspect where different methods are used to locate and recover the mark, to visualise the detail and to provide the optimum conditions for comparison of detail from the control item (known) with the scene mark (unknown).

The second and equally important aspect of any examination is the assessment of the evidential significance of the result of the comparison. The fundamental concept of any marks examination is that the detail transferred from one object to the other may show a degree of association between the two objects. The evidential significance of the degree of correspondence or 'match' depends on the quality of the mark, that is, the amount and definition of the detail present, how the detail was originally produced, and if it has been modified by wear or damage. The approach undertaken in reaching an opinion on the evidence should always be based on the four guiding principles of balance, logic, transparency, and robustness. These principles are discussed later in the chapter.

It is a critical aspect of a comparison to understand the manufacturing processes associated with the control item. There are three classes of features that an examiner will need to consider:

Class features – Intentional or unavoidable features that repeat during the manufacturing process and are common to all items of a particular type.

Subclass features – Features that are not unique to one particular item but allow some discrimination between groups of items with the same class features. They arise during manufacture but are not necessarily introduced deliberately.
Individual features – Characteristic features that arise at random during the manufacturing process or through normal use.

In any crime scene there is a potential for recovering footwear marks and these are the most commonly recovered marks after fingermarks. Where premises have been entered by force, there are likely to be tool marks at the point of entry. A marks examination can also provide intelligence when there is no control item. The make and model of a shoe may be identified to assist an investigation, and possible links can be identified between different crime scenes. The first part of the chapter discusses the fundamental processes that can be applied to any marks examination.

8.2 The Fundamental Process

The fundamental processes in a marks examination are the same as any forensic examination. A common structured approach can be applied and is summarised here under a number of general headings:

Search and recovery – The search of a crime scene to recover the items of interest. This may involve the use of specialist equipment and techniques. In the case of footwear marks, the scene examiner will use different lighting techniques to search for marks. Footwear marks may also be located and enhanced using lifting techniques and chemical treatments. In the case of tool marks, these are located in a visual search and the original item recovered, or the mark is cut out or cast. Further search and recovery may be done at the laboratory.
Analysis – The assessment of a mark to determine suitability for comparison with control items. Factors considered include the following:

(1) Assessment of the case information provided and the questions being asked. This assessment includes whether further information is required before commencing the examination.
(2) Consideration of other relevant forensic science aspects such as the recovery of DNA, fingermarks or other trace materials.

(3) Development of an examination strategy derived from an assessment of the quality and level of detail present in the mark and how the detail should be enhanced to visualise it more clearly.
(4) Deciding what test marks are required.

Comparison – The mark is compared with the control item to determine the level of agreement, and establish detail or features that correspond or are different. This may involve making test marks with the control item, with the objective of replicating the suspect mark. Traditionally, marks examinations are done by a visual and microscopic comparison of the mark and control items. Matching detail may be recorded electronically but the comparison itself is not electronic.

Evaluation – The formulation of a conclusion from the findings by weighing up the prosecution and defence propositions. The interpretation of the findings should take into consideration aspects such as the degree of correspondence between the control item and the mark, the quality of the mark and any limiting factors such as the substrate, the time between offence and recovery of the control items, and hence their potential modification through further use.

Verification – Marks examinations are subjective, and although an opinion may be based on established guidelines or convention, there are many examinations that are not straightforward and this can potentially lead to a misinterpretation of the evidence. As with all subjective areas of forensic science, an independent review by another expert, often termed a 'critical findings check', is of enormous benefit. The essential aspects of the verification process are to check that:

(1) The examiner has followed the appropriate documented examination process and applied the relevant scientific methodology and techniques.
(2) The work and findings of the examination are reflected in the conclusion reported. The results must support the conclusion and clearly there should be no understatement or overstatement of the findings.
(3) The maximum evidence has been obtained; nothing has been overlooked and there are no other marks that might change the outcome.
(4) The specific questions asked have been fully addressed.

The second examiner should have no prior knowledge of the findings of the first examiner and this verification process forms part of the approach to mitigate the risks of cognitive bias.

8.3 Footwear Marks

8.3.1 Introduction

The vast majority of footwear marks are recovered from volume crime scenes and they are usually made by training shoes. Since the 1980s there has been a significant increase in the number of brands and models of training shoe available. There are thousands of different sole patterns, and millions of training shoes are distributed within the UK every year. Shoe designs are constantly being added while others are discontinued. Figure 8.1 shows a selection of training shoes which all have the same sole pattern, but with different uppers.

The sole is likely to be made of rubber or a synthetic material such as polyurethane and is usually composed of a midsole and an outsole. The midsole is positioned between the upper and the outsole to provide cushioning and support. The bottom of the outsole comes into contact with the ground. In this chapter, for simplicity, the patterned bottom surface or tread will usually be referred to as the 'sole' rather than the outsole. As discussed previously, to be able to interpret and evaluate features found in a mark, the examiner must have a good understanding of the manufacturing processes for footwear.

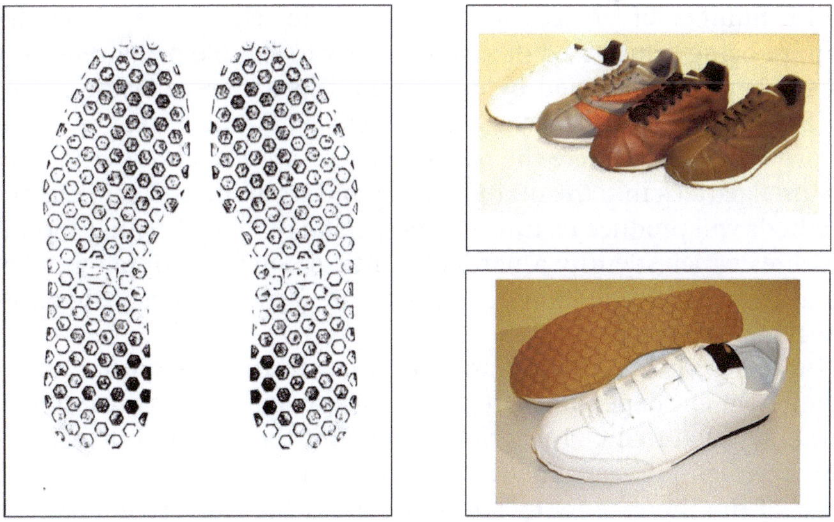

Figure 8.1 Training shoes with the same sole pattern and different uppers.

8.3.2 Manufacture of Footwear

The soles of most modern footwear are mass produced using a number of different moulding processes including compression, injection and open-pour moulding. In compression moulding a block of the sole material is placed in a metal mould; the mould is closed and then subjected to considerable heat and pressure. This is the most commonly used process for the manufacture of natural and synthetic rubber soles. In injection moulding, the liquid sole material is injected into the closed mould. Both midsole and outsole can be done as one injection or two separate injections, if the midsole is a different colour for example. Soles can be attached directly to the lasted uppers as part of this process, or just the sole units can be manufactured, to be attached later in the process. In open-pour moulding the polyurethane is poured into the mould and then the mould is closed. There is a pre-moulded outsole already present and the liquid midsole is poured into the open mould; while still liquid, the lasted upper is lowered in.

For any given model of footwear, there will be at least one mould, and sometimes several, for each size. A textured pattern known as 'stippling' can be added to areas of the mould surface. There are two main methods. Hand stippling is done by physically striking the mould with a steel die; the tip of the die is small and it is struck on the mould numerous times. The second method uses acid to etch the texture onto the mould. The texture pattern is computer designed, and a number of processes are then conducted that result in this pattern being transferred in a random way to soluble paper. Pieces of this paper are burnished by hand onto selected areas of the mould surface. The mould is placed in an acid bath which eats away the unprotected areas where the paper has been applied and this etches a textured pattern into the mould. Both hand stippling and acid etching methods will produce texture effects specific to that particular mould. Features which identify a particular mould, such as stippling, allow some discrimination between groups of items and are therefore subclass features.

Unintended defects and flaws can occur in the sole during manufacture which, depending on their cause, result in class features or randomly acquired characteristics. Some of these mould defects are only revealed by wear. The first step of rubber processing is to combine all the ingredients including fillers. If the mixture is not completely blended, then the filler particles may remain in pockets within the sole and are only exposed when the footwear is extensively worn;

they then drop out of the sole leaving odd shaped spaces in the sole pattern. These are random characteristics and therefore unique to that sole. Air bubbles have also been seen in polyurethane soles, produced during the manufacturing process; care must be taken in the interpretation of such features as the position of such bubbles may not be entirely random.

The size of the shoe is determined by the manufacturer and there is not a strict relationship between the shoe size and the length and width of the sole. Any experienced examiner would be reluctant to give a specific size for a particular mark even if most of it appears to be visible (unless the shoe size label on the sole is also present in the mark!). In certain circumstances, it may be possible to look at the slight pattern variations that occur between different sizes. Hence, it may be possible to determine the shoe size for a very small area of mark but may not be possible for an apparently full-length mark.

8.3.3 Types of Footwear Mark

Footwear marks are normally classified as two- or three-dimensional. The vast majority of recovered marks are two-dimensional. As a shoe is worn, it accumulates material on the sole, acquired from the surfaces in which it has been in contact, such as soil or grease. This material can then be deposited elsewhere to leave a mark; the mechanism by which this happens is complicated by the background surface on which the mark was made, known as the 'substrate', and the nature of the deposit on the sole. The substrate can be porous, non-porous or a combination. Non-porous surfaces include glass, ceramic tiles and uPVC window frames; porous surfaces include paper and card items.

8.3.3.1 Two-dimensional

The footwear makes contact with a surface which does not yield permanently and the sole leaves a mark. Such marks may be visible or invisible (latent). Examples have been found on every conceivable type of surface including floors, doors, furniture, glass, paper and fabric. The substrate will normally determine the quality of detail seen in the mark. The quality may also be affected by the nature of the deposit left by the footwear in making the impression. A muddy mark on a heavily textured surface such as a thick pile carpet is likely to result in any fine detail being obscured.

8.3.3.2 Three-dimensional

The footwear makes contact with a surface which yields permanently such as soil, sand or snow. The sole sinks into the soft material and the impression may contain details from the uppers of the footwear as well as the sole. Such marks usually contain less fine detail than two-dimensional marks, although excellent results have been obtained from marks made in wet concrete.

8.3.4 Search and Recovery at the Scene

8.3.4.1 Light Sourcing and Photography

Volume crime scenes are usually burglaries at either domestic or commercial premises, and the scene will be attended by one scene examiner with limited time. The point of entry and areas that are disturbed will be examined. Where the crime is a serious one such as murder, a team of examiners may be used and a more extensive examination will be undertaken. A forensic scientist may be requested to attend a serious crime scene. A scene is initially examined for visible marks using the natural or artificial light at the scene and different light sources that the examiner brings to the scene. The principle of a visual examination is to use lighting to provide as much contrast as possible between the mark and the background. For initial detection of any marks present, this is done by trying different lighting angles and intensities, for example, oblique white light illumination is used to find marks in dust. Portable Nd:YAG (neodymium-doped yttrium aluminium garnet) and semi-conductor diode lasers have been in use for many years, and marks can be searched for or viewed at different wavelengths. Typically light in the blue, green or yellow parts of the spectrum is used to excite fluorescence. Contaminants present in the mark, such as grease, may excite fluorescence and some surfaces such as natural fibres found in paper or wood may exhibit background fluorescence; this can improve the contrast of marks in blood or dirt. Blood absorbs ultraviolet (UV) light and appears dark against the light fluorescing background.

Once the mark has been located, different lighting techniques can be used to optimise the imaging of the mark if it cannot be removed, for example, the mark is on a fixed floor. The basic aim of mark photography is background removal. The photography should be carried out by an individual who has been suitably trained to get the best possible result – in focus, correctly lit, scale in the correct position and no distortion. If the mark has been poorly photographed, then any subsequent comparison will not achieve the best evidence.

If at all possible, the object bearing the mark should be recovered from the scene to be examined at the laboratory, for example, if the mark is on paper, cardboard or broken glass. The object must be protected in suitable packaging so that the mark is not damaged.

8.3.4.2 Lifting Techniques

A dusty mark, that is a mark left by dry footwear, can be recovered by electrostatic lifting. This is the least destructive method available and unsuccessful attempts will not prevent other techniques being subsequently used. It is particularly useful on patterned or textured surfaces that are difficult to photograph satisfactorily. Electrostatic devices are now very small and simple to use. They work by charging a plastic film placed over the mark. The film consists of a black side and an aluminium-coated side. A high voltage charge is applied to the aluminium and this draws the black side of the film onto the surface bearing the mark. The dust particles in the mark are drawn to the film and adhere to it. After the unit is turned off, the film retains a charge and hence the particles of the mark. The lift is examined using a high-intensity light source at an oblique angle. Electrostatic lifting can be used to recover visible marks, but perhaps more importantly will also recover latent marks. In serious crime scenes, a floor surface may be completely lifted by using long strips of lifting film which can be screened at the scene for suitable marks. The lifts can be rolled to preserve them, although this can result in some transfer of material. Smaller lifts are retained in card folders to preserve them. Electrostatic lifting can also be used to clean up a mark to then allow it to be lifted with a further electrostatic lift or a gelatine lifter.

Gelatine lifters work well on marks that were originally wet and have dried. They are normally black but are also available in white or transparent. The lifter has a pliable backing and the gelatine is able to deform to the substrate contours during application and smoothing in place. Some of the loose particulate matter and any grease on the surface of the mark will be transferred to the surface of the gelatine which is slightly tacky. The lift must be photographed promptly once it has been submitted to the laboratory, as any latent marks progressively degrade. If the gelatine lifter does not initially recover much of the mark, a second lifter can be left on the mark for a longer period, say 20 minutes, and a better mark may be recovered. Black gelatine lifts absorb light so strongly that

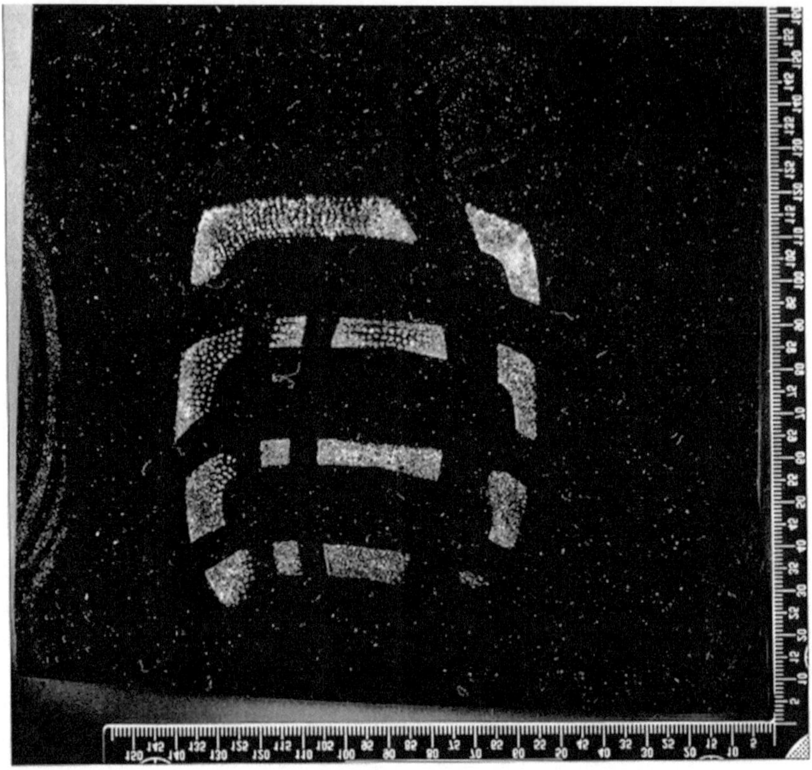

Figure 8.2 Faint heel mark recovered from a pamphlet with a black gelatine lifter.

even faint marks can be photographed satisfactorily with strong overexposure. Figure 8.2 shows the heel mark recovered from a pamphlet using a gelatine lifter. The mark was extremely faint on the original item.

Originally wet marks can also be brushed (dusted) with black or aluminium fingerprint powder and lifted onto a sheet of clear adhesive film which is then pressed onto a clear plastic sheet to give an adhesive lift. A mark made by dry footwear is likely to be destroyed by over-brushing. Dusting marks with powder and the use of adhesive lifts was common practice before gelatine lifts became universally used by scene examiners.

8.3.4.3 Chemical Treatments

Most marks found at a volume crime scene will be in dirt or grease of some type, or they could be in a material which has been spilled at the scene. Negative impressions are also found that have resulted from the removal of material from the substrate by the footwear. At a serious crime scene, marks may be in blood and these will have considerable importance in any investigation.

Marks made with bloodstained footwear are commonly treated with chemicals that react to different properties of blood. Marks in blood vary considerably, depending on the viscosity and quantity of blood on the sole. The best marks are usually those where excess blood has been tracked off the sole; heavily bloodstained marks may obscure detail. The two main techniques used for enhancement are protein stains and peroxidase reagents.

8.3.4.4 Protein Stains

The protein stain acid black 1, commonly known as 'amido black' has been in use since the early 1960s and stains the blood blue–black. It reacts with the proteins in blood but is not specific for blood. The mark is fixed using a solution of 5-sulphosalicylic acid in water. This precipitates the basic proteins and thus prevents diffusion of the mark and loss of detail. The mark is then treated with the acidic protein stain and this dyes the precipitated basic proteins to give a coloured product. Before applying the stain, an area of the background away from the mark is tested to ensure that there is no heavy background staining. A washing stage is required post-staining to remove excess dye on non-porous surfaces. On porous surfaces, it acts as a destainer, removing dye absorbed by the background.

Many other protein stains have been proposed for use in the enhancement of both fingermarks and footwear marks over the years. None of these have been found to be as effective as acid black 1. Both acid violet 17 and acid yellow 7 have been recommended by the UK Home Office Centre for Applied Science and Technology (CAST) for use in addition to acid black 1, where the choice of stain is made depending on the colour and porosity of the background.

8.3.4.5 Peroxidase Reagents

Peroxidase reagents are less sensitive than protein dyes and rely on the property of haemoglobin to catalyse oxidation reactions in the presence of hydrogen peroxide.

Diaminobenzidine (DAB), a colourless reagent, is oxidised to give a dark-brown colour. It can be used on both porous and non-porous surfaces. 5-Sulphosalicylic acid is again used as the fixing solution.

DAB is known to be potentially hazardous and has only been used by specialist teams. Colourless leuco crystal violet has also been used to enhance marks and is oxidised to purple crystal violet. This is known to be hazardous and therefore, large-scale spraying at scenes is undesirable.

Luminol reacts with haemoglobin to produce a blue glow. The process is known as 'chemiluminescence'. Specific areas are sprayed with luminol and the reaction must be observed in darkened conditions. The reaction lasts a relatively short time and must be photographed immediately. Luminol is not specific for blood and reacts with other oxidising reagents such as bleaches (containing sodium hypochlorite) which may have been used to clean up the scene. Luminol is useful for recovering marks on surfaces such as carpets and on dark surfaces.

Other chemicals such as ninhydrin have occasionally been used to enhance footwear marks in blood. Ninhydrin reacts primarily with amino acids and gives an intensely coloured product known as 'Ruhemann's Purple'. This renders detail readily visible on background materials, permitting sufficient contrast. These techniques should only be used at the scene when it is not possible to recover the object bearing the mark from the scene for treatment at the laboratory.

8.3.4.6 Three-dimensional Marks

Such marks should always be carefully photographed and certain marks, for example those in soil, can be recovered by casting with a dental stone casting material. Water is added to the material to produce a pourable mix and this is then carefully poured onto the impression. It is important to photograph the mark first as the casting material may not flow properly into all areas of the mark leaving some detail obscured. The photograph and the cast can provide complementary information. In the laboratory, the loose material is cleaned off the cast and then the cast is soaked in a saturated potassium sulphate solution to harden it. It can then be cleaned under cold running tap water.

8.3.5 Laboratory Examination and Comparison

8.3.5.1 Analysis of the Mark

Footwear marks have been found on many different objects and can range from a small partial mark of a few studs on a post-it note to a large piece of wooden flooring bearing numerous overlapping marks of different patterns. The principles for examination and comparison are the same. A detailed assessment of the quality and level of detail in the mark is made. Issues that might affect the clarity of the mark are identified such as movement, the substrate on which the mark was made or the type of deposit left by the footwear. Pattern elements are identified; this could be a small area of blocks, seen as a component in many different pattern types or an almost complete mark where the examiner can identify the likely brand and model of the footwear. The level of wear and any possible mould or damage features are noted. Where there are multiple marks which are apparently made by the same footwear, the examiner will look for features that are reproduced in more than one mark.

The examiner tries to get the best representation possible of the mark by enhancing it where appropriate to improve the clarity of the mark. Clearly if the submitted footwear has a different pattern, further enhancement techniques would be pointless. Porous items such as paper or card permit the most enhancement techniques and such items will always be electrostatically lifted to identify any latent marks. The mark should always be photographed before any destructive techniques are applied.

Many different treatments have been developed for recovery of fingermarks on both porous and non-porous surfaces by forensic science practitioners, researchers and academic groups over the years, and often these treatments are applied in a sequence. Some of these techniques have also been usefully adopted by footwear mark examiners, depending on the substrate and the deposits within the mark. The examiner makes an assessment of which techniques are likely to produce further enhancement of the mark.

Some of the chemical treatments for blood marks have already been discussed; in addition, in the laboratory, 1,8-diazafluoren-9-one (DFO) can be used. Like ninhydrin, it reacts with amino acids, and DFO produces a fluorescent product under appropriate illumination. This is especially useful for development on backgrounds where a conventionally coloured product would be hard to visualise. DFO is only recommended for use on paper and other porous surfaces. DFO and ninhydrin can be applied in sequence.

Physical Developer (PD) is a complex mixture including silver nitrate and it reacts with certain components to form a grey–silver deposit. It has been used on paper items where the footwear mark is oily or waxy. An acid pre-wash is used to stop darkening of the background, caused by the alkali fillers in the paper. PD can be used on porous items and is always the last stage in any sequential treatment. The mark is photographed wet and as soon as possible.

Electrostatic Detection Apparatus (ESDA) used for detecting indented handwriting impressions, has also been used for marks on paper. The paper has to be humidified first to increase sensitivity to the electric charge. This technique was effectively abandoned once electrostatic lifting became universally used.

8.3.5.2 Marks on Clothing and Bodies

Clothing is examined under different lighting conditions. Any marks observed will be photographed and can be lifted where appropriate. The marks are often distorted or broken up by creases in the garment. Where marks are faintly visible under certain lighting conditions, enhancement has been observed with both DFO and ninhydrin. In such cases, marks have appeared on the inside of the garment rather than the outside, and are likely to have resulted from the transfer of skin debris or sweat from the victim when stamped on or kicked. A series of publications by Farrugia and Nic Daeid discusses the use of a variety of chemical enhancement techniques for footwear marks in blood on dark clothing.

Marks on skin resulting from kicking or stamping on the victim are usually caused by bruising or abrasion and will often have diffuse edges. The marks may take some time to develop and a second viewing after a day or two may show an improvement. The yielding nature of skin means that detail in recessed areas of the sole may be seen and also other areas such as the laces have been observed in such marks. These marks may provide corroboration of the circumstances but are very unlikely to give a strong link to a particular control item. The skin is generally not a good medium to record detail.

8.3.6 Examination of the Footwear and Test Impressions

As discussed, the mark is always analysed first but decisions on further enhancement are made as the comparison with the submitted footwear progresses. Before test impressions are taken, any necessary contact trace examination is undertaken; this could be sampling for blood or removal of glass or paint particles, for example.

Test impressions can be static or dynamic. The sole of the control item is dusted with black or aluminium fingerprint powder, the excess is removed, a sheet of clear adhesive applied to the sole and then pressed onto a clear plastic sheet to produce a transparency. Most comparisons can be undertaken using static test marks, however, in some cases dynamic marks are required. These can be produced by a colleague with the appropriate shoe size wearing the footwear and can be walking, running, kicking or stamping, as appropriate. A kickboard is useful where kicking marks are required. Test marks can be made in sand or mud or the sole wetted to illustrate a particular point.

8.3.7 Comparison

The comparison is a step-wise process using pattern, mould, general and specific wear.

Pattern – Most recovered marks are partial only with possibly a relatively small area of the sole pattern present. If the pattern or pattern components of the submitted footwear are different from the mark, then the footwear is clearly excluded as a possible source.

Mould – If the footwear has the same pattern, then the configuration of the pattern components is examined. Different moulds may be differentiated on this basis. For different sizes of a particular pattern, manufacturers can take various approaches. The pattern may just be expanded or additional components can be added or they may be rearranged. For an examiner to decide that the mark and the footwear came from the same mould, then mould features would need to be observed such as stippling detail or mould defects.

General wear – The term 'general wear' is used for the overall change that takes place when a sole wears down with use. It can lead to the obliterating of parts of the sole pattern. Clearly, the areas that contact the ground with every step receive the most wear, whereas areas that are not routinely in contact with the ground receive little or no wear. General wear is affected by the walk of the wearer and their habits, the type of footwear and the composition of the sole material. Where general wear appears to be different, it is important to have information on the time elapsed between the mark being left at the scene and the seizure of the footwear.

Feathering – 'Feathering' detail can be seen on some soles. This consists of microscopic detail that develops due to abrasion of the sole surface resulting in very fine wear. This appears like friction ridge detail and is highly individual. Any additional abrasion will alter this feathering detail in a very short time frame. Feathering has been seen on natural rubber soles but not on synthetic soles like polyurethane. Where the feathering is clearly defined, a few millimetres square is sufficient to show that it was made by a particular shoe. This feathering is also known as 'Schallamach Patterns' after a researcher in the rubber industry. Where a feathering comparison is undertaken, both mark and test prints are enlarged to two or three times life size to enable a detailed comparison.

Individual wear – Specific wear characteristics are acquired by footwear constantly while being worn. They can be cuts, gouges, scratches, acquisition of small stones or other debris. These random wear marks are acquired purely by chance and can be very large or minute. Individual characteristics are evaluated according to their position, size, shape and how clearly they are defined. In those instances when characteristics are not clear enough to correlate well, consideration must be given to the possible or known reason for the limited clarity, for example, the mark is in sand or soil or the mark is wet.

The test print is overlaid on the photograph of the mark which has been printed to life size and matching features are marked up. There will never be complete correspondence between the mark and the test print. The expert must look at the matching and non-matching features and account for differences. The sole is examined microscopically to confirm that identified characteristics are random damage features, resulting in loss of part of the sole. Figure 8.3 shows a partial mark which has been left as a narrow strip, where the complete pattern components cannot be seen but there are characteristic damage features present.

8.3.8 Evaluation

Marks comparisons have traditionally been reported subjectively where the expert comes to a decision on the level of evidence based on his or her knowledge and experience of footwear marks and gives an expert opinion. The footwear can be excluded if the pattern, mould or wear—for example, the shoe is clearly less worn than the mark—is different. If there is very good agreement of pattern, mould, general wear and clearly defined specific damage features, the expert may conclude that the footwear was responsible for the mark.

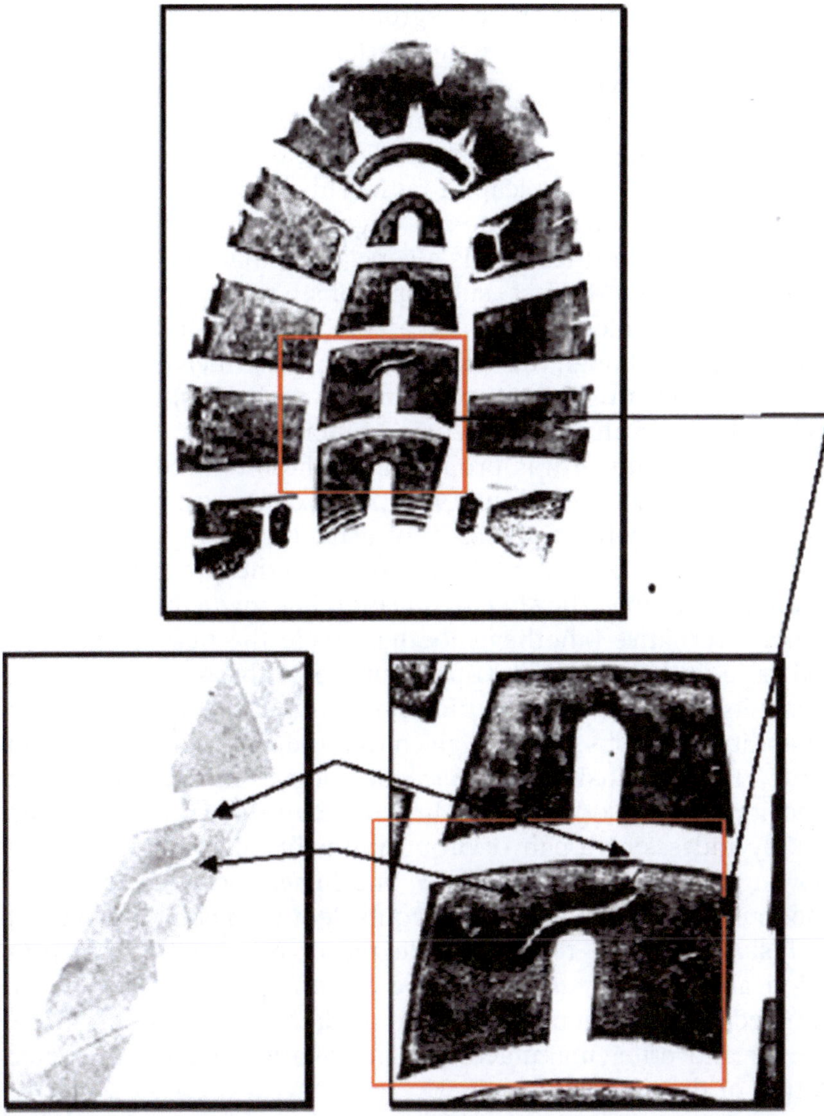

Figure 8.3 A scene mark (bottom left) compared to a test mark. Corresponding detail indicated by the arrows.

This conclusion is termed an identification. The expert has concluded that the mark could not possibly be duplicated by chance and all other sources are eliminated. For a significant proportion of cases, the level of correspondence will fall somewhere between exclusion and identification. There may be enough detail to determine that the pattern matches but no unique features are present, or the background interferes so any features cannot be relied upon. Where the mark is of poor quality, the result may be inconclusive.

The categorical opinion of certainty cannot be rationalised logically. In 1991, Stoney suggested that identification can be considered somewhat analogous to a leap of faith. This leap occurs without any statistical foundation and there is ongoing discussion at this time about whether 'identification' is a conclusion that may be scientifically sustained. Nevertheless, such opinions have long been accepted by courts as valuable evidence and are permitted in relation to many evidence types although this acceptance is under increasing scrutiny.

With the advent of DNA profiling, a probabilistic model was developed for evaluating forensic evidence using the Bayesian approach. This model was proposed for the interpretation of footwear marks by Evett *et al.* in 1998. The expert gives an evaluative opinion based on the estimation of a likelihood ratio. The expert addresses a pair of propositions based on the prosecution proposition (for example, the shoe in question has made the mark) and a defence alternative (another shoe has made the mark). The likelihood ratio is the ratio of two probabilities – the probability of the observations given the prosecution proposition is true divided by the probability of the observations given that the alternative proposition is true. The conclusion expresses the degree of support for one proposition or the other depending on the value of the likelihood ratio. By addressing a pair of propositions in this way, the expert is following the principles of balance and logic. The value of the likelihood ratio is related to a verbal equivalent on a standard range of conclusions used to reflect the degree of correspondence. Information about the frequency of the pattern, the frequency of the estimated size and general wear from data collections of suspect footwear and other manufacturing information is used in the calculation of the likelihood ratio.

A direct survey of some relevant population of footwear is difficult to achieve although some research using citizen science based approaches have begun in this area. A number of data sources can be considered of value regarding the distribution of footwear patterns. Information on sales figures from manufacturers is useful, but it represents total sales and does not necessarily reflect what is being worn in a given area at a given time. The information from footwear submitted to a forensic laboratory more accurately reflects the type of shoes being worn by offenders, but care must be taken. Regional variations can be seen in UK collections, certain patterns will be very popular in different areas of the country and not in others. The frequency of a pattern in a particular area can be dependent on fashion, the choice of a particular group, a discounted price, local availability, the length of time footwear is kept and counterfeit copies of a pattern. Information on counterfeit copies is clearly difficult to obtain. Laboratory data collections are based mainly on volume crime and, therefore, great care must be taken if such data is used for serious crime offences. It is also very important to remember that it is about how common the pattern in the mark is, not the pattern of the footwear that has been submitted. As discussed previously, the same pattern components can be used on many different shoes.

In Appeal Court Judgement, R *vs.* T, (2010) concerns were expressed about the size and relevance of the database used and the Court concluded that an expert could consider such data when expressing an opinion but should not use a mathematical formula (likelihood ratio), as the data was unreliable. Any experienced scientist recognises that such data must be treated with caution and its limitations should be recognised. There have been many publications on this matter post the Appeal Court Judgement. The two remaining principles of transparency and robustness require the expert to demonstrate how the conclusion was reached, and that the evidence must withstand scrutiny by other experts and cross-examination.

8.3.9 Footwear Databases and the Use of Footwear for Intelligence

Data collections of both crime marks and footwear have been in use for many years. In the early 1970s, marks from unsolved burglaries and other scenes were submitted to the Metropolitan Police Forensic Science Laboratory (MPFSL) and details were kept in a card index. A scheme for coding sole patterns was developed based on a set of rules, and covered aspects such as the shape of pattern components and their location on the sole. Coding the marks enabled the index to be searched for possible links. At that time, the index consisted almost exclusively of marks made by the soles of Doctor Martens type footwear and Dunlop Green Flash shoes. During the 1980s, there was a move away from the centralised index and local indexes were set up by scene examiners. Today, computerised collections of local crime marks are maintained for possible scene linking based on pattern type and other information. A large number of volume crimes are committed by repeat offenders and it is common for burglars to commit a number of offences in the same day. As it would be unusual for an offender to discard footwear between crimes, timely identification and matching of marks allows different crime scenes to be linked. However, any intelligence possibilities must be considered in context, as the same footwear pattern type at two different scenes may not indicate a link, if the sole pattern is very common.

The other type of database is that of sole prints of submitted footwear. Again since the 1970s, such collections have been maintained in forensic science laboratories, in addition to information obtained from manufacturers. The database of footwear patterns has been used for intelligence in suggesting the brand and model for an unknown mark left at a murder or other serious crime scene. As already discussed, databases provide information on the frequency of pattern types, different sizes and moulds for the same pattern, and general wear patterns. All of this information must be considered in context.

Many police forces now maintain their own computerised footwear database and use a national coding scheme, as part of the National Footwear Reference Collection (NFRC). There is also a national capability for collating and searching scene marks. Figure 8.4 shows a screenshot of an NFRC search. The search shows a page of coded patterns where the manufacturer is known for some of the patterns, Gola and Fila in this case, and 'Named' patterns which have been coded but where the manufacturer has not been identified. Commercial companies have also been involved in providing computerised reference collections and coding capability. These commercial systems provide regular updates through the year on major brands of footwear and are sold on the basis of identifying unknown marks at crime scenes.

8.4 Tool Marks

8.4.1 Introduction

Tool marks can be left at any crime scene and are often present at scenes involving forced entry into premises. They may be found at the point of entry on a door or window and are also found inside the premises where maybe a desk drawer or safe deposit box has been forced open. Almost any kind of tool can be involved in a crime. Probably the most common are tools used for levering such as screwdrivers, case openers also known as 'wrecking bars' or 'jemmies', and chisels. Cutting tools such as bolt cutters may be used to snap the shackles of padlocks or to cut wire fences. Knives have been

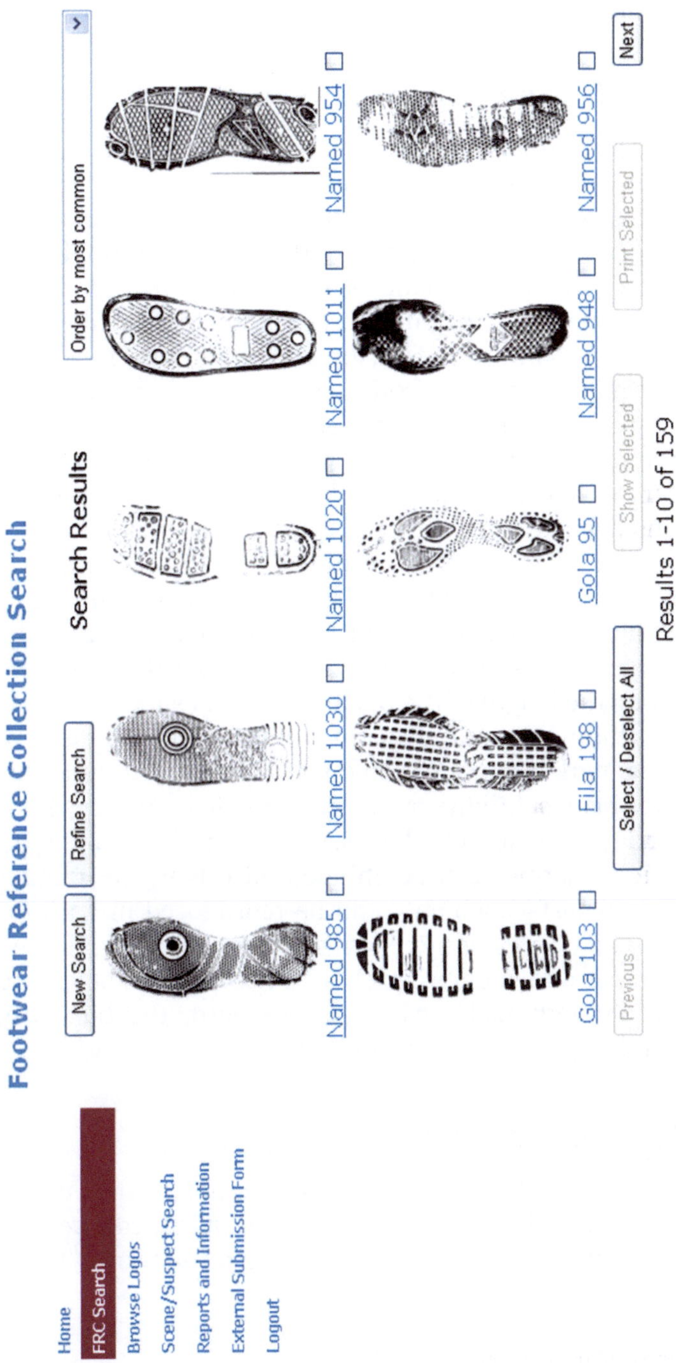

Figure 8.4 Screenshot of National Footwear Reference Collection search.

used to cut cables or stab tyres. Marks made by hammers, saws, drills and many other tool types have been seen.

8.4.2 Manufacture of Tools

There are many different types of tools which require different properties depending on their intended use. Hence, there is a large variation in the processes used in their manufacture. Tools need to be hard-wearing, and the working part of the tool will normally be made of steel. Usually, a rough shape of the tool, known as a 'blank' is formed first from the stock material. The metal may be forged, that is, shaped under the application of force at different temperatures, or tools may be cut or stamped from metal sheets or rods.

When the basic tool has been formed, several processes may be used to finish the tool for use. A coating may be applied such as metallic plating, paint or varnish, and some have a heat-treated tip. Grinding and linishing detail are found on many tools. These are processes used mainly to remove flash or other unwanted features resulting from the primary shaping or cutting process. The tool is held against a moving abrasive surface and depending on how the tool is held against the surface; grinding marks can be horizontal across the blade or longitudinally along the length of the blade. Figure 8.5 shows the magnified view of longitudinal grinding detail on a slightly rusty screwdriver blade. Milling marks are often seen on bevelled tools such as chisels or on bolt cutter blades. Milling is carried out using a cutter with several evenly spaced teeth. Once a tool is used, it begins to acquire individual characteristics. Coatings can be chipped off leaving an irregular outline on the tool surface which could be reproduced in an impressed mark. The tip and edges become less sharp through use; the tool may acquire scratches and gouges or dents. If, for example, bolt cutters are used to cut through metal which is too hard, the blades can be damaged, and Figure 8.6 shows such damage.

8.4.3 Types of Tool Mark

When a tool comes into contact with another object, then evidence of that contact may be left behind. Useful tool marks can be produced when a hard surface comes into contact with a softer one. Tool marks generally fall into two main types:

(1) Levering/force marks
(2) Cutting marks

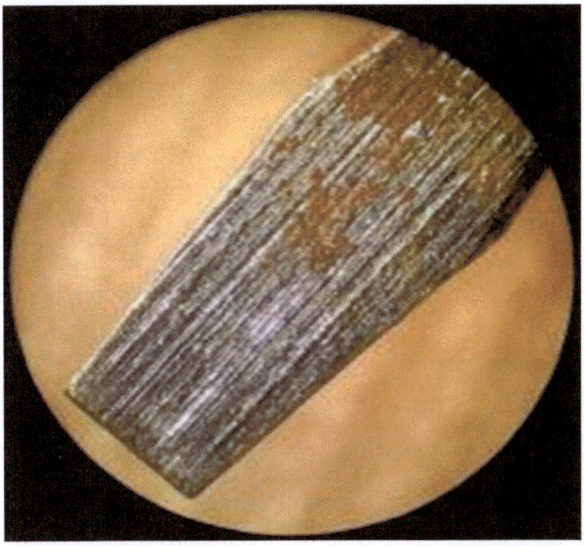

Figure 8.5 The slightly rusty surface of an old screwdriver blade with grinding detail running the length of the blade.

Figure 8.6 The damaged blades on a pair of bolt cutters.

There are two ways by which detail in a mark can be produced, known as 'impressed' and 'dynamic' detail. If there is no additional movement when the tool and object are forced together, then an impressed mark may be left. This impression reproduces detail on the surface of the tool, for example, if a screwdriver is used to lever open a window and the tip of the tool is pressed into the surface of the window frame. If there is movement by the tool across the surface of the object, then a dynamic mark results, consisting of parallel lines known as 'striations'. These striations are produced by damage and other uneven points along the tip of the tool. Dynamic marks also result when a cutting tool is used to sever an object or if an object is stabbed with a tool, for example, a knife stabbing a tyre. In many cases a mark will consist of a combination of impressed and dynamic detail.

The examples of tool marks given so far are all marks produced in the commission of a crime, but tool marks are also produced during the manufacture of tools and other mass-produced items. We normally refer to this type of mark as a 'manufacturing' mark and these are dealt with later in the chapter.

8.4.4 Recovery of Marks from the Scene

The detail required to identify a tool from a mark is microscopic, it is not possible to determine at the scene if a mark is likely to contain useful detail or not. Where possible the entire item bearing the marks is retrieved and packaged carefully. If this is not possible, the marks are carefully cut out, for example, marks in a window frame. All marks should be retrieved from the scene, not a selection, as additional marks may show further or reproducible detail, and indeed more than one tool may have been used. What makes a mark obvious is usually the extent of the mark and the damage associated with it. Very damaged marks are often of poorer quality and contain the least amount of detail. It is important to detail the position of the marks at the scene in relation to each other and to provide orientation information for the examination of the tool mark, for example, ease of access for the tool. If it is not possible to recover the original marks, then they are cast. Casting materials usually consist of a liquid silicone material and a hardener which is added to set the cast.

There are occasions when hand tools are not used in a crime but much larger tools are suspected. Where the scene is an illicit drugs factory, tools may be used to press tablets or produce blocks of drugs. To produce fake credit cards or fake coins, moulds and other equipment may be present. Where the equipment is very large, this may have to be examined *in situ* by the scientist.

Marks may also be recovered from a person (alive or dead). This could be as the result of an assault using a weapon or perhaps a mark left by a ring during a punch and will require an experienced photographer, and an appropriate scale must be used. Some marks will develop more clearly over a few days. Such marks may also be clearer under different lighting conditions, for example, UV or near infrared.

Other marks that can be recovered are those from a dead body where the victim has been stabbed or slashed with a tool such as a knife, axe or machete. Such marks will usually be removed by the pathologist. It is also possible that a body may have been sawn or cut up for disposal.

Tools may be abandoned at the scene; these could belong to the suspect but also could have been taken from the scene, for example, from a garage or garden shed, or from a previous scene for use by the suspect. Where tools are recovered directly from the suspect, they are also preserved for possible contact trace evidence. Tools are packaged carefully to avoid any subsequent damage to the tool.

8.4.5 Laboratory Examination and Comparison

Tool marks are initially examined with a stereomicroscope at low magnification with a flexible light source. As with footwear marks, a detailed assessment of the quality and level of detail in the mark is made and issues that might affect the clarity of the mark are identified, such as the substrate. A mark in highly splintered wood may well lack detail and marks in tyres may be distorted by the rubber.

8.4.5.1 Impressed Marks

If the mark is a levering mark and shows impressed detail or a mixture of impressed and striation detail, it may be possible to determine class characteristics such as tip width, tapering or not of the sides, or whether the tool has a bevel or forked end. It may also be possible to see potential damage such as an uneven tip or a damaged corner, chipped paint or plating. Where there are several marks, potential characteristic features can be checked to see if they appear to be reproducible and therefore, are reliable characteristics of the tool responsible.

Although levering marks are the most common impressed marks seen, other types are seen in impact marks, for example, a hammer used to strike a door to gain entry, or grip marks, for example, where mole grips have been used to bend and break a lock.

8.4.5.2 Dynamic Marks

Levering marks can also consist of mainly striations with possibly a small area of impressed detail where the tool initially made contact with the substrate. This is also true of cutting marks such as those made by bolt cutters. Where an item has been cut, the tool normally leaves a characteristic profile, and this can often be seen without the aid of a microscope.

8.4.5.3 Specialised Marks

8.4.5.3.1 Saw Marks

Marks left by a saw are different from the other tool marks discussed here. There are many different types of saw but they are essentially a series of teeth along the blade. By the nature of the sawing action, each mark made by an individual tooth is modified or obliterated as the saw is used. If there is an initial or final cut, it may be possible to determine the tooth spacing to give teeth per inch (tpi) and the set of the teeth, for example, alternate, raker or wavy.

8.4.5.3.2 Alphanumeric Punches

Two types of punch marks have been seen in tool mark examinations. A set of individual metal punches can be used to strike characters into metal. Such marks used to be frequently seen in the 'ringing' of vehicles where engine numbers or vehicle identification numbers were altered or erased and replaced with new numbers using such punches. In the manufacture of these punches, the relevant character is engraved onto each face of a steel hob. The steel punch is cut to length and one end is forced against a face of the hob to impress the character into the end of the punch. The punch will then be finished off and may be plated. The detail on a new punch will relate to the manufacturing process but, as a punch is used, microscopic damage occurs and the plating chips. The examiner has to ascertain what detail relates to the manufacture and what is unique.

The second type of punch is used in card-embossing machines in the production of fake credit cards or similar. These machines have two punches that work in unison; one leaves a recessed negative impression of the character, while the other leaves a raised positive impression of the character. The examination of suspect cards is best made by making test cards from the same type in the machine. Figure 8.7 shows the comparison of casts taken from two credit cards, showing corresponding detail on a particular digit.

8.4.5.3.3 Drill Marks

Where an item has been completely drilled leaving only the hole, little can be done with the mark other than the dimension of the drill bit. However, if there are incomplete marks or swarf produced by the

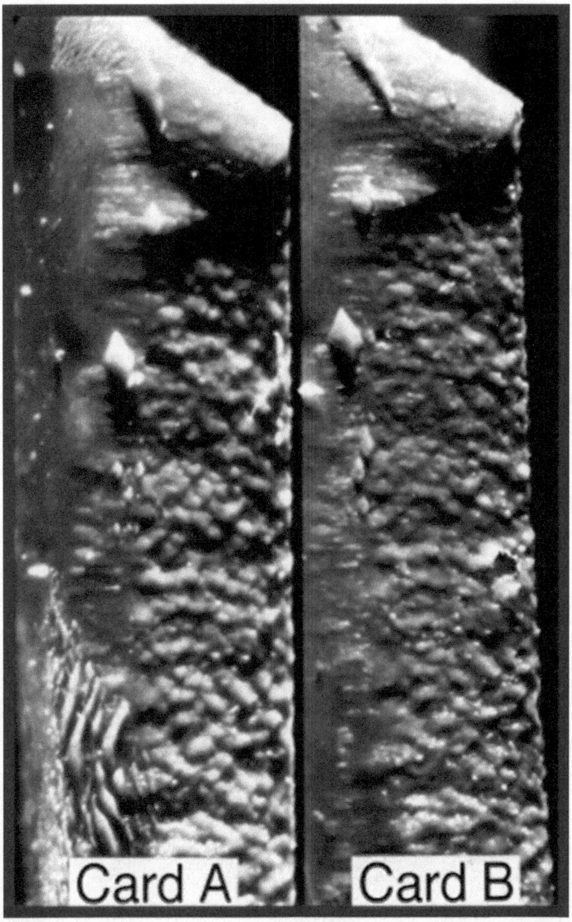

Figure 8.7 Casts taken from two credit cards. The degree of corresponding detail demonstrates that the digit 1 has been made by the same punch.

drilling process, such marks could show the profile of the drill tip. Drill bits wear down during use and the profile will change.

8.4.6 Test Marks and Comparison

Over the last 25 or so years there has been a considerable improvement in the microscopic equipment used for tool mark comparison, and this has led to an increased use of casting materials that allow better visualisation of the marks. Many of the casting materials were originally designed for dental use but in recent years materials have been specifically designed or modified for forensic use. Most current materials are silicone rubbers which are applied as liquids and flow into the mark. These are available as a two-pack system where the examiner mixes the hardener with the base and therefore can vary the consistency and setting time of the mixture, or a one-pack system where the base and hardener are automatically mixed in a 'gun' when the plunger is depressed. One of the most important features of any material is the ability to cast and be recast without loss of detail. Different viscosities of casting materials are available, with higher viscosity materials being used for vertical surfaces. Polysulphide casting materials have been produced which allow casting of silicone rubber.

Detailed comparison of the scene mark and test marks made with the tool is undertaken using a comparison microscope. This consists of two stages with two sets of objectives but only one pair of eyepieces, effectively two microscopes joined by a bridge. The field of view is divided into two so the object on each stage fills only half the field of view and allows the user to view both images side-by-side; they can also be superimposed. The stages can be rotated, tilted and moved independently in the x–y and z planes. Both objects should be examined under the same conditions and be illuminated from the same direction and angle. The comparison must be undertaken with two positives or two negatives. A tool cannot be compared with the mark directly as this would be a positive with a negative, so the mark is cast.

For an impressed mark, it may be possible to directly compare the relevant area of the tool with a cast of the mark, but due to lighting problems and lack of maneuverability of the tool, it is much more likely that a replica cast of the relevant area of the tool will be compared with the cast of the mark. A replica cast is made by casting the relevant area of the tool and then recasting this cast with a different material to produce a replica of the original tool. If a cast of the mark has been submitted, it is also likely that this cast will be cast and recast to give a replica of the original cast. This is due to the better contrast produced by certain casting materials under a comparison microscope. It may be preferable to make a test mark with the tool in modelling wax and cast it for comparison.

Where the mark is dynamic, test marks are made, and this is the most difficult aspect of a tool mark examination. A small change in the angle or pressure of the working edge of the tool alters the points of contact with the surface and hence, alters the appearance of the striations. Where the mark is a sliding mark, it may have been possible to identify which face of the tool is the more likely from the initial examination. Test marks are then made at various angles to attempt to reproduce the detail. Where it was not possible to decide on the likely face, test marks are made with both faces of the tool. The test marks

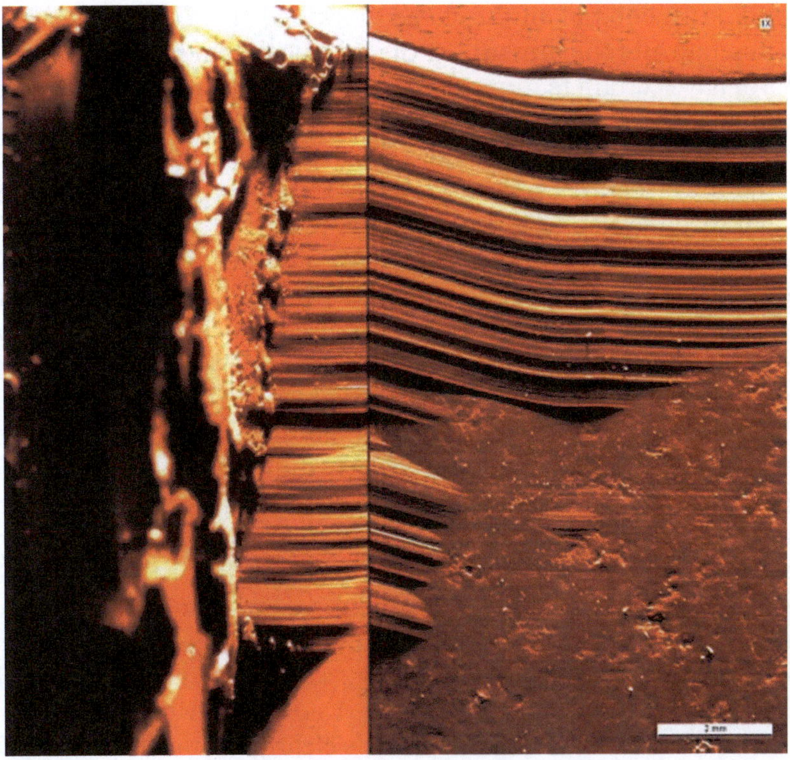

Figure 8.8 A comparison of dynamic striation detail. The mark on the left is a cast from a scene mark, and that on the right is the cast of a test mark made using the suspect screwdriver.

are nearly always made in lead or modelling wax as these materials are soft and will not damage the tool. An examination of casts of the initial test marks against the cast of the scene mark enables the examiner to decide what further test marks may be needed. Figure 8.8 shows dynamic striation detail present in the cast of a scene mark on the left and in a cast of a test mark in lead using a screwdriver on the right.

For a cutting tool such as bolt cutters, test marks are made in lead rod; here the issue is to identify the likely part of the blades used to make the scene cut. There may be an indication on the blades to assist with this, such as contact trace material, or the blades may appear shiny in a particular area.

8.4.7 Evaluation

In evaluating a scene mark and a tool, the examiner has to decide on the significance of matching and non-matching features. Much will depend on the quality of the scene mark and this can be affected by the substrate and the pressure of the tool when making the mark. The tool will have certain features on the surface and some of these will be reproduced in marks made by the tool but some may not. Where features are reproduced in the mark an assessment is made on whether these features are unique to that tool, for example, damage to the tool through use or if the features are part of the manufacturing process and therefore, could be present on other tools.

Where there are apparent differences but some corresponding detail, the examiner has to decide if the discrepancies can be explained by subsequent damage to the tool, for example, if there has been a delay in seizing the tool, or perhaps the test mark materials used do not properly reflect the substrate in which the mark was found.

As with footwear marks, the expert addresses two propositions based on the prosecution proposition (for example, the tool in question has made the mark) and a defence alternative (another tool has made the mark). There is no reliable data regarding the commonality of tools or features on tools and, therefore, it is not possible to determine a numerical result or likelihood ratio. The expert gives his or her subjective opinion using knowledge and experience of tool marks.

Any comparison is limited by the amount and quality of detail present in the mark. Where the mark is clearly defined with sufficient characteristic detail, and the expert decides that the detail in the mark corresponds to features on the tool which have been randomly acquired, the expert concludes that the tool made the mark. Where the detail appears different, for example, the tip width and tapering in an impressed mark is clearly different, then the expert will exclude the tool. Conclusions falling between these two extremes provide a level of support for either the prosecution or defence proposition, expressed on a verbal scale. Where the detail in the mark is limited, it may not be possible to exclude the tool and such an examination is reported as inconclusive, offering no support for either the defence or prosecution propositions. The results from tool marks are very often identifications, exclusions or inconclusive due to lack of detail.

8.5 Manufacturing Marks

8.5.1 Introduction

Manufacturing marks can be found on a wide range of mass-produced items, resulting from either the detail on the machine tools and other equipment used to produce the items or the material used to make the product. Such evidence can be used where there is a requirement to link items in the suspect's possession or their premises to a scene.

The range of items that are mass produced is enormous and methods of production involve an incredible variety of processes and techniques, including moulding, casting, stamping, cutting and grinding. The importance of understanding the manufacturing process has already been mentioned in previous sections. The scientist will consider aspects such as:

(1) What manufacturing detail is present on the items?
(2) Do different machines produce different detail?
(3) Does the detail change over time and, if so, what is the rate of change?
(4) How much of this material or these items has been produced?
(5) How are the manufactured items packaged and distributed?

Clearly it will not always be possible to have answers to all of these questions, but it is often possible to obtain important information from various sources including published research, the manufacturer of the item, which may include a visit to the factory, and samples from the actual process, and retail packets containing samples of the item in question. The range of items that could be examined and compared is huge but items seen in case work have included cable ties, nails, screws and buttons. It would be impossible to cover all of the potential production processes involved in the manufacture of such items, and the main focus in the following section will be to use plastic film to illustrate various aspects. The most frequently encountered items are indeed packaging materials such as plastic bags and sacks, cling film and tape. These materials have been most frequently seen in drugs distribution but have also been encountered in the production of devices in terrorist cases and in the disposal of bodies. To interpret the detail seen on such items, the examiner must understand the manufacturing processes used.

8.5.2 Manufacture of Plastic Film

Plastic bags are usually manufactured using the blown film extrusion process. Polymer granules are heated until molten and then the plastic melt is extruded through an annular die slit, usually vertically, to form a tube. While hot, the tube is blown by compressed air to a larger diameter which produces thinning of the tube wall. The air is introduced *via* a hole in the centre of the die. A high-speed air ring is mounted on top of the die and this blows on the hot film to cool it. The tube of film continues upward, continually cooling, until it passes through nip rolls where the tube is flattened to create what is known as a 'lay-flat' tube of film. The lay-flat tube is then taken back down the extrusion tower through a series of rollers. The lay-flat tube can then be converted into bags by heat sealing across the width of the film and cutting or perforating to make each bag. Alternatively, the tube can be slit along the edges to produce two flat films which are then wound up onto reels.

The width and thickness of the film are controlled by the volume of air in the bubble, the output of the extruder and the speed of the haul-off. This process can produce extrusion detail on the plastic, a series of bands that run parallel to the direction of the production. The bands are a result of the flow characteristics of the polymer through the die and are affected by build-up of polymer around the die. Cleaning of the equipment involves changing the die and this removes the signature; the build-up will be different following each clean. More modern lines produce less of a signature. The rollers may produce fine scratches on the surface of the film due to defects on their surfaces.

As part of the process, bags can have handles and gussets incorporated, they can be perforated to allow them to be torn from a roll, or be printed. Resealable bags have a grip seal and this is produced by having grooves in the sides of the die of the right profile to form the sealing strips. Heat seals are produced in two ways, an electrically heated strip is pressed against the film to give a flat seam, usually found at the base of bags. The second type of seal is a vertical hot wire which both cuts and seals the bag, giving a narrow seal seen on the sides of bags. Perforating is done by a notched blade piercing the extruded film.

Cling film is also made by the blown film process and here the film is cut to form a single sheet and wound onto a cardboard former and then cut to form individual rolls. There are many types of self-adhesive tapes available including clear, masking, insulating, duct and brown packaging tape. The most commonly seen tapes are clear and brown packaging tape. The film used in such tape is usually polypropylene and has a primer coat applied to one side and the adhesive coating is applied to this side. The other side has a release coating so the tape does not stick to itself. Where bags, tape or cling film are recovered from the suspect's premises, wherever possible, all the materials should be seized, not just samples.

8.5.3 Laboratory Examination and Comparison

8.5.3.1 Bags Examination

As before, an assessment is made of the suspect bags. The general appearance, colour, size and construction of the bags is noted. The method of construction of a bag can be deduced from the arrangement of the folds and heat seals on an individual bag. The initial comparison of the reference bags with the suspect bags will consist of comparing the physical appearance. Many bags can be excluded on this basis given the huge variety of bags now available. If the reference bags are not excluded at this point, then manufacturing detail can be examined.

Transparent polythene bags can show coloured bands when they are viewed under polarised light and the patterns usually align along the extrusion direction of the film. When bags have very distinctive patterns of polarisation colours they can be sorted into a matching sequence. By using a large table fitted with circularly polarised light, large numbers of bags can be examined at the same time and polarisation colours can be seen in all orientations. The colours are determined by the birefringence and the thickness of the plastic and therefore reflect changes in one or both properties. These colours can be continuous through several bags or taper off within one bag. Figure 8.9 shows a number of transparent grip seal bags which have

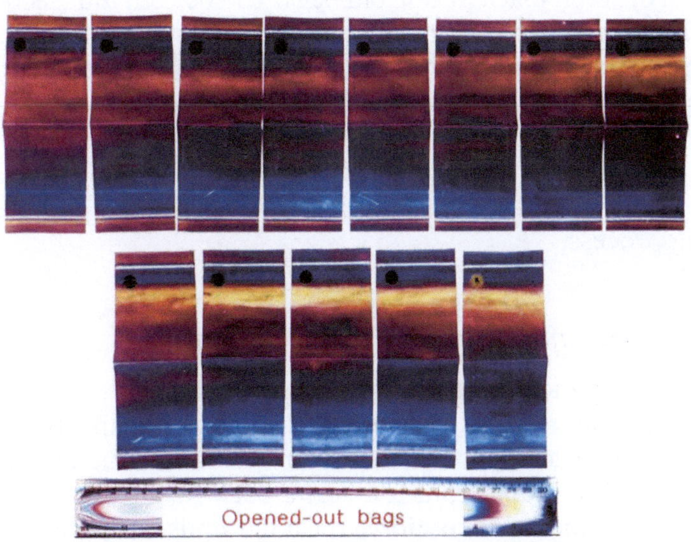

Figure 8.9 A number of grip seal bags that have been opened out by cutting down each edge and then viewed under circularly polarised light.

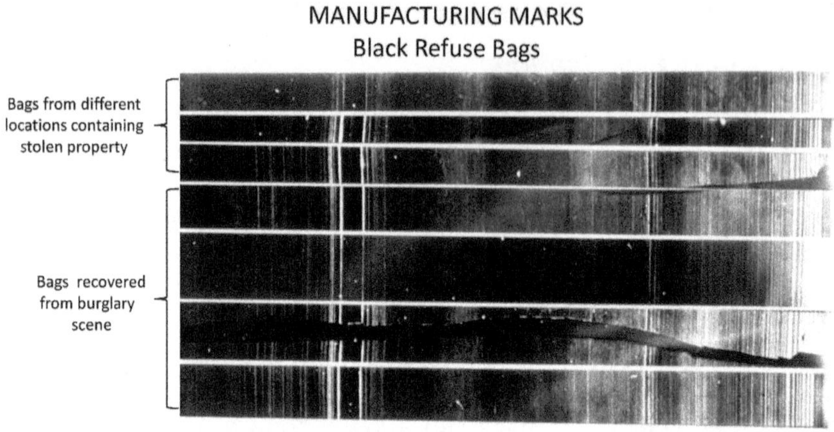

Figure 8.10 Strips from seven black plastic bags showing extrusion detail. The bottom four sections are from a burglary scene. The top three are from different locations and contained items of stolen property.

been opened out by cutting down the sides and then viewed under circularly polarised light. The bags have been sorted into two sequences based on the changes in the coloured bands.

For black or highly coloured bags, a light table is used with a high-intensity white light source and the bags are viewed in a darkened room. The manufacturing detail is seen as a random pattern or a variable pattern of parallel lines. Figure 8.10 shows thin strips removed from each of seven black plastic bin bags. The bottom four sections are bags recovered from a burglary scene; the top three are from different locations and contained items of stolen property. The matching of extrusion detail shows the association of these bags.

Transmitted light microscopy can be used to view scratches on the surface of bags. Heat seals and printing detail on the bags can also be examined and compared for possible features relating to the manufacturing process.

8.5.3.2 Tape and Cling Film Examination

In a tape examination, the initial comparison will be of general appearance, colour, width and surface texture of the tape. If the roll of tape cannot be excluded, the examiner will then usually look for a physical fit of the tape from the scene back to the roll or to other linked scenes. In the laboratory, the tape ends are often stuck to sheets of clear plastic so they can be overlaid. Manufacturing detail and surface scratches may be used to confirm the fit.

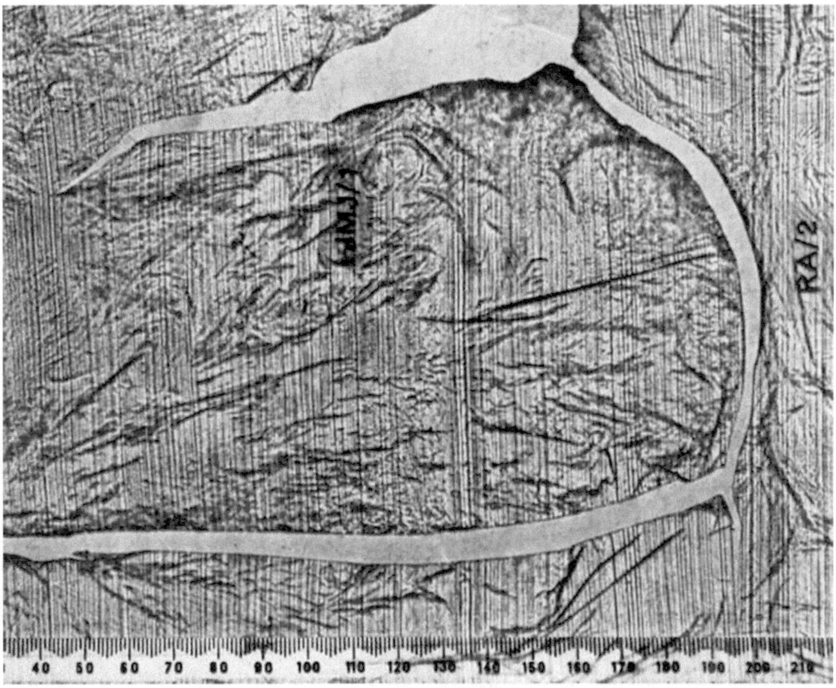

Figure 8.11 The comparison of a large piece of cling film and a roll of cling film.

In a cling film comparison, this again will usually be to look for physical fits which may be confirmed with manufacturing detail using the shadowgraph technique. In this technique, a point source of white light is used, the discontinuities in the film deflect the light and this is recorded. Cling film is easily distorted by creasing and stretching, which can make these examinations difficult. Figure 8.11 shows the comparison of a large piece of cling film with the end of a roll of cling film. There is excellent agreement of the manufacturing detail.

8.5.4 Evaluation

Knowledge of the manufacturing process permits the identification of various marks on mass-produced plastic bags. Sequentially made bags can be related to each other by marks present on their cut ends or sides. When all class characteristics agree, and consecutive marks correspond on two bags, it can be concluded that they were originally one piece of plastic. Owing to the large numbers of bags produced by an individual production line, care has to be taken in the assessment of the evidential value of those which are not consecutive but share the same manufacturing source. Generally, this depends on the variation of the features examined, which has to be judged by the examination of the bags in the case. This may be impossible in those cases where only a few bags have been recovered.

8.6 Physical Fits

Physical fits, sometimes referred to as 'mechanical fits' can be encountered in a wide variety of circumstances. Such an examination is required when it is necessary to show that two or more pieces of material which have been broken, torn, cut or in some way separated, were once joined together. In a very simple way, these can be separated into two groups depending on the complexity of the detail, and the knowledge and skill of the examiner. The first group can be described as 'jigsaw fits' and, as the name suggests, these are broken items that can be obviously fitted back together due to the characteristic profile of the broken edges or detail on the surface such as a pattern or texture. Examples include fitting together large pieces of broken patterned glass or torn pieces of printed paper from a magazine.

The second group requires a level of knowledge and skill associated with the material being considered. It may require manipulation of very small fragments of material, an examination of surface detail or the fractured surface using a microscope and the use of casting and recasting to enable a comparison to take place. Examples include pieces of tape as shown in Figure 8.12, a tool tip that has broken off, or a knife blade which has become detached from its handle. There are examples of where a bottle or weapon has been broken at the scene and it is necessary to reconstruct the item to link the scene with a suspect or to determine what happened at the scene, for example, reconstruction of the broken pieces of glass from a petrol bomb.

8.7 Conclusion

This chapter has covered the main areas of marks and impressions work, explaining the fundamental requirements of any examination and comparison, following a clear, structured approach. Marks and

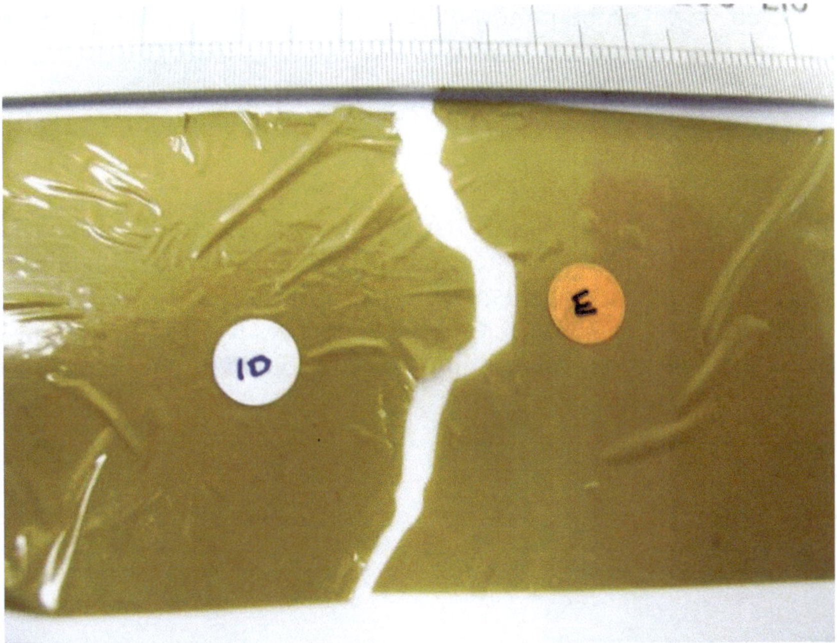

Figure 8.12 Physical fit of a piece of tape back to a roll.

impressions have been a critical area of forensic work for many years and has been routinely used in nearly all offence types, from criminal damage to murder and terrorist incidents. It remains one of the few areas where the forensic scientist can express the evidential result as conclusive, based on their own subjective expert opinion with all of the associated challenges that this brings.

Further Reading

D. Baldwin, J. Birkett, O. Facey and G. Rabey, *The Forensic Examination and Interpretation of Tool Marks*, John Wiley & Sons, Ltd, West Sussex, 2013.

W. J. Bodziak, *Footwear Impression Evidence*, CRC Press, Boca Raton, London, 2nd edn, 2000.

Home Office Centre for Applied Science and Technology, *The Fingerprint Source Book*, London, 2012.

Association of Forensic Science Providers, Standards for the Formulation of Evaluative Forensic Science Expert Opinion, *Sci. Justice*, 2009, **49**, 161–164.

I. W. Evett, J. A. Lambert and J. S. Buckleton, A Bayesian Approach to Interpreting Footwear Marks in Forensic Casework, *Sci. Justice*, 1998, **38**, 241–247.

K. J. Farrugia, K. A. Savage, H. Bandey and N. Nic Daéid, Chemical enhancement of footwear impressions in blood on fabric – part 1: protein stains, *Sci. Justice*, 2011, **51**(3), 99–109.

K. J. Farrugia, K. A. Savage, H. Bandey, T. Ciuksza and N. Nic Daéid, Chemical enhancement of footwear impressions in blood on fabric – part 2: peroxidase reagents, *Sci. Justice*, 2011, **51**(3), 110–121.

K. J. Farrugia, H. Bandey, K. Savage and N. NicDaéid, Chemical enhancement of footwear impressions in blood on fabric – part 3: amino acid staining, *Sci. Justice*, 2013, **53**(1), 8–13.

9 Trace and Contact Evidence

Sarah Jacob

Formerly of Forensic Access, Aspect House, The Quadrangle, Grove Business Park, Wantage, OX12 9FA, UK
Email: sarah_ruth_jacob@yahoo.co.uk

9.1 Introduction

The discipline of trace evidence refers to the recovery and examination of minute samples of material other than those which are biological in nature. The trace material under consideration may be particulate or liquid. Some trace examinations, for example glass, paint or fibre examinations, are routine and subject to standard operating procedures in United Kingdom (UK) forensic laboratories. However, any trace material can, in theory, become the subject of a forensic examination and more unusual trace evidence types would be considered an infrequently encountered test.

Trace evidence examinations can involve identification of the nature of the trace material present or they can involve comparison of the trace material in question to a control sample recovered from a crime scene. Trace evidence comparisons rarely provide a conclusive link because the items for comparison tend to be mass-produced. However, such examinations regularly provide strong evidence of a link or a conclusive elimination, both of which could be extremely useful in an investigation. Trace evidence comparison also has the advantage that, given enough background detail, it can provide

Crime Scene to Court: The Essentials of Forensic Science, 5th Edition
Edited by Niamh NicDaeid and Peter C. White

Published by the Royal Society of Chemistry, www.rsc.org

information at activity level. That is to say, the interpretation of the trace evidence can address the activity which may have caused the transfer of the trace material rather than source level which would merely consider the possible origin of the trace substance in question. For example, for a glass case, given enough information it is possible to address the issue of whether or not Mr X broke the window rather than just whether or not any recovered glass from his clothing originated from the broken window in question.

In this chapter, the general principles of trace evidence examination in broad terms including transfer, persistence, contamination and case assessment will be considered and the analytical techniques a forensic scientist has in their armoury to analyse trace evidence samples will be discussed. Finally, there follows a more detailed look at various different types of trace evidence routinely and not so routinely examined at the laboratory.

9.2 Basic Principles of Laboratory Examination

The laboratory examination of the various types of trace evidence follows the same fundamental principles as described below.

9.2.1 Case Assessment

Upon initial receipt of a trace evidence case at the forensic science laboratory, the case is assessed prior to the commencement of any laboratory examination. Case assessment will consider the expectation of finding the trace evidence in question on the examined item within the framework of case circumstances. This will be dependent on factors such as case scenario based on the scene examination and witness statements; the suspect's version of events; transfer; persistence and time delays. It may be that given the case circumstances, the forensic scientist will come to the conclusion that the finding of any trace evidence will be unlikely and recommend that no further examination for trace materials is undertaken. Alternatively, the scientist may consider that the finding of any trace evidence will not assist the investigation given the case circumstances and again recommend that no further work relating to these evidence types is carried out. If it is the forensic scientist's recommendation to go ahead with the full laboratory examination, the case strategy will be decided at this stage including the order of examination of the items. If the case is multi-disciplinary, for example, items may also be required for examination of biological fluids, liaison with other forensic scientists will occur at this stage to determine the over-arching case strategy.

9.2.1.1 Transfer

The likelihood of transfer of trace material will depend on case circumstances including the nature and degree of contact with the source material. Forcible contact between an item and the source, for example a tool coming into contact with a painted window frame or used to break a window, may result in the acquisition of embedded or impacted material by the item. An example of paint crushed and embedded on a wrecking bar used to force entry *via* a window is shown in Figure 9.1. The likelihood of an item acquiring embedded fragments will also be dependent on the nature of the surface of the item.

For material thrown or sprayed over a person's clothing, for example glass from a breaking window, paint sprayed from a graffiti can or particles of metal from an Automatic Teller Machine (ATM) being attacked with an angle grinder, the amount transferred will be dependent on the proximity of the person to the source. Direct transfer of material to an item through contact will be dependent on the degree and length of contact and the 'shedability' of the source.

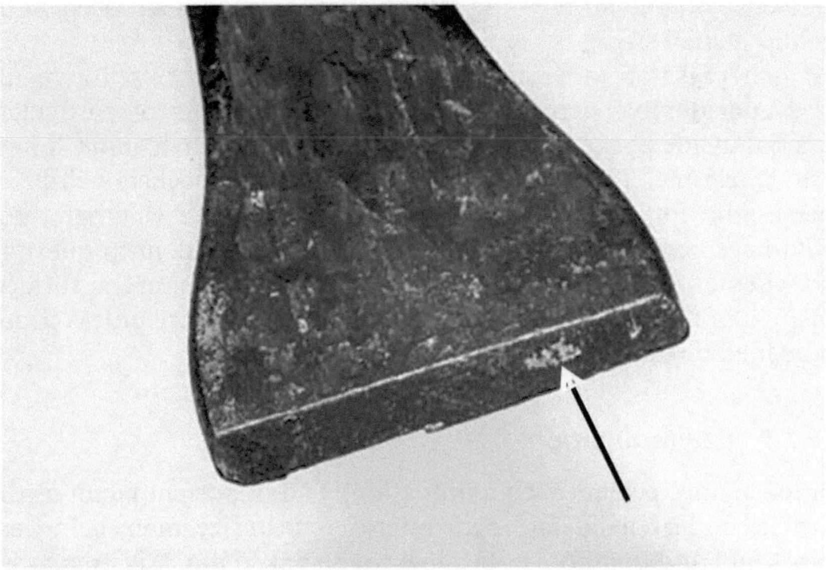

Figure 9.1 Crushed paint on the end of a wrecking bar. The paint is the light area at the right-hand side of the bevel (identified by the arrow).

For example, a flaking paint surface will likely transfer paint fragments to an item of clothing whereas a pristine freshly painted surface will not.

9.2.1.2 Persistence

Impacted and embedded material on an item together with liquids which have wetted the item and then dried may well persist for some considerable period of time depending on what that item is subsequently used for.

Loose fragments of trace material will be lost quickly from the surface of an item. The persistence of several loose particulate evidence types on clothing has been studied in more detail, including glass and fibre evidence. It has been shown that loose fragments of glass and fibres on clothing are lost very quickly within a matter of hours to leave a residual amount. The rate at which this occurs will depend on the surface characteristics of both the receiving and donor materials, the degree of movement and retentiveness of the item. For example, for an item of clothing worn by a suspect, the persistence of transferred trace materials will depend on the degree of activity of the wearer. A person running will most likely shake fragments off the surface of their clothing more quickly than a person moving slowly. A rough-surfaced item of clothing, such as a woolly jumper, will retain fragments on the surface for longer than a smooth-surfaced item, such as a shell-suit.

If an item is not worn extensively then it may retain loose fragments for a considerable period of time. Loose fragments of particulate material inside pockets will likely be retained for considerably longer periods of time. The finding of fragments inside pockets will be of note as the forensic scientist will need to consider whether these could have been in the pockets from before the incident in question and whether they could be a source of any fragments on the surface. Washing of items will most likely remove trace evidence unless firmly embedded or impacted onto the surface of an item.

9.2.1.3 Contamination

For loose trace evidence, contamination is an important factor as the particles under consideration are microscopic in size, many being less than 1 mm and therefore not visible to the naked eye. It is imperative that the forensic scientist can say that the fragments of material were acquired at the time of the incident in question and not subsequently, as a result of contamination. During case assessment, the forensic scientist will flag up potential contamination of the items prior to receipt at the laboratory. For example, when examining a suspect's clothing, the forensic scientist will seek to identify any contact between the suspect and an individual or item having prior contact with the source of the trace material in question.

Transfer of loose fragments of particulate material directly from one item to another is known as primary transfer. Studies for some types of trace evidence, including glass and fibres, have shown that it is also possible to transfer particles between items *via* an intermediate person or object. This is known as secondary or indirect transfer. The number of fragments transferred *via* secondary transfer is generally much less than by primary transfer. Therefore, generally the finding of a large quantity of loose fragments on the surface of an item will be indicative of them having been acquired by primary transfer rather than by contamination through secondary transfer even if this is a possibility.

9.2.2 Recovery of Trace Material

The technique used for recovery of any trace material will be dependent on the nature of the material in question as described below.

9.2.2.1 Loose Particulate Material

Loose particles on items such as clothing are recovered by shaking and brushing the surface of the item and transferring any debris collected into a labelled storage receptacle such as a petri-dish or paper fold. A small paint brush or toothbrush can be used to brush the item. The debris can be collected for transfer to the storage container in either of two ways. The item can be held over a sheet of brown paper, the surface brushed and then the piece of brown paper carefully lifted and the debris poured into the storage receptacle. Alternatively, the item can be hung over a piece of apparatus called a cone. This is a large steel funnel with a highly polished surface. A labelled petri-dish is placed underneath the hole at the bottom. Material on the surface of the item is swept into the cone and then brushed into the petri-dish at the bottom. An image of the cone in use at the laboratory is shown in Figure 9.2.

Debris collected from the surfaces of an item can then be searched for any particles of interest under a low power microscope. It is important to note that it is not possible to say where any loose debris collected from

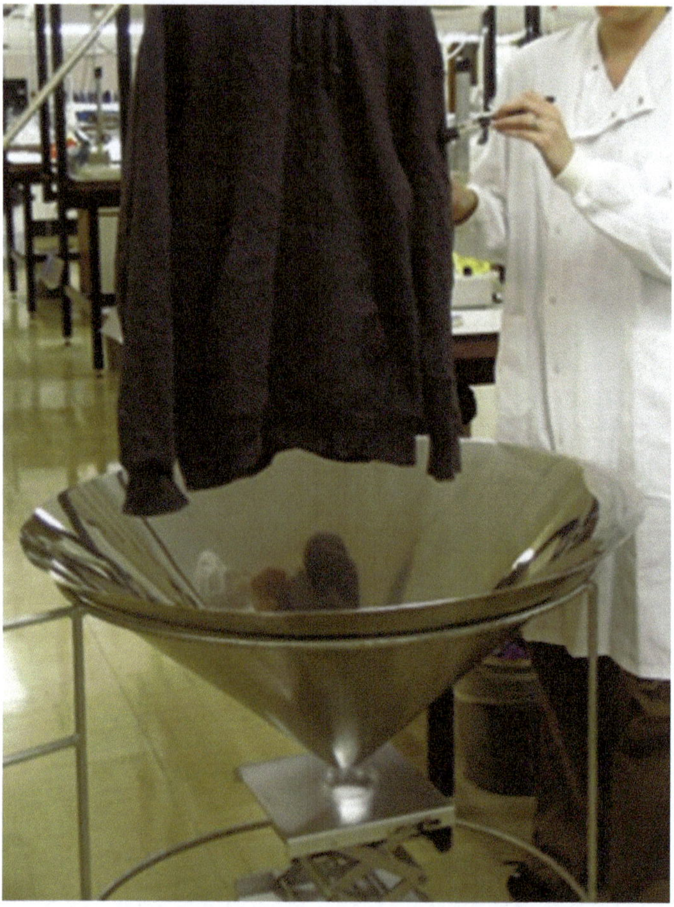

Figure 9.2 A stainless steel cone used for collection of debris from clothing. An item is suspended above the cone and any particles adhering to the item are brushed from it and collected at the base of the cone.

the surface of an item in this manner originated from on the item. For items having pockets or compartments such as clothing or bags, the individual pockets or compartments can each be brushed separately for transfer to separate individually labelled storage containers.

In some instances, the position of a particle on an item will be important to the case interpretation, particularly fibres cases. For example, the position of extraneous fibres on clothing could provide support for contact with another individual in an assault case alleged to have happened in a certain manner. Extraneous fibres are recovered from the surface of clothing or other items such as car seats by taping. Taping conventionally uses lengths of clear tape wrapped around a roller. The tape is rolled across a small area of the item. The tape is then removed from the roller and placed flat onto a sheet of clear acetate and the sides sealed with lengths of clean clear tape. The area of the item which the tape relates to is clearly labelled on the acetate and a diagram of the item. This process is repeated until the entire surface of the item has been taped. Care should be taken so as to not overload the tape with extraneous material from the surface. Lengths of tape which are overloaded in this manner will be difficult to search for particulate material. Tapings can then be searched under a low power microscope for any relevant particulate material. More recently bespoke taping kits have been developed for the recovery of extraneous fibres from the surfaces of items.

9.2.2.2 Liquid Stains and Impacted Material

Impacted, crushed or smeared materials adhering to items, including those which have been transferred between items during forcible contact or those which have been acquired wet by an item and which have subsequently dried, can be searched for using a low power microscope. For the examination of large items, for example, car body parts for the presence of impacted paint relating to a road traffic accident, specialist low power microscopes exist which are on a long, flexible arm to facilitate the examination. Any particulate material found can be removed using a scalpel and stored in a clean, labelled container such as a petri-dish. Areas of liquid staining can be cut out of an item and extracted using a suitable solvent for further analysis.

9.2.2.3 Liquid Traces and Fine Particles

Liquid traces, such as condom and other sexual lubricants and cash-box dyes, can be sampled using a standard swab. This technique is also suitable for the sampling of very fine particulate material which cannot be recovered using a scalpel, for example, possible traces of lead on the hands of an individual suspected of the theft of lead from a roof.

9.2.3 Techniques of Examination

The forensic scientist has a number of techniques in their armoury which they can call upon to examine and compare trace materials. The choice of techniques will be dependent on the nature of the material in question. Forensic science laboratories in the UK are busy places dealing with a high volume of cases which are examined under contract with set prices. Examination techniques must therefore be cost-effective and suitable for a high throughput of cases. Techniques should also be non-destructive where possible so that the examined material is available for future re-examination on behalf of the defence. Techniques utilised at the laboratory are fully verified for use and standard samples are examined at the same time as casework samples to confirm the accuracy of the results obtained.

9.2.3.1 Random Sampling

Where a small number of loose particles or areas of staining or impacted trace material is found on an item, it will be possible to examine all the recovered fragments or areas. However, where a large number is found, it is not practical to examine them all. Therefore, examination of areas or fragments of interest is based on the principle of random sampling. A representative number of areas or fragments will be chosen at random from the item.

9.2.3.2 Examination Techniques

Low power microscopy with a magnification of up to $\times 50$ is used to identify the nature of particulate material and examine overall morphology and appearance. High power microscopy with a magnification of up to $\times 500$ can be used for detailed examination of particles. It can be used in conjunction with different lighting, for example, blue light and ultraviolet (UV) light, to compare how particles or different areas of the particulate material behave. High power microscopy is utilised, for example, in the examination and comparison of fibres and paint. Interference microscopy is used for the examination of original glass surfaces to determine whether the surface in question originates from a flat or non-flat source of glass.

Glass Refractive Index Measurement (GRIM) can be used to differentiate between glass samples originating from different sources. The technique works by immersing the glass fragment in an oil which has a known linear correlation of refractive index and temperature. The glass itself will only show a negligible change of refractive index with temperature. By heating and cooling the glass fragment in the oil and noting the temperature at which the glass matches the refractive index of the oil, it is possible to determine the refractive index of the glass.

Scanning Electron Microscopy (SEM) can be used to identify the elemental composition of a particulate sample. It does this by subjecting the sample to a beam of electrons which will interact with elements in the sample to release X-rays. The energies of the X-rays produced are indicative of elements present in the sample. This technique is suitable for identifying elements present in glass, paint and other particulate samples such as metal alloys.

Fourier Transform Infrared (FTIR) spectroscopy is used to determine the organic component of a sample. This technique subjects the sample to a beam of infrared light and measures the wavelengths which are subsequently absorbed. These will be dependent on stretching, bending and rocking motions of the bonds making up the structure of the sample. This technique is routinely used to identify the binder component in paint samples as well as for the analysis of other organic samples such as condom lubricants and adhesives.

Chromatographic techniques are used to separate out and identify components present in mixtures. Gas Chromatography–Mass Spectrometry (GC–MS) can be used to separate and identify components in complex organic mixtures. The components of the mixture are separated as they pass through the Gas Chromatography (GC) column and identified by comparison of the retention time and mass spectrum to known reference samples. Thin Layer Chromatography (TLC) can be used to separate out the individual dyes making up a coloured dye or ink sample. A sample of the dye is placed on a chromatography plate and one end of the plate is immersed in a suitable solvent. As the solvent moves up the plate it separates out the individual dye components present. It is suitable for the analysis of liquid samples such as extracted dyes from fibres, security cash-box dyes and pen inks.

Microspectrophotometry subjects the sample to a beam of visible light and measures the wavelengths of light which are absorbed. It is routinely used for the analysis of extracted dyes from samples such as fibres.

Other analytical techniques could, in theory, be used for the analysis of trace materials. However, techniques of analysis are required to be fully validated prior to use (a time-consuming and costly process) and forensic laboratories are functioning within a budget to provide cost-effective, rapid, non-destructive analysis of samples. Routine analysis of trace samples will therefore be limited to a number of techniques suitable for the analysis of the broadest range of samples possible.

9.2.3.3 Contamination

For loose particles of trace evidence, contamination at the laboratory is an important consideration. Stringent steps are taken at the laboratory to eliminate the possibility of contamination. Control samples are opened after suspect items have been examined to ensure there is no accidental contamination from the control. Control samples are examined in a separate room from suspect items. When examining items for loose particulate material, the outer packaging is blanked prior to opening. In other words, debris is collected from the outer surface of the packaging and searched for any relevant trace material to confirm that none is present. This is to ensure there is no transfer of trace material from the outside of the bag to the item as it is being removed from the packaging. Cones and benches are cleaned before and after use. Cone and bench blanks are also carried out. The examining forensic scientist would wear a clean laboratory coat and gloves as standard.

9.2.4 Interpretation

The evidential significance of the findings will depend on the likelihood of finding the trace material in question on the examined item. In very general terms, factors for consideration are:

- The commonality of the trace material – this can be assessed against a relevant collection kept at the laboratory or be based on the knowledge and experience of the forensic scientist.
- The amount of trace material found – a large number of areas of staining or fragments may add to the significance of the findings depending on case circumstances.
- The presence of non-matching sources on the examined item – this may increase the chance of a coincidental match which will reduce the evidential significance of the findings.
- The nature of the trace material given the case circumstances, for example, is the trace material under consideration loose fragments, crushed material on a surface or areas of staining on an item. The appearance or position of the trace material may support a particular version of events which may add to the evidential significance of the findings.

Trace evidence in the UK is interpreted using the model of Case Assessment and Interpretation (CAI).

The prosecution and defence viewpoints are defined at the outset. They are expressed within a hierarchy of propositions, which address source, activity and offence level propositions. Source level propositions will consider the origin of the material in question. For example, prosecution and defence propositions for a glass case interpreted at source level may be:

Prosecution viewpoint: The glass on the T-shirt originated from the broken window.
Defence viewpoint: The glass on the T-shirt originated from some other source.

Activity level propositions will consider how the trace material in question was acquired by an item. Typical prosecution and defence propositions for a glass case interpreted at activity level may be:

Prosecution viewpoint: The wearer of the T-shirt broke the window.
Defence viewpoint: The wearer of the T-shirt did not break the window.

Interpretation of the case at activity level will require sufficient background information in the form of information from the scene or witness statements to formulate the two propositions. If this is not available, the case should be interpreted at source level.

Offence level propositions, for example, whether or not the wearer of the T-shirt committed the burglary, are never considered by the forensic scientist as this is the domain of the court.

Trace materials are mass-produced. There will be other sources of the material under consideration and the evidence will not provide a conclusive link. An exception to this is paint evidence where a sufficient number of sequential layers in a sample can make it unique. In the absence of a conclusive link, a conclusive elimination or an inconclusive result, the strength of evidence in the UK is articulated by the verbal scale:

Limited support; moderate support; moderately strong support; strong support; very strong support; extremely strong support.

The verbal scale can be used to express evidential significance in support of either the prosecution or defence proposition. For example:

- The findings provide strong support for the proposition that the wearer of the jumper broke the window.
- The findings provide limited support for the proposition that the wearer of the jumper did not break the window.

9.2.5 Bayesian Approach to Interpretation

Forensic evidence can be interpreted using Bayesian theory. A likelihood ratio is calculated which is the probability of obtaining the evidence if the prosecution viewpoint is true divided by the probability of the evidence if the defence viewpoint is true. The numerical value obtained correlates to points on the verbal scale. The Appeal Court Judgement in the case of R *v* T states that Bayesian theory can only be used where there is robust data on which calculations can be based.

There is sufficient background data to permit the use of Bayesian theory in the interpretation of glass cases. At source level, the likelihood ratio will be dependent on finding the glass type in question by chance. The background data which provides this information is the frequency of occurrence of the glass type amongst a control glass reference collection retained at the forensic science laboratory. The calculation becomes more complex at activity level and must take into account transfer and persistence of the glass on the item in question. The transfer of glass fragments to items of clothing has been the subject of considerable research. This has led to the development of equations for the calculation of a likelihood ratio at activity level if a population or populations of glass fragments are found. In very general terms, the calculation considers the probability of finding a certain number of groups (types) of glass on clothing along with the probability of finding a particular number of glass fragments within a group of interest. The values for these terms are taken from information provided by background surveys of glass fragments on clothing.

The calculation also considers the likelihood of transfer of a certain number of glass fragments to the item in question given the case circumstances and in particular, how the glass object was broken. A value for this is based on information from studies relating to the transfer of glass to clothing. Lastly, the calculation will consider the frequency of occurrence of the glass type based on information from the control glass reference collection.

For all other trace evidence types, the background data is not yet sufficient to permit the use of Bayes theorem and the interpretation of evidence is based on the skill and experience of the reporting forensic scientist.

9.2.6 Bias

Cognitive bias is a recognisable behaviour in the Human thought process. It happens when an individual is influenced either consciously or sub-consciously by outside factors when drawing conclusions. There is a risk of cognitive bias in trace evidence examinations. Background information is required when defining case strategies, for example, how a window was broken in a glass case or how an assault occurred in a fibres case. Opinions are formed in light of such information which could include contextual bias. Many of the tests performed during trace evidence examination are subjective including microscopic examination. Empirical findings based on instrumental methods are less prone to bias but assessment of the correlation between analytical findings may be liable to a degree of subjectivity. This could lead to expectation bias where the expectation of what an individual will find directly influences actual findings; confirmation bias where the scientist tests for confirming evidence rather than potential conflicting evidence; role effects where a scientist aligns themselves with either the prosecution or defence team rather than remaining impartial.

To mitigate against these risks of bias, it is good practice for the forensic scientist to adopt the CAI approach where both prosecution and defence viewpoints are considered. The expectation of the findings in light of each scenario is defined at the outset during pre-assessment of the case. To guard against contextual bias, the forensic scientist should only be party to sufficient information as is necessary for case assessment. Laboratory examination should be based on standard operating procedures with fully validated methods where possible. Where appropriate, use of statistical evaluation should be adopted, for example, assessment of the correlation in refractive index between recovered and control glass samples. Forensic scientists should be well-trained and undergo regular competency testing. Casework findings, particularly those which are critical to the interpretation of evidence or which are subjective in nature, should be independently peer-reviewed without influence from the reporting scientist. Pre-assessment of the case and the interpretation of findings should also be independently checked.

9.3 Glass Evidence

Glass encountered in casework may originate from flat window glass, patterned glass or from non-flat container glass such as drinking glasses or bottles. Flat window glass may be single-paned, double-glazed or from a laminated source. Laminated glass comprises two layers of glass each of which is bonded either side of a layer of clear plastic. It is a type of safety glass that holds together when shattered and is used for vehicle windscreens and shop windows. Toughened glass is another type of safety glass which forms small cubes when it breaks rather than pointed shards. It is typically used in vehicle side windows. The modern method of flat glass manufacture is the float process. The constituent components of the glass are melted together, and the molten mixture poured onto a bath of molten tin and allowed to cool. The surface of the glass sheet in contact with the tin bath will absorb tin into the uppermost layer.

Offences where glass evidence can be of use include burglaries involving a broken window; assault cases involving glassing; road traffic accidents where glass from a broken car window can be used to determine whether there is evidence to support the view that a car was involved, or an individual was inside the vehicle at the time. Types of glass encountered in forensic casework are microscopically sized loose glass fragments and embedded or impacted glass.

Loose glass fragments can be acquired by an item when a window or other glass object is broken. When a window breaks, many glass fragments will travel away from the breaker but some small particles of glass will be showered back towards them in a process known as backward fragmentation. These fragments can travel up to a distance of approximately 3 m.

9.3.1 Recovery of Glass Control Samples

Glass control samples should be recovered from all broken sources of glass. Even if two windows or glass objects look the same, they may be differentiated at the laboratory. Glass should be recovered from the item itself rather than off the ground. Six pieces of glass should be recovered for examination. These should be from different points around the break if sampling a window in order to get a range of analytical readings for comparison. Full thickness pieces of glass should be taken where possible to allow examination of original surfaces. Both sides of laminated and double-glazed windows should be recovered and labelled appropriately as this information may be relevant to the interpretation of the findings.

9.3.2 Case Assessment of Glass as Evidence

Case assessment will consider the likelihood of transfer of glass to an item. For loose glass fragments originating from a broken window, the forensic scientist will consider the likelihood of transfer of glass to the clothing or hair of the breaker. This will depend on how the window was broken which will dictate proximity of the breaker to the window. For example, the likelihood of transfer of glass fragments from a window broken using a tool with the breaker standing close by will be greater than to a person throwing a stone from a distance to break the window. If an individual subsequently reaches or climbs through a broken window, one might expect direct transfer of glass to their clothing if it comes into contact with the broken glass. The size of the window and size of the break will affect how much glass is ejected from the window. The position of the break will influence case strategy and what item is examined first. For a break in a window which is high up, the forensic scientist will examine hair combings or upper body clothing first. For a window which is low down, the forensic scientist may view kicking as a potential method for the window having been broken and examine the suspect's shoes for embedded glass.

The type of glass will influence how much glass is ejected as an object breaks. A laminated glass window will break less easily than a single-paned window and therefore eject fewer small particles of glass. The case assessment process will also consider the persistence of loose glass fragments. In particular, the forensic scientist will assess the activity of the wearer and the retentiveness of the fabric from which clothing items are made. Related to this, the forensic scientist will consider the time delay between the incident having occurred and suspect items being seized. If the time delay is a matter of days and the clothing items are thought to have been worn during the interim period, it is unlikely that any glass examination of clothing items will take place. Potential contamination issues will also be considered.

For embedded or impacted glass fragments, the forensic scientist will consider the nature of the contact between the item and the glass object and the type of surface on the item. The forensic scientist will also consider whether the item may have been used during the time between the offence and seizure of the item and whether the item has been packaged correctly to preserve any traces which may be present. These factors will influence the likelihood of finding embedded or impacted glass and determine case strategy, specifically which areas of the item to examine.

9.3.3 Laboratory Examination of Glass

Debris is collected from items and searched for any loose glass fragments of a freshly broken appearance. Freshly broken glass fragments

Figure 9.3 Freshly broken glass fragments typical of those encountered in forensic casework with a pinhead for size reference.

will appear clean and sharp-edged. Figure 9.3 shows freshly broken glass fragments typical of those encountered in forensic casework with a pinhead for size reference.

Debris is collected from pockets of clothing items to assess whether fragments of glass recovered from the surface could have originated from the pockets during transit. Items examined for embedded or impacted glass are searched under a low power microscope.

Recovered glass fragments are compared against the control on the basis of random sampling. A representative number of fragments from those recovered will be selected for GRIM. If a group of glass fragments are found to correspond to the control sample of glass in this respect, at least one of these will be selected at random for elemental analysis using SEM.

Further tests can be carried out on the group of corresponding fragments to determine whether there are other points of similarity with the control sample. Fragments bearing an original surface, indicating that they originated from the surface of the glass item from which they came, can be examined using interference microscopy. This will determine whether they originate from a flat or non-flat source of glass in line with the control sample. Original surfaces can also be viewed under UV light to determine whether they fluoresce which is indicative of the presence of tin and manufacture by the float process. Recovered fragments can also be annealed, a heat treatment process involving heating and subsequent cooling of the glass fragment, and the refractive index remeasured. A change in refractive index following the annealing process will indicate the glass fragment has originated from a source of toughened glass.

9.3.4 Interpretation of Glass Evidence

Glass evidence will not provide a conclusive link; glass is a mass-produced item and there will be other sources of the glass in question. However, the finding of a population of corresponding glass fragments is significant and may be sufficient for strong or very strong evidence. If there are several populations of glass corresponding to multiple controls, the evidence can provide extremely strong support.

The significance of glass evidence at source level *i.e.* whether or not the recovered glass originated from the glass source in question, be it a broken window or other glass item, is assessed on the basis of the commonality of the glass type. This is gauged by comparison of the refractive index, elemental composition, and other general characteristics against the collection of information retained at the laboratory for previously examined control glass samples. This information should be used as a guide only as the collection is limited in the number of glass samples it contains and is also biased in that it contains glass samples seized in relation to a crime. These will mostly originate from broken windows and will not be representative of broken glass in the general environment.

If there is sufficient information from the case circumstances to indicate how the glass may have been acquired by the item, the case can be interpreted at activity level. For example, whether or not a tool was used to break a window or whether or not the wearer of the T-shirt broke the window. The significance of glass evidence at activity level is assessed on the basis of:

- the commonality of the glass type
- the number of glass fragments found
- the number of distinguishable controls having corresponding glass fragments
- the number of non-corresponding sources
- the appearance of embedded or impacted glass.

The evidential strength will increase if there are several populations of glass matching multiple controls. Conversely, the evidential strength will decrease if there are fragments from non-corresponding sources since this will suggest habitual contact with broken glass which will in turn increase the chance that any corresponding fragments are present coincidentally. There is sufficient background information to permit the use of Bayesian inference when interpreting glass cases at both source and activity level.

Studies have shown that it is possible to break a window and for the clothing of the breaker to not acquire glass. The evidence as to whether or not a person has broken a window may therefore be inconclusive in the absence of corresponding glass fragments.

9.3.5 Case Scenario

A hit and run incident occurred in which a pedestrian was knocked down and killed whilst crossing the road. A short while later, the owner of a vehicle contacted the police to report his car as stolen. It was subsequently found parked in a nearby supermarket car park. The car had a broken windscreen from which police recovered a control sample of glass. The owner of the vehicle was arrested and his T-shirt seized. The owner denied being the driver of the vehicle at the time the accident occurred.

The items were submitted for examination at the laboratory. Case assessment showed that there was no contamination of the items prior to their receipt. Debris was collected from the T-shirt and searched for the presence of freshly broken glass fragments. Twenty fragments were found.

The control sample of glass was examined. It was found to be laminated; the inner and outer sides of the windscreen had been labelled. The two sides of glass were distinguishable from one another on the basis of appearance. The outer side was tinted pale green whilst the inner pane of glass was colourless. Both were examined under UV light and found to be manufactured by the float process. The refractive index and elemental composition of both panes of glass were measured and they were found to be distinguishable from one another in both respects.

A representative sample of eight recovered fragments was examined by GRIM. Seven were found to correspond to the control sample originating from the inside of the windscreen whilst the remaining fragment corresponded to the outer pane. A representative fragment from the seven corresponding to the inner pane and the fragment corresponding to the outer pane were chemically analysed. Both were found to correspond to the relevant pane of glass in chemical composition. One of the recovered fragments had an original surface. This was examined microscopically and under UV light. It was found to have originated from a flat source of glass manufactured by the float process in keeping with the control.

As a representative portion of the glass fragments was examined, there were likely to be other fragments of glass corresponding to the control samples amongst those not sampled. Comparison of the two types of glass in the control sample to the control glass reference collection at the laboratory showed that both had been rarely encountered. The finding of both types of glass increased the evidential strength of the evidence. The majority of glass fragments corresponding to the pane of glass on the inside of the windscreen was in keeping with expectations had the wearer of the T-shirt been inside the vehicle at the time the accident occurred. In conclusion, there was very strong support for the view that the wearer of the T-shirt was inside the vehicle at the time the accident occurred rather than not. It was not possible to comment on whether or not the wearer of the T-shirt was the driver of the vehicle as the glass would likely have sprayed throughout the inside of the vehicle as the window was breaking.

9.4 Paint Evidence

Paint is a coating added to items to improve their appearance and/or protect the item. Paint is often added to a surface as part of a layered sequence incorporating primer, undercoat and topcoat. Items may be subsequently repainted which builds up the layer structure present. In general terms, the constituent components of a paint are the binder, solvent, pigments, extenders and other additives. The binder is the film-forming component of the paint. It is the vehicle that carries the pigments and other constituent components and dries or cures to hold them in place. Binders are organic in nature, examples being alkyds, polyurethanes, acrylics and polyesters. The solvent is a liquid substance added to paint to give it the necessary consistency for application. It evaporates from the paint coating once applied. Pigments are small particles which add colour to the paint layer. Extenders are particles of inorganic substances added to paint to improve its mechanical performance. Other additives are added to impart certain other desirable characteristics to the paint layer if required, for example fire retardants, anti-fungal agents and UV stabilisers. An undercoat is designed to protect the surface and will typically contain more extenders *versus* a topcoat which is decorative and will include pigments.

Paint encountered in casework may originate from any painted surface including doors, windows, motor vehicles or tools. Offences where paint evidence can be of use include burglary where a person has levered open a painted door or window frame with a tool or climbed through a painted window frame in gaining entry; road traffic accidents including examination for vehicle-to-vehicle transfer of paint, or search of clothing from a victim of a road traffic accident for paint from a motor vehicle; criminal damage involving wet paint or sprayed graffiti paint. Types of paint encountered in forensic casework include microscopically sized loose paint fragments, impacted crushed paint or paint which has gone onto an item wet and subsequently dried. Paint originating from spray paint cans will appear as microscopically sized balls from the paint aerosol produced by the can.

9.4.1 Recovery of Paint Control Samples

Paint control samples should be recovered from as close to the area of damage or point of transfer as possible. Paint from other locations on the item may show subtle differences in the appearance of the paint due to weathering effects or inconsistencies in the paint layers present *e.g.* a car may have been touched up over a small area following an accident. Samples should be taken down to the base layer so as to capture all layers present.

9.4.2 Case Assessment of Paint as Evidence

Case assessment will consider the likelihood of transfer of paint to an item. The transfer of paint will depend on the degree and nature of the contact between the item and the paint source. It will also depend on how readily paint fragments are shed from the painted surface. Case assessment will also consider the persistence of the paint on the item. This will depend on the subsequent use and movement of the item and its surface texture. For loose paint on clothing items, persistence will depend on the activity of the wearer and the retentiveness of the fabric from which the item is made. For an item such as a tool, the forensic scientist will consider whether the tool may have been used during the time between the incident having occurred and the item being seized. Related to this, the forensic scientist will consider the time delay between the incident and seizure of the items. Depending on case circumstances, it may be that the time delay is too long for any examination for paint fragments to be viable. In the case of dried or impacted paint traces, the forensic scientist should consider whether the packaging of the item is sufficient to preserve any paint traces which may be present. The case circumstances will influence case strategy, dictate the order of the examination if there are multiple items and also determine whether there are any potential contamination issues.

9.4.3 Laboratory Examination of Paint

Debris is collected from the surfaces and pockets of items using brown paper or a cone and searched for any relevant paint fragments using a low power microscope. Items examined for dried or impacted paint are searched under a low power microscope and any relevant areas sampled.

Any loose paint fragments or samples of dried or impacted paint found are compared against the control on the basis of random sampling. A representative number of the fragments or samples are compared microscopically against the control. This will include high power microscopy up to a magnification of $\times 500$ under different lighting conditions including blue and UV light. The overall appearance of the paint and the density and appearance of individual pigment and extender particles within the paint are compared. Fragments composed of layers are examined sideways up to facilitate a cross-sectional comparison of each of the individual layers.

If a population of paint fragments is found corresponding in microscopic appearance to the control, representative fragments can be selected for further analysis. FTIR spectroscopy can be used to compare and identify the type of binder present in the paint along with some types of extenders. SEM can be used to compare and identify inorganic pigments present in the paint. If there are a sufficient number of corresponding paint layers between an item and a control sample, there is the potential for a conclusive link on the basis of microscopic analysis alone.

9.4.4 Interpretation of Paint Evidence

The strength of paint evidence is dependent, in general terms, on the number of layers present and the colours of the layers. Some colours, such as white, may be deemed commonly encountered whereas other colours, for example orange, may be viewed as rarely encountered. Paint of a more unusual colour will be more evidentially significant than a paint colour which is commonly encountered. The greater the number of corresponding layers present the higher the evidential strength. If an item has been repainted, the colour of additional paint layers will in many cases be down to an individual's personal choice. The adding of more paint layers to an item in this manner will eventually make the paint layer sequence unique with the potential for a conclusive link. A motor vehicle with a factory finish will have a paint layer sequence common to all vehicles made on that production line. The evidential strength of this paint will be low. A vehicle that has been resprayed will have a paint layer sequence becoming more evidentially significant the greater the number of paint layers that have been added. Cross-sectional images of fragments of paint taken from a vehicle with a factory finish and a vehicle having been resprayed twice are shown in Figures 9.4 and 9.5, respectively. For a source of paint comprising a single colour, be that paint from a spray canister, paint which has been acquired wet or impacted paint on an item, the evidential strength will be corroborative only. Two-way paint transfer will increase the evidential significance of the findings.

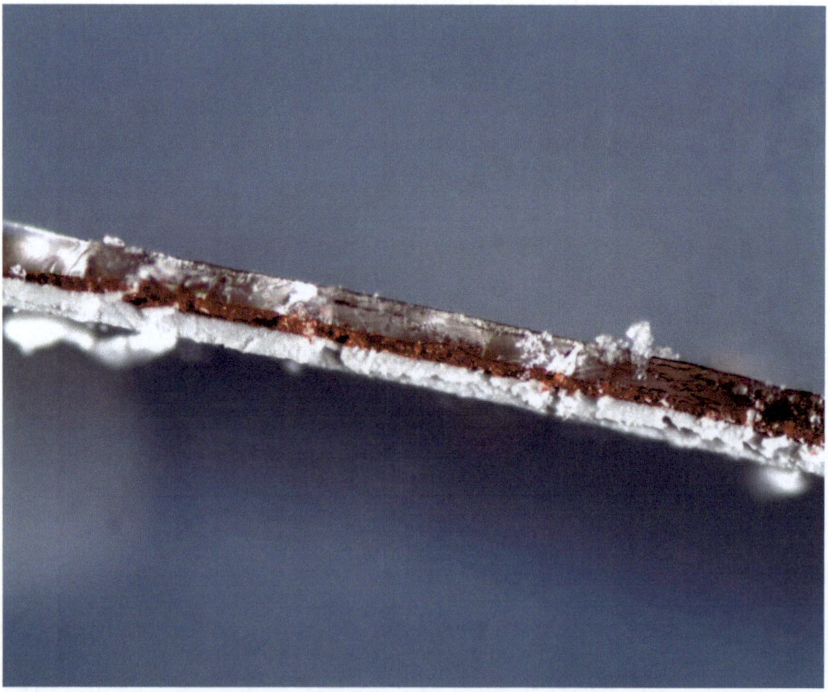

Figure 9.4 Microscopic examination at ×100 magnification of a cross section of a fragment of paint taken from a vehicle with a factory finish.

Figure 9.5 Microscopic examination at ×100 magnification of a cross section of a fragment of paint taken from a vehicle that had been resprayed twice.

If there is sufficient information about the case circumstances to indicate how the paint may have been transferred to the item, the case can be interpreted at activity level. The appearance of the paint may also add evidential significance. For example, the finding of crushed paint adhering to a tool as typified previously in Figure 9.1 will provide evidence to support the view that the item was used to lever open a painted window frame.

9.4.5 Case Scenario 1

A motor vehicle collision occurred and the driver of one of the vehicles was killed. The second vehicle fled the scene. Following a police investigation, an individual was arrested and his motor vehicle seized. Both vehicles were examined for the presence of any extraneous paint associated with areas of damage to the bodywork and samples taken. The extraneous paint found on one of the vehicles is shown in Figure 9.6. Control samples of paint were also taken from each of the cars. Both vehicles were metallic blue in colour. In cases

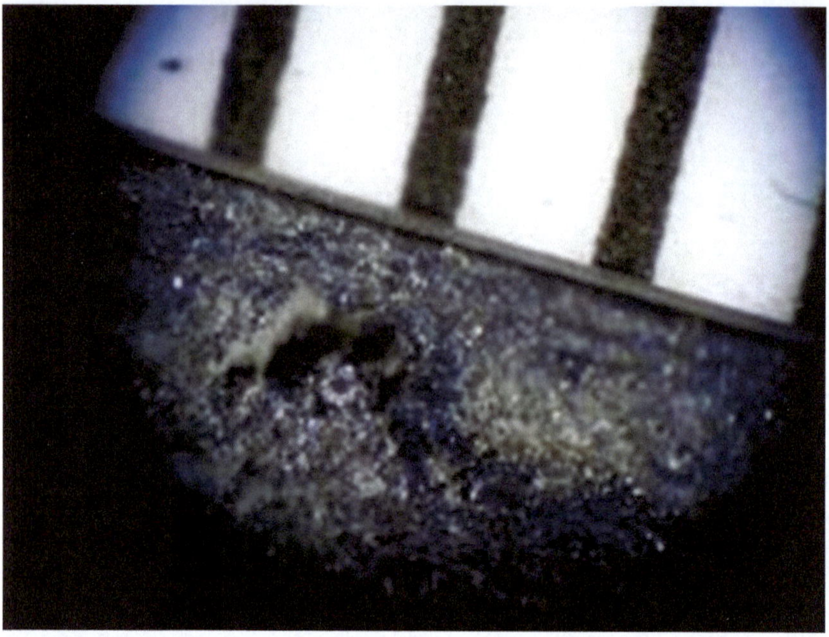

Figure 9.6 Microscopic examination at ×50 magnification of extraneous paint associated with damage to a motor vehicle that had been involved in a collision with another motor vehicle.

where two items in forcible contact are the same nominal paint colour, it is prudent to compare control samples at the outset to confirm they can be differentiated from one another at the laboratory. This was done and both were found to consist of a primer, metallic blue paint layer and clear lacquer topcoat. Both were likely the original factory finish. The paint layers were distinguishable between the two samples.

The extraneous paint from each of the vehicles was compared against the control sample of paint from the second motor vehicle. Both extraneous paint samples had the appearance of having been crushed which supported the view that the two cars had been in forcible collision. Both were found to consist of a metallic blue layer and a clear lacquer layer. The layers in both corresponded in the results of microscopic comparison and chemical analysis to the relevant layers of the compared control. The strength of the evidence depended on the likelihood of finding crushed paint corresponding to the control from the other car on both motor vehicles coincidentally. The findings were not conclusive as the paint on the two vehicles was likely the original factory finish. However, the cross-transfer of paint between the two vehicles increased the evidential strength of the findings as did the crushed appearance of the paint. Overall, the findings provided very strong support for the view that the two vehicles had been in forcible collision.

9.4.6 Case Scenario 2

A couple were found murdered in their own home following a presumed burglary which went wrong. Police examined the scene and identified the point of entry as a skylight on the top floor of the property. The paint around the skylight frame was white in colour. It was noted to be in poor condition and flaking away from the surface. A sample of the paint was taken from the window frame. A suspect was subsequently identified, and his clothing seized. Debris was collected from his top and searched for the presence of loose fragments of white paint. Eight such fragments were found. Three of these were compared microscopically against the control sample. At first sight the control sample of white paint had the appearance of a single layer. White paint is ubiquitous and, as such, will likely only provide limited evidential strength. However, examination of the control sample using high power microscopy under UV light revealed the presence of six layers. The appearance and density of pigment and extender particles within each of the layers was distinguishable under UV light. The cross-sectional images of the white control paint under the high-power microscope using white and UV light are shown in Figures 9.7(a) and (b) respectively. These six layers were replicated in the three fragments of white paint selected at random for examination from the eight fragments recovered from the top. The finding of

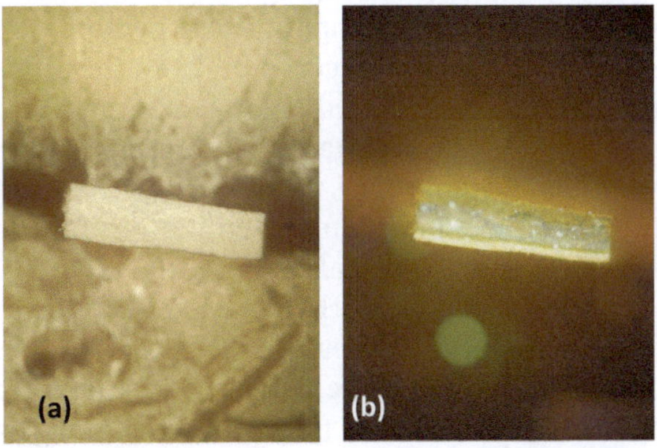

Figure 9.7 A cross section of a white paint fragment at ×100 magnification under white light – (a) compared with examination under UV light – (b), which shows the presence of six layers of paint.

six layers corresponding microscopically between the recovered fragments and control was sufficient to provide a conclusive link without the need for further chemical analysis.

9.5 Fibre Evidence

Fibres are ubiquitous in the environment originating from fabric items be they clothing items, car seats, furniture items or soft furnishings. Fibres may be natural or synthetic. Natural fibres can originate from plant sources (for example cotton, linen, hemp) or animal sources (for example wool, silk). Synthetic fibres include polyester, nylon, spandex. Fabric items are often composed of a mixture of different fibre types, for example, polyester/cotton blended fabrics. Fibres are dyed to impart colour to the item. Some fabric items are multi-coloured and therefore will be obviously composed of fibres of varying colours. Other fabric items may appear to be single-coloured but will in fact also be composed of fibres of different shades.

Offences where fibres evidence can be of use include burglary where an individual has left a clump of fibres on a window frame in gaining entry to the property; assault where contact between two individuals has resulted in the cross-transfer of fibres between their clothing; car theft where fibres recovered from a car seat can provide evidence that an individual has sat inside the car.

9.5.1 Recovery of Fibres

Extraneous fibres are recovered from an item by taping. This can be done at the scene or items can be recovered from the scene for taping at the laboratory. Tapings are placed onto acetate sheets. Care should be taken to seal the edges of the tapes so that any subsequent contamination of tapings cannot occur. In addition, the area of an item to which a taping relates, be it clothing or an item of furniture, should be carefully recorded as in some cases the position of the fibres may be critical to the interpretation of the evidence.

Clumps of fibres may be recovered direct from the scene. When comparing such fibres with specific damage to an item, control samples should be taken from as close to the area of damage as possible. There may be differences in fibres across an item if, for example, they originate from different batches having a different combination of dyes.

9.5.2 Case Assessment of Fibres as Evidence

Case assessment will consider the likelihood of fibre transfer to or between items. The transfer of extraneous fibres between items will depend on the degree and nature of the contact. Additionally, it will depend on how readily fibres are shed from the source. It will also be dependent on the nature of the receiving surface, specifically how readily extraneous fibres will adhere to it. Case assessment will consider the persistence of extraneous fibres. As with other trace evidence types, fibres will be lost from items over time to leave a residual amount. Related to this, the forensic scientist will consider the time delay between the incident having occurred and suspect items being seized. The case circumstances will influence the examination strategy, dictate the order of the examination if there are multiple items and also determine whether there are any potential contamination issues.

9.5.3 Laboratory Examination of Fibres

Fibre tapings are examined using low power microscopy and any fibres of interest are circled. Prior to any further examination, fibres on tapings must be removed from their tape and acetate backing sheets. The fibre of interest is cut out of the backing sheet using a scalpel and soaked in solvent to dissolve any remaining glue present.

Recovered fibres are compared against control fibres on the basis of random sampling. Representative numbers of recovered and control fibres are compared microscopically including the use of high power microscopy. This will compare the shape and morphology of the fibres along with their colour.

Representative fibres from among those found to correspond in microscopic appearance are selected for further analysis using FTIR spectroscopy, microspectrophotometry and/or TLC which will compare the combination of polymers and dyes present.

9.5.4 Interpretation of Fibre Evidence

The strength of fibres evidence will depend, in general terms, on the colour and type of fibre along with the number of those fibres present. Some fibre types and colours are deemed more common than others. For example, white and blue cotton fibres are very common due to the prevalence of white T-shirts and blue jeans compared to fibres which are orange or green in colour. The finding of fibres of a more unusual type and colour will be more evidentially significant than a fibre type/colour which is more commonly encountered. If there are multiple fibre types in a control sample, the transfer of fibres from this item has the potential to become more evidentially significant as the number of corresponding fibre types increases. Two-way fibre transfer will increase the evidential significance of the findings.

If there is sufficient information about the case circumstances to indicate how the fibres may have been transferred, the case can be interpreted at activity level. The position of the fibres may add evidential significance depending on the case circumstances. For example, the finding of fibres in a particular location in an assault case.

9.5.5 Fibre Plastic Fusion

Fibre plastic fusion can occur if a fabric item is in forcible contact with a plastic surface. The heat and pressure generated when this occurs may cause the plastic to melt and fibres to become impacted into the melted surface. This scenario can occur in a road traffic accident if the clothing of an occupant makes forcible contact with plastic components inside the car such as the dashboard or steering wheel. The examination of fibre plastic fusion can be useful in instances where an individual is denying being the driver of a car involved in a road traffic collision.

9.5.6 Case Scenario 1

An elderly lady was found murdered in her home. She had been sexually assaulted. Fibre tapings were recovered from her clothing *in situ* prior to removal of her body. A population of extraneous fibres was noted when these were examined at the laboratory. An individual was considered a suspect in this case. However, he had destroyed all of his clothing items prior to arrest. It was therefore not possible to identify any clothing items for comparison. In the absence of clothing, fibre tapings were recovered from the driver's seat of his motor vehicle. These were examined and several different fibre types were found to be present. Representative fibres of each were compared microscopically and using chemical analysis with fibres recovered from the clothing of the victim. The morphology and shade of fibres were found to correspond under high power microscopy. The results of FTIR spectroscopy and TLC were also found to correspond indicating a correlation in the constituent polymers and dyes present.

In this instance, there was no potential source of the fibres. Rather the proposition under consideration was whether or not the fibres were transferred to the clothing of the victim and car seat from the same item or items. In assessing the significance of the findings, the number of extraneous fibres found was considered as were the types, colours and commonality of the fibres. Corresponding fibres found to be present comprised blue cotton, in keeping with denim, and two types of red polyester. The number of fibres found was significant and could be considered in keeping with direct transfer from the originating item or items. The blue cotton fibres had some significance but could be considered reasonably common, for example, if they had originated from a pair of jeans which had been mass-produced. The two shades of red polyester fibres were more significant but also likely to have originated from a mass-produced item of clothing. Much greater evidential significance came from the fact that these fibres were found in combination with one another. Overall, the findings provided strong support for the view that fibres on the clothing of the victim and the car seat originated from the same source.

9.5.7 Case Scenario 2

Fibres have on occasion been used in an investigative capacity. A series of rape cases occurred in the London area perpetrated by an unknown individual dubbed The M25 Rapist. Forensic scientists found brightly coloured fibres during the examination of a victim's clothing. These were examined and found to correspond to fibres in high visibility jackets worn by railway workers. The individual subsequently convicted of these attacks, Antoni Imiela, worked on the railways. Other fibres found on a victim's clothing also corresponded to fibres that originated from a pair of curtains found at Imiela's home address.

9.6 Condom and Other Sexual Lubricants

Examination of items for traces of condom and other sexual lubricants can be useful in the investigation of cases of sexual assault and rape. Many of the condoms available in the UK have lubricant added to their surfaces. Silicone based condom lubricant otherwise known as polydimethylsiloxane (PDMS) is by far the most frequently encountered condom lubricant. Aqueous based glycols (for example, polyethylene glycol based lubricants) are also in use. Typically between 100–500 mg of lubricant are applied to a condom. A small percentage of condoms are unlubricated. Other substances encountered as lubricants in cases of sexual assault include products sold specifically as sexual lubricants such as KY Jelly; baby oil; petroleum based products such as Vaseline; body lotions and creams.

Research has shown that during intercourse with an individual wearing a lubricated condom, residues of the condom lubricant may be transferred to both the vagina and the penis. Penetration with an item bearing a lubricant may also result in the transfer of the lubricant to the vagina or anus. Transfer of lubricant traces may subsequently occur to the underwear of an individual.

Direct transfer of lubricant from an item such as a condom to a surface such as the vagina is primary transfer. Lubricant can also be transferred *via* secondary transfer through an intermediate surface or object, for example, from a condom to the vagina and then subsequently to the underwear of that individual. The amount of lubricant transferred through secondary transfer is generally much less than by primary transfer.

The persistence of lubricant in the vagina or anus is dependent on numerous factors including the type of lubricant in question. Lubricant residues will disappear over a period of time through a combination of normal body processes including urination and defecation and through washing actions so that all evidence of the lubricant is likely to be lost. In general, oil based lubricants including PDMS condom lubricant will persist for significantly longer periods of time than aqueous based lubricants. PDMS lubricant can potentially persist in the vagina for several days after contact. Washing or wiping is likely to remove traces of lubricant.

Items typically examined in cases of sexual assault include intimate swabs, condoms, items of underwear, items it is alleged have been used to penetrate a victim of assault.

9.6.1 Case Assessment of Lubricants

Case assessment will consider the likelihood of transfer of lubricant based on the prosecution and defence propositions. For example, consider a case in which a victim is claiming rape using a condom and a suspect is claiming consensual intercourse without a condom. Intimate swabs taken from both can be examined for the presence of condom lubricants, the finding of which would support the prosecution viewpoint. In such a case, it would be prudent to ascertain whether either individual has had intercourse using a condom in the days prior to the alleged incident. In other words, whether there is another explanation for the traces being present. For intimate swabs, case assessment will also consider the persistence of lubricants. This will depend on the activity of an individual, for example, whether or not they had a wash or used the toilet following an alleged incident. Related to this, the forensic scientist will consider the time delay between the incident having occurred and items being seized. For PDMS condom lubricants, a time delay of up to several days is acceptable for the examination of intimate swabs whereas for aqueous based lubricants the acceptable time delay is much shorter and only up to a matter of 12 hours. The case circumstances will dictate the order of the examination if there are multiple items and also determine whether there are any potential contamination issues.

9.6.2 Laboratory Examination of Lubricants

Items of underwear can be examined for any relevant areas of staining. Any staining found can be cut out of the item and extracted using an appropriate solvent. Other items, including those believed to have been used to rape an individual, can be swabbed to recover any traces present. Swabs can be extracted by removal of the head from the handle and extraction of any lubricant traces using an appropriate solvent. The solvent extracts are chemically analysed for the presence of any materials of interest. FTIR spectroscopy can be used to determine whether any traces of condom lubricants or some general lubricants are present. GC–MS can be used for the separation and identification of components in more complex mixtures.

9.6.3 Interpretation of Lubricant Evidence

The finding of lubricants may support one scenario over another if both prosecution and defence viewpoints are clearly defined. In such cases, the findings have the potential to be reported at activity level. However, it is important to note that since some types of lubricant are lost very quickly from the body, including aqueous based condom lubricants, the absence of lubricant does not necessarily mean that a condom or lubricant has not been used and the findings should be regarded as inconclusive.

In the absence of enough information, the forensic scientist will consider the findings at source level and whether or not the lubricant of interest may be present. They will not make any comment at activity level regarding the mechanism of transfer of any relevant trace material.

The forensic scientist should consider the potential for other sources of the trace material in question. For example, a drawback in the interpretation of PDMS condom lubricant evidence is that PDMS is a component of many other products in daily use, for example, cosmetics, shampoos, skin creams, deodorants, polishes and general lubricating oils. Similarly, glycols are found in other products including cosmetics and toiletries, hand wipes, feminine hygiene products. It is not possible to distinguish one source of PDMS or glycol from another by identification of the presence of PDMS or glycol alone. Materials used as lubricants will be mass-produced. They may be found in other formulations and it may not be possible to distinguish one source from another.

9.6.4 Contamination in Lubricant Cases

Contamination is a recognized problem in condom lubricant cases as the traces are often not visible to the naked eye. In addition, considering PDMS specifically, historically there have been problems with swabs having been contaminated with PDMS during manufacture. Blank swabs originating from the same batch as those used for sampling should also be examined to determine whether there might be a background level of PDMS present.

9.7 Noxious Substances

Noxious liquids, such as ammonia, acids, and drain cleaner, are sometimes used to attack individuals in cases of assault. Canisters of lachrymators, specifically *o*-chlorobenzalmalononitrile (CS), chloroacetophenone (CN) and oleoresin capsicum (OC) often referred to as pepper spray, are also used during assaults. Containers of liquids and canisters seized in relation to such incidents can be examined to determine if the contents can be defined as a noxious substance. Items of clothing and swabs taken from the skin of individuals can be examined for traces of noxious liquids.

The persistence of traces on items depends on the volatility of the substance under consideration. In particular, ammonia and CS are extremely volatile and any traces on items will evaporate in a matter of hours. Items suspected of bearing traces of CS or ammonia should be packaged inside a nylon bag with a swan neck seal. As the substance evaporates, the gaseous traces will be trapped inside the bag and these should be analysed prior to opening the packaging of the item.

The method of analysis used here is known as headspace analysis whereby a gas syringe is used to withdraw a sample of the gas(es) present in the nylon bag. The contents of the syringe are then injected into a GC instrument to identify any volatile components that may be present.

If there is a suspicion that acid has been used to assault an individual, items should first be sealed inside a polythene bag prior to being placed inside a police evidence bag. This is because the acid will attack any paper packaging with which it comes into contact.

9.7.1 Case Assessment of Noxious Substances

Case assessment will consider the likelihood of transfer of a noxious substance based on the prosecution and defence propositions. The alleged circumstances of an attack will dictate what items are examined. The nature of the substance in question will also determine how the examination is carried out. In particular, if the substance is suspected to be ammonia or CS, the atmosphere inside the packaging bag of an item thought to bear traces of the noxious material should be sampled prior to opening for any evaporated traces of the substance. With this in mind, if a suspected source of the noxious substance has been found, including any container or canister used to facilitate the attack, this should be examined first to determine the identity of the substance. An empty container is suitable for examination as it will likely contain enough traces of material for identification. If no potential source has been found, the victim or medical professionals may be able to give an indication as to what the substance may have been based on smell and the nature of the injuries.

9.7.2 Laboratory Examination of Noxious Substances

Bottles of suspected noxious substances can be examined using wet chemistry to determine the identity of the material present. Indicators can be used to determine the acidity or alkalinity of the substance. Specific tests can be utilized to identify the presence of particular ions, for example, sulphates or nitrates in the case of sulphuric and nitric acid respectively. Titration can be used to determine the concentration of the substance if there is sufficient material present. Empty bottles can be swabbed and the recovered traces used for identification. Canisters of lachrymators can be sampled and the contents chemically analysed using GC–MS to determine the identity of the substance present.

When considering volatile substances, specifically ammonia or CS, if items for examination have been packaged correctly inside a nylon bag with a swan neck seal, the atmosphere inside the bag can be sampled prior to opening using headspace analysis in conjunction with GC–MS.

Items of clothing are examined visually for any relevant areas of staining or damage. Items are also examined using UV light as some substances used in assault include a fluorescent component. Any such areas found can be cut out of the item and extracted using an appropriate solvent. The solvent extracts can then be chemically analysed for the presence of any materials of interest using GC–MS or wet chemistry as appropriate.

9.7.3 Interpretation of Noxious Substance Evidence

The findings may support one scenario over another if both prosecution and defence viewpoints are defined, including the appearance and position of relevant staining on clothing. In this case, the findings have the potential to be reported at activity level. However, it is important to note that traces of ammonia and CS in particular are lost very quickly to the atmosphere. In addition, an attack may occur using a noxious substance but traces of the substance may not fall onto the victim's clothing. The absence of traces does not necessarily mean that a noxious substance has not been used and the findings should be regarded as inconclusive.

In the absence of enough information, the forensic scientist will consider the findings at source level and whether or not any noxious substance is present. The forensic scientist should also consider the potential for other sources of the trace material in question.

9.8 Other Types of Trace Evidence

Other more unusual trace evidence types can be encountered during forensic casework. As these types of evidence are not routine, there will not be any specific laboratory procedures for their examination. Instead, the forensic scientist will use their experience along with the analytical techniques at their disposal to tackle and interpret these types of cases. Some examples follow.

9.8.1 Glitter

During the investigation of a case of sexual assault, it transpired that the victim was wearing make-up containing particles of glitter. Debris was collected from a top taken from a suspect and this was searched for glitter. More than one hundred particles of glitter were recovered of various shades. All were hexagonal in shape. They were compared microscopically against the glitter in the make-up of the victim. The shape and combination of shades of glitter particles were found to correspond. The strength of the evidence depended on the commonality of the glitter and the likelihood of finding it on the clothing of the suspect by chance. In order to assess the variation in glitter in make-up and toiletry products, a number were purchased and samples of each examined microscopically. It was apparent that all were distinguishable on the basis of the size and shape of glitter particles and combinations of shades of glitter present in each. The finding of glitter on the top of the suspect corresponding to that in the make-up of the victim was therefore deemed significant although not conclusive as make-up products are mass-produced. The forensic scientist concluded that there was strong support for the glitter on the suspect's top having originated from the make-up worn by the victim. The particles of glitter recovered from the clothing of the suspect illustrating their variation in colour, size and shape are shown in Figure 9.8.

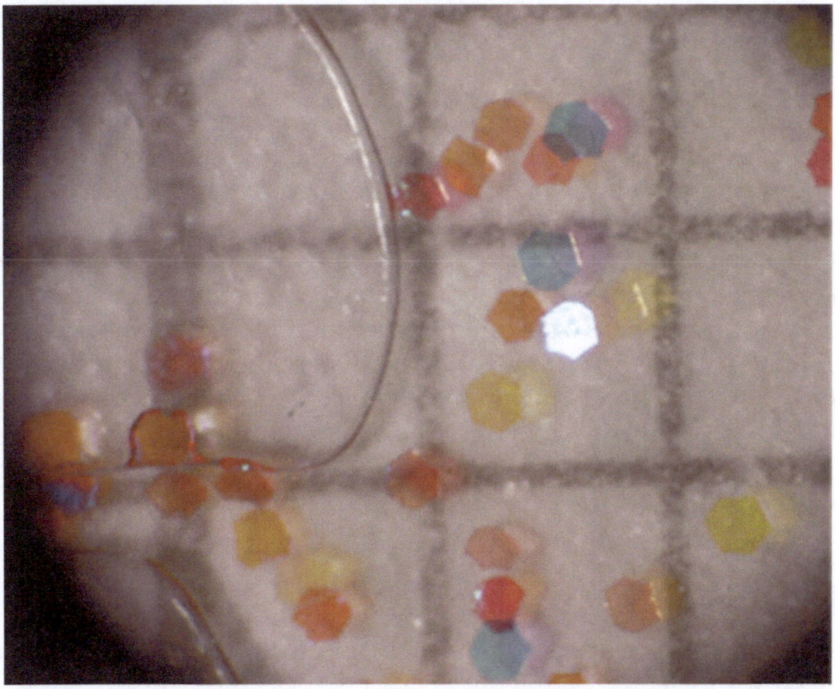

Figure 9.8 Microscopic examination at ×50 magnification of particles of glitter recovered from clothing in a case of sexual assault showing their variation in colour, shape and size.

9.8.2 Glue

A woman being stalked by her ex-partner found that her car locks had been glued up. Samples of glue were recovered from the locks and the ex-partner was arrested. A tube of glue was seized from his car by police. The recovered glue samples and a control sample of glue from the tube were compared. They were found to correspond in appearance under both low and high power microscopy. They were also found to correspond in the results of FTIR spectroscopy which showed that both were cyanoacrylate, aka superglue. This type of glue is a mass-produced item and readily available. Since there were many other potential sources of the glue in the car locks, the findings provided limited support for the glue in the car locks having originated from the tube of glue from the suspect. Although the evidential strength was limited, in combination with other evidence, this result was useful to police and helped secure a conviction.

9.8.3 Pen Ink

Writing in gold ink was found on pieces of a high-end motor vehicle which it is believed had been stolen to order and stored in a garage. The owner of the garage denied knowledge of the origins of the vehicle and the writing present. A pen containing gold ink was recovered from his office and this was compared against samples of the writing recovered using a swab. The writing was found to correspond to the ink in the pen on the basis of both low and high power microscopy and analysis using TLC.

The strength of the evidence depended on the commonality of the ink in the pen. Gold inks are less commonly encountered than many other colours of ink. To get an indication of the variation in shades of gold inks, twelve different gold ink pens were purchased and the ink from each compared to that in the pen recovered in this case. All were found to be distinguishable from the pen in question. On the basis of this, the forensic scientist concluded that there was strong support that the writing on the car had originated from the recovered pen.

9.8.4 Nail Polish

During the investigation of a murder case, the forensic scientist was tasked with examining the surfaces of a rock, the believed murder weapon, for nail polish from the victim which may have transferred there as she held her hands up to defend herself. The surfaces of the rock were examined using a low power microscope and minute fragments of a pink material were found. A representative sample of these were compared against a control sample of pink nail polish recovered from the victim's home address. A correspondence was found between the recovered and control samples on the basis of low and high power microscopy and chemical analysis using FTIR spectroscopy and SEM. The forensic scientist concluded that the findings provided strong support for the proposition that the recovered fragments originated from the bottle of nail polish. The correspondence did not provide a conclusive link as nail polish is a mass-produced item and there would be other bottles of the same type of nail polish in existence.

9.8.5 Metal Particles

It is believed that during a robbery an angle grinder had been used to break into the ATM. The forensic scientist examined a top from a suspect for small particles of metal which may have originated from the ATM as it was cut open. Debris was collected from the top. This was searched under a low power microscope and more than one hundred small balls of metal were recovered. Several of the balls were chemically analysed using SEM and the alloy present was found to correspond to a control sample of metal taken from the ATM. The alloy itself was not unique; there would be many other ATM and other items composed of the same alloy. However, much of the strength of the evidence came from the appearance of the metal balls. These were typical of having been acquired from a metal item which had been cut at high temperature using an angle grinder. The tiny droplets of molten metal produced during this process having subsequently cooled down quickly to form small balls of metal. To find this type of metal particle (shape and metallic composition) on a member of the public chosen at random from the general population would be rare and the forensic scientist concluded that the findings provided strong support for the proposition that the wearer of the top was close to the ATM as it was cut.

9.9 Conclusion

There is scope for the improvement of trace evidence examination and interpretation through an increase in research in this area. For example, more detailed studies of the transfer and persistence of a range of trace evidence types which would allow a more robust interpretation of the evidence. The future of trace evidence should also feature research into other methods of chemical analysis to provide more cost-effective and rapid examination of samples.

In recent years, forensic laboratories in the UK have seen a decline in the case submission levels of many of the traditional forms of trace evidence including glass, paint and fibres as police rely more heavily on other forms of forensic evidence. An exception to this is the submission of cases for the examination of noxious chemicals. However, as we have seen, trace evidence has a distinct advantage in that it can provide evidence at activity level and routinely provides evidential strengths of strong or very strong support. Dependent on case circumstances, traces may be the only form of forensic evidence at a scene available for examination. Trace evidence therefore retains an important place in forensic science and it is imperative that the skills of trace evidence examination and interpretation are maintained.

Websites

https://www.researchgate.net/publication/273698920: Case Assessment and Interpretation of Expert Evidence: Guidance for Judges, Lawyers, Forensic Scientists and Expert Witnesses: Jackson G. *et al.*

Further Reading

R v T, Court of Appeal – Criminal Division, EWCA 2439, 2010.

J. M. Curran, T. N. Hicks and J. S. Buckleton, *Forensic Interpretation of Glass Evidence*, CRC Press, Boca Raton, 1st edn, 2000.

B. Caddy, *Forensic Examination of Glass and Paint: Analysis and Interpretation*, CRC Press, Boca Raton, 1st edn, 2001.

J. Bentley and G. P. A. Turner, *Introduction to Paint Chemistry and Principles of Paint Technology*, CRC Press, Boca Raton, 4th edn, 1997.

M. M. Houck, *Identification of Textile Fibres*, Woodhead Publishing, Cambridge, 1st edn, 2009.

J. Robertson, C. Roux and K. Wiggins, *Forensic Examination of Fibres*, CRC Press, Boca Raton, 3rd edn, 2021.

10 Identification of the Individual

Stephen Bleay,*[a] David Charlton[b] and Clive Reedman[c]

[a] London South Bank University, 103 Borough Road, London SE1 0AA, UK;
[b] Formerly of Surrey and Sussex Police, Church Lane, Lewes,
East Sussex BN7 2DZ, UK; [c] Formerly of Go4IT Consulting, Jubilee Road,
Portslade, East Sussex BN41 1SU, UK
*Email: bleays@lsbu.ac.uk

10.1 Introduction

Each human body is unique, and distinguishable from others of the human population by several means. Some of these means are more widely used than others, and have wider applications including use as biometrics in confirming identity during day-to-day interactions with other members of the population, and in forensic investigations.

The unique structures of the human body occur at a number of levels working from the interior outwards. The attributes of some of these 'layers' are potentially useful for forensic and biometric purposes, and in some cases records of them can be captured and stored in searchable databases. For an attribute to be useful as a means of identification/individualisation, it is important to demonstrate that the attribute is persistent, preferably throughout life, or that the changes that occur through ageing can be readily accounted for.

Crime Scene to Court: The Essentials of Forensic Science, 5th Edition
Edited by Niamh NicDaeid and Peter C. White

Published by the Royal Society of Chemistry, www.rsc.org

Starting from the innermost 'layer' of the human body and working outwards, some of the features that may be used for biometric and forensic identification include:

- **Skeleton:** Including the hard bones of the body and the teeth. Dental records and skeletal structures have long been used to identify human remains, and the impressions left by teeth on other surfaces have been used in forensic investigation.
- **Veins:** Vein structures/pathways within the body differ for each member of the population. Vein patterns have begun to be used as biometric identifiers, and their use in forensic investigations has also been pursued.
- **Organs:** Most organs of the body are not exposed at the surface, but both of those that are (eyes, skin) have characteristic features that may enable them to be used for identification. In the case of the eye, the iris has patterns that have found use as a biometric identifier. The skin has a characteristic texture over its entire surface and may also have individual surface features (moles, *etc.*). The texture in certain regions of the body, especially the fingerprint patterns found on the fingertips, has become an essential tool in both biometric identification and forensic investigation.
- **Bodily secretions/excretions:** Such secretions and excretions (which include sweat, body fluids, faeces, mucus, earwax, scent) originate from inside the body and pass to the outside. Many of these secretions/excretions (*e.g.* body fluids) contain DNA, which may be used as a means of identification in its own right. Others, such as sweat, may form a layer on the surface of the skin that may be transferred to other surfaces to leave a representation of the surface texture of the body. Others still (such as scent) may have a chemistry and/or biology that may be characteristic to the individual, but are currently insufficiently researched to be used as a reliable means of identification.

The use of the features in any of these 'layers' as a biometric identifier or in forensic investigation generally relies on the initial capture of a reference image or dataset that can be used in comparison. This comparison may either be with a mark/impression recovered from a crime scene, or with another image captured from the same individual at a different point in time. It is usually only the patterns of the skin and other externally visible features (*e.g.* iris patterns) that can be readily captured by standard photography, either by direct capture of features *in situ*, or by transferring an impression of them to another medium (*e.g.* 'inked' fingerprints). However, it is possible to capture images of subsurface features such as veins, bones, *etc.* by use of specialist imaging technologies (*e.g.* infrared reflection, X-rays, computed tomography scanning, magnetic resonance imaging).

As has been described, there are many surface and subsurface features that can be used for both biometric and forensic identification and it is important to highlight the differences between the two applications. Biometric applications utilise features that have been captured from the individual under controlled conditions, and compares them to sets of features that have been captured from a wide range of individuals under similarly controlled conditions and that are held on a database. This comparison enables the identity of the individual to be confirmed (provided that the subject has already been enrolled on the database). Fingerprints are commonly used as biometric identifiers but there are also a range of features (iris, vein patterns) that may be regarded as 'subsurface' but are still useful as biometrics.

Forensic investigation typically utilises what are typically referred to as 'marks', *i.e.* representations of the outer surface of the body that are left as a result of contact with other surfaces. The contacts that occur between the body and other surfaces may either result in the transfer of material, or in the formation of impressions in the surface. These 'marks' are created under uncontrolled conditions and will be imperfect – either incomplete, or affected by movement, distortion or contamination. Although marks left by physical contact are the features most commonly used in criminal investigation, the increase in the use of digital imaging means that other body features may occasionally be captured in images (*e.g.* of abuse) in sufficient biometric detail to enable comparison with a database or with a suspect in a criminal investigation.

10.2 Fingerprints in Criminal Investigation

Throughout this chapter, the term 'fingerprint' is used to describe representations of the ridge patterns on the fingertips that have been taken under controlled conditions, for example inked or digitally enrolled ten print sets. The term 'fingermark' is used to describe the marks left by contact between the finger and a surface under uncontrolled conditions, which includes those recovered from articles or surfaces at crime scenes.

Of all the different types of marks that may be encountered at a crime scene, fingermarks are the easiest to utilise. There are several characteristics of fingermarks that make them highly suited for criminal investigation, including:

- They contain features of a scale that are readily discernible.
- They are deposited by the regions of the body that are used to manipulate/touch articles and surfaces.
- They can be classified on a coarse scale by human examiners by using simple, easily recognisable patterns.
- They also contain finer scale features that can be recognised and searched both by human examiners and by an automated database.

Although marks may be deposited by the skin present on many other parts of the body (*e.g.* lips, ears, back of hand, elbow, cheek), none of these offers the same range of characteristics that make them as useful for comparison purposes as fingermarks.

10.2.1 A Brief History of Fingerprint Identification

An effective and comparatively inexpensive forensic discipline, the examination of friction ridge skin from the fingers, palms and even feet of individuals is considered one of the strongest of forensic sciences because of the discriminatory power of the evidence. After all, no two fingerprints of any two different individuals, even those of identical twins, have ever been found to have the same configuration of features present.

Friction ridges form before birth and are incredibly robust and resilient. Even in death it is possible to obtain valuable post-mortem fingerprint impressions that can be used to identify deceased in accidents, criminal activity and of course natural disasters. This was never more typified than during the attempts to identify victims of the Asian Tsunami in 2004, Figure 10.1.

The distinctive patterns of friction ridge skin found on the tips of the fingers have been utilised by humans for many thousands of years. Fingerprints have attracted the attention of artists, examples including 6000 year old cave paintings, 'signatures' on Roman pottery, and their use by Thomas Bewick as frontispieces in illustrated books in the late 1700s–early 1800s. From the 1600s scientists including Nehemiah Grew and Jan Purkinje also began to

Figure 10.1 An example of dermal/epidermal separation during efforts to identify victims of the 2004 Asian Tsunami.

study these characteristic patterns and began to consider how to classify them.

In the mid to late 1800s pioneering work by Sir William Hershel, Dr Henry Faulds and Sir Francis Galton established important principles that are still core to the use of fingerprints in criminal investigation, including the fact that ridge patterns are permanent and do not change for an individual over time, fingerprints and the marks left by them can be used to identify individuals, and a classification scheme can be developed for fingerprint patterns to enable individuals to be searched against a database. Sir Edward Henry, while Inspector-General of Police in Bengal in the late 1890s, developed what is known as the Henry Classification system for the storage and retrieval of fingerprint records with his assistants Azizul Haque and Chandra Bose. On his return to the United Kingdom to become Assistant Commissioner at Scotland Yard in 1901, Henry established the Metropolitan Police Fingerprint Bureau based on the principles of this system. In Argentina Juan Vucetich had independently developed an alternative classification system, and by the early 1900s fingerprints were becoming widely used as a tool in crime investigation. As a consequence, for over 100 years fingerprints have been core to the way that law enforcement agencies have sought to both solve crime and to verify the identity of individuals coming to their notice.

10.2.2 Principles of Fingermark Comparison and Identification

Friction ridges are not uniform in configuration, nor do they necessarily flow in straight lines. There are fractures and interruptions within the structure of these ridges as well as a general directional flow. Friction ridges also contain finer features such as pores that can also be highly distinctive in shape and distribution. The patterns, the unique configuration of the way features and interruptions present themselves are the basis of how expert examiners can differentiate between fingermarks found at crime scenes and the fingerprints recovered from donors, either voluntarily or through criminal processing.

Every feature and characteristic in a sample of friction ridge skin is readily recognizable and identifiable and individualisation is possible so long as there are sufficient characteristics of acceptable quality present. It is possible to match the marks left by very small amounts of friction ridge skin at a crime scene against a suspect, so long as there is sufficient clarity upon which to rely upon the permanence and reproducibility of visible features. The features that may be utilised during comparison include the general pattern formed by the ridge flow (termed '1st level detail'), shown in Figure 10.2, the characteristics formed by interruptions to the ridge flow (termed '2nd level detail') shown in Figure 10.3, and the finer features such as pores (termed '3rd level detail') shown in Figure 10.4.

When an examiner compares ridge detail between a crime scene mark and an exemplar from a ten-print form, a generic methodology has been adopted known as ACE-V. In order to establish whether there is any basis upon which to determine that the ridge detail found at a crime scene can be individualised against that of a known

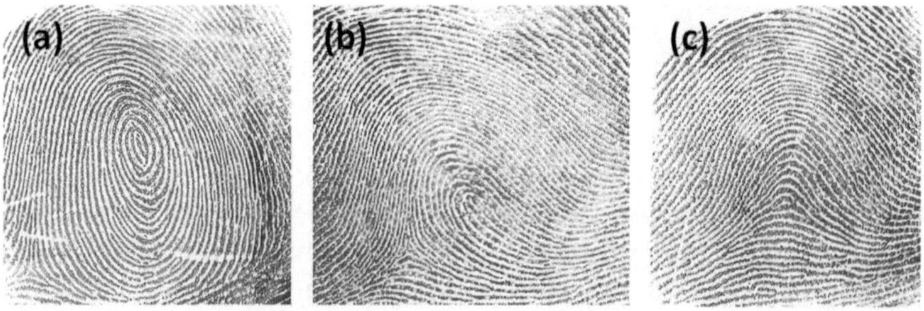

Figure 10.2 Examples of the principal fingerprint patterns used to describe '1st level detail' in fingerprint comparison, (a) whorl, (b) loop, and (c) arch.

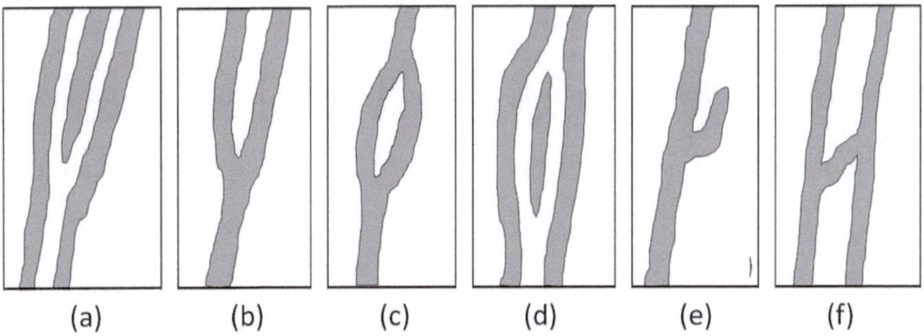

Figure 10.3 Schematic diagrams showing the principal '2nd level detail' features and how other features are defined by combinations of them, (a) ridge ending, (b) bifurcation, (c) 'lake' (two opposing bifurcations), (d) 'island' (two opposing ridge endings), (e) 'spur' (a bifurcation and ridge ending) and (f) 'crossover' (two linked bifurcations).

Figure 10.4 Examples of '3rd level detail' features, in this case pores present within the ridge.

exemplar, one must first **assess** the friction ridge detail in the crime scene mark so that useful reference points can be identified in the next phase, when there will be a **comparison** of the mark features with that of features within the reference print. Subsequently, the examiner will make an **evaluation** of the findings, from which conclusions as to identity will be derived. Finally, as in all scientific endeavours, there must be an element of peer review, known as **verification,** to repeat the ACE process independently to hopefully reach the same conclusions.

An **assessment** of a fingermark by a skilled examiner will include consideration of the following factors:

- Any distortion present (*e.g.* has the finger slipped or twisted during contact?)
- Development methods and medium used to visualize latent material (*e.g.* has the mark been visualised using powdering or a chemical process?)
- Deposition pressure (*e.g.* has high pressure been applied during contact that could lead to additional distortion in certain regions?)
- Anatomical attributes (features)
- Clarity (*e.g.* is ridge flow distinct in all regions of the mark or are some areas blurred?)

During the **comparison**, the examiner will look for similarities in:

- Pattern (1st level detail)
- Ridge path (2nd level detail)
- Ridge shape (3rd level detail)
- Pore positioning (3rd level detail)

During the **evaluation**, the examiner will look to form an opinion:

- Can the mark be eliminated as it belongs to someone having legitimate access to the crime scene?
- Is there sufficient information available to individualize (or exclude)?

In **verification** of any findings, an independent assessment of the casework is undertaken using the ACE process described earlier in order to ensure that the conclusions reached are consistent with the original findings. In the UK, for a crime scene mark to be confirmed as identified, the comparison was historically undertaken three times. One check was made by the original examiner, then verification by two further experts who would hopefully reach consistent conclusions independently of one another. However, this requirement is not universally adhered to any longer and many fingerprint bureaux have adopted a two check approach. While further longitudinal research is always of benefit to ensure the accuracy and reliability of examiners and processes adopted, the reduction from three to two checks has had no obvious impact on practitioner accuracy and there is no basis in science (where traditionally repeatability of findings is a matter of peer review based on one other assessor) to support maintaining a three check strategy. Thus the reduction in checks from three to two brings fingerprint examination in line with broader scientific principles around peer review and critical findings checks and it could be suggested supports adherence to best practice around ISO 17025 accreditation.

Until recently, fingermarks were generally regarded as an 'absolute' form of evidence, with experts being able to state that a mark was an identification with an implied certainty of 100%. The findings of the Fingerprint Inquiry Scotland in 2010 stated that fingerprint evidence should be regarded as opinion evidence, not fact, and as such assessed on its merits. There has also been a growth in the research and development of probabilistic approaches to fingerprint comparison, using the number of features found in agreement to calculate 'likelihood ratios' for the mark and exemplar print originating from the same source. Similar approaches are used for other forms of forensic evidence and this would make fingerprints consistent with them. To date, although several models have been developed there is no universally agreed system in use and implementation of such a system in the UK is several years away.

10.2.3 Contextual and Cognitive Bias in Fingermark Identification

The 2009 US National Academy of Sciences report suggested that fingerprint examiners, in addition to other forensic science disciplines, were vulnerable to cognitive and contextual bias. The Fingerprint Inquiry Scotland also recommended that steps were taken to ensure that potential contextual bias in fingerprint comparison was minimised. Research has highlighted over many years now that decision making in all walks of life, including the analytical sciences, can be influenced by the context in which evidence is presented to the practitioner. Because of research by Dror, Charlton and others, it is now widely accepted that the lower the quality of a fingermark, or the more ambiguous the information is within a fingermark, the more likely it is that decision thresholds as to how ACE-V is applied and conclusions drawn may change over time. In some cases actual case decision outcomes can be affected by the level of context within a case that is available at the time of the examination. It is widely documented that the same fingerprint examiners can make different decisions over time when presented with evidence they have seen previously, if the context in which that evidence is presented is changed.

Fingerprint examiners are experts in interpretation. They are able to seek targets of information, often clouded by distortion, by superimposition of ridge detail, and they are able to use their experience to make value judgements on what they see and how such data is interpreted. However, often, when asked, experts, all experts, not just those engaged in the examination of fingerprints, find it incredibly hard to both explain and to justify their thought processes. The reason for this is quite simple. The human mind is a machine. Like all machines it has an optimum operating capacity before it becomes inefficient. Such 'cognitive overload' can lead to the mind becoming overwhelmed with information and the person would be unable to function. The mind copes with this bombardment of information by selecting what it needs to process immediately and what it can afford to 'archive'. It tends to archive things it believes it already knows and instead concentrates on what it believes is new information. It is therefore quite easy to carry out a task that has been done many times before without much thought going into it. Cognitive shortcuts make it easier for our brains to cope and compartmentalise information. The truth is our brains decide to reassign cognitive power to concentrate on what it sees as more useful and valuable tasks. This is the very nature of expertise and makes experts incredibly efficient people, but it can also make them extremely vulnerable to ignoring important new information if their minds perceive that through experiences it can afford to miss out vital clues that the human may find useful in completing a task accurately.

Fingerprint experts can be influenced in their decision making by external factors such as the context in which a particular piece of evidence is presented to an examiner for processing. An illustration of the issue of contextual bias can be seen in Figure 10.5 where visual interpretation of the central figure in the chart *i.e.*, is it B or 13 will depend on the visual context placed before the examiner. It is possible for practitioners to unavoidably and unconsciously lose objectivity and hence analyses may be biased in favour of a desired outcome.

Crime scene examiners will be aware of physical contamination when they gather evidence from a crime scene. It is entirely possible to conceive that it is not just physical evidence that can be contaminated, but also the scientists themselves. It is this 'cognitive contamination'

Figure 10.5 A demonstration of the issue of contextual bias. Visual interpretation of the central figure in the chart above will depend on the visual context placed before the examiner.

that is as much of a risk in modern laboratories as more traditional forms of evidential contamination.

Cognitive contamination in the fingerprint domain can manifest itself across a wide range of situations and tasks. This may be as simple as labelling errors, for example a right palmprint presented to an expert as a left palm, meaning that if this information is taken at face value an incorrect conclusion regarding whether the print is a match will be made. However, a full examination of the print, without taking cognitive shortcuts, should show an expert from ridge flow, *etc.* that a right palm is more likely and enable correct conclusions to be drawn.

As highlighted in studies over the past ten years, it is also possible to contaminate human decision making by providing the examiner with too much case information. Many agencies use what are known as case management systems. This technology makes it possible to track evidence through the chain of custody, record findings and facilitate the communication of results. While such tools greatly aid the continuity of evidential process, it can be argued that such systems are also detrimental to accurate decision making. In research conducted it has been shown that it is possible to get fingerprint examiners to change their original decision based upon the context in which they receive the evidence for analysis. For example, a study looked at examiners and real case work. Each examiner was presented with casework that they had also analysed 5 years previously. It was discovered that where a strong context (*e.g.* 'this is the mark from a brutal murder') was placed before the examiner, that it was possible to influence their decisions to a point where they changed their mind in certain cases. While the majority of changes observed were to the more cautious end of the spectrum, whereby previously identified evidence was now left unidentified, there were a few examples where an initial negative decision was later changed to that of an identification simply based on the strong context in which it appeared to the examiner. Many other studies by researchers in the domain have produced similar findings. It also seems that the more ambiguous, or difficult the comparison task, the more likely it is for such context effects to manifest themselves.

One of the major areas of cognitive contamination revolves around the perceptions of the examiners when carrying out an analysis of a fingerprint. Their experience and past bench work will influence them in both their interpretation of the evidence and how they may choose to report on their findings.

This 'base rate bias' is a source of concern when an examiner assesses fingerprint evidence arising from an Automated Fingerprint Identification System (AFIS) match, or indeed from the need to conduct a check of crime scene marks against a suspect. The way AFIS systems are designed, with scoring thresholds and suggested respondents, it is easy for examiners based on their experiential learning to begin to apply false expectation rules when conducting their analysis. For example, they may place too much weight on a respondent appearing number one in the list. This is especially true when the vast majority of AFIS matches appear as number one on the list. Indeed, the examiner, as has been shown in other studies, may begin to 'lose interest' lower down the respondent list and actually miss a match because they did not expect to see it so far down the order of comparison.

So far, many of the cognitive contamination issues have been part of the human condition, with an understandable desire to conform. These biases are subconscious and impossible to avoid. It is only possible to put measures in place to minimise the impact of cognitive contamination and to lessen the risk of influence on decision outcomes. But cognitive contamination can also arise from the way an organisation delivers a forensic service and the way in which it places expectations on the scientists working for that organisation. There are many examples of what can be termed corporate or cultural bias:

- Examiners working for law enforcement agencies risk a bias in that the overarching culture is to catch criminals rather than to do what should be the primary focus of any forensic scientist, namely to carry out objective and unbiased analysis.
- The way in which practitioner performance is measured may be considered a biasing factor in the decision making of examiners. A common scenario in most agencies would be to measure the performance of a fingerprint practitioner by the number of

identifications they make, which creates an atmosphere whereby the examiner feels under pressure to perform and find identifications.

- Many agencies may choose to adopt a hierarchical approach to verification of finding through peer review procedures. In some agencies the verification, or peer review, is carried out by a senior officer or manager. In such instances, the ability to challenge or discuss findings is harder for the practitioner.

The risk of cognitive contamination is always possible in a forensic domain where it is the human who is essentially the tool of the science. That is to say the examination of fingerprints is a fundamentally human process of analysis and comparison. The human eye is the primary method for deciphering the complex information presented to it and it is for the human mind to filter out the noise from the real data in order to reach valid and hopefully reliable conclusions.

It could be suggested based on what has been discussed already that the only way to proceed is to eliminate all context and to ensure that the examiner is not exposed to any information that may bias their decision making reliability. This sounds reasonable. However there is a real world element to this argument that should not be ignored. Namely that not all context is bad context. Sometimes context and information pertaining to a case can actually enhance the ability of a fingerprint practitioner (or any other forensic scientist) to make a valid decision. For example, a fingerprint examiner will try to make value judgements about the orientation of a finger, the interpretation of information within the print, as well as assess the pattern, 2nd level minutiae and 3rd level information like sweat pores and any distorting factors. To make this process easier and less prone to misinterpretation, it is entirely reasonable to expect that relevant contextual information may be made available to the examiner to aid them in their work. For example, information about the processes used to enhance a latent print, or even whether the mark was deposited in blood or other such circumstances will greatly enhance the ability of an examiner to make a more accurate evaluation. As a consequence, a knowledge of how different processing methods can impact upon image clarity and other characteristics seen in the mark can be included and accounted for in the evaluation.

Cognitive contamination has been largely misunderstood by fingerprint practitioners who see the concept as a direct challenge to their impartiality and professional integrity. Research to assess not just strengths, but also potential vulnerabilities has traditionally been met by the fingerprint profession with both scepticism and distrust. Some commentators within the profession have even ridiculed the importance of such research and denied that bias is even a real threat. However such assertions are not now universally held within the fingerprint profession and there is broad acceptance of both the academic validity of cognitive contamination, as well as a realisation that the fingerprint profession and wider forensic sciences must do more to understand where cognitive contamination is a risk to the accurate processing of evidence and to seek better processes and technologies to minimise the risk of bias in the workplace. Recent publications from the UK Forensic Science Regulator further underpin the perception that the need to address biasing factors in the forensic domain is gaining wider credibility.

Many agencies are proactively looking at ways to introduce ways to 'sequentially unmask' context and information to the practitioner to minimise the risk of biasing factors influencing decision making while still enabling the examiner access to information that may help the analysis of the evidence. While the ACE-V process is defined as a linear process, it is considered appropriate to look upon the sequential unmasking process as a linear process also. In this way appropriate information might be made available at various stages of the ACE-V process to facilitate the examination. While the linear process implies a need to maintain order to the unmasking process, it is recognised that there will always be in a visual and subjective process like fingerprint examination a need to revert to the source print to aid the interpretation of the crime scene fingermark. It is accepted that the crime scene mark should always be the first to be assessed prior to any comparison with a known print, but on occasions it might be helpful to refer to the known to aid the interpretation of features within the crime scene mark. This is not strictly allowed in a linear unmasking process, however, in the real world this might be a realistic option. In such circumstances documented and details contemporaneous notes of the examination would be appropriate whereby any changes to the original interpretation of the mark after further analysis would be noted.

Other ways to ensure that contextual information is minimised is to seek ways to ensure that there is a limitation on the potential for one examiner to favour a particular other to verify their conclusions. An easy way to avoid this issue is to introduce blinded case processing and blinded verification.

Modern forensic science must consider how to do more with fewer available resources. More than ever, the power of the human mind must be understood with regard to the comparison of fingerprints and other physical trace evidence. New processes and technologies, including any reduction in verification checks or lights out processing of evidence without any human intervention, must be introduced with knowledge of how such changes might influence human decision making in the future.

Only further research to better understand these issues will support any changes introduced.

10.2.4 Formation of Fingermarks

Before fingermarks can be utilised in comparison, they must first be located, visualised and imaged in a way that enables both a human examiner and AFIS to find and highlight all features of relevance. The objective of those involved in the investigation of crimes is to maximise the number of fingermarks that are located and visualised. In order to do this it is useful to have an understanding of how the different types of fingermarks that may be encountered are formed, and of the properties that can be utilised to visualise them.

Fingermarks are formed as the outcome of a finger making contact with a surface. In such an interaction there are several possible outcomes, and one or more of these may be expected to occur in any given interaction:

- Material present on the surface of the finger (including sweat and other contaminants) may be transferred to the surface.
- Material present on the surface (including dust and other contaminants) may be transferred to the finger and physically removed from the surface.
- The surface may deform due to the force applied by the finger (which in some cases may leave a permanent impression in the surface).
- The finger may deform against a rigid surface (although this deformation will be reversible).

The marks that are left as impressions in the surface and by removal of material from the surface can be often be located relatively easily by use of different lighting techniques which produce contrast between the surface and the local disturbance left by the contact of the finger. However, by far the most common outcome of the interaction during contact is the transfer of material from the finger to the surface. In some cases the material deposited by the finger is visible to the unaided eye (*e.g.* blood, dirt, ink), and the resultant mark is called a 'patent' mark. In the majority of cases the material transferred to the surface is not readily visible by eye and the resultant marks are termed as 'latent' marks. Latent marks require some form of enhancement or the use of specialist imaging techniques to make them sufficiently visible for images of them to be captured.

Although the amount of material transferred from the finger to the surface is typically only a few microgrammes in weight and the height of the resultant surface features only a few micrometres, the residue has a highly complex chemical composition and also a range of other properties that can be utilised to visualise it. Some of these properties are summarised in Table 10.1.

The complex chemical mixture forming the fingermark exists as an emulsion which is heterogeneous on a microscopic level. The distribution of individual chemical constituents may also be highly localised within the fingermark, for example the constituents of eccrine sweat may be concentrated around the pores that secrete them.

It should also be noted that the material deposited in fingermarks can produce local modifications in the substrate they are deposited on, for example by initiating corrosion, by changing dielectric constant, or by changing the chemical composition of the surface. In some cases it may be possible to detect the presence of the mark by detecting its effect on the substrate as opposed to utilising the properties of the mark itself.

Table 10.1 The range of different properties of fingermarks that can potentially be exploited in their visualisation.

Generic property	Examples
Chemical composition	Eccrine sweat (*e.g.* water, amino acids, salts) Sebaceous sweat (*e.g.* fatty acids, wax esters) Contaminants (*e.g.* foodstuffs, cosmetics, blood)
Physical properties	Adhesion Refractive index Reflectance/absorption Luminescence Electrical charge Dielectric constant Topography
Biological properties	Resident microbial flora Transient microbial flora DNA (*e.g.* in shed skin cells)

The nature of the substrate the fingermark has been deposited on will also affect what material is available for enhancement. On non-porous surfaces, the fingermark residue remains on the surface, but progressively dries out and the surface features reduce in height as volatile components are lost. On porous surfaces, the fingermark may diffuse into the interior of the substrate, leaving very few of the constituents remaining on the surface. The properties of the fingermark will also be influenced by the environment the fingermark and surface are exposed to after deposition, for example high humidity environments may diffuse or even dissolve water soluble constituents.

The wide variety of properties that fingermarks possess means that there is an equally wide range of processes that can be used to visualise them. This distinguishes fingermarks from many other forms of forensic evidence because there may be multiple opportunities to visualise the mark – if initial attempts to target one property or chemical constituent are unsuccessful, the mark may be enhanced by switching to a process that targets another property or chemical constituent. It is often possible to utilise sequences of processes, progressively moving from those that are non-contact and non-destructive to those that may result in permanent changes to the surface, in order to maximise the chances of visualising the fingermark.

Examples of some of the methods that have been used to visualise fingermarks are given in the following section and in the Bibliography.

10.2.5 Visualisation of Fingermarks

10.2.5.1 Optical Methods (Utilising Physical Properties Including Refractive Index, Reflectance, Topography, Luminescence)

10.2.5.1.1 Visual (White Light) Examination

The use of a range of white light sources providing different lighting conditions can be a very powerful means of detecting fingermarks. Because fingermark residues are likely to have a different refractive index and different reflectance/absorbance properties to the surface they are deposited on, it is often possible to maximise the contrast between the mark and the surface by varying the angle of illumination. On transparent surfaces it is the refractive index that is likely to be more important, whilst on opaque surfaces the reflectance/absorbance will dominate. The topography of the mark is also a factor, especially on

Figure 10.6 Friction ridge detail on the screen of a mobile phone revealed by visual examination.

smooth surfaces, where the presence of the mark may result in increased scattering of incident light making the mark visible, as illustrated in Figure 10.6. Visual examination is often used as the initial stage in a processing sequence because it is non-contact and non-destructive, meaning that the mark is unaltered by the process and other processes can be used subsequently.

10.2.5.1.2 Fluorescence Examination

Fingermarks may contain constituents that are fluorescent. Some of the naturally occurring constituents of sweat (*e.g.* the amino acid tryptophan) are fluorescent, but fluorescent contaminants may also be present on the fingers from sources such as foodstuffs and cosmetics and these are likely to be more abundant. In fluorescence examination, a high intensity, coloured light source is used to illuminate the fingermark. If the illuminating light is of an appropriate wavelength (or wavelength range), then energy is absorbed by the fluorescent constituents of the fingermark and then re-emitted as light of a different, longer wavelength. Normally, this re-emitted light is not readily visible because it is of lower intensity than the reflected light and is 'swamped' by it. However, by viewing the surface through a filter that blocks the lower wavelength, reflected light and transmits the higher wavelength fluorescence, the fingermarks can be seen.

Fluorescence examination may also reveal fingermarks by another mechanism. Fingermarks may also contain contaminants (*e.g.* blood) that strongly absorb light of certain wavelengths. If they are present on a surface that is fluorescent, the fingermark may become visible as dark ridges against a bright background.

In fluorescence examination, the wavelength and power of the light source are important considerations, as well as the characteristics of the filters used to view the resultant fluorescence.

10.2.5.1.3 Ultraviolet Reflection

It is also possible to utilise electromagnetic radiation outside the visible region of the spectrum to visualise fingermarks. The most widely utilised of these methods is ultraviolet reflection. Ultraviolet (UV) reflection can be effective on both porous and non-porous surfaces, especially where the surface is smooth. On smooth porous surfaces, some fingermark constituents strongly absorb UV radiation and thus appear as dark ridges against a bright background. On smooth non-porous surfaces, the topography of the fingermark, even though features may be less than 1 μm in height, can still strongly scatter the short wavelength radiation and make the fingermark visible as bright ridges against a dark background. Although both short wave and long wave UV light sources are readily available, specialist optics, sensors and filters are necessary to directly view the reflected radiation and distinguish the marks. In the broader context of forensic investigation and identification, it is important to recognise that short wave UV radiation is destructive to DNA and this process should not be used if DNA recovery may be important. Long wave UV reflection is less destructive to DNA, but less effective than short wave UV reflection in detecting fingermarks.

10.2.5.2 Physical Methods (Utilising Properties Such as Adhesion, Electrical Charge, Dielectric Constant)

10.2.5.2.1 Powders

There are many different types of powder available for fingermark enhancement, ranging in nature from metal to organic, with morphologies from flake to granular, and secondary properties from magnetic to fluorescent. The fundamental principle behind all of these types of powder is the same, when they are applied to the surface with an appropriate applicator (usually a brush), the powder particles preferentially adhere to the fingermark deposits as shown in Figure 10.7 and not to the surface. The mechanisms of adhesion are likely to differ between

Figure 10.7 Fingermarks revealed on a glass bottle by powdering with aluminium flake powder and lifted from the surface using transparent lifting tape.

different powder types – it is known that certain types of powder will develop marks not detected by others. The mechanisms that may influence powder adhesion include attraction to the residual electrical charge on the fingermark, the capillary action of the water content of the mark, or the 'sticky' nature of viscous sebaceous constituents or contaminants.

10.2.5.2.2 ESDA

Although ESDA is now firmly associated with revelation of indented writing during document examination, it was originally investigated as a means of enhancing fingermarks on paper (but found to be ineffective on marks more than a few days old). When used for fingermark enhancement, ESDA utilises differences between the electrical properties of the region of the paper touched by the finger and the adjacent, untouched region. These electrical differences may arise from either the residual electrical charge on the fingermark, and/or from local modification of the dielectric constant of the paper as a result of the fingermark residue diffusing into it (although neither mechanism has been conclusively established). Modification of dielectric constant may arise from filling of porosity in the paper with solid material, or from the dielectric constant of the constituents of the residue. Any local surface differences in electrical properties are visualised by covering the surface in a thin polymer layer that retains electrical charge, passing a high voltage corona source across the surface to develop charge patterns on the polymer sheet, then passing a development powder over the surface that selectively adheres to the regions of highest charge on the polymer sheet.

10.2.5.3 Chemical Methods (Utilising Chemical Constituents of Fingermarks)

10.2.5.3.1 Amino Acid Reagents

Amino acids are one of the most abundant constituents of eccrine sweat, and are utilised to visualise marks by several reagents including ninhydrin, 1,2-indandione and 1,9-diazofluoren-9-one (DFO). These reagents react with the amino acids present in the fingermark to produce reaction products that are either highly coloured (*e.g.* ninhydrin – Figure 10.8a) or fluorescent (*e.g.* 1,2-indandione, DFO – Figure 10.8b). These reagents are most suited for use on porous substrates that absorb the eccrine sweat constituents and ninhydrin is probably the most widely used reagent for this purpose worldwide. Interactions can occur between amino acids and the cellulose present in paper, meaning that the amino acids migrate very little from the position they were originally deposited in. As a consequence, marks in excess of 20 years old have been successfully visualised using amino acid reagents. The principal application of amino acid reagents is to enhance marks on porous surfaces. The amino acid reagent is generally used as a solution which is used to 'wet' the porous surface, and the solvent allowed to

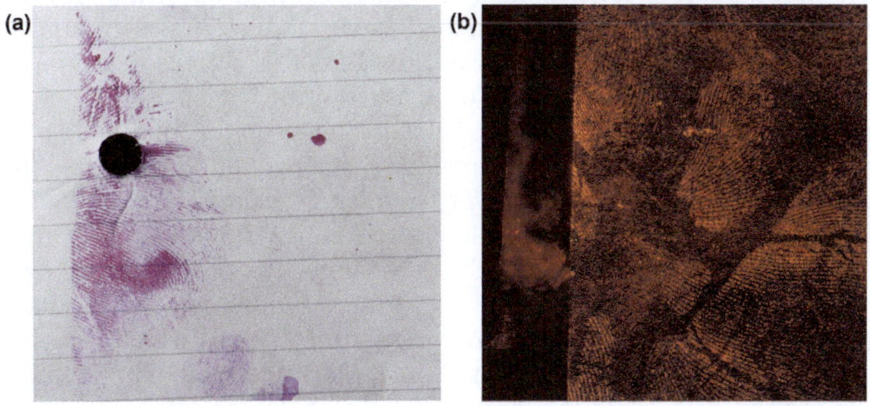

Figure 10.8 Friction ridge detail revealed by processing with amino acid reagents, (a) a palmar mark on lined notepaper developed using ninhydrin, and (b) a palmar mark on a brown envelope developed by 1,2 indandione.

evaporate before the item is exposed to elevated temperature (and in some cases humidity) to enable the reaction to proceed.

10.2.5.3.2 Silver Nitrate

Before the introduction of amino acid reagents in the 1950s, one of the oldest and widely used reagents to visualise fingermarks on porous surfaces was silver nitrate. Silver nitrate targets the chloride constituents of eccrine sweat, undergoing a reaction with them to form silver chloride which is then converted to silver by exposure to light (with short wavelength, UV light giving the most rapid results). Marks become visible as the invisible deposits are converted to a brown or grey product. Disadvantages of the process are that residual silver salts in the background progressively convert to silver, obscuring the developed marks, and the fact that the small chloride ions can readily migrate within porous substrates, especially when humidity is high. As a consequence, marks older than a few weeks developed using silver nitrate may appear diffuse with no distinguishable ridge detail. Silver nitrate is applied to the surface as a solution, and the surface allowed to dry in the dark before being exposed to light to progressively develop the mark.

10.2.5.3.3 Iodine

Iodine is probably the oldest fingermark visualisation reagent and was observed to develop fingermarks on paper as early as the 1860s, before fingermarks had been proposed as a means of criminal identification. Iodine can be used to visualise fingermarks in two ways, by the sublimation of solid crystals of iodine, or by the application of an iodine solution containing a fixing agent. Iodine is preferentially absorbed by several fingermark constituents (*e.g.* water), developing brown/purple marks if iodine is used as a fuming reagent or dark blue marks if a fixing agent has been used. The strongest interaction is observed between iodine and squalene (which is abundant in freshly deposited fingermarks). However, squalene decomposes rapidly in the days after deposition and therefore iodine (in both fuming and solution forms) tends to be relatively ineffective in enhancing fingermarks more than a few days old. Iodine fuming is conducted by placing items in a cabinet containing iodine crystals which may be gently heated to cause them to sublime. Marks develop as the fumes are progressively absorbed by the fingermark. In the iodine solution method, a solution containing iodine and fixing agent is applied to the surface, and marks are revealed rapidly as the iodine is absorbed by the mark and the carrier solvent evaporates.

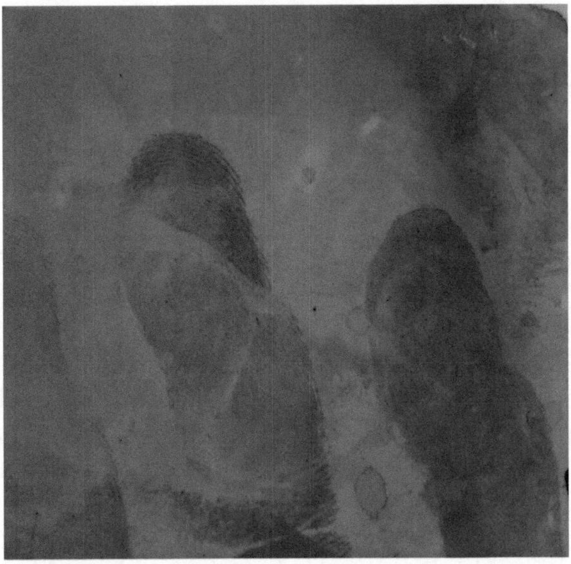

Figure 10.9 A sebaceous fingermark on a ceramic tile enhanced by Solvent Black 3.

10.2.5.3.4 Oil Red O, Solvent Black 3

Oil Red O and Solvent Black 3 are examples of reagents that selectively stain fats, and find uses in fingermark visualisation because they can stain both constituents of sebaceous sweat and certain greasy contaminants such as butter and vegetable oil. Although these substances are generally less abundant in fingermarks than eccrine sweat, they may in some cases be the dominant material present. By using these reagents in sequences with reagents targeting eccrine constituents, more marks may be detected overall. Oil Red O is a reagent designed for use on porous surfaces and stains the fatty constituents in the fingermark a deep red colour, whilst Solvent Black 3 is used for grease contaminated non-porous surfaces resulting in the mark being stained a deep blue–black as shown in Figure 10.9. Both reagents are used in similar ways – the surface being treated is immersed in or wetted with the reagent solution until the dye has stained the fats present and then the excess solution is washed from the surface using water.

10.2.5.3.5 Basic Violet 3

Basic Violet 3 is similar to Solvent Black 3 in that it stains certain fatty constituents of fingermarks. However, Basic Violet 3 also strongly stains skin cells and can be a useful reagent for enhancing fingermarks on the adhesive side of adhesive tapes, where many skin cells may be pulled from the surface of the finger. Basic Violet 3 is applied by immersing the solution in the dye solution, then washing the excess dye from the surface using water. Marks are stained a violet colour, but the dye is also fluorescent and additional marks may be detected by conducting a fluorescence examination.

10.2.5.3.6 Acid Dyes

Acid dyes are principally used for the enhancement of fingermarks contaminated with blood. None of the reagents are blood-specific, but all are protein stains and therefore capable of interacting with many of the proteinaceous constituents of blood. However, it should be recognised that there are many protein containing contaminants that can give 'false positive' reactions and therefore the reagents should be used with other confirmatory tests to demonstrate that blood is actually present. The acid dyes used include Acid Black 1 (which stains marks a deep blue -black), Acid Violet 17 (which stains marks a deep violet) and Acid Yellow 7 (which stains marks a pale yellow and is also highly fluorescent). The reagents are applied in three stages, immersion or wetting with a fixing solution to fix the basic proteins in the mark, followed by application of a staining solution to selectively stain the fixed proteins, and finally a wash solution to wash the excess dye from the surface.

10.2.5.3.7 Haem Reagents

Another set of reagents used for enhancement of fingermarks contaminated with blood are reagents that utilise the peroxidase reaction with the haem molecules present in blood, converting the active chemical from a colourless to a coloured form to make the mark visible. These include leuco crystal violet (colourness – violet), leuco malachite green (colourless – green-blue) and diaminobenzidine (colourless – brown). These are applied to the surface as colourless solutions (some incorporating fixing agents) and the marks appear as coloured products over a period of a few minutes. Note that although luminol is a reagent of this type, and produces a chemiluminescent reaction product, it should not be used for fingermark enhancement because it does not contain a fixing agent and ridge detail may be diffused by repeat applications. Haem reagents are more specific to blood than the acid dyes, but there are still several substances that can produce false positive reactions.

10.2.5.3.8 Cyanoacrylate Fuming

Cyanoacrylate fuming enhances fingermarks by means of a polymerisation reaction, with certain constituents of the fingermark (including water and some amino acids) initiating polymerisation

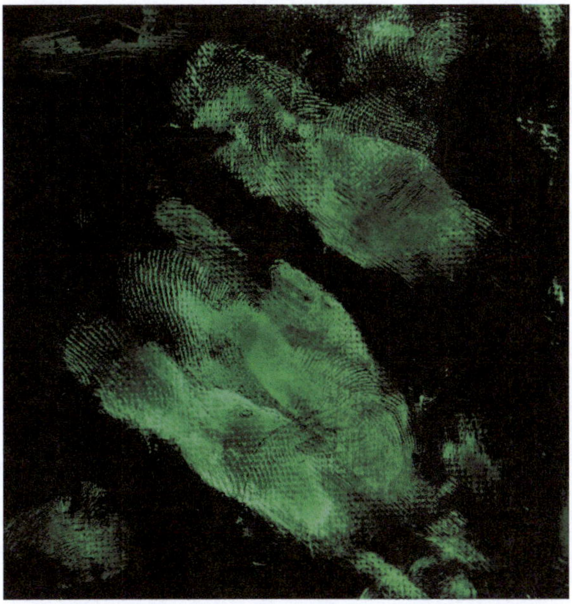

Figure 10.10 A fingermark on a plastic document wallet developed using cyanoacrylate fuming and dyed with the fluorescent dye Basic Yellow 40.

of cyanoacrylate vapour and resulting in the growth of fine polycyanoacrylate 'noodles' on the fingermark. On some surfaces this alone may be sufficient for the mark to become visible, but further enhancement can be obtained by immersing the surface in solutions containing basic, fluorescent dyes which selectively stain the polycyanoacrylate, making the marks readily visible by fluorescence examination. An example of a fingermark enhanced by this procedure is shown in Figure 10.10. Cyanoacrylate fuming is best conducted within cabinets with controlled humidity levels, allowing the salts present in the mark to reabsorb moisture from the atmosphere to assist with initiation of the polymerisation reaction. The cyanoacrylate vapour is produced by placing liquid cyanoacrylate in a dish placed on a heated plate, although chemical initiation of fuming using sodium hydroxide and sodium carbonate has been considered in the past.

10.2.5.4 Other Methods (Utilising Non-specific Differences Between the Fingermark and the Background)

10.2.5.4.1 Vacuum Metal Deposition

Vacuum metal deposition is a process that visualises fingermarks in a way analogous to breathing on a mirror or glass to reveal previously

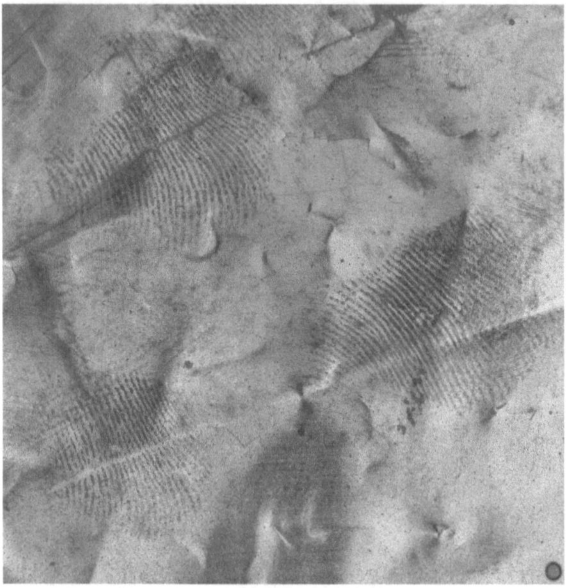

Figure 10.11 Fingermarks on a plastic bag revealed by gold/zinc vacuum metal deposition.

existing marks. On a mirror, these marks become visible because of the different sizes of water droplets that form on the surface – vacuum metal deposition operates in a similar way but utilises the deposition of metals from heated filaments at high levels of vacuum to produce metal clusters on the surface. The characteristics of the surface (surface energy, surface chemistry, abrasion) all influence the size and distribution of the clusters that are formed. In the more widely used gold/zinc process, a small amount of gold is deposited to produce the different distribution of gold clusters on the surface, followed by zinc to produce differential growth rates on the different regions and to visualise the mark as seen in Figure 10.11. Alternatively, silver can be used as a single deposition metal, although the resultant marks are more difficult to see and more prone to handling damage.

10.2.5.4.2 Powder Suspensions

Powder suspension is a very simple process but one where the mechanism has not been conclusively established. It involves the mixing of a dry powder (*e.g.* titanium dioxide, iron(II/III) oxide or carbon) into a relatively thick paste with a water/surfactant mixture. This is then applied to the surface using a wetted, soft brush, then the excess washed from the surface. It is thought that a micelle structure is formed around the powder particles, and this is disrupted by certain constituents of the fingermark to result in preferential deposition of powder on the ridges. It is known that the process does not work well on purely sebaceous marks, nor on marks that have been freshly deposited, but can still develop marks on surfaces that have been wetted. It is known to work well on faint traces of blood in addition to latent marks.

10.2.5.4.3 Physical Developer

Physical developer is a process adapted from the photographic industry, and results in the selective growth of silver particles on the ridges of the fingermark. It is the most effective process for enhancing fingermarks on porous surfaces that have been wetted. For paper items that contain alkali fillers such as calcium carbonate, it is first necessary to immerse the item in an acidic solution such as maleic acid to neutralise the fillers and prevent them acting as preferential deposition sites for the silver particles. The item is then immersed in a developer solution containing silver and iron ions in a fine equilibrium maintained by a combination of surfactants that create stable micelle structures around silver nuclei. The presence of constituents present in the fingermark, or charges developed on them by the acidic developer solution, results in the reaction being tipped in favour of the breakdown of micelle structures and deposition of silver nanoparticles in the locality of the fingermark ridges. Further silver deposition results in the growth of spherical particles on the fingermark ridges, and the developed marks appear dark grey. The item is then washed in water until all residual traces of the developer solution have been removed.

10.2.6 Capturing Images of Fingermarks

10.2.6.1 Capture Requirements

Visualisation of fingermarks using the processes outlined above is only part of the process. Developed marks are very rarely examined and compared *in situ*, and therefore an image of the fingermark must be captured in order for a comparison to be made. Such images must be fit for the purposes of comparison by a fingerprint identification specialist, and also to be loaded and searched on an AFIS system. All the effort that has been expended in visualising as many fingermarks as possible may be negated if the images that are captured of them are not suitable for an expert to use them in comparison.

Almost all images of fingermarks are now captured digitally, either by digital cameras or in some cases by scanners. Important factors to consider when capturing an image of a fingermark that is fit for identification purposes are:

- Inclusion of a scale. Both identification specialists and AFIS systems rely on the fingermark being at the same scale as the fingerprint it is being compared to. If the image is being printed, it is important that it is output at 1:1 scale. By including a graduated scale, or marker of known dimensions in the image it is possible to ensure that the image is correctly scaled.
- Resolution. The image needs to be captured at a resolution that is sufficient to reproduce all of the detail relevant to the identification process. Where only second level detail (minutiae) are present, 500 pixels per inch (ppi) is generally sufficient but where third level detail such as pores and the shapes of ridge edges need to be captured, resolutions of 2000 ppi may be required. Most modern digital cameras have large sensors and filling most of the frame with the fingermark will result in an image that is at a sufficient resolution to capture all detail that may be required. The image may need to be rescaled to a particular resolution (500 ppi or 1000 ppi) if it is to be uploaded to an AFIS system.
- Distortion. Although both human identification specialists and AFIS systems can compensate for some distortion in images, it is desirable that this be kept to a minimum. Fingermark images should be captured with the camera perpendicular to the mark, and not at an angle. If this is not possible then the image should include an 'L' shaped scale so that any distortion can potentially be adjusted for during processing.

In many cases, even images captured with all these factors taken into account may not be easy to identify because the flow of the fingermark ridges may be difficult to discriminate against the background, for example because the background is highly patterned, multi-coloured, textured or obscured by reflections from the surface. In these situations it may be possible to use specialist imaging or image processing techniques to improve the contrast between the fingermark and the background. An outline of some of the methods that may be used for these purposes are described in the following sections.

10.2.6.2 Specialist Image Capture Techniques

10.2.6.2.1 Infrared Reflection

Infrared reflection in the near infrared (700–1100 nm) region of the electromagnetic spectrum can sometimes be used to suppress

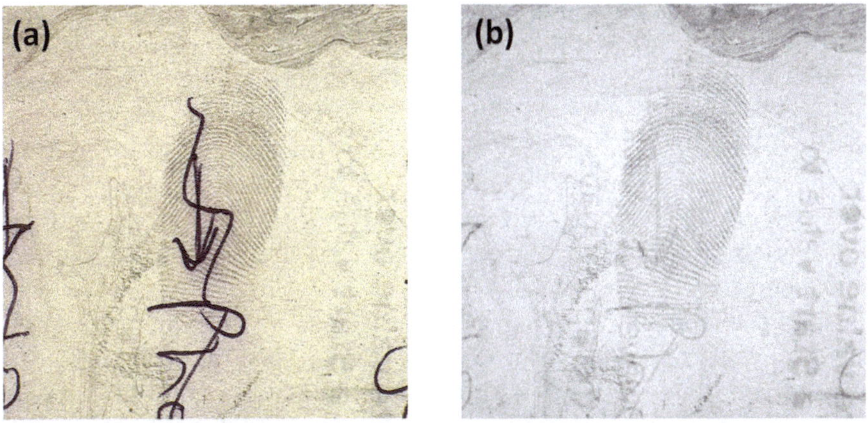

Figure 10.12 A fingermark developed using physical developer and overwritten with blue ballpoint pen, (a) as seen under white light, and (b) viewed in infrared reflection using a filter to block radiation below 830 nm wavelength.

background colours and patterns because many of the organic pigments used in printing inks are infrared transparent, and surfaces that appear highly patterned and/or coloured under white light may appear devoid of printing when viewed using infrared reflection.

The infrared reflection process requires an infrared sensitive imaging system, a source emitting infrared radiation and a filter that blocks visible light and transmits infrared radiation. The surface is irradiated using the infrared source and the reflected infrared image captured using the infrared sensitive imaging system, using the filter to block any reflected visible light. Marks developed using deposition processes incorporating inorganic or metallic materials including vacuum metal deposition, powders and powder suspensions either absorb or scatter infrared radiation more than the background. Thus, developed marks remain visible when imaged in the infrared. As outlined above, many surface patterns printed using organic dyes 'disappear' or fade when viewed using infrared reflection, making a developed mark, such as the one shown in Figure 10.12, easier to discriminate from the background.

10.2.6.2.2 Multispectral Imaging

There are several technologies that can be used in multispectral imaging, each enabling the capture of a digital image providing both spatial and spectral information, with each pixel having a reflectivity or fluorescence spectrum associated with it.

One common approach is to use a monochrome sensor array in combination with a tunable filter or a series of light sources with different, narrow band outputs. Multispectral imaging requires the capture of a series of images of the surface and fingermark over a range of wavelengths. This can be achieved by using a broadband light source and tuning the filter to vary capture wavelength, or by progressively illuminating the surface and fingermark with different wavelength light sources. The series of monochrome images thus collected are known as an 'image cube' (or 'image stack') and can be interpreted by software to give an RGB (red–green–blue, *i.e.* colour) representation of the surface.

Once the 'image cube' has been obtained, a range of processing techniques can be applied to the spatial and spectral information to extract the desired information. These include simple selection of regions with the desired colour spectrum (*e.g.* ninhydrin) and deselection of regions of unwanted background colour/pattern. Other approaches such as Principal Component Analysis can be utilised to identify the major spectral responses from the fingermark and background. In all cases the output is a false colour image that more clearly discriminates the fingermark from the background than the original colour representation.

10.2.6.2.3 Mass Spectrometry Imaging

An area of growing interest is the use of mass spectrometry techniques with an imaging capability, including Matrix Assisted Laser Desorption Ionisation (MALDI) and Secondary Ionisation Mass Spectrometry (SIMS). These techniques produce scans of the surface with each point of the scan producing a mass spectrum of the molecular species present. Not only can these techniques produce an image at a resolution suitable for comparison, but they can also identify some of the substances present (*e.g.* metabolites and contaminants) and map their distribution in the fingermark. This information may add contextual information to an investigation, although validation work is required before wider application of these techniques on casework.

10.2.6.3 Image Processing

The transition from traditional wet photography to digital image capture has significantly increased the number of image processing methods that can be applied to enhance the contrast between fingermark and background post-capture.

Some of these methods are simple (such as contrast and brightness adjustment) whilst others are more complex and may need some expertise to apply. One example of this is the use of Fast Fourier Transform algorithms to separate out the components of the image associated with a regular, interfering background pattern from the irregular components of the fingermark. By masking out the image components associated with the regular background and reversing the transform, the fingermark can be clearly seen.

When conducting any image processing, it is essential to first save a 'master copy' that represents the fingermark in its 'as captured' state (this can effectively be regarded as the 'digital negative'). Any processes then applied to the image should be applied to a 'working copy' and the order and nature of those processes saved as an audit trail. This is so that if the origin of any working copy image is called into question, then it can be repeatably demonstrated that by applying the processes in the audit trail to the master copy, the same result can be obtained.

10.3 Fingerprints and Other Biometrics in Identification

10.3.1 Introduction to Biometric Systems

The concept of the use of biometric information (including fingerprints) for personal identification has been touched upon in Section 10.1. The aim of this section is to provide a view as to the role that such biometrics currently play in law enforcement and how they may in the future offer new opportunities (and threats) to the work of the crime scene examiner.

There is a very simple definition of biometrics, namely, "the automated identification, or verification of identity through repeatable measurements of physiological, or behavioural traits." Perhaps the most important word in the definition is 'automated'. It is automation that provides the greatest benefit of biometrics and separates it from its roots in the card-based Anthropometric Identification System devised by Alphonse Bertillon back in the 19th century.

Our human abilities outstrip the accuracy of biometric matching engines in almost all circumstances. We often read about the incredible 'new' technologies that are shaping the future of human identification, when in reality our natural ability to use physical and behavioural traits in this way have been around as long as humans themselves. Our ability to accurately identify other people without employing technology 'external' to our own selves is phenomenal. Our own eyes are an incredibly efficient 'sensor' and our brains the most advanced matching and image storage devices that will ever exist. The 'communications network' that links the sensors, processors and image stores together does so with lightning speed and is able to outstrip the performance of any computer, even to the point of being able to self-repair damaged or worn out parts automatically.

Even employing standard 'binning' methods a fingerprint expert could use these natural abilities to perhaps manually search twenty to thirty reasonable quality scenes of crime (latent) marks a day against collections of rolled, inked impressions. Sooner or later though, the limitations of the human come into play and the examiner will become tired, bored, or distracted. Concentration levels will fail and mistakes will be made. However, we also have the ability to just 'know' that the person in the crowd is somebody we are acquainted with, or have seen before. We can do this even after the shortest of glimpses and often at a considerable distance. What is it that makes us so sure? The answer is that our own brains are already able to achieve the 'holy grail' of biometrics, namely 'recognition at a distance'. Without any apparent effort our data stores fuse the information that our 'sensors' collect, make comparisons and report results; the program written, updated and executed in our own mobile super-computers. However, we cannot know everybody, or expect others to. Technology is a tool. Where it comes to identification, this tool provides many advantages that are recognised through automation.

To consider how a biometric system actually works we need to take a generic view that encompasses all of the requirements of the definition given above. In this, it is useful to maintain the 'human biometric processor' analogy. To either identify somebody we know, or to come to the conclusion that the person we are looking at is in all probability one of those people, we have first to 'enrol' an image of that person in our own personal super-computer (our brain). Just how a human being achieves this is still something of a mystery, but in all probability our brains do much the same as a biometric system. We 'encode' the image that we view (or hear) using sensors (our eyes and ears) in such a way that we can store away a physical image as a 'template', *i.e.* an encoded representation of a person that can be added to the millions that we will process in our lifetimes.

A biometric enrolment therefore requires that an image, be it a face, a fingerprint, an iris pattern, a voice, or any other human characteristic, is first captured. This original image may be captured at a high resolution to ensure the features of interest are recorded, and can be many megabytes in size. That image is then encoded; a process that turns the image into a mathematical reference 'map' of that image. This encoded information is what is known as a 'biometric template'. This template is usually very small in size (often only around 4 kilobytes) and it is often the case that more than one template is produced during the enrolment session, so as to enhance the overall performance of the system. It is important to note here that once a template, or set of templates, has been produced it may not then be necessary to keep the original image, which can be discarded as there will be no requirement to display that image again in the automated world of biometrics (although some systems, such as fingerprint databases, may retain low resolution images as an aid to human examiners checking the outputs of an automated comparison). It is also important to note that it is not possible to reconstruct the original image of a face, iris pattern, or whatever human characteristic was originally captured from the biometric template produced during the encoding process.

Once the template(s) exist, we need to consider how are they used to either identify a person, or verify their identity. We can again use the 'human biometric processor' analogy. There is now a representation of an image (a biometric template) stored in our brains against which we miraculously compare many new such templates that we produce every time we see a face, hear a voice, or even note how a person walks. In a fraction of a second we capture that image, we encode it and we compare it against the data we already have stored in our personal cranial databases. We then come to some conclusions, *i.e.* "I know this person", "I don't know this person", or perhaps, "that person may be somebody I have come in contact with in the past, but I'm not completely sure of that". This is in effect exactly what the biometric system does, but in many ways it is more efficient than a human in that it can keep templates that don't 'degrade' over time, *i.e.* its memory is always sharp, it doesn't go 'face blind' and of course it can make comparisons against millions of templates in a fraction of a second, 24 hours a day, seven days a week. But it also has drawbacks when compared with the human brain processor. It works based on a computer program written by humans and it cannot (to date) apply a great deal of intelligence to a comparison. We learn from experience and in some ways so do computer programs. But we have an innate ability to adjust our decisions based on many factors, such as the environment we are in and (for facial comparison) accounting for the likely ageing effects on the image we are seeing now, as opposed to the image we originally processed, *i.e.* we compensate for ageing in a manner that is difficult to emulate in a biometric algorithm.

10.3.2 Biometrics in Forensic Crime Scene Investigation

The growth in the use of biometric systems in our everyday lives offers new opportunities and threats to their use in forensic crime scene investigation. There are many that feel that we live in a world where biometric systems are adding a further layer of surveillance that is intrusive, or beneficial to our security, depending on personal viewpoints. The reality is that in the near future (perhaps only a few years), all mobile devices manufactured in the world will have biometric capabilities built in as standard. We are already becoming habituated to the use of biometrics in our personal lives, be it to log into our mobile phone 'apps', or make an online, or in-store payment. But we must distinguish between this 'civil' use of biometrics and that employed by governments for law enforcement, or other purposes such as voter registration, identity cards, or passports.

In most cases our use of biometrics is, or soon will be, in what is known as the 'one to one' verification mode. In this scenario the person who has enrolled their face, voice, fingerprint, iris, vein pattern, *etc.* does not need to be identified against a large database and it is usual for the biometric template to be stored on the device that remains in the possession of the end user. All that is required is to answer the question "is the person using this device, in all likelihood the person who is entitled to do so and so may gain some form of benefit from verifying their identity?" In truth the same applies in many government applications. The biometric passport is now standard in much of the world, but again, the question is the same and what happens is almost identical in the civilian and governmental application; a one to one verification.

However, biometric databases against which templates can be compared on a massive scale do exist and their number is growing rapidly. There is growing opposition to the holding of biometric data on civilian systems and in truth, there is often no good reason to do so, because the decision as to identity has been made and all that needs to be communicated to a bank, travel operator, shop, *etc.* is that the user of a device is probably the person that has the right to whatever benefit they are seeking. This is in practice no different in many respects from that person having used a password, or PIN, or any combination of 'traditional' identification techniques. All that has happened, or is happening, is that now we can add the 'something we are' proposition to the existing 'something we know' (a password, *etc.*) and the 'something we have' (a token, bank card, *etc.*). The benefit of this to a person is that their biometric characteristic is not easily stolen, compromised, or otherwise used illegally and that the biometric can replace, or augment the existing security methods available. So from a law enforcement perspective there are very limited forensic investigation benefits accrued over and above what would have been available before this particular mode of biometric usage was available.

It is also important to note that the biometric algorithms adopted in any given application may be provided by different commercial companies and that the templates produced are not usually suitable for use in any other commercially supplied system. The lack of interoperability is and will remain for some time an issue. The human biometric matching algorithm is hard wired into us at birth and we all apply it, as humans, in much the same manner. Computer systems and the programs they run are diverse and despite some similarities (such as programming languages), they are designed to meet commercial needs and thus far, the sharing of that data across differing applications supplied by different producers is rare and the constraints of commercial advantage and data protection legislation will continue to prevent the widespread adoption of biometric data sharing.

Law enforcement employs three distinct, but not inseparable types of identification in the prevention and detection of crime. These can be classed as, person-to-person, latent-to-suspect and biometric 'image'-to-suspect. This book covers much, if not all of the first two law enforcement scenarios but in summary; in the vast majority (if not all) of countries throughout the world there is now only one principal method employed for the 'measurement' of a person to establish whether that person is known to the authorities, or is who they are suspected of being. That method is of course, fingerprints. The taking of a set of 'ink on paper' and now increasingly 'livescan' prints from arrested persons is normal. These fingerprints are routinely used to establish one of four things, namely that;

(1) A person is known.
(2) A person is not known.
(3) A person is known and is who he is suspected of being,
(4) A person is known, but is not who they were suspected of being.

Governments have invested large amounts of money in AFIS and it is unlikely that their predominance as the principal person-to-person identification method will be challenged in the near future.

However, the emergence of the new biometric technologies (particularly iris and facial recognition) and their increasing accuracy and falling costs, could challenge this supremacy in the longer term. It may not be unreasonable to soon start considering the taking of an iris pattern from an arrestee. Certainly, the 'maths' behind iris recognition technology seems convincing when compared with fingerprints. It is possible now to conduct high penetration searches using iris recognition technology, so why are we not clamouring to start immediately on the road to replace fingerprints for person to person identification? There are a number of reasons, including the protection of existing investment and the risks surrounding single supplier sources. It is also possible that legal recognition will not be easily obtained.

The first two areas of identification discussed above (person-to-person, latent-to-suspect) are not new. Emerging biometric technologies do affect them both, but the general processes of these forms of identification are tried and trusted over many years. However, there is one area of identification (biometric 'image'-to-suspect) in the law enforcement arena that stands to benefit from the emergence and improvement of biometric technologies in a more immediate and dramatic way.

The potential of automated face recognition has been long recognised as a way of overcoming the difficulties involved in searching hours and hours of CCTV footage for a single individual, assisting in the almost impossible task of trying to match a poor quality facial image on a bank's video tape with an unidentified armed robber, and many other scenarios. Similarly, voice recognition could aid in analysis of audio recordings of kidnappers, or hoaxers who have never been identified. All of these possibilities represent a scenario where there may be no 'physical' trace evidence, or any idea who the individual may be at all. Manually attempting to find a 'face in the crowd', or identify a suspect from pictures of known offenders is a notoriously difficult task, as well as a very costly one in terms of an investigators time. Observing a single video monitor in a local council's control room for hours on end waiting for a particular individual to appear for a second or two, will soon demonstrate the concept of 'face blindness' and why a biometric system may be more suited to the task in hand. There are already on the market numerous products that can scan hours of CCTV footage in seconds, extract faces, encode them and search them against databases of one, or multiple suspects. Such products have been proven, in some scenarios, to be far more accurate in their ability to find the 'suspect' than any human.

There are however a number of obstacles that have to be overcome before the effective and efficient identification of such biometric images becomes a reality. Firstly, we have to consider the state of the biometric technologies themselves at this current point in time and the chances of them improving in the foreseeable future. Face in the crowd and automated witness album searching are with us now, but what levels of failure in terms of false accept rate and false reject rates are deemed acceptable by the operators? If a national database of facial images were to be instigated, a multitude of issues would need to be considered. These include how a national facial image database would be used and the business case behind it, legal and civil liberty implications and any changes in the law necessary to make such a database a workable proposition, national standards around facial image capture, image compression, and many others. All of these and many more questions need to be answered before any further move is made towards the establishment, or expansion of such law enforcement based databases and systems.

Biometric technologies offer many advantages to the law enforcement community. These advantages fall into two distinct areas, but are not mutually exclusive. Firstly, the business, process, cost and security advantages that biometrics bring to any organisation can be applied throughout any criminal justice system. Secondly, there are the direct advantages to law enforcement itself as applied to the two major objectives of the prevention and detection of crime. Identification always plays a part in crime investigation and conviction in a fair and just system. Clearly, there is a wealth of experience and expertise already applied in these fields and technology is enhancing them at a rapid pace. However, biometrics opens up new possibilities and challenges, which may increase dramatically the effectiveness of the law enforcement community. We may now have to consider how we apply Locard's Principle to evidence of a purely digital nature. It is a certainty that we will see a rise in cybercrime and that increasingly, such crime will involve an element of biometrics.

Consider a scenario where a multi-million pound trading fraud is investigated that involves a particular trader having signed onto a trading system using biometrics, but who denies having done so. Digital images of abuse may contain biometric features that can be matched to suspects, such as fingerprints, vein patterns or irises, and indeed several successful prosecutions have proceeded on this basis. However, it is likely that the additional forensic crime scene investigation aspects that may be required in future have not yet been fully considered.

Biometrics are already enhancing our daily lives, both as security and convenience applications and we are fast becoming reliant on them, but there are issues that have to be considered from the investigatory perspective. Principal amongst these issues are the lack of understanding regarding biometrics that currently exists amongst the forensic investigation community and the very diverse nature of the technologies themselves. There are tremendous benefits to be gained from biometrics, but also a very large potential for misuse and criminal activity, and it is essential that crime scene investigators and the judicial system as a whole begin to understand the impact that the mass adoption of biometrics will have.

Further Reading

D. Ashbaugh, *Quantitative-qualitative Friction Ridge Analysis: An Introduction to Basic and Advanced Ridgeology*, CRC Press, Boca Raton, 1999.

Fingermark Visualisation Manual, ed. H. L. Bandey, Home Office, London, 2nd edn, 2022.

I. E. Dror, Practical Solutions to Cognitive and Human Factor Challenges in Forensic Science, *Forens. Sci. Pol. Manage.*, 2013, **4**, 1–9.

Advances in Fingerprint Technology, ed. R. Ramotowski, CRC Press, Boca Raton, 3rd edn, 2012.

Strengthening Forensic Science in the United States: A Path Forward, National Research Council, National Academies Press, Washington D.C., 2009.

11 Body Fluids

C. Crawford* and C. Jarman*

Forensic Access Ltd, Aspect House, The Quadrangle, Grove Business Park, Wantage, Oxfordshire, OX12 9FA, UK
*Emails: caroline.crawford@forensic-access.co.uk; clare.jarman@forensic-access.co.uk

11.1 Introduction

Body fluid testing and DNA profiling are invaluable tools in the investigation of crime and can form crucial evidence within the judicial process. Arthur Conan Doyle's Sherlock Holmes described the importance of body fluid identification back in 1887 in 'A Study in Scarlet'. Sherlock stated "*Criminal cases are continually hinging upon that one point. A man is suspected of a crime months perhaps after it has been committed. His linen or clothes are examined, and brownish stains discovered upon them. Are they blood stains, or mud stains, or rust stains, or fruit stains, or what are they? That is a question which has puzzled many an expert, and why? Because there was no reliable test.*" Fortunately, we now have robust, validated methods for the detection of not only blood but also a range of other body fluids including semen and saliva.

The detection of a body fluid assists in the investigation of crime, but far greater value comes if we can determine from whom that body fluid may have originated. Blood staining on the clothing of a suspect in an assault case may be of interest, but only if that blood is from the complainant rather than from the suspect themself. Similarly, semen

Crime Scene to Court: The Essentials of Forensic Science, 5th Edition
Edited by Niamh NicDaeid and Peter C. White

Published by the Royal Society of Chemistry, www.rsc.org

on the intimate swabs of a complainant in a sexual assault case is of high importance if it could have come from the suspect, but potentially of no importance if it came from a consensual sexual partner. Historically there were no tests available to determine from whom a particular body fluid might have originated. Later, blood grouping tests allowed limited information to be gained about the potential source of a body fluid. However, it was the advent of ever more discriminating DNA profiling tests in the late 1980s and 1990s that revolutionised this area of forensic science.

DNA is a complex chemical found in almost all the cells of the human body. DNA acts as a blueprint or code determining everything about us, from how our body functions to our physical appearance. Our DNA is usually the same throughout our body, so the DNA profile obtained from blood is the same as that obtained from semen or saliva. If we could analyse the whole of a person's DNA, it would, with the exception of identical twins, be unique to that person. However, this is neither practical nor necessary and so DNA profiling involves testing targeted areas of DNA which are known to vary considerably between individuals. This produces what is known as a DNA profile. If a DNA profile obtained from a crime scene or exhibit (collectively termed a 'crime stain') matches the reference DNA profile of an individual, the crime stain could have come from that person. If the profiles do not match the crime stain could not have come from that individual.

Whilst extremely discriminating, DNA profiling on its own is often not enough. Body fluid testing and interpretation can further assist in addressing what may have occurred during an incident, rather than just who was present. Take the example of an allegation where a female states that a male bit her breast over her clothing. The male states that the two spent the evening together. They hugged and kissed but at no time did he bite her. DNA transfer between both individuals would be expected given either scenario and so DNA evidence on its own would not be of use. However, the finding of a distinctive area of saliva staining on the chest area of the female's clothing, along with DNA profiling of that staining showing significant amounts of DNA that could have come from the male, may start to build a picture and begin to provide evidence to support the complainant's version of events over that of the defendant and build a picture of what may have occurred.

The role of the forensic biologist is to use scientific techniques to detect and identify body fluids on items and at scenes; to employ DNA profiling to help address from whom that body fluid may have originated; to then use the results to comment on how that body fluid or DNA may have been deposited and finally, where possible, to comment on whether their findings may lend support to one version of events in a case over another.

11.2 Overarching Principles Within Forensic Biology

11.2.1 Case Assessment and Interpretation

In approaching any examinations involving forensic biology techniques it is essential that the scientist adopts an approach that:

(1) Maximises useful evidence recovery
(2) Preserves the item as far as is possible for other evidence types and future examinations
(3) Prioritises items and examinations that may provide most value to that particular investigation
(4) Aims to deal with the central questions of interest to the investigators and the court
(5) Utilises testing in a way which gives best value whilst taking into account budgetary constraints

Cognitive bias (which is explored elsewhere in this book) can have an effect in forensic biology where subjective interpretations and comparisons are undertaken. Safeguards against cognitive bias in biology include training scientists to have an awareness of the issues, following standard operating procedures, independent checking of results and the use of a Case Assessment and Interpretation (CAI) model.

The CAI model was developed in the late 1990s by a group of scientists from the Home Office Forensic Science Service to provide a robust framework and best practice approach to the assessment and interpretation of forensic casework and to protect against cognitive bias. This approach is widely adopted within forensic biology in the UK today. For the purposes of this chapter, we focus on specific aspects of the CAI approach and its application to forensic biology – more information is provided in the Further Reading section.

When using the CAI model, casework is divided into investigative and evaluative aspects. Investigative is often, though not always, in the early stages of the case, where the scientist conducts tests and makes inferences from their findings that may then assist with the investigation – in essence the scientist uses their techniques and expertise to investigate what may have happened. Examples include identifying a body through DNA profiling or using the detection of semen on intimate swabs to infer that sexual intercourse has occurred, then using DNA profiling and the National DNA Database to identify a potential suspect. The evaluative phase is usually later in the case when the scientist evaluates their findings in light of competing prosecution and defence accounts – this aligns naturally with the accusatorial court process. The CAI approach is largely based on Bayes Theorem and is explored in more detail elsewhere in this book. As discussed elsewhere, the scientist deals only with the likelihood ratio part of Bayes and evaluates their evidence within this format.

$$\text{Likelihood Ratio (LR)} = \frac{\text{PrE}|H_\text{p}}{\text{PrE}|H_\text{d}} \tag{11.1}$$

Crucially here, the scientist compares the probability of a particular piece, or combination of pieces, of scientific evidence given the prosecution hypothesis, H_p – what the prosecution say occurred, against the probability of that same evidence given the defence hypothesis, H_d – what the defence say occurred.

If the evidence is equally probable given H_p or H_d then an LR of 1 is obtained – this is in essence neutral – the findings can equally be accounted for by the prosecution or defence accounts. An example of this might be the detection of a suspect's blood at the scene of the crime. H_p is that his blood was present because he committed the crime, however H_d is that he was at that location earlier in the day and was bleeding at the time. The forensic evidence is equally probable given either account. An LR of above 1 indicates that the evidence provides support for H_p over that of H_d. An LR of below one indicates that the evidence provides support for H_d over H_p.

The likelihood ratio is intrinsic to all areas of forensic biology, from employing validated data to undertake statistical analysis of DNA profiles resulting in a numerical LR, to a more subjective approach based on experience for the evaluation of body fluid presence and distribution.

11.2.2 Case Strategy

It is imperative at the onset of any case that the scientist develops a considered case strategy. This may evolve through the life of the case, incorporating new information and results of examinations. A robust case strategy will set the foundations of that investigation and feed into the final interpretations and evaluations. It also ensures that proper consideration is given to the best items for examination, most appropriate tests to be performed, the order of those tests and the potential usefulness of other evidence types.

11.3 Detection of Body Fluids

The detection of any body fluid usually begins with a visual search using strong white lighting. Some body fluids such as blood may be relatively easily observed as red/brown stains, although very weak, small or dilute blood stains and stains on certain backgrounds may be more difficult to see. Other body fluids, such as semen and saliva, may only be visible if they are present in high quantities or on contrasting backgrounds. Biochemical screening tests are therefore required to locate body fluids. Even where a visible stain has been found these tests are used to ensure that the stain is a body fluid rather than some other substance with a similar appearance.

Some of the biochemical tests are confirmatory. They will only give a positive result in the presence of a specific body fluid. Others are known to have a small number of false positives and with these, in the presence of a stain with the appearance of the body fluid AND a positive reaction we might *presume* the body fluid is present, the test being termed a 'presumptive test'. Some tests have so many false positives that they can only give an indication the body fluid is present. These tests are still useful to locate stains, sometimes for further testing, and give some information about whether the body fluid may be present. A combination of tests may be used depending on case circumstances.

Prior to any destructive biochemical tests which may alter or remove body fluids (or other evidence) from a surface, non-destructive techniques such as the recovery of hairs, fibres and debris, or sampling of areas for 'touch'/'wearer' DNA must be considered. Consideration must also be given to the order of examinations to maximise recovery of evidence that would best address the issues in the case. Finally, evidence types within other disciplines and the effect of body fluid testing on these should also be considered, for example fingerprints or footwear marks. This is usually assessed using a CAI approach and detailed in the case strategy.

11.3.1 Alternative Light Sourcing

Alternative light sources (ALS) can be used to locate body fluids at scenes and on items in the laboratory. This is particularly useful where there is a source of body fluid or DNA (such as 'touch DNA') which is colourless and would not be found during a search using white light (a white light is what we would think of as a 'normal' light, albeit a very bright one). Alternative light sources use different wavelengths of light. The light is passed over the surface or item to be examined based on the premise that some body fluids and areas of skin contact may fluoresce under certain wavelengths. ALS is particularly useful at crime scenes where portable light sources are used to screen large areas that cannot reasonably be recovered for laboratory examination. Body fluids such as semen, saliva, urine and vaginal material can fluoresce when Ultraviolet (UV) light is applied, however, background surfaces may also fluoresce so multiple light filters can be used depending on the body fluid and background material. Body fluids will not always fluoresce under ALS and so a lack of fluorescence is not an indication that the body fluid is absent.

Some laboratories have successfully used ALS to locate areas of clothing that have been touched or handled in order to recover 'touch DNA'. This can be useful where it is likely that contact has been made but the exact location is not known – for example where the victim of a robbery states that they were grabbed on the shoulder – ALS could be used to narrow down the area of their clothing for DNA sampling. Again, this has to be treated with caution and used in combination with case information.

11.3.2 Blood

Blood is the typical body fluid one thinks of when imagining a crime. It can provide shocking images when presented in crime dramas and photographs. However, the scientist must be careful that what they are looking at is actually blood rather than some other substance. Some scenes, particularly kitchens, and exhibits such as dirty clothing or rusty knives, may have numerous stains that have an appearance that is similar, or even visually indistinguishable to blood but are not actually blood and so it is imperative that we are able to differentiate blood from these.

Blood is composed of plasma, which accounts for 55% of the total volume, with the balance being made up of cells. Plasma is around

90% water and is a straw-coloured, almost clear liquid containing dissolved plasma proteins, metal ions and organic substances.

The most common type of cell is the red blood cell, or 'erythrocyte'. There are about 5 million red blood cells in 1 mL of blood. These have no nucleus (and therefore no DNA) and are the only cells to contain the pigment haemoglobin, which gives blood its red colour. Their outer cell membrane, in common with all other body cells, carries biochemical markers which were historically used by the forensic scientist for blood grouping.

White blood cells, or 'leucocytes', are part of the immunological defence of the body and are present in much lower numbers than red cells, around 3000–11 000 per mL. White blood cells have a nucleus and this contains DNA.

The first step in blood identification is a thorough search, usually using a strong white light source, and sometimes a low power microscope (LPM). Depending on the item in question and case circumstances it may be necessary to spend hours searching an item, for example black footwear in which there are seams, treads and laces requiring the item to be manipulated under the LPM to search for minute traces of blood. This was the scenario in the reinvestigation of the murder of Vikki Thompson. Mark Weston was originally tried for her murder and found not guilty but following changes to the double jeopardy law the case was re-opened. No blood had been found on Mark Weston's boots during the original investigation in 1995 but at that time DNA profiling was significantly less sensitive so there was less emphasis on searching for minute traces of blood. With advances in the sensitivity of DNA profiling, smaller traces of blood could now be analysed. During the reinvestigation, which began in 2005, a lengthy search of the boots over several days resulted in the detection of small blood stains which gave a DNA profile matching Vicky Thompson. Mark Weston was subsequently convicted of her murder in 2010.

Other lighting techniques such as infrared have also been used to visualise blood.

If a blood-like stain, usually red–brown in colour, is found it may be tested to determine whether it is blood or some other substance. Several such tests utilise the enzyme-like activity of haemoglobin which, in the presence of hydrogen peroxide, catalyses the release of oxygen. Certain dyes such as Leuco Malachite Green (LMG) or phenolphthalein in the Kastle–Meyer (KM) test, have the characteristic of existing in two different states, a reduced form, which is colourless, and an oxidized, coloured form. If a drop of the dye in solution, kept in its reduced state, is applied to a blood stain, no colour should be observed. However, if a drop of hydrogen peroxide is added, the haemoglobin in the blood will catalyse the breakdown of the hydrogen peroxide. This results in oxidization of the dye with an associated colour change (Figure 11.1). Usually, in casework, a piece of folded filter paper will be rubbed onto the stain to recover a small portion of blood and the dye added to the filter paper so that minimal disruption is caused to the stain itself. This allows the bulk of the stain to be retained for blood pattern analysis and DNA profiling. KM is the most common blood detection test used in UK forensic laboratories. A positive KM reaction is denoted by a pink colour and LMG by a green colour. If a stain has the appearance of blood and gives a positive reaction to the KM or LMG test, it is presumed to be blood.

If no visible blood staining has been detected, a 'general screen' may be undertaken. Filter paper is methodically rubbed over the item in small sections and then tested with KM or LMG to determine

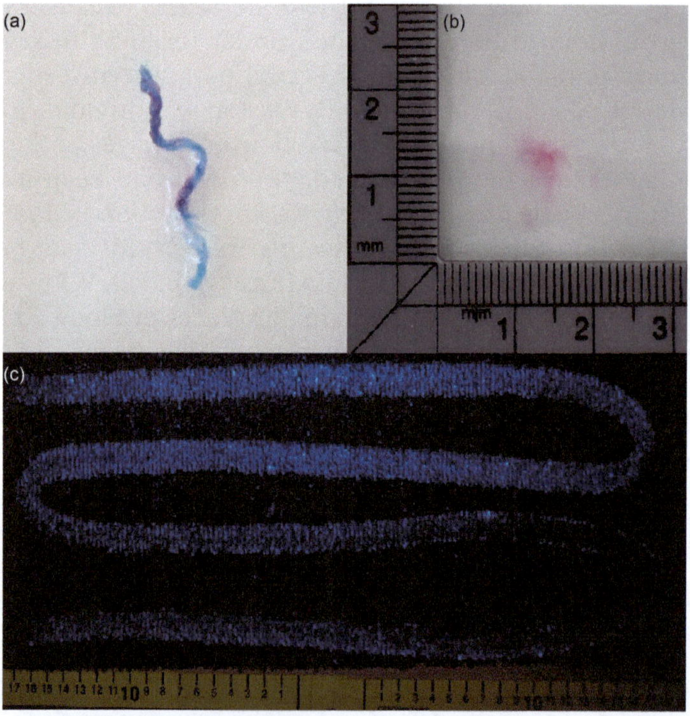

Figure 11.1 (a) KM positive reaction on a piece of green thread that has been 'directly' tested. (b) KM positive reaction on a piece of filter paper that has been rubbed against a blood stain. (c) Luminol positive reaction on a piece of carpet that a blood stained swab has been wiped over to mimic a drag mark.

whether any traces of blood staining can be detected chemically. If a positive reaction is obtained the scientist can examine that particular area in more detail. A positive reaction in the absence of visible staining can be said to indicate the possible presence of blood but cannot confirm its presence.

If a stain has the appearance of blood but is very light and/or may be dilute, a negative reaction to the presumptive test may be obtained, either because the stain is not blood, or because there is insufficient blood present to give a positive result. In this circumstance a 'direct' test can be employed. A thread of the fabric of the garment can be cut out and tested directly with the presumptive test allowing the chemical to react with any blood present within the thread. This results in a more sensitive test (Figure 11.1).

Luminol testing for blood has been used since the 1960s and goes in and out of fashion. Its uses include following trails of blood-stained footprints, where there has been a clean-up of blood at crime scenes or blood that has been washed from clothing, or where many years have passed since blood was deposited. The luminol solution is usually applied directly to the surface to be tested as a spray. As with the blood tests described above, haemoglobin in blood catalyses the reaction – this time causing a release of luminescence (light) rather than a colour change. Luminol is extremely sensitive and can work particularly well on old, degraded and dilute blood staining. As a positive reaction is denoted by chemiluminescence rather than a colour change, luminol can be good at detecting blood on very dark or patterned backgrounds.

There are disadvantages to luminol use. It must be applied in a darkened environment and the chemiluminescence time is short such that photographic recording of reactions can be problematic, though advances in digital photography have greatly improved this. There are considerable issues with its use on non-porous surfaces because blood staining can become wet and run. Even on porous surfaces, luminol can cause some distortion of the pattern of any blood present. A further drawback of luminol is that there are many substances other than blood which can produce a positive luminol reaction and so further chemical testing is always necessary where a positive reaction has been obtained. As luminol is applied directly to the surface, it is a potentially destructive technique and should only be applied after other examinations and following a visual blood search. It does not replace a thorough search for visible bloodstaining.

Another presumptive test, also very good at detecting very low levels of blood is leucocrystal violet (LCV) which is used to detect non-visible bloodstains or enhance marks in blood. LCV is similar to KM and LMG producing a purple colour when oxidized, but as with luminol it is applied directly onto the surface to be tested. LCV has the advantage of containing a fixative, sulfacylic acid, and so can be used on non-porous surfaces without significantly altering the blood distribution. LCV also reacts positively with considerably fewer other substances than luminol. LCV has associated health and safety considerations leading to it currently being a predominantly lab-based technique.

Hemastix® are dip sticks that were originally developed to detect blood in urine. In forensic examinations the yellow pad is wetted and pressed onto suspect stains, or a swab of the stain pressed against the pad. The activity of haemoglobin catalyses an oxidation reaction of 3,3′,5,5′-tetramethylbenzidine to the oxidized coloured form which turns the pad green. Hemastix® are useful for scene work as there are no liquid chemicals to apply, however as a one-step test can produce more false positives.

Confirmatory tests for the presence of blood are also available however, these are generally less sensitive and/or require more test material. These include the Takayama Crystal Assay which results in the formation of distinctive haemochromogen crystals when a blood sample is treated with pyridine.

The tests described above react with human and animal blood. However, immunochromatographic tests are available which detect antihuman monoclonal antibodies. Within a test cassette (similar to those used for Covid or pregnancy testing) these form antigen–antibody complexes in the presence of human blood resulting in a positive test line. These cassettes can therefore indicate that blood is human in origin although other higher primate and ferret blood samples have also given positive reactions. This type of test was particularly useful in the case of a missing person who made fake snuff movies in his spare time. An examination of his home for signs that he had been murdered led to a fridge being found, the inside of which was covered in blood. The use of an immunochromatographic test quickly showed that the blood inside the fridge was not human.

11.3.3 Semen

Most semen examinations are undertaken in relation to sexual assault cases, predominantly on intimate swabs and clothing. The finding of semen and issues in relation to its transfer and persistence are the remit of the forensic biologist.

Semen consists of seminal fluid which usually contains spermatozoa or 'sperm'. The sperm is the male sex cell, or gamete, and its biological function is to pass genetic information from the father to the female gamete, the egg or ovum. Human gametes carry a set of 23 chromosomes, one from each of the 23 pairs that make up the normal human complement of 46. 1 mL of semen from a healthy adult male contains around 60–1000 million sperm cells and a normal ejaculate contains approximately 2.5–6 mL. Most of the volume of semen comes from the accessory glands comprising the prostate gland, the seminal vesicles, Cowper's glands and the glands of Littre. The prostate gland produces a slightly acidic fluid containing high levels of acid phosphatase (AP) which is used extensively for semen testing.

The definitive test for semen is the identification of spermatozoa cells. These can be extracted from material such as intimate swabs (*e.g.* vaginal or anal) or clothing and viewed microscopically. Historically a swab may have been wiped directly onto the microscope slide in order to transfer any spermatozoa present. This progressed to the swab head being mixed with water and centrifugation used to pull the spermatozoa cells out of the fabric through a small hole or spin basket. These methods also extracted any nucleated epithelial cells (NECs) within the sample. NECs line body orifices and, if recovered from intimate swabs in large enough numbers, can mask the presence of spermatozoa. Today, optimised methods using buffers and enzymes are used which increase the yield of spermatozoa and have the additional benefit of separating spermatozoa from NECs. These techniques utilise the different cell properties to preferentially digest NECs separating them from the spermatozoa in the sample. This is particularly useful where there is a significant number of NECs but only a low quantity of spermatozoa, for example vaginal swabs containing a trace (a very small amount) of semen.

In order to visualise spermatozoa a portion of the extract is dried to a microscope slide. Common histological staining chemicals such as Haematoxylin and Eosin or Christmas Tree (nuclear fast red and picoindigocarmine) are applied to the stain. The spermatozoa head is oval shaped and the chemicals differentially stain the nucleus and cytoplasm giving a distinctive bipolar appearance. Occasionally the tail is also visualised but usually the extraction process causes any remaining tails to be broken off. Where buffers and enzymes were not used in the extraction process, microscopic searching for spermatozoa can be hindered by the presence of NECs and other debris, particularly from intimate swabs. In such cases the use of digestion enzymes such as proteinaseK can be used to differentially digest epithelial cells from, for example, the lining of the vagina while retaining spermatozoa.

An estimate of the quantity of spermatozoa present in a sample can be a useful tool in determining how and when semen was deposited, particularly in relation to intimate swabs. Studies are available which give information as to the persistence of semen within the vagina, anus and mouth following sexual acts. However, care is needed to avoid the overinterpretation of evidence in relation to quantities of spermatozoa. Differences in the number of spermatozoa males produce, activities of the complainant after sexual activity (such as bathing, urinating, defecating and tooth brushing), sampling methods, extraction methods and stochastic effects can all have an effect. As such, the number of spermatozoa observed may assist in determining, for example, whether vaginal intercourse occurred a day ago rather than six days ago but would not be able to assist with more narrow time frames such as two days *versus* three. These studies address retention of spermatozoa in living, active individuals. If a person is incapacitated or dead, then these time frames will no longer apply.

Some semen samples may be 'azospermic' containing no spermatozoa, or 'oligospermic' containing few spermatozoa. This may be due to factors such as vasectomy, illness or drug use.

While the identification of spermatozoa will confirm the presence of semen, in many instances we must first locate the staining, for example on bedding or clothing, and it is not practical to extract the whole item. The acid phosphatase (AP) test is a useful and long-standing screening method which detects the enzyme acid phosphatase, often present at high levels in semen.

This test involves wetting the item and laying blotting paper over it. The paper is pressed down and where semen is present, water-soluble AP within it transfers to the paper which is subsequently sprayed with the brown coloured AP reagent (Figure 11.2).

AP reagent contains sodium alpha naphthol phosphate and brentamine fast black potassium salts. When applied to a surface containing acid phosphatase the naphthol is freed from the sodium alpha naphthol phosphate. The naphthol then combines with the brentamine fast black potassium salts causing the production of a purple azo dye. The change in colour on the sprayed paper from brown to purple denotes an area of possible semen. The paper can then be realigned over the item and the location of the purple staining marked on that item. The marked area may be sampled, extracted and microscopically examined for spermatozoa.

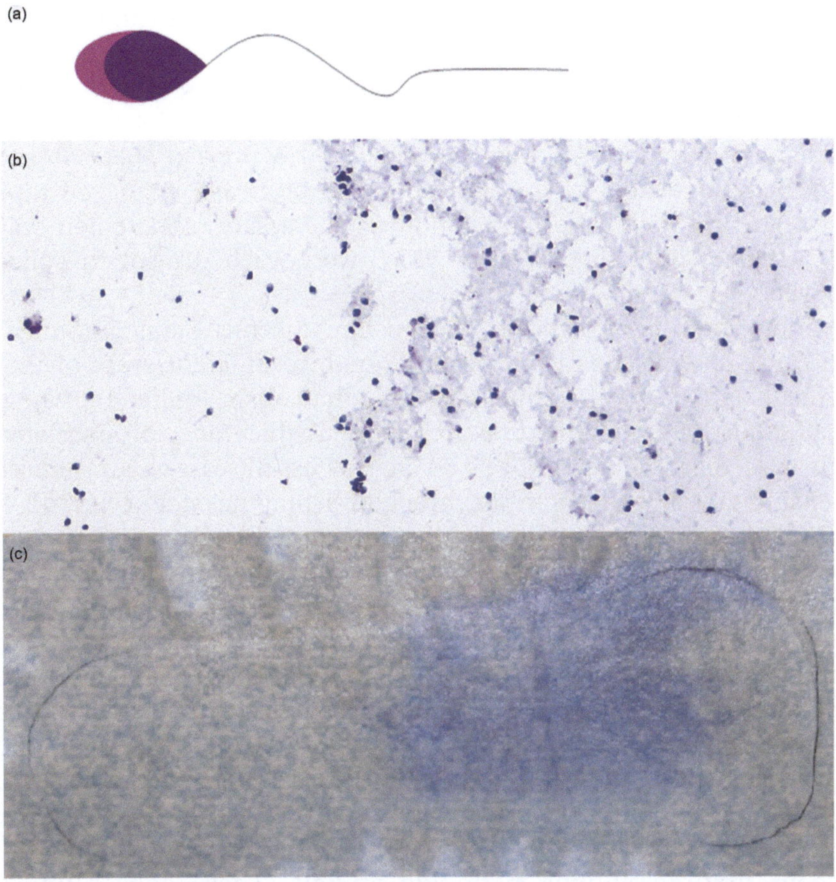

Figure 11.2 (a) Bipolar staining of a spermatozoa with tail still attached. (b) Numerous spermatozoa in one field of view under microscopy. (c) AP positive reaction obtained from a pantyliner (test undertaken on Phadebas® paper).

Other body fluids, including vaginal material, also contain AP and other substances such as perfumes are known to sometimes give false positive reactions. These are usually weaker reactions, and some may be a slightly different colour – vaginal material can give a pink reaction for example. Therefore, a positive AP reaction can only indicate the presence of semen and extraction for the presence of spermatozoa is required in order to confirm its presence. The speed and strength of the AP reaction are recorded as this can give an indication of whether semen may be present and at what quantity or whether there may be a false positive reaction.

As AP is water soluble its detection is of limited use where items have been wet, particularly if they have been thoroughly washed.

In these cases, the detection of AP may not be possible and instead areas of fabric, such as from the crotch of knickers, may be extracted for spermatozoa without prior testing. Spermatozoa are able to persist in low numbers through the washing process and so whilst washing a garment may remove AP, it will not necessarily prevent spermatozoa from being detected. This might be useful in a case of alleged rape where the knickers were worn following the assault and semen may have drained onto the knickers which were washed prior to police recovery.

Various studies have shown that washing a semen-stained garment can cause spermatozoa to redistribute both to different areas of that garment and to other garments within that wash. Therefore, where small amounts of spermatozoa are detected which are not associated with an AP reaction, care must be taken to assess the case circumstances to ensure that innocent transfer through washing has not occurred.

Where a strong AP reaction has been obtained but no spermatozoa identified, further tests should be considered to help address whether azospermic or oligospermic semen is present. The Florence Iodine test, also known as the Choline test, reacts with choline from the seminal vesicles. Choline is water soluble and so this test cannot detect semen after an item has been washed. It is not as sensitive as AP and cannot be used to screen a whole item. Instead, the supernatant from an area of extracted material is dried onto a slide. The Florence Iodine reagent is passed over the dried stain and, in the presence of semen, the choline within it combines with iodine in the reagent to form choline periodide crystals. These appear under the microscope as thin brown needle shaped crystals.

Research is ongoing into newer, easier screening tests for the presence of semen, particularly by commercial biotech companies, however, validation of these in UK forensic casework are still in early stages and it is yet to be seen whether any are as reliable as the standard tests used.

11.3.4 Saliva

Saliva testing might be undertaken in many different types of cases: biting and spitting can occur during assaults; kissing and/or licking often occurs during sexual assaults; when a car crashes and airbags are deployed the saliva of the driver may be deposited onto their airbag. Any of these may result in the transfer of saliva which can then be detected. Saliva can also transfer to drinking vessels, cigarette ends and other items which are naturally in contact with the mouth, however saliva testing is not necessarily undertaken on these items as they may be assumed to contain saliva if a significant amount of DNA is recovered from them.

Saliva is a watery fluid secreted by the salivary glands into the mouth. It contains mucin, which assists the passage of food into the oesophagus, and salivary amylase that digests starches. Saliva also contains cells and hence DNA. These are predominantly NECs shed from the inside surfaces of the cheeks. The salivary amylase can be used to locate and identify the possible presence of saliva.

Items, such as clothing, can be screened for the presence of amylase using paper tests. One of these tests involves the use of Phadebas® paper which consists of homogeneously interlinked starch polymers which have a blue dye covalently bound to them and are insoluble in water. In the presence of amylase, the starch polymers are hydrolysed and the blue dye is released. Thus, a blue colour would indicate the presence of amylase. The screening technique is similar to that of AP screening whereby the item is wetted and the Phadebas® paper placed, dye side down, onto the item. This is then weighted to ensure good contact. Any amylase present will leach into the paper causing the blue dye to be released in that area. If required, the Phadebas® paper may subsequently be screened for acid phosphatase in an attempt to detect any semen present.

Following extraction of swabs (in a similar way to semen extraction) the supernatant may be examined for the presence of saliva using the Phadebas tube test. This is useful where a surface has been swabbed to recover any possible saliva staining – for example in biting, licking or spitting cases. Phadebas® (starch linked with blue dye) is added to the supernatant and incubated. In the presence of amylase the starch is broken down and the blue dye released causing the liquid to turn blue (Figure 11.3).

A similar method to the Phadebas® paper method, Seretec® Amylase-Paper, uses a paper impregnated with starch. Amylase digests starch so that a starch–iodine complex cannot be formed and this results in colour variation on the amylase paper.

As amylase is present in other body fluids, albeit at lower levels, these tests are not confirmatory. They can still be powerful, for example detecting saliva for subsequent DNA profiling in a spitting case where the assailant is unknown to the complainant. However, where intimate samples and underwear are being examined, or other items where false positives may be present, care must be taken to assess whether the reaction is more likely due to saliva being present or because some other false positive is present.

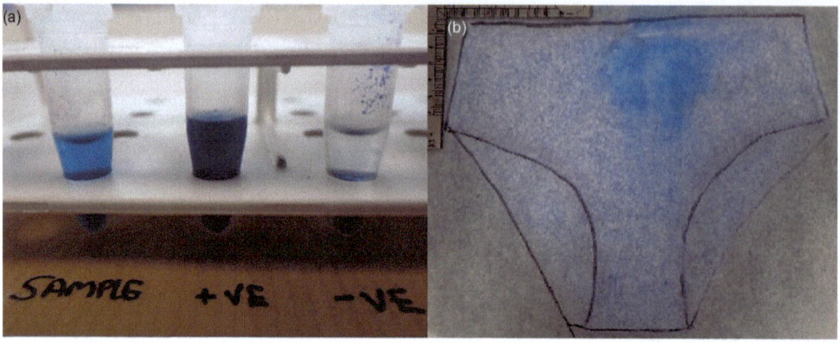

Figure 11.3 (a) Phadebas® tube test results showing a positive reaction to the test sample (left), strong positive reaction to the positive control (middle) and negative reaction to the negative control (right). (b) Phadebas® paper test on a pair of knickers with a positive reaction on the centre front.

11.3.5 Urine and Faeces

Urine and faeces testing, whilst not as commonly used, are none the less important within the arsenal of the forensic biologist. The detection of faeces may be an important aspect in alleged anal rape cases and urine testing may be required in, for example, cases of commercial contamination.

Several tests can be utilised to examine for the presence of faeces. Stains are checked for the odour of faeces and portions can be microscopically examined for the presence of undigested foodstuffs. A chemical test for the presence of urobilinogen, a component of faeces, can also be carried out. Thin layer chromatography can be used to supplement this test, and may give additional information on origin of sample, *i.e.* herbivore, omnivore/carnivore. These tests are not confirmatory.

Urine also can have a characteristic odour and the *para*-dimethylaminocinnamaldehyde (DMAC) chemical test can be used to detect the presence of the component urea. Again, this test may indicate the presence of urine but is not confirmatory.

While it is sometimes possible to obtain DNA profiles from faeces and urine, neither are good sources of human DNA as neither intrinsically contain DNA. Both are waste products and any DNA present is by virtue of being transferred as the product leaves the body. That said, in toxicology cases where an illegal or other banned substance has been detected in urine, the individual may question whether the urine was actually theirs and DNA profiling can help address this in some cases.

11.3.6 Vaginal Material

Particularly in cases of alleged sexual assault, the detection of vaginal material on the penis, fingers or boxer shorts of a suspect may be of use. However, NECs from the lining of the vaginal canal cannot be differentiated from those of other body orifices such as the mouth. As such, whilst it may be possible to ascertain that DNA of a female is present, it is not possible to specifically attribute it to vaginal material.

11.3.7 The Future of Body Fluid Testing

Additional body fluid tests are available, and others are being developed all the time. These range in specificity, sensitivity and ease of use, as do the tests described above. It is the role of the forensic biologist to determine which tests are most appropriate for their laboratory and to validate accordingly, then to select the relevant tests and order of application for each item. That said, the tests most commonly used in UK laboratories today, such as the AP test for semen and KM test for blood have been the main techniques for decades with relatively little change to their application.

DNA is contained within the nucleus of cells (see the sections below for more information). Messenger RNA (mRNA) is a single stranded RNA molecule which carries information from the cell nucleus to the protein synthesising ribosomes. The RNA molecules are different depending on the protein for which they code and thus RNA relating to different body fluids will contain different markers. Efforts are being made to exploit the RNA markers from specific body fluids to confirm their presence. Success in this area might allow the identification of vaginal material for example, or may assist in differentiating menstrual blood from trauma blood. This would be particularly useful if the RNA could be co-extracted with DNA for a single analysis which might detect multiple body fluids with high specificity and sensitivity.

11.4 Contact DNA

While not a body fluid, contact DNA testing has become a significant part of the forensic biologist's work. 'Touch', 'handler' or 'wearer' DNA are descriptions of DNA recovered from items that have been touched, handled or worn *i.e.* DNA transferred through contact. These terms describe the reasons for sampling a particular area rather than a specific type of cellular material. There will rarely be any visual staining and no specific tests are available to indicate where high levels of DNA are present, although alternative light sourcing can be helpful to direct scientists to potential areas of contact. Usually, the scientist will sample from areas that appear well worn or handled or would be expected to come into prolonged contact with the wearer or handler such as the inside of a collar of a jacket or the handle of a knife. Areas of importance can also be dictated by case circumstances, for example in a robbery where the complainant stated their upper sleeve was grabbed by the offender, or an area of a victim's clothing that is believed to have been torn by the offender.

Contact DNA is powerful in a wide range of cases and forms a significant portion of the DNA profiles analysed in modern casework. It can provide evidence of contact between individuals, contact with certain surfaces such as stolen cars or at scenes of crime, the wearing of clothing stained with a victim's blood, or the handling of tools or weapons used in a crime. In the case of an office worker who was beaten to death in his office, samples were recovered from his hands in an attempt to recover the DNA of anyone he had recent contact with. Over 50 other workers were originally identified as suspects and comparison of their reference DNA profiles with the samples from the deceased's hands resulted a match between one of the workers with samples from the deceased's nails. When the evidence was put to the individual whose DNA matched, he confessed to the crime, giving details of what had occurred and implicating another worker who had been with him.

When interpreting and reporting touch DNA profiles, the scientist must make clear the limitations of their findings. When dealing with such results they must emphasise:

(1) It is not possible to determine when any of the DNA detected was deposited, other than potentially since the item was last thoroughly washed.
(2) Where there are multiple contributors of DNA to a mixed result it is not possible to determine the order in which these different contributions of DNA were deposited. It is not possible to say which was most recent or whether any were deposited at the same or different times. Just because someone has contributed more of the DNA does not mean they have contributed more recently.
(3) It is possible to handle or wear an item and not leave detectable DNA behind. The longer an item is handled or worn the more one might expect detectable DNA to be deposited but there is no robust way of quantifying this.

DNA can be deposited directly onto an item through handling or wearing. However, DNA can also be deposited indirectly onto an item *via* another person or surface. This is sometimes termed secondary transfer. This involves a person depositing their DNA onto a surface. That surface then makes contact with the item in question, thus transferring DNA of the person to an item they have never directly contacted.

Direct transfer can result in small or large quantities of DNA being transferred to the recipient surface. Indirect transfer usually results in only small quantities of DNA being transferred unless the DNA is in the form of a high DNA yielding body fluid such as blood, semen or saliva or transfer *via* a regularly used item such as a glove or other clothing, bags or bedding as these items have the potential to build up significant amounts of DNA over time. With each potential transfer step, the amount of DNA available for transfer will be reduced.

DNA transferred *via* another person's hands usually requires a fairly short time gap between the DNA being deposited on the intermediate's hands and that intermediate coming into contact with the surface in question. This is because DNA is often quickly lost from hands through washing, wiping or touching other surfaces.

The forensic biologist is often asked to comment on whether a DNA profile is more likely due to direct or indirect transfer. In assessing this the scientist will consider the amount of DNA detected and the number of possible contributors to that result along with the examination results, for example how large an area was sampled, any known body fluids present within that area and what results were obtained from other areas of that item. They will also carefully consider the case circumstances and any accounts given by prosecution and defence. In some instances it may be that the findings favour one method of transfer over another but often, and particularly when dealing with small amounts of DNA, the findings may be wholly explainable by direct or indirect routes. It is worthy of note that for indirect transfer to occur, a surface bearing that person's DNA must be available to transfer. If that surface is available for testing this should be recommended.

11.5 DNA Recovery Methods

In order to perform DNA profiling on a sample it must first be recovered from the item, person or scene. Various methods are available for body fluid/DNA recovery. Swabbing using a wet, sometimes followed by a dry swab (similar to a cotton bud), has traditionally been used for non-porous surfaces and various swab types are available. Minitaping involves repeatedly pressing a small piece of adhesive tape onto a surface. This is usually, though not always, used for DNA recovery from porous items such as clothing. Where a body fluid has soaked into fabric, a sample of the stained fabric is often cut out and submitted directly for DNA profiling.

For historic cases where adhesive tapes were used to recover hairs and fibres, there may be visible skin flakes present on the tapes which can be cut out for DNA profiling, or in some cases the whole surface or sections of the tape may be cut up or swabbed for DNA profiling, or for further testing, for example spermatozoa extraction in sexual assault cases.

11.6 Blood Grouping

Historically, blood grouping systems were used in forensic science to address who body fluids may have originated from. The ABO system detected variations in proteins on the surface of cells. Phosphoglucomutase (PGM) made use of protein variants that have small differences in their chemical composition, and hence slightly different electrical charges when in solution. ABO and PGM are independent of each other and analysis of both could give rise to 40 distinct combinations. The advent of DNA profiling with its vastly increased sensitivity and specificity means blood grouping markers are no longer used in standard casework.

11.7 DNA and Its Analysis

In 1982 a young black man brutally raped a young white female in Virginia, USA. During the assault the perpetrator told the victim that he had a white girlfriend. On that basis, Marvin Anderson, who lived with a white woman, was arrested and later convicted of the assault. Swabs had been recovered from the victim, but it was not until 2001 that DNA profiling of these samples showed that Marvin Anderson could not have been the offender. He had spent more than 15 years in prison. Mr Anderson is one of a growing number of people who have been exonerated by DNA evidence. DNA evidence has helped convict countless more.

11.7.1 Genetics

Deoxyribonucleic acid (DNA) is the hereditary material in most organisms. Genes are made up of DNA and these genes code for specific proteins. Within a cell the DNA is wrapped tightly into chromosomes with each chromosome containing many genes. Traits are inherited from parents through these genes. Variations of these traits, or 'polymorphisms' exist in the population. Some of these are obvious and visible, for example eye colour, others are more subtle and/or may not be visible, for example different blood groups. The location of a gene in the genetic content of a cell is called the 'genetic locus', and different versions of the genetic information at a genetic locus may co-exist within a population. The different versions are called 'alleles'. It is these variations that are of use in forensic science.

11.7.2 Structure of DNA

All human cells, with the exception of erythrocytes and gametes (sperm and ovum), contain 46 chromosomes that comprise one pair of sex chromosomes plus 22 pairs of chromosomes. The sex chromosomes are called X and Y. Males have one of each and females have two X chromosomes.

One member of each chromosome pair is inherited from each parent. The pairs of chromosomes will carry the same genes, but they may contain different alleles. For example, both carry genes for eye colour but one may have a blue eye allele and the other may not. In this case where different alleles are inherited from each parent the genetic locus is described as heterozygous. If they were the same *i.e.* both with the same blue eye colour gene, they would be described as homozygous. Incidentally blue eye colour is a recessive gene so an individual must be homozygous to have blue eyes.

Gametes carry only a half complement of DNA *i.e.* 23 chromosomes, so the result of a combination of gametes is a zygote – the cell that will develop into an embryo – that has a complete complement *i.e.* 46 chromosomes. The inheritance of one chromosome out of each pair is random, so each gamete contains a random one of each of the 23 pairs of that parent and each new baby possesses a shuffled collection of half of the chromosomes available from each parent. Therefore, if two polymorphisms occur on different pairs of chromosomes, they can be regarded as being inherited independently from each other. All the cells of a human arise from the subsequent cell divisions the zygote undergoes. Therefore, all the cells of the body have identical sets of DNA (although in some instances somatic mutations may occur).

Each one of the 46 chromosomes in human cells includes a double stranded piece of DNA which is wound around itself into a double helix. These are coiled and supercoiled around a supporting structure of protein enabling a vast length to be present within the microscopic length of a nucleus.

The structure of DNA is well documented elsewhere and can be found in any basic genetics book. In essence DNA is composed of a backbone of alternating sugar and phosphate molecules. The sugar is a five-carbon-ring deoxyribose which has one of four nitrogenous bases attached. These bases are adenine (A) and guanine (G) which are purines, and cytosine (C) and thymine (T) which are pyrimidines. Each strand is assembled from individual units called nucleotides which comprise a deoxyribose sugar, a phosphate and a base. The sugars are linked *via* the phosphate group from the fifth, or 5′, carbon on one sugar to the third, or 3′, carbon on the following sugar.

The two strands of the double helix do not carry the same information but complementary sequences with the strands being held together by hydrogen bonds between the T and A bases and the C and G bases. The backbone of the DNA strands run in opposite directions due to the carbon-to-carbon linkages, for example, the sequence 5′-GGCTATAAT-3′ on one strand would be matched by 3′-CCGATATTA-5′ on the other strand given the A–T and C–G pairing.

The sequence of the nucleotides on the DNA strand encodes genetic information. However human DNA has been found to consist mostly of base sequences that carry no genetic information. These 'non-coding' regions are useful for forensic identification because they are not subject to selection pressure and so possess greater variation.

11.7.3 DNA Profiling

DNA was first considered in a forensic context by Alec Jeffries in 1984 when he noted similarities and differences between the DNA of his lab technician's family members. He realised that these techniques could be used to produce a 'DNA fingerprint' of an individual. This was first used in a forensic context in the Colin Pitchfork case where two young girls had been murdered. A DNA profile was obtained from semen from each girl and the profile matched that of Colin Pitchfork. The DNA profiling evidence in this case gave far more discriminating evidence than previous blood grouping could. At the same time as implicating Colin Pitchfork, it exonerated another suspect who had previously admitted to killing one of the two girls.

Today multiple regions of DNA (loci) are simultaneously targeted and amplified during DNA profiling. Each of these loci vary between people and it is this variability combined with the testing of multiple loci that gives DNA profiling its discriminatory power. Today DNA profiling follows the following stages:

(1) Extraction
(2) Quantification and amplification
(3) Separation, detection and analysis
(4) Interpretation

11.7.4 Extraction

DNA must be extracted from the sample before it can be amplified and analysed. There are various extraction methods and the choice depends on the sample type as well as ease of use, cost and DNA yield. These processes involve extraction of the cellular material from the substrate (*e.g.* swab, minitape or fabric), cell lysis (the disruption of cells to release DNA into the lysate), and isolation of the DNA from the rest of the sample. Some lysis methods use chemicals such as SDS and enzymes such as ProteinaseK to disrupt cell membranes and denature proteins. DNA can be purified (isolated from the other material) by processes such as binding DNA to magnetic particles or membranes before washing steps to remove the remaining particles. Many other methods are available and most forensic laboratories use multiple extraction methods depending on the type of sample to be processed. Most also offer multi step reactions using enzymes and buffers to separate cell types enabling spermatozoa cells to be processed separately from other cell types. Many of these extraction methods use automated technologies.

11.7.5 Quantification and Amplification

Following extraction, the quantity of DNA is usually still too small to be analysed and so the areas of interest are targeted and amplified (copied many times). The advent of the polymerase chain reaction (PCR), a technique that uses polymerases to replicate sections of DNA exponentially and thus amplify those sections, has led to vastly more sensitive forms of DNA profiling. Theoretically a DNA profile can be obtained from only one cell; however, in practice larger numbers of cells are needed.

In the bulk of forensic casework PCR is used to target and amplify specific regions of DNA called short tandem repeats (STRs). These are tandem repeat sequences of DNA that are 2–6 base pairs long. There are large numbers of STR regions in the human genome but only some have been selected for use in human identification. The majority of STRs have between 6–30 repeat units and it is this variability which makes them useful for identification of individuals.

The replication of these STR sequences by PCR is carried out within an aqueous solution in a small tube with a typical reaction volume of 25–50 μL. The reaction is controlled by a programmable thermal block called a thermocycler which heats and cools the sample allowing the following process to occur:

Denaturation: The block heats to 90 °C causing the hydrogen bonds between the two strands of DNA to break apart rendering the DNA single stranded.

Annealing: The temperature is lowered to around 55 °C. At this temperature synthetic primers (or oligonucleotides) anneal to each end of the target DNA. These primers are around 20 base pairs long. There are two for each locus in order to bind to either side of the STR. This target STR sequence which lies between the two primers is used as a template for the amplification.

Extension: The temperature is raised to 75 °C. Within the solution are the four types of nucleotides (A, C, T and G) and polymerase which is an enzyme used by the body for DNA replication. At this temperature the polymerase catalyses the addition of the nucleotides to the 3′ end of the primers where they bind to the template DNA strand. The type of nucleotide added is determined by the base on the template strand with complementary bases being added each time. The polymerase will continue to add bases to extend the primers until it is interrupted.

Once a full cycle of denaturation, annealing and extension is complete the process is repeated. After each cycle there should be double the number of template strands because the new strands from the last cycles will themselves act as templates for the subsequent cycle. In this way exponential amplification of those target areas should be achieved, although in practice the full potential is not realised. When a new strand is used as a template the 5′ end terminates with the primer and does not continue as the original sample

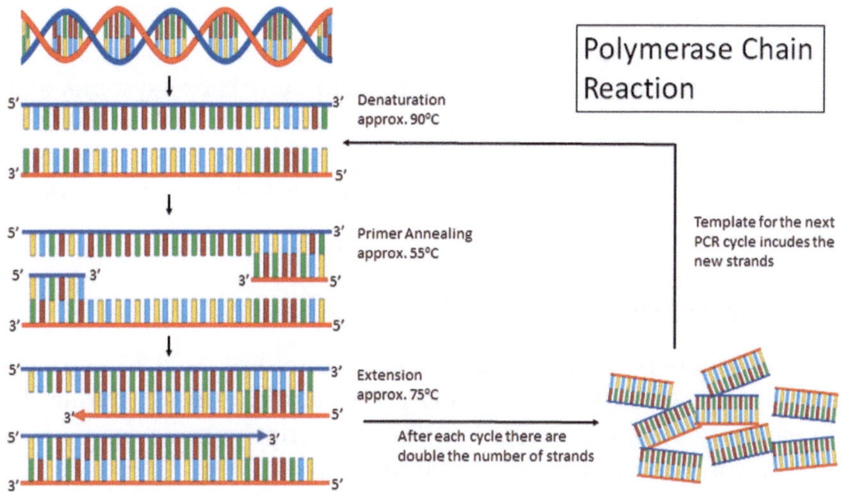

Figure 11.4 Polymerase chain reaction.

DNA would. In this way most of the DNA in the original sample is not copied. Instead, it is the greater number of these shorter units that continue to be copied. The PCR reaction is shown in Figure 11.4.

Using a reaction mixture that contains primers for multiple genetic loci, several STR loci can be amplified at the same time. This is called multiplexing and the combination of an individual's independently inherited STRs is their DNA, or STR, profile.

Traditionally the PCR cycle was repeated 28 times. 'Low Copy Number' (LCN) profiling, a specialist technique devised by the Forensic Science Service, used 34 repeat cycles which resulted in the amplification of such small quantities of DNA that specialist examination and interpretation methods were used. Today 30 cycles are usually used as standard, which coupled with better extraction and detection techniques, allows the detection of extremely small amounts of DNA.

During PCR one member of each primer pair has a fluorescent label attached. Therefore, each of the PCR products will also carry this label. This allows for detection during electrophoresis (see below).

If too little template DNA is added to the PCR cycle no, or a low quality DNA profile may be obtained. If too much is added the profile will be 'overamped' and may contain so many artefacts that it is uninterpretable. For this reason, the DNA in the sample is quantified prior to the PCR process. Today quantification usually utilises 'real-time PCR'. This is a PCR system that incorporates fluorescently labelled primers. As the PCR cycles progress, more PCR products accumulate. This causes an increase in fluorescence which can be monitored. The result is compared against a standard curve to allow the quantity of template DNA to be determined. These systems also quantify the amount of DNA from male donors, and the amount of inhibiting substances in the sample. The quantification data is used as a guide to determine how much template DNA will be required for PCR.

11.7.6 Separation, Detection and Analysis

Today automated electrophoresis instruments are used to separate the amplified STRs through narrow glass or silica tubes (capillaries). This capillary electrophoresis (CE) uses multiple capillaries so that multiple samples can be run simultaneously.

DNA is negatively charged and during capillary electrophoresis the PCR product, which contains a mixture of amplified STRs, is loaded into the capillary and then moved through a separating medium by the application of an electric field. The smaller fragments move faster than the larger fragments and so the mixture is separated by size into groups of comigrating molecules. These amplification products, labelled with the fluorescent dye, are detected at the other end of the capillary by a photosensor. The data collected by the photosensor is displayed in a chart called an electrophoretogram (epg). This is a plot of the time taken for a fragment to pass through the capillary (retention time on the x-axis), *versus* the detector response (y-axis) measured in relative fluorescence units (rfu). The quantities of the fragments comigrating in groups in the capillary column are in approximate proportion to the quantity of DNA in the sample that has been amplified.

The size of the fragments are measured in base pairs (bp) and this corresponds to the number of repeating STR units within the allele and the length of each of those STR units. Each fragment size is compared against an internal size standard – a collection of fluorescently labelled DNA fragments of known bp size that are added to the mixture for this purpose. Once the size of the test sample fragments is known they are compared against an allelic ladder – an artificial mixture of the most common alleles in the population, also run through a capillary and sized against an internal standard. The base pair size of the test fragment can be compared to the ladder to ascertain which allele on the ladder it corresponds to. The use of multiple fluorescent dyes and loci of differing sizes ensures that there is rarely overlap of different loci in the same location in the resultant epg.

If the donor has inherited the same allele from both parents, they are homozygous at that locus and will show only one peak (as both alleles are the same size and will have migrated through the capillary at the same speed). If they have inherited different alleles from both parents, they are heterozygous and will show two peaks. When an optimal amount of DNA is tested heterozygous pairs of peaks are usually of a similar height.

Multiplex kits contain a marker that indicates the sex of the donor. Females have two X chromosomes and males have an X and a Y. This marker (called amelogenin) is located on both the X and Y chromosomes but is six base pairs shorter on the X chromosome. Thus, a male donor will show two peaks at amelogenin representing the two different sized fragments and a female will show only one because both of their amelogenin fragments are the same size.

Positive and negative extraction and PCR controls are reviewed to ensure that the tests have worked and there is no indication of gross contamination.

The first multiplex DNA profiling kits (Quad) in 1994 detected only four STR loci. In 1995 Second Generation Multiplex (SGM) kits based on six STR loci and amelogenin were introduced. In 1999, SGMPlus, which detected ten loci (including the six included in SGM) and amelogenin was introduced. Since 2014 labs in England and Wales have used DNA17 kits. These detect 16 STR loci (including the ten included in SGM+) and amelogenin. This means that a complete DNA17 profile contains 34 alleles – two at each of the 17 areas. An example of a DNA17 epg is provided in Figure 11.5.

11.7.7 Interpretation

Whilst analytical software largely designates the alleles within a profile, a scientist must still review this. Analysis of DNA profiles can be complicated by the presence of artefacts, some of which may fall into allelic positions. Common artefacts include:

Stutter: A small peak, most commonly one repeat length (usually −4 bp) shorter than the parent allele, caused by slippage of the polymerase during PCR. Depending on the locus, amount of DNA present and the specific processes of the forensic laboratory −2, −8 and +4 bp stutters may also be seen.

Spikes: Narrow peaks due to fluctuations in voltage or minute air bubbles in the capillary during electrophoresis.

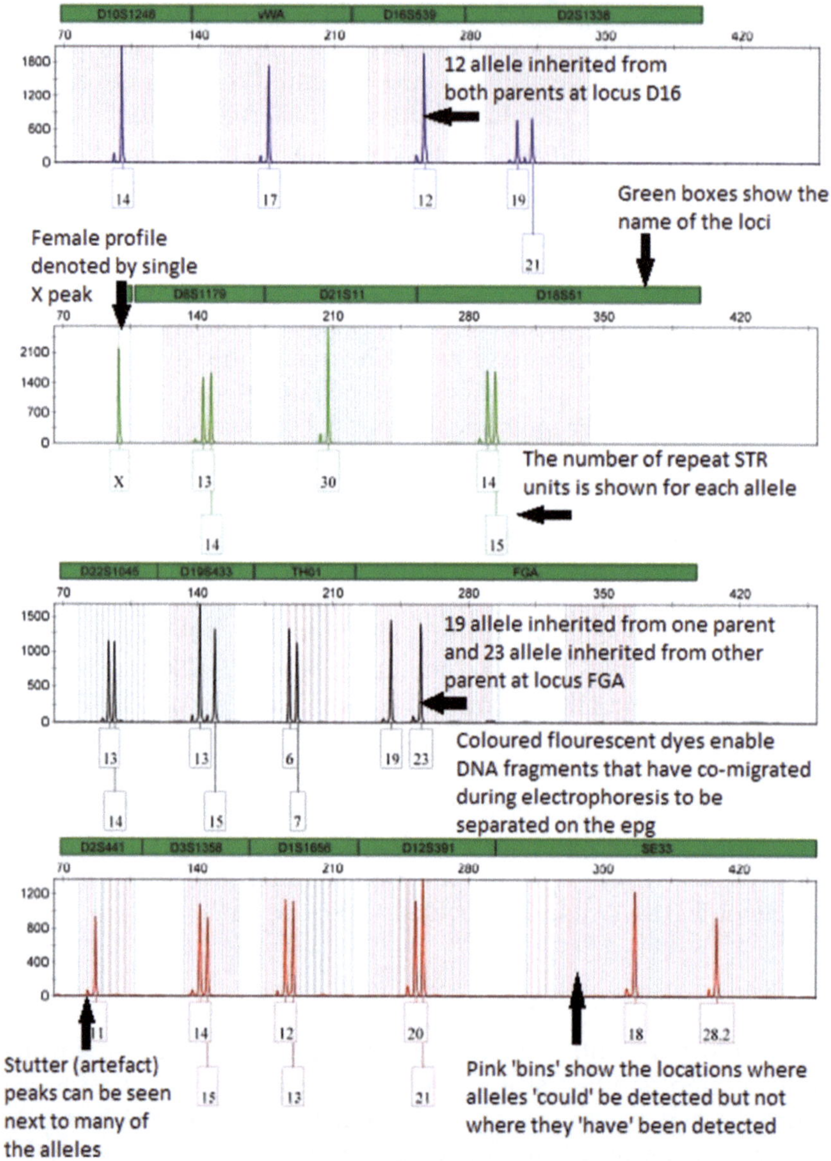

Figure 11.5 Electrophoretogram (epg).

Pull up: Additional peaks on the epg due to a failure of analysis software to discriminate the different dye colours.

Drop in: The presence of an additional DNA peak (or two) due to a fragmented DNA molecule contaminating a consumable used in the test.

11.7.8 Single Source DNA Profiles

A full or complete DNA profile is one where all of the DNA components have been detected and this DNA appears to be from one person.

Where less than the optimal amount of DNA is present alleles may 'drop out' of the sample as there is insufficient DNA for all of the alleles to be detected. In samples with low quantities of DNA stochastic variation means that different alleles within the sample may be detected each time the sample undergoes PCR. These partial, or incomplete, DNA profiles may range from 1–33 DNA components. When degraded DNA is detected it is the larger DNA components which will tend to drop out first because degradation effects the larger DNA molecules more than the small ones. There are a number of mechanisms that might degrade DNA, such as the presence of bacteria, humid conditions, extreme heat or UV light.

Inhibition may also cause a reduction in the number of components detected. This is when another substance present in the sample may be inhibiting the PCR reaction thus preventing the DNA from being amplified. Inhibition can be caused by various substances including some clothing dyes, wood and a component of blood, but modern extraction methods have mostly overcome these.

Where weak, partial DNA profiles are obtained containing limited DNA components it may not be possible to determine whether the DNA is from only one person or may be a mixture of low levels of DNA from more than one. Repeating the PCR with another aliquot from the extract is advisable to provide a fuller picture of the sample under test.

11.7.9 Mixed DNA Profiles

Many samples contain DNA from more than one person. This may be due to mixtures of body fluids such as semen and vaginal material in a sexual assault case, a body fluid such as blood combined with the DNA of the wearer of an item of clothing from which the blood was recovered, or mixtures of touch DNA from an item that has been handled by more than one individual. There are many more examples, and the sensitivity of DNA profiling today is such that complex, mixed DNA profiles are commonplace.

A mixed DNA profile may contain DNA from a minimum of two people but there is no maximum, although there are limits to the usefulness of a DNA profile with numerous contributors. A single source DNA profile will have a maximum of two allelic peaks at each locus. A mixture of DNA from two individuals will have between one and four alleles at each locus (depending on how many alleles the contributors have in common). A mixture of DNA from three individuals would have between one and six alleles at each locus, and so on. As the number of contributors grows so do the number of DNA components. The more DNA components there are the more likely that an individual will have matching components by chance and so it becomes more difficult to exclude individuals from having contributed their DNA.

The height of peaks can also be used as a guide to the minimum number of contributors of DNA as where DNA components are shared a higher peak is produced. For example, in a mixture where there are two contributors in roughly equal amounts whose alleles at a locus are 17, 17 and 17, 18 only two peaks would be seen because the three 17 alleles would only form one peak. That peak would be significantly higher than the 18 peak and would suggest that DNA from more than one person may be present.

11.7.10 Major/Minor DNA Profiles

Where DNA from one individual in a mixed result is present in significantly greater quantities than the other contributor(s) a 'major' DNA profile can sometimes be determined. In this situation the peaks relating to the major DNA components are significantly bigger than those relating to the minor and a full profile of the major contributor may be determined. This can be compared to reference DNA profiles as if it were a single source DNA profile and single source statistical analysis applied (see below).

11.7.11 Complex Mixed DNA Profiles

The paragraphs above describe relatively simple mixed DNA profiles but there are many instances where the number of contributors is high, or unknown, or there are low levels of DNA meaning that the absence of components cannot necessarily be used as evidence that an individual's DNA is not present. Experienced forensic scientists are required to interpret these profiles in a meaningful way. These interpretations take account of the potential number of contributors and thus the potential representation of matching alleles even if an individual's DNA was not present, the location of matching alleles and whether there would be a high expectation of seeing those components at the lower molecular weight loci with potential drop out at the high molecular weight loci, the potential masking of DNA components by artefacts such as stutter and the comparative heights of the peaks present on the epg. Some mixed DNA profiles become so complex that they are not suitable for meaningful comparison.

11.7.12 Attributing DNA to a Body Fluid

Once a DNA profile has been obtained, consideration must be given to whether it can be attributed to a particular body fluid or type of cellular material. In order to do this the scientist will consider the strength and nature of the DNA profile itself, the tests performed on the item prior to sampling and any other results from the item. Where a body fluid such as blood or semen has been profiled, a strong DNA profile would normally be expected because those samples contain high amounts of DNA. In this situation it would be reported that the DNA originated from that body fluid.

If there is insufficient DNA in the sample to give a clear, strong DNA profile, for example where only a trace of spermatozoa has been detected, it becomes difficult to confidently determine what type of cellular material the DNA originated from because we would not necessarily expect the samples to contain much DNA. Therefore, any profile obtained might have come from that body fluid or alternatively from something else, such as contact DNA present on the surface that the DNA was recovered from. Similarly, if a very small weak blood stain was sampled and a low level DNA profile was obtained, an assumption cannot be made that the DNA detected definitely originated from the blood. Other considerations should be made such as whether the DNA could have originated from the surface on which the blood was present, with no profile being obtained from the blood. In relation to touch DNA samples it is rarely possible to determine the specific source of the DNA. It may be skin cells shed from the body, but it may contain traces of saliva or other body fluids.

11.7.13 Reference DNA Profiles

In order to consider whether or not a particular individual may have contributed DNA to a sample, it is necessary to have a reference DNA profile of that individual for comparison. As DNA is the same throughout the body, a reference DNA profile taken from, for example, a buccal (cheek) swab, can be used for comparison against a DNA profile obtained from any biological material. Buccal swabs provide an excellent source of DNA and can be collected in a relatively inobtrusive manner. A blood sample can alternatively be taken and often this or deep muscle tissue are recovered from cadavers. Pulled head hair can also be recovered though this is relatively rare.

The National DNA Database (NDNAD) contains a variety of reference DNA profiles taken from individuals on arrest. The oldest are known as pre-PACE and these include SGM and some SGMPlus profiles. These are not suitable for use as evidence in court as they were taken under older legislation. They can be used for intelligence comparisons, but a new reference sample and resultant profile must be obtained before the evidence can be used at trial. SGMPlus PACE samples taken more recently are suitable for use in court but only have information at 11 loci rather than the current 17. Any PACE reference profiles loaded in recent years will be DNA17. It is possible to compare an SGMPlus reference profile to a DNA17 crime stain profile but only at the 11 loci the two share in common. Often this may be sufficient, particularly when dealing with full or major crime stain profiles, however when dealing with comparisons to complex mixed results, it is advisable to obtain a new DNA17 reference profile to allow a fuller comparison.

Volunteer or elimination mouth swab samples may be taken by police for use in a particular case. They are commonly taken from complainants and/or those whose DNA may be expected to be on the item in question, for example the legitimate partner of a complainant in a rape case. These may be submitted to the laboratory for comparison in that particular case but are not included on the NDNAD.

Surrogate reference samples may be taken when the individual in question cannot give a reference sample, for example a missing person. Items are selected that are likely to bear large amounts of the person in question's DNA and little or no DNA of others, for example a toothbrush or razor. Comparison of the resultant profile with close relatives such as parents or children can be of use in increasing confidence that the profile obtained can be assumed to be from the person in question.

11.7.14 Comparison

Once a reference profile has been obtained it can be compared to the crime stain. In the case of a single source crime stain profile, if any alleles differ then the DNA could not have come from that person. If all of the alleles match then the DNA could have come from that person, or anyone else who by chance has the same DNA profile.

In the case of a mixed DNA result, comparisons are more complicated. The scientist will compare the reference DNA profile to each locus of the mixed result and consider how well represented that person is. In doing this they will take into account the nature and complexity of the profile. The forensic scientist will then conclude whether the person could be considered a possible contributor, whether they do not appear to have contributed, or whether due to the nature of the result it is not possible to determine.

11.7.15 Assumed Contributors of DNA

It may be possible, depending on the results of the comparison and the circumstances of the case, to assume the presence of the DNA of a particular individual in a sample. In doing so, the scientist should be satisfied that the individual could have contributed both by means of direct comparison and that case circumstances dictate their DNA is likely to be present on that item. Examples include the donor of an intimate swab, the wearer of a garment or the known driver of a car.

11.8 Statistical Analysis

Where an individual could have contributed DNA to a sample the scientist must also consider the possibility that they have not contributed, and their reference DNA profile matches by chance alone. Statistical analysis centred in the likelihood ratio of Bayes Theorem allows a robust, quantitative evaluation of whether the result is more likely if the DNA came from the person in question rather than if it came from another unrelated person (*i.e.* a coincidental match).

This process, and the mathematical principles behind it, are detailed in length in various textbooks. In summary, the scientist first assesses the probability of the evidence (a DNA profile matching the person in question) if the prosecution hypothesis (H_p), that the DNA did indeed come from that person, is correct. This based on the First Law of Probability, is given a probability of 1. This is the numerator of our likelihood ratio.

The next, and crucial, part of this evaluation is determining the probability of the evidence if the defence hypotheses (H_d) is correct. This is the probability of obtaining this matching DNA profile if it did not come from the person in question but instead came from someone other than, and unrelated to, that person – in essence how often that profile will be seen (its frequency) in the general population.

To do this, proven statistical formulae and validated frequency databases which represent the distribution of alleles within the main racial groups of the general population are used. The Hardy Weinberg formula and validated frequency databases are employed to estimate the frequency of the specific pair of alleles at each individual locus of the DNA profile in question. Then, again utilising basic laws of probability, this time the Third Law of Probability, these individual frequencies can be combined to give an assessment of the frequency of that whole profile within the general population. Factors are incorporated into these calculations to allow for under representation of less common alleles within the frequency databases (known as the size bias correction factor), and the possibility of individuals within a community sharing more DNA in common than when compared to a wider population (known as the F statistic or Theta). This final frequency is the denominator of our likelihood ratio and, when considering full profiles, is always considerably smaller than 1 in a billion (or <0.000000001).

The final step is to calculate the overall likelihood ratio by dividing our numerator of 1 by our denominator of <0.000000001. Thus, where a complete matching DNA profile is obtained using a DNA17 system, it has been demonstrated that even if a person has the most common alleles at each locus, the final LR will always be over a billion (a billion is one thousand million).

Forensic Regulator guidance, and accepted protocol within the UK, is to never report above the level of one billion. Therefore, where a full DNA profile matching that of a reference DNA profile is obtained, there is no need to calculate a specific LR and instead the result can be reported as: '*The DNA profiling result is over a billion times more likely if this DNA came from Mr X rather than from someone other than, and, unrelated to him*'.

11.8.1 Continuous Models for Statistical Analysis

Traditional methods of statistical analysis (as described above) do not allow for complex mixtures of DNA, or many of the uncertainties within a DNA profiling result such as imbalances of height between DNA peaks, drop in and out of peaks and uncertainty with regards to whether peaks may be due to the presence of DNA components or artefacts of the DNA profiling process. Advanced, or 'continuous' software programs, have been written to allow analysis of such complex results. These programs use biological modelling, statistical theory, computer algorithms and probability distributions to calculate likelihood ratios and/or infer genotypes for the DNA results. They build millions of conceptual profiles and grade them against the result observed. Like traditional, or 'binary' models, these continuous advanced statistical models are reported as a likelihood ratio.

Such evaluations require certain assumptions to be made. The most subjective of these is the number of contributors to that mixture. As discussed previously, due to allele sharing and dropout the scientist cannot be certain of how many contributors there may be to a particular mixture, instead they state a minimum contributor number. However, for statistical evaluation, an assumption of contributor number has to be made. This becomes particularly difficult as the contributor number increases. Research shows that allele sharing increases considerably with increased contributor number, leading to a very real risk of under representation of contributor number. Where there is uncertainty it may be advisable to run the analysis based on for example three contributors and then again as four. As with single source statistics it is assumed that all contributors are unrelated to each other and to the person of interest.

DNA profiles with more than five contributors are currently unsuitable for statistical evaluation. Forensic Regulator guidance states that where statistical analysis is not possible, it is uninformative for a scientist to tell the court that an individual '*might have been a contributor*' as they alternatively might not have been a contributor and in the absence of statistical analysis the court has no evidential weight in favour of either of those propositions. If an assessment of evidential weight is not possible the scientist should inform the court that they can give no guidance with regard to probative value.

11.8.2 National DNA Database

The National DNA Database (NDNAD) was created in 1995 and is run by the Home Office on behalf of UK police forces who use it to identify who may have left a body fluid stain or DNA sample in a particular case. The NDNAD comprises reference DNA profiles of known individuals and crime stain profiles of unknown origin. As of 31st March 2023, the NDNAD held 7 045 155 reference profiles (some of which are duplicates with an estimate of a total number of individuals retained of 5 911 703) and 670 817 crime stain profiles.

Crime stain DNA profiles that do not match any of the reference profiles in a criminal investigation and are deemed to be of interest may be loaded to the NDNAD. Each night the NDNAD compares all crime stain profiles to each other and to all reference profiles and notifies the relevant police force of any matches.

In order to load a crime stain DNA profile to the NDNAD, it must meet certain criteria – these include having four complete original SGM loci and for the provider loading the profile to be confident that the DNA is from one person. The less information there is within the profile, the more likely it is that adventitious matches will occur. Profiles loaded to the NDNAD are held indefinitely and will be searched against profiles already held and others as they are added.

A DNA profile that is not suitable for loading to the NDNAD, for example it contains too little information and/or the scientist is not confident the DNA is all from one person, can be submitted for a one-off speculative search. This undergoes a search against profiles currently held on the NDNAD but is not loaded permanently.

Other functions of the NDNAD include:

Familial searching – based on the premise that related individuals may share more DNA in common than unrelated and thus looks for similar profiles within the NDNAD rather than wholly matching ones. A list of possible profiles ranked in order of similarity is issued for intelligence purposes.

Missing Persons DNA Database (MPDD) – separate to the main NDNAD and holds DNA profiles obtained from personal items and/or family members of missing persons to assist with identification of any unidentified bodies.

Vulnerable Persons DNA Database (VPDD) – similar to the MPDD this is separate to the main NDNAD and holds DNA profiles of individuals considered at risk of harm.

11.9 Other DNA Profiling Techniques

The DNA profiling described above details 'standard' nuclear DNA profiling which is routinely used for the vast majority of forensic samples. However, where there are samples that contain insufficient or badly degraded DNA, or where male DNA is mixed with significant amounts of female DNA, other techniques may be of assistance. The technique used will depend on what is required for the sample, what tests have already been undertaken and how much sample is available.

11.9.1 Y-STR Profiling

Y-STR profiling utilises loci on the Y chromosome which are amplified using PCR in the same way as autosomal DNA. Y chromosomes are only found in males and so this technique will only detect male DNA. The Y chromosome is inherited in whole by a son from their father and so the combination of Y-STR alleles found in one individual is normally inherited without change from father to son. This means that all closely related paternal line male relatives (*e.g.* father, brothers, paternal grandfather, paternal uncles, cousins and nephews related through the paternal line, *etc.*) are likely to share the same Y-STR profile. The only exception to this may be if a mutation occurs between generations that could result in a difference to the profiles.

When matching DNA profiles are obtained following Y-STR analysis, the significance of a match may be assessed by reference to an online global Y-STR database called the 'Y Chromosome Haplotype Reference Database' (YHRD). The number of matching Y-STR profiles observed in a relevant population dataset on the YHRD is used to provide an estimate of the frequency of that Y-STR profile in that population. Due to paternal inheritance of Y-STR profiles this method of assessing the significance of results is different to that used in standard DNA profiling, and the resulting figures are much lower than are usually seen with DNA17 profiles.

Y-STR profiling can be useful in assessing relatedness between male individuals. It is also useful where a mixture of small amounts of DNA from a male, and large amounts from one or more females is obtained, for example in digital penetration cases. If an offender penetrates the complainant's vagina with his fingers, only small amounts of his DNA may be transferred to the vagina. During DNA profiling this can easily be swamped by the large amounts of vaginal DNA present. By only testing areas on the Y chromosome it is possible to detect the male's DNA without interference from the female(s) DNA.

Resultant profiles are not comparable with standard DNA profiles. Therefore, the samples are not suitable for searching on the National DNA Database.

11.9.2 Mitochondrial DNA

Mitochondria are present in almost all human cells, their function being to provide energy within the cell. There are 100–1000 mitochondria in each cell, each with two to three copies of its own mitochondrial DNA. Each cell may therefore have several hundred copies of mitochondrial DNA but only one copy of nuclear DNA. This means mitochondrial DNA can be detected in samples where little or no nuclear DNA is present, for example hair shafts or highly degraded samples. Mitochondrial DNA is therefore particularly useful in disaster victim identification or investigation of mass graves.

Within all the cells of a human, all of the copies of mitochondrial DNA have the same sequence which consists of around 16 500 base pairs. In contrast to nuclear DNA, almost all of the mitochondrial DNA sequence is used for coding and only a relatively small portion called the 'D' (Diversity) loop, or the control region, has no specific coding function. The control sequence includes a point at which the mitochondrial DNA replication always starts. This is called the 'origin of replication' and is numbered 1, continuing through the chain to around 15 500. Two regions have been found to exhibit a high degree of variation between individuals. These hypervariable regions (HVRs) 1 and 2 occur within the control region. HVR 1 and HVR 2 are linked and so mitochondrial DNA cannot have the same discriminating power as nuclear DNA. Mitochondria are inherited through the maternal line and can therefore be useful where relationships through the maternal line are important. This was demonstrated when bones, alleged to be the remains of Tsar Nicholas II, the Tsarina, their three daughters, the family doctor and two servants were discovered by amateur historians in 1991. The Tsar and his family were murdered in 1917 during the Russian revolution. STR and mitochondrial sequencing was performed on the bones. STR profiling allowed a tentative distinction of a family group, with genotypes consistent with a mother, father and three children. The mitochondrial analysis was of particular importance because, through this, individual skeletons could be linked to the maternal lineage of the Russian royal family. The maternal descendants included HRH Prince Philip Duke of Edinburgh. The skeleton identified tentatively as being that of the Tsar matched the mitochondrial HVR sequences of living maternal descendants perfectly (except for a single base mismatch) which, taken in the context of other supporting evidence, strongly supported the claims for the remains when taking mutation rates into account.

11.9.3 Single Nucleotide Polymorphisms

Single nucleotide polymorphisms (SNPs) are genomic variants at a single base pair and are the most common type of genetic polymorphism in humans. Panels of SNPs can be tested using much shorter DNA sections than mitochondrial or STR analysis and have been invaluable in mass disaster events such as the disaster victim identification effort following the attack on the World Trade Centre in September 2001 where nearly 3000 people died.

11.9.4 Next Generation Sequencing

Next generation sequencing (NGS) is a massively parallel sequencing technology that allows millions of fragments of DNA to be sequenced in parallel. In many cases this may be used for medical purposes but in the forensic context it would provide highly discriminating evidence, if not full identification of an individual. NGS may also be used to target mitochondrial DNA, STRs and SNP markers including on the Y chromosome. The detection of STRs means that NGS profiles could be compatible with current DNA databases. This could be hugely helpful in disaster victim identification and other cases where samples are degraded. Results are produced without the need for electrophoresis vastly increasing the potential throughput of samples. The increased genetic information provided by NGS technologies could also provide information about physiological characteristics of individuals as well as their ancestry. Potential uses in the future may also include epigenetics – genetic modifications that impact gene activity without changing the DNA sequence. Changes such as methylation (the addition of methyl groups to the DNA without changing the DNA sequence) over an individual's life could be used to predict the sample donor's age. NGS can also distinguish between identical twins which standard STR technologies cannot.

NGS is not yet in routine use in forensic science, but it is an exceptionally powerful tool which may prove invaluable in future cases.

11.9.5 Anti-contamination

As DNA profiling has become more sensitive the need to minimise and detect contamination has increased. Unwanted transference of DNA can occur from individuals to exhibits, between cases examined within the laboratory and, perhaps most crucially, between items within a case.

Contamination can be reduced with the use of Personal Protective Equipment (disposable lab coats, masks, hair nets, double gloves – where a first pair of gloves is worn throughout and the top pair changed or cleaned frequently); regular cleaning of laboratories and equipment; wiping exhibit packaging on entry to the laboratory; use of disposable equipment, laboratory specific clothing and equipment that is not moved between laboratories; and separation of items from different scenes or individuals into individual laboratories. Despite best efforts contamination does occur but this can often be detected by searching of DNA profiles against staff, casework, police and consumable manufacturer elimination databases, the use of bench logs to determine what items may have been examined at the same time, and batch tracking of consumables. DNA laboratories must also undertake environmental monitoring to monitor background DNA within the laboratory environment.

11.10 Interpretation and Evaluation of Biological Evidence

Once body fluid tests have been performed and DNA profiling and associated interpretations undertaken, the next step is to determine what the scientific findings mean in the overall context of the case. Forensic Regulator guidance states that: '*Part of the role of expert witnesses is to assess the probative value of their observations and to present their evidence in a way that conveys its strengths and limitations in a clear, complete, correct and consistent manner*'.

The extent of this assessment in any particular case will depend on the specific findings and the case information available. This final step is underpinned by the principles of CAI and the Likelihood Ratio with the scientist evaluating their evidence within a robust framework of a set or sets of competing propositions. They first consider how likely are the scientific findings, given what the prosecution allege has occurred (H_p), and compare this against how likely are the scientific findings, given what the defence allege has occurred (H_d) – in essence the likelihood ratio within Bayes Theorem. The results of such an evaluation may result in the findings: supporting one account over the other; being equally explained by both; or in some cases, being explained by neither, instead pointing to an entirely different scenario.

In instances where the findings favour one account over another it may be possible to quantify that support. This is often verbalised using a widely recognised scale of support with a numerical guide which relates to the LR: Neutral (1), Weak/limited (2–10), Moderate (11–100), Moderately Strong (101–1000), Strong (1001–10 000), Very Strong (10 001–1 000 000), Extremely Strong (>1 000 000).

It is crucial that in employing this approach, the scientist ensures they have the following:

(1) A clear H_p and H_d – 'no comment' cannot be used as H_d and the scientist must be careful not to infer a H_d that maximises the prosecution evidence.
(2) Sufficient case information.
(3) Sufficient data and experience to robustly evaluate the probability of the evidence given each account.
(4) Sufficient scientific results to allow such a comparison to be made.

The forensic scientist must also be aware of the potential of hypotheses changing and the effect this can have on any evaluations made and make this clear within their report. The final conclusions rely heavily on the specific propositions considered. If one was to alter as new or different information becomes available this could significantly alter the findings. Consider a case where small amounts of semen matching the defendant are detected on an external vaginal swab of a complainant. The complainant states that vaginal intercourse with internal ejaculation occurred five days prior to the swabs being taken. The defendant initially states that no sexual activity occurred between the two at all. The evidence clearly supports the complainant's account over that of the defendant. However, what if the defendant later says that the two engaged in sexual activity a day prior to the swabs being taken and he may have touched the vagina with a semen stained hand? The LR alters considerably. Now the evidence can be explained by either account and does not favour one over the other.

11.11 Conclusion

The detection of body fluids in combination with DNA profiling within a robust CAI framework provides the justice system with invaluable investigative and evaluative information. Many of the body fluid identification techniques are long proven, having been used for decades. A revolution occurred in forensic biology with the advent of DNA profiling in the 1990s. DNA profiling techniques continue to grow in sensitivity and discrimination but with this come more complex, mixed DNA results and associated challenges in interpreting and evaluating these. The advent of new and exciting techniques such as messenger RNA analysis to identify body fluids and next generation sequencing (NGS) analysis of DNA may again revolutionise this area of forensic science.

Abbreviations

CAI	Casework assessment and interpretation
H_p	Prosecution hypothesis
H_d	Defence hypothesis
LR	Likelihood ratio
ALS	Alternative light sourcing
UV	Ultraviolet
LPM	Low power microscope
LMG	Leuco malachite green
KM	Kastle–Meyer
LCV	Leucocrystal violet
AP	Acid phosphatase
NECs	Nucleated epithelial cells
PSA	Prostate-specific antigen
DMAC	*para*-Dimethylaminocinnamaldehyde
mRNA	Messenger RNA
DNA	Deoxyribonucleic acid
PGM	Phosphoglucomutase
A	Adenine
G	Guanine
C	Cytosine
T	Thymine
PCR	Polymerase chain reaction
STRs	Short tandem repeats
LCN	Low copy number
CE	Capillary electrophoresis
rfu	Relative fluorescence units
bp	Base pairs
epg	Electropherogram
PACE	Police and Criminal Evidence Act
NDNAD	National DNA Database
YHRD	Y Chromosome Haplotype Reference Database
SNPs	Single nucleotide polymorphisms
NGS	Next generation sequencing
MPDD	Missing Persons DNA Database
VPDD	Vulnerable Persons DNA Database

Websites

https://www.gov.uk/government/organisations/forensic-science-regulator
https://yhrd.org, Y Chromosome Haplotype Reference Database

Further Reading

M. R. Cummings, W. S. Klug, C. A. Spencer, M. A. Palladino and D. Killian, *Concepts of Genetics*, 12th Global edn, ISBN 9781292265322.

J. E. Allard, The collection of data from findings in cases of sexual assault and the significance of spermatozoa on vaginal, anal and oral swabs, *Sci. Justice*, 1997, 99–108.

R. Owers, G. Davidson, A. McDonald, R. Morgan and P. O'Rourke, Time since intercourse (TSI) data from a large-scale casework study of penile-vaginal penetration allegations using Sperm Elution™, *Forensic Sci. Int.*, 2018, **288**, 10–13.

D. J. Balding and R. A. Nichols, DNA profile match probability calculation: how to allow for population stratification, relatedness, database selection and single bands, *Forensic Sci. Int.*, 1994, **64** (2–3), 125–140.

A. Rafferty, N. NicDaeid, J. Butler, J. Chalmers, G. McVean and G. Tully, Forensic DNA analysis, A primer for the Courts, The Royal Society and Royal Society of Edinburgh, 2017, https://royalsociety.org/-/media/about-us/programmes/science-and-law/royal-society-forensic-dna-analysis-primer-for-courts.pdf.

G. Jackson, C. Aitken and P. Roberts, *Case Assessment and Interpretation of Expert Evidence: Guidance for Judges, Lawyers, Forensic Scientists and Expert Witnesses*, Royal Statistical Society, 2015, https://rss.org.uk/RSS/media/File-library/Publications/rss-case-assessment-interpretation-expert-evidence.pdf.

12 Digital Forensics

John Tuer*[a] and Benjamin Donnachie[b]

[a] University of Central Lancashire, UK; [b] School of Computing and Communications, The Open University, Milton Keynes, UK
*Email: JGTuer@uclan.ac.uk

12.1 Introduction

The approach to digital evidence at a crime scene has undergone more dramatic changes in the last decade than in the entire 50 plus years that preceded it. Until relatively recently, the challenges that faced digital crime scene investigators from year to year were driven, to some extent, by the well-known Moore's law (about transistor density and hence processor speed doubling every two years) as well as the lesser-known Kryder's law which gives similar predictions about digital storage capacity increasing. Typically, an investigator at a crime scene where digital artefacts need to be acquired, could easily identify the computers needed by simply ensuring that they brought sufficient empty, portable storage to image (copy) all hard disks on any computers present. The main challenge as disk capacity increased, was the time to copy the data and sometimes, to bring large enough empty disks.

Early 21st century technology presents many new challenges to the digital crime scene investigator. The increasing storage capacity problem remains but pales almost into insignificance compared to

Crime Scene to Court: The Essentials of Forensic Science, 5th Edition
Edited by Niamh NicDaeid and Peter C. White

Published by the Royal Society of Chemistry, www.rsc.org

newer problems such as:

(1) Recognizing where the computing technology is; laptops, desktops and servers are no longer the only computing devices as smartphones, tablets, games consoles, smart home technology, wearables and even smart vehicles become people's primary computing devices.
(2) Evidence is more widespread; previously the primary evidence source was stored locally on disk. Now evidence may be in memory, passing briefly through routers, to be stored on remote cloud-based servers before apparently disappearing. Google Drive, OneDrive and Dropbox are just a few examples of the many locations that masses of data are now stored.
(3) Encryption and access to devices. Where previously investigators dealt with unencrypted disks they could remove from a device, now smartphones and cloud-based services do not give up their data so easily, besides ordinary users are more aware of privacy risks, thanks to whistleblowers such as Edward Snowden; plus, manufacturers provide complex measures to protect users' data.
(4) Piecing together a vast number of artefacts from disparate sources and possibly located in different jurisdictions. Email, instant message and web browser logs are no longer the primary software sources of evidence as a changing landscape of social media and other communication sites and app stores' digital traces, both on the owner's computers and on third-party cloud sites.
(5) The fact that just as digital devices have permeated our lives, digital evidence has done the same. Seldom is any crime committed, any dispute litigated, or any case enacted where digital evidence is not a central element of that case. The 'Wagatha Christie' libel case between Rebekah Vardy and Coleen Rooney, in 2022, relied heavily upon digital evidence and is a prime example.

Fortunately, the basic methodology for investigating and reporting on digital evidence remains the same: identify, capture, analyse, and report. In this chapter we will discuss how this methodology is applied in practice both for the classic, physical crime scene and the 'virtual crime scene'. An overview of the laws and regulations which both constrain and enable investigators is presented and, although examples from an England & Wales and UK perspective are cited, most other Western jurisdictions use a similar framework. Best practice guidelines designed to prevent evidence contamination are discussed as well as how they can be adapted in situations where the only way to acquire evidence is to interact with it. We will also look to the future of these guidelines and how they may change in line with development of technology and of the understanding of digital evidence by the legal profession.

The initial 'look' at any digital investigation, is by considering the classic artefacts and how to apply the basic investigation methodology within the legal framework and guidelines. However, this can be affected by problems introduced from newer technologies such as mobile and ubiquitous computing as well as social networking applications. For each of these, some of the acquisition and analysis techniques used will be discussed, together with the kinds of results that can be achieved alongside the challenges they present, such as the importance of digital triage and the identification of appropriate digital evidence.

In approaching any modern digital evidence problem, gathering intelligence is very important and some of the ways to legally acquire the intelligence necessary to carry out an investigation are discussed. Effectively reporting results is the final part of the digital forensics process and some of the styles of reporting for different audiences are presented. This chapter concludes with a look ahead to how we, as practitioners, might approach emerging technologies, such as the so-called 'internet of things', embedded/wearable technology and what the future for digital forensics might look like.

12.2 Laws, Regulations and Guidelines

12.2.1 Introduction

This section covers some of the laws and regulations that both enable a digital investigator to do their job as well as those that the investigator might easily fall foul of during the course of an investigation. There are many news stories covering old crimes that are easier to commit using new technology (such as fraud or the distribution of indecent images of children) as well as new crimes that have required new laws to be created. Readers interested in a comprehensive overview of the laws, and crimes relating to information technology in the UK and cybercrime, are referred to texts cited in the Further Reading section.

12.2.2 Hacking Laws

Cybercrime is a regular feature in the news cycle whether it is foreign governments trying to spy on people or companies, criminals using malware to steal data or money from people, or political activist groups trying to take down websites of organizations they disagree with. An article by the British Computer Society summarised the biggest cyberattacks in 2022:

(1) Attacks linked to Russia against Ukrainian infrastructure such as the electricity network, internet infrastructure, banks government administration and military.
(2) Costa Rica declared a national emergency following a ransomware attack that caused major financial disruption.
(3) A criminal group known as Lapsus$ stealing data from Nvidia, Ubisoft, Samsung and Microsoft before leaking it online.
(4) Major outage to NHS emergency services in the UK following a ransomware attack.
(5) Attack against Iranian steel companies causing machines to malfunction sending molten steel across the factory floor.

The media use the term 'hacker' to cover all of these activities although, amongst computer scientists and software engineers, the term refers to a particularly skilled (possibly benevolent) programmer. With apologies to our computing colleagues, we will use the popular media meaning here. Malevolent hacking has been around since computers began to do useful financial work, either by disgruntled employees wanting to get revenge, dishonest people using it as a tool to steal money or those curious to test or understand security mechanisms. While many countries tried to apply old laws to these situations, in the UK the Computer Misuse Act 1990 was passed into primary legislation and an examination of the early drafts demonstrates how far advanced the UK was in dealing with cybercrime. Subsequently, the Budapest Convention on Cybercrime in 2001 sought to standardise cybercrime legislation across Europe and coordinate responses then, by the mid-2000s, most European counties had enacted specific cybercrime laws.

The UK Computer Misuse Act (CMA) 1990 is fairly representative of these laws in that it forbids unauthorized access (*e.g.*, logging into a system with stolen credentials, amongst other methods), committing crimes using unauthorised access, altering data having gained unauthorised access and preventing a system from operating properly (*e.g.*, a denial of service attack). A more recent (and controversial) addition to the CMA, is a prohibition on the making, distributing or possession of hacking tools. This last section can be a problem for the digital investigator because many of the tools in their toolbox, such as a password cracker, can also be used to commit hacking offenses. Fortunately, the UK's prosecuting body, the Crown Prosecution Service (CPS) recognised this and set out guidance as to how the legislation should be used and, most helpfully, a means by which such 'tools' could be classified as malicious or benign.

The key, common feature of all the anti-hacking laws is the prohibiting of unauthorized access to data. Any digital investigator, whether in a civil or criminal investigation, could be guilty of one of these hacking laws if they gain access to a machine to investigate it, without the proper authority. Being told by your superiors to access a machine may not be a sufficient defence if it is clear that they do not have the authority to grant access. DPP v Bignall is a clear example of such a situation. Personal data owned by an employee or on equipment owned by them, brings with it more layers of complexity such as necessity and proportionality, particularly where these devices are used in an employment context. In the next section you will see that the digital investigator must be aware of relevant laws protecting individual privacy so that they stay on the correct side of the law.

12.2.3 Privacy Laws

Aside from the risk of violating a criminal law in accessing data while performing a digital forensic investigation, the next most likely mistake is violating a person's civil rights. This could be the rights of the person being investigated or even a digital bystander, that is, someone unrelated to an investigation whose data is captured by a device owned by the person under investigation. In the United Kingdom, this is enshrined in the Human Rights Act, where the individual is protected from the actions of those acting on behalf of the state. Whereas at an individual level the Data Protection Act 2018, derived from the General Data Protection Regulation (GDPR), deals with protecting and accessing personal data. Even though these laws are designed to protect citizens from intrusive and unnecessary access to their data, many (wrongly) interpret these as absolute rights. The use of these acts, and subsequent access to data, is subject to strict controls that prevent both the state and individuals from unnecessary and unrestricted access to personal data. Specific legislation is also available to prevent the interception of communications, such as the Investigatory Powers

Act 2016 (previously the Regulation of Investigatory Powers Act 2000) in the UK. This legislation was used in relation to the News of the World scandal where journalists unlawfully accessed the voice mail accounts of over 4500 individuals, leading to the jailing of senior newspaper staff. Of course, the right to privacy in all these jurisdictions is not absolute; for each there is a test which allows a limited violation of the right to privacy. For criminal matters, this requires the authority of a senior police officer or, in certain circumstances, the Home Secretary but even for an internal investigation within a company on company owned equipment, individuals may have privacy rights which must be handled properly by an investigator.

One of the basic tests which applies in most jurisdictions is that the need to examine an individual's private data must be both necessary and proportionate. Any investigator asked to examine private data has a duty not only to obey the relevant privacy laws but also an ethical duty to ensure that their actions are necessary and proportionate or that they have written assurance from the person commissioning the investigation, that this has been considered.

12.2.4 Guidelines

In addition to the various constitutional laws, treaties, and bills of rights that protect individuals, a digital investigator must follow good practice guidelines if, the evidence they collect or analyse, is to be acceptable in a civil or criminal court, employment tribunal or even an internal, corporate, disciplinary procedure or investigation. A number of national justice organizations have produced guidelines governing the collection and investigation of digital evidence. Notable among these are the UK Association of Chief Police Officers (ACPO) *Good Practice Guide for Digital Evidence* and the US National Institute of Justice publication, *Forensic Examination of Digital Evidence: A Guide for Law Enforcement*. The most detailed guidelines at the time of writing were in the ISO standard ISO/IET 27037:2012, *Guidelines for Identification, Collection, Acquisition and Preservation of Digital Evidence*.

The ACPO guidelines are relatively concise and generalizable compared to the other documents and the principles which are relevant for many investigations in most jurisdictions and hence described in further detail here. (Note that at the time of writing, ACPO had rebranded itself as the National Police Chiefs Council (NPCC) and so are in the process of re-issuing the guidelines, therefore you may find them labelled differently but with very similar content.)

The guidelines are built around four simple principles:

(1) **No action** taken by law enforcement agencies, persons employed within those agencies or their agents, **should change data** which may subsequently be relied upon in court.
(2) In circumstances where a person finds it necessary to access original data, that person **must be competent to do so** and be able to give evidence explaining the relevance and the implications of their actions.
(3) An audit trail, or other record of all processes applied to digital evidence, should be created and preserved. An **independent, third party** should be able to examine those processes and **achieve the same result**.
(4) The **person in charge** of the investigation has **overall responsibility** for ensuring that the law and these principles are adhered to.

The first version of the guidelines was published in 1998 and, despite calls for the principles to be updated and expanded to include robust testing and validation of methods, arguably their simplicity has led to widespread adoption nationally and internationally. Only in 2020 were the first meaningful amendments published. Put simply, digital evidence must be handled in such a way that the collection and analysis does not alter the data. As will be seen in the next section, this can be proved mathematically using a technique called 'hashing'. This makes use of a one-way mathematical function which produces a unique digital signature (known as hash value) for a given digital file and can be equated to the digital equivalent of DNA. No matter how large the file is (even if it is an entire disk size measured in Gigabytes or Terabytes), should a single bit (the smallest piece of data that can be stored) in the file (or disk) be altered, the number produced by the hash algorithm will appear radically different. It is best practice to calculate and record a hash value for every piece of digital media as soon as it is seized so it can be demonstrated that the evidence produced in court, is exactly identical to that seized. This proves it cannot have decayed or been modified by the digital forensic analyst during analysis.

As will be seen in the next section, life isn't always quite as simple for the digital investigator. ACPO Principle 1 covers situations when the investigator is handed a disk that is switched off and not connected to a running computer therefore assuming that the disk is not protected by strong encryption. If, for example, the investigator finds a disk connected to a computer that is running, and they know that the disk is encrypted and thus unintelligible once the computer is switched off, then Principle 1 becomes meaningless: there will be no evidence unless the investigator interacts with the computer in some way, even if just to confirm that the disk can be decrypted later. In doing so it is inevitable that the analyst will make changes to the data. Here Principle 2 applies, and extra care must be taken, so that it can be made absolutely clear later which changes are due to the actions of the investigator retrieving the evidence.

Finally, ACPO Principles 3 and 4 apply as much to digital evidence as they do to evidential procedures for any of the forensic sciences: principle 3 compels everyone involved in the evidence chain to keep contemporaneous notes at all stages, ensuring it is clearly explained what has been done with that evidence, and how the results presented to the court were obtained. By doing so, it can be established that the work is reproducible. Principle 4 says that there can be no buck-passing. The person in charge of the investigation takes full responsibility for making sure that evidence is handled and analysed in a principled and legal manner.

12.3 Identification and Acquisition

12.3.1 Introduction

Computers do not generate or store evidence for investigators. Digital evidence is essentially a representation of something created by a computer that can be connected to the actions of an individual. In the days when desktop computers were the subject of our investigations, evidence was fairly easy to locate and identify. In this current age, we find computing devices embedded in objects all around us, either storing data in tiny local disks or, more likely, transmitting them to a cloud server which is difficult to identify, let alone access. In the following sections we will look at how to approach, identify, and acquire relevant data from both a physical crime scene and a virtual crime scene spread across uncountable machines, cloud servers and social media sites.

12.3.2 Preparing and Approaching a Physical Crime Scene

The digital crime scene is an incredibly complex environment. In addition to all the other aspects of management seen in traditional crime scenes, there are several facets that are unique to the digital world. Unlike traditional crime scenes, at a blood scene for example, there are many ways in which the digital scene can change right before the investigator's eyes. For example, a wired or wireless network connection can facilitate the remote amendment or, in the worst-case scenario, destruction of evidence.

A running computer is also a changing environment, with running processes potentially making changes to files. The process of switching a computer on alone changes over 2000 files. The simple solution to this is just to turn the computer off; yet this simple action is, in itself, fraught with danger. Shutting down a system running Windows in the normal manner runs the risk of activating unknown applications which might encrypt the entire drive, or worse still, erase it. Traditionally, the advice to circumvent this included cutting the power to the machine, however many operating systems (OSs) now automatically encrypt their disks (bitlocker in Windows and FileVault on MacOS) making the files inaccessible once the OS is not running or the user is logged out. To increase performance, and avoid making users wait, computer systems temporarily write to a dedicated 'buffer' in main memory before periodically transferring the contents to disk, a background process known as 'flushing'. However, if the power is suddenly removed the contents of main memory will be lost and so any data not flushed, will also be permanently lost. Clearly, getting this process wrong could mean that all of the data is permanently lost or perhaps the defence lawyer will allege that, the lost data from the buffers, was the very piece of evidence that would exonerate his client.

Therefore, it is important that a significant amount of thought and consideration goes into any investigation strategy or plans where computer systems may be seized. The first consideration is the intelligence picture. Who is the investigator dealing with? What are they suspected of? What are the suspect's technical capabilities? All of these will affect how to approach the scene. Take, for example, the terrorist who has rigged his computer with a tilt switch and explosives; is it reasonable to have the machine X-rayed *in situ*? Would the investigator want to enter this property by knocking on the front door? Or would a dynamic, forced entry be more appropriate? What about the computer scientist deploying military grade encryption on his machine such as was the case in the FBI's arrest of Ross Ulbricht, creator of the Silk Road, the Dark Web's most notorious online marketplace? How would an investigator approach the suspect's machine so that they can make sure that the suspect does not get the chance to activate the encryption and lose all of their evidence.

Do these scenarios sound far-fetched? They are not; the authors have first-hand experience of incidents just like these plus they have had to provide advice to senior investigating officers to make these critical decisions.

Although these are extreme examples, they do indicate the complexity of the challenge facing digital investigators. Almost every serious crime investigation has an important digital element to it and, with the ubiquity of smartphones, this will only become more common.

Once the intelligence picture is understood and the scene secured, the situation does not get any simpler; the next decisions are just as complex. Firstly, there is a need to identify and document the context of all digital evidence. When considering the amount of data that can be stored on a micro-SD card, (which at the time of writing was measured in Terabytes) the significance of locating these cannot be overstated. In the most serious of cases, fingertip searches lasting days are not uncommon.

Any items found must be documented *in situ,* to provide context to the find, ideally by photograph. This is often significant in attributing the item to an individual, particularly where the chances of another person being concerned with it are reduced by the circumstances. For example, a USB 'thumb drive' full of evidence, found with a credit card receipt relating to the suspect's account, is quite compelling even if the computer to which it has been connected cannot be found.

This process is also vital in showing what the investigator was confronted with and what was happening with the computers at that moment in time. In extreme cases, it can become the only evidence. For example, if a photograph is taken of the screen of a computer or smartphone showing a significant email, just before encryption is activated and locks out all the data on the computer, then the photograph itself may be the only proof that an email existed. Current training, under the Police Education Qualification Framework scheme, teaches officers to make best use of body worn video to document these scenes as first responders. Enhancements in police issued technology allows live images of the scene to be transmitted, allowing instant access to expert advice.

Once the scene is documented, the seizure phase commences. It is important here that the digital investigator is fully aware of the powers under which they are operating and how their actions will interact with that of traditional forensic crime scene examiners. The law around accessing data unlawfully is, quite rightly, widely constructed. It is therefore important that the investigator is fully aware of the limitations of the authority to investigate that they have. This could be under the protection of a search warrant or civil search order. In the UK, it could be under any one of a number of powers conferred by the Police and Criminal Evidence Act 1984 relating to searching of, and seizure from, premises or even Section 50/51 of the Criminal Justice and Police Act 2001 which relates to examination of material deemed too excessive to be examined at the scene. Although some of these laws were not designed specifically for computer examination, it can be seen how easily they can be shown to apply to digital investigations, potentially to the detriment of the investigator and investigation.

Any breach of the relevant rules can mean that evidence is deemed unfairly obtained and is at risk of being ruled inadmissible in any future court proceedings. Furthermore, the investigator can find themselves personally liable if they deliberately access data that they have no entitlement to.

When originally drafted, Section 1 of the Computer Misuse Act dealt with unauthorised access to computers. In 2006, this was amended to cover programs and data (note the computer is not mentioned explicitly). This is a much wider interpretation. Consider the situation where a USB drive is unlawfully seized, when the warrant only allows seizure of computers, due to poor wording. Clearly there is no authority to access this data. Any subsequent search of that drive could be a breach of Section 1 of the Computer Misuse Act for which the examiner could be personally and criminally liable.

Having documented the scene, ensured all is well and that the examiner is acting lawfully, the actual seizure and collection of evidence will commence. Often there is only one chance to collect the evidence, particularly where running systems are concerned (see Section 12.3.4). It is therefore important that the correct equipment and tools (both hardware and software) are available. What an investigator needs to take with them will of course be dependent on the intelligence picture available to the investigator and the jurisdiction involved. This can range from basic plastic evidence bags and exhibit labels to Faraday boxes which isolate wireless devices capable of accessing mobile data from the network. Significant high-capacity storage media are also required. It is sensible to take far more than one would anticipate using rather than be in a situation of not having enough room to copy the password protected cloud drive that could lock out at any moment. Care must also be taken when crossing jurisdictions with evidence; some countries forbid the import or export of encrypted drives.

The watchword with properly capturing a digital crime scene is to be prepared, to anticipate the unexpected and to plan for the worst-case scenario. It is far better to waste an hour loading and unloading the entire contents of your laboratory than lose the case because you were not prepared.

12.3.3 Classical Digital Forensics (Dead Machine)

Although there can be situations where visiting the scene of the crime will be necessary, the reality of digital forensic work is that more often than not, the work will come to the investigator. Generally, this is in the form of a seized computer packaged and exhibited. The following will provide an overview of what happens next and why attention to detail is vitally important.

Any item received should be an exhibit. This means that it has been properly seized, packaged and continuity of evidence (or "chain of custody") is maintained. Often this takes the form of an exhibit label. Examination of this should show the last signature on it is that of the person who hands it to the investigator. This might not always be the case. In civil investigations, exhibits can come by post or by other registered means. These should be accompanied by documentation to allow the investigator to satisfy themselves that the exhibit is exactly as it was when it left the sender. Indications that the exhibit has been tampered with can include changed seal numbers, rips to the bag and any variety of physical damage to the exhibit. Should anything like this be discovered, then it is important that it be properly documented.

Once satisfied that the exhibit's integrity is intact, only then it is appropriate to conduct a physical examination of the exhibit. Care should be taken to identify and record all component parts of the device so that evidence can be adduced, if necessary that it was actually capable of performing the acts alleged. For example, it is unlikely a device without network connectivity would be used in a hacking case. In many cases, this evidence is just as important as any file fragments that may be recovered from the hard disks inside.

Time stamps on files are a vital part of digital forensics and it is important that they are as accurate as possible. Incorrectly calculated time stamps can, at worst, convict the innocent or provide a digital alibi for the guilty.

BIOS time (here referring to the computer's internal clock) is used by the operating system to calculate the time it believes that files have been created, modified or accessed (collectively known as 'MAC times'

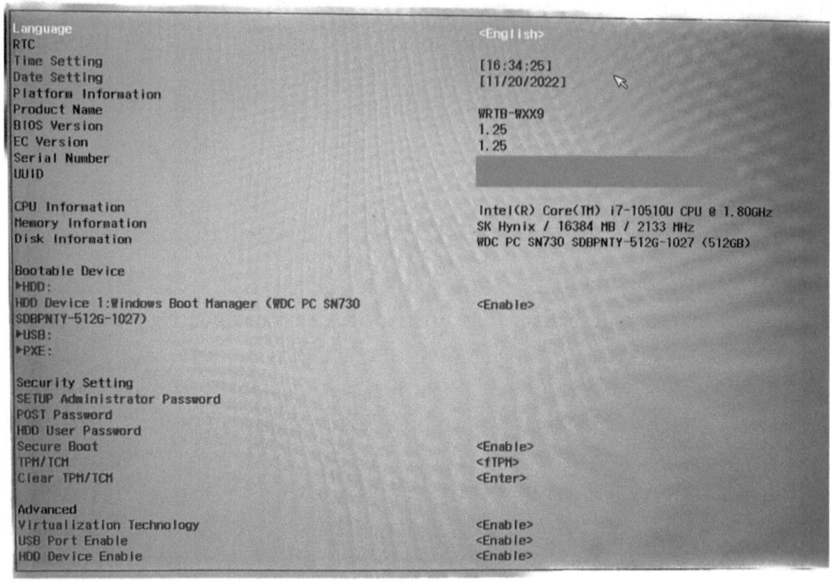

Figure 12.1 BIOS clock.

and discussed further in Section 12.4.2). The BIOS is the basic program used by a computer when it first starts and also contains configuration settings often accessed using F12, ESC or DEL resulting in a display similar to the one shown in Figure 12.1. The time held in the BIOS is then augmented by a variety of settings within the operating system to provide the times recorded; for example, some operating systems adjust the BIOS time for daylight savings while others will apply the adjustment internally. Often this BIOS time is not accurate, so it is important that one of the first steps taken during the examination of the computer hardware, is to note this time and compare it against a known accurate source such as a radio time signal clock. Any difference can be applied to system calculated times to create a true and accurate time at which the files were updated. Modern systems update their time automatically using internet time servers, where highly accurate clocks make the time available online. So, while the BIOS time may be inaccurate, care should be taken to confirm with any logs whether the time was automatically adjusted.

Digital evidence is incredibly fragile and susceptible to the slightest interactions with a running computer. For example, over 2000 files are touched and have amendments made to their timestamps upon normal start up. Imagine the effect on the case and personal credibility if the defence can show files were amended whilst in the investigator's possession. Of course, ACPO Guideline 2 may be relevant here and this serves to highlight the value of keeping rigorous notes as required by ACPO Guideline 3.

The ACPO guidelines for handling digital evidence and their four basic principles were outlined previously (Section 12.2.4). The first of these principles stated that no action shall be taken that changes the original evidence. In practice this means that a copy is taken of the drive in a manner that demonstrates the act of taking that copy, did not change the contents of the drive at all, a process called imaging. All subsequent investigative work is then done on this copy (or image) which can itself be replicated many times without reference to the original.

The process of obtaining this copy, however, is a complex matter and requires the use of two tools. The first of these is called a write blocker with its purpose being to prevent an operating system making any changes to the suspect hard drive while the copying is taking place. Hardware write blockers are considered the 'gold standard' where a physical device intercepts commands that would change the content of the hard drive and then discards them before they get the option to do so – Figure 12.2 shows a hardware write blocker connected to a hard drive.

Software write blockers are also available that force the computer to treat the device as read-only but in some cases modifications may occur. An example the reader could test for themselves is Thumbscrew,

Figure 12.2 Hardware write blocker.

a free download for Windows systems, that can be used to demonstrate the principles of write blocking.

Once the drive has been secured, it is important to be able to demonstrate that the act of taking the copy has not changed the contents. This is done by applying a mathematical 'hash function' to the original, as described previously (Section 12.2.4). Two of the most popular functions historically have been MD5 and SHA-0, each of which produce a 'unique' hexadecimal number for any file; there are many other types of hash that can be used in the same way.

Once a hash has been calculated and recorded in contemporaneous notes then the drive itself can be copied or to use the more correct term, imaged. Initially, this process involves making an identical bit for bit copy of the entire disk into a single file using a Unix tool called dd so the disk image file created often has the suffix 'dd'. This file would be exactly the same size as the entire capacity of the disk (even if the disk was empty) because every value on the disk was copied. For large investigations, this obviously became untenable, so the format of one of the popular commercial digital forensics packages has become the *de facto* standard namely, Expert Witness Format (EWF) which was developed by Guidance Software for their popular EnCase tool. It has a file suffix ending in E01 (or in the latest version, Ex01) with the ability to compress disk images, embed additional integrity protection and spread them over multiple files, as well as to store metadata about the disk and the investigation.

Regardless of the format used to store the disk image, once it has been created the hash function can be applied to this file and the result recorded in contemporaneous notes where it should be identical to the value for the suspect drive.

If these conditions can be met, then it can be said that a forensically sound image has been obtained. This can then be duplicated as many times as is necessary. Any tests or processes the examiner wishes to apply that may change time stamps or file content, can be undertaken safely in the knowledge that they can start afresh, with a new copy, within minutes.

While this situation is fairly typical of the day-to-day work in a digital forensics' laboratory, there are occasionally situations where it is necessary to work in a fashion that will, without doubt, change the data on the original drive, a breach of the ACPO Principle 1. Fortunately, the principles accept this situation can arise and ACPO Principle 2 explains that, where it is deemed necessary to act like this, then that is permissible as long as the examiner is competent and is capable of giving evidence as to the effect of their actions.

The circumstances when it is deemed necessary to access live data are not defined but they generally are due to either a time constraint or a technical reason. One very common situation where it may be necessary to immediately examine a computer at the home address, is when a missing child's life is at immediate risk. Clues to their whereabouts could be found on their laptop therefore the delays in conveying the laptop to the lab, increase the risk to the child significantly.

Another situation where data must be captured live, is when corporate servers are involved. These are seldom switched off to obtain digital evidence because the cost to the business far outweighs the value of the evidence. In this case, compromise solutions of obtaining an image of a selective part of the system is obtained while the system is still running. Clearly data will change as the process is ongoing but as this is the best available option, it is acceptable and justifiable. Certain embedded computers, such as those on smartphones, make it impossible to separate the unencrypted disk from the rest of the system consequently changes to the system have to be made to allow access to the data.

Whatever the reason for examining a live system, as long as it is reasonable under the circumstances, the person doing it is qualified, able to show how their actions can be separated from those of the suspect and there were no other options, then it is likely to be accepted by a court. Here, it is important that the circumstances, context and options available at that time are accurately recorded, along with the decision making process that led to the conclusion that this was the only viable option available. (The National Decision Model – is an effective way of making and recording decisions such as this.) A simple test here is: could I lose evidence if I follow ACPO Guideline 1? If the answer is yes, then relying on the protections afforded by ACPO Guideline 2 is reasonable.

12.3.4 Digital Triage and Live Forensics

The astute reader will have noticed a growing problem with the imaging process described above. Less than 15 years ago, the average hard drive size found in a home computer was around 40 GB; today drive sizes are measured in TB, an increase of more than 25-fold. Although the cost of storage is relatively cheap, increasing drive sizes have brought their own problems to the field. For the examiner who has to retain data for many years, possibly decades after the examination, there is the issue of cost and space for storage. However, the main issue is performing the actual investigation in the first place.

As hard drives have grown in capacity, the more users will fill it. Whether it is photographs, office documentation, video or the like, it will be stored. With larger and larger hard drives, the incentive to periodically weed out and remove material no longer required, is no longer there. This means that the digital 'haystack' of irrelevant data has become so large that a methodical search of every part of it for the virtual 'needle,' may be impractical.

In the recent past (*ca.*, 2010) it was commonplace that drives were searched in their entirety, and without too much focus, simply because the amount of data made this a feasible task. Attempting to perform the same style of forensic investigation now, is an almost impossible task making backlogs and delay common place in all investigative institutions.

Approaching this problem has required some creative techniques and a change of investigative mind-set from the 'just in case' approach to a much more focussed, intelligence led, proportionate and 'risk aware' type of investigation.

Understandably, no investigator wants to miss a piece of evidence that would allow a paedophile to go free or undetected and, for that reason, even cases submitted for investigation of fraud would be investigated for indecent images. Today, without intelligence to suggest that such an image is likely to be found, the investigation would be tailored to meet the needs of the submitting investigation team hence would focus on appropriate searches to look for the evidence that will prove or disprove their particular case. Speculative, *'just in case'* searches will not be undertaken.

In the current era of cheap hardware, it is not unusual for multiple computers to be recovered from a single crime scene. For example, looking around his office, as shown in Figure 12.3, the author can see approximately 40 devices containing over 10 TB of data that could potentially need to be examined if it were a crime scene. So how does the investigator set about determining which key devices to prioritise are and how to approach it?

Digital triage is a process that has been developed to allow investigators to have an automated (and forensically sound, by complying with the ACPO principles) 'quick look' at devices to determine whether they contain material of relevance. In the previous section, the importance of not changing the data by working on an image so that the original data was not used, was emphasised. In a triage situation, the emphasis changes and the time-consuming process of imaging a drive, is initially substituted for a targeted search of the drive which determines whether or not imaging is worthwhile. The

Figure 12.3 Office crime scene.

most common way of doing this, and ensuring the drive is not changed during the process, is to boot the computer with a specially adapted Linux 'live disk'. This disk is adapted to act as a software write blocker. Usually, a pre-programmed USB thumb drive will be used in conjunction with that disk to identify and record files that are of specific interest to the investigation.

One way this technique can be used is to identify which machine is of interest to the investigation. In one eBay fraud case known to the author, while executing a warrant to search a house, eight computers were seized. Traditionally, all the drives on these computers would have been imaged and investigated. The imaging process alone would have taken at least two days, probably more, to allow the laboratory to continue functioning on other higher priority investigations. The investigation process would have required probably another two weeks to complete.

In this case all eight machines were triaged simultaneously, overnight, for traces of the known eBay user ID. When the examiner returned the following morning, only one machine was identified as containing traces of the relevant user ID. This machine was imaged and investigated, and the rest of the devices pended. In total, the investigation concluded in around four days rather than three weeks, providing evidence that led to a conviction.

Unfortunately, such an approach is not without risk therefore it is important that these risks are understood and accepted. In doing so, a proportionate and reasonable approach to the sheer volume of data, is possible.

12.3.5 Mobile Devices

Alluded to above were some of the special problems presented by smartphones and devices where the computer is embedded in a device. Getting data from a mobile phone has always been a specialist area, requiring proprietary, commercial software and often special hardware connections for each model of phone. At the time of writing, it was still possible to buy a feature phone (non-smartphone) and with the right equipment it is possible to be able to extract the address book, call and text history. However, the two dominant types of smartphone, Google Android and Apple iOS (iPhone), present special problems from a digital forensic viewpoint. The operating systems are specifically designed to resist tampering, or a forensic investigation. Consequently, traditional imaging is not possible unless the phone can be 'rooted' or the OS by-passed to give the investigator root (superuser) access to the phone. This would of course violate ACPO Principle 1 thus special care needs to be taken and only performed by a suitably qualified investigator, as the provisions of ACPO Principle 2 will apply.

If a smartphone is seized while powered off then there is not much that can be done, other than bagging it, as with other dead computing devices, and leaving it for the specialist mobile forensics examiner. In the case of Android phones and devices, however, it may be possible to perform some traditional forensics if there is an external, removable memory card. This can be imaged and examined just like any other disk and some data, such as photos, may not be encrypted. Encryption of mobile phones is a particular issue, with even 6 numerical codes taking months, if not many years, to brute force (a technique where every possible combination of password is checked sequentially). In this case, the ideal situation would be the suspect to provide you with the access (PIN) code. Of course, there are very obvious reasons why they may not wish to do so. The law recognises this as obstruction and penalises it heavily with up to two years in prison for the wilful refusal to disclose a decryption key (an unlock code) following a formal request made under S49 of the Regulation of Investigatory Powers Act.

If a smartphone or tablet is seized while it is on and unlocked, then a number of choices may need to be made relatively quickly. Most devices will be set to automatically lock after a certain period of inactivity and may be set to erase after 10 password attempts or *via* a remote command. If the investigator does have the password or fingerprint (if using fingerprint recognition) of the owner, then the first priority is to isolate it from radio signals. This could be done with a special Faraday (radio wave secure) bag, by putting the device in flight mode or by removing the SIM card. If you need to keep it unlocked then you will have to find a way to do this and keep the device powered, which can be challenging if using a Faraday bag. As mobile device forensics is a rapidly changing and specialist area, how one proceeds will be guided by the circumstances of the case and the advice of the mobile forensics expert. For example, there may be a need to perform a live examination of the device. However, this has to follow a strict protocol, with a video recording all of the actions taken, so it can be shown which artefacts are due to the suspect and which are due to the investigation.

12.3.6 Virtual Crime Scene

So far, the discussions have been related to seizing physical devices which have some kind of long-term storage (disks) which have digital artefacts stored which can usually be directly related to the actions of an individual person. Until the early part of this century, digital forensics dealt mostly with examinations of this type of device. The advent of always-on, ultra-cheap, fast wireless network connections has made online storage both attractive and practical. The physical devices that might be seized now could merely be input/output devices which only temporarily buffer the digital artefacts. The artefacts may now be stored on many disks, duplicated on unknown, highly secure servers around the world, only accessible to a very small number of people in one company. No one is ever going to image one of these disks. The best the investigator could hope for, if able to obtain a court order that the company recognised, would be a report from the company in reply to specific questions from the investigator about an individual account.

However, in the case of data from social media or websites, there may be no need to image disks or try for difficult court orders. If the suspect has posted in a publicly available forum, such as Twitter or a public posting in Facebook, LinkedIn or Google Plus, then it is available on every computer in the world: a simple screenshot saved with a hash of the image and entered in the contemporaneous notes, may be sufficient. Should additional data be required from a social media platform, then it may be necessary to use international, mutual, legal assistance treaties. This, however, is very time consuming and a balance should be struck between this resource-extensive process and the seriousness of the offence.

Having secured the evidence, all that is now required is simply a way to tie the account that posted the message with the individual. As you will see later, this however, is not as easy as it may seem. The app or webpage used to post messages, will often leave artefacts at least indicating the account logged in, so an image of the suspect's computer or smartphone may provide this evidence. Depending on how long ago the post was visible, evidence may be found in the Internet Archive (www.archive.org, also known as the Wayback Machine). This is a service that attempts to keep an archival record of content on the internet since the mid-1990s; however, not all sites, and not all updates, are recorded. Figure 12.4 shows an example of the Wayback Machine archiving the BBC News website from 1st December 1998.

Since most social media introduced privacy controls by about 2010, it is now rare (but not unknown) that someone will make a public incriminating statement. If a posting is limited however, then the options for obtaining evidence about it are also limited. The investigator would either have to convince someone in the suspect's network to allow access to their account to view postings, or create a fake account then have that account accepted into the network and view the postings. The investigator might also need to communicate with the suspect to obtain information or to become trusted enough to be accepted. In each of these cases, there are rules about what can and cannot be done. In England and Wales, these are covered by two rules: Covert Human Intelligence Source (CHIS), also known as a covert internet investigator, and Directed Surveillance. Both of these are dealt with in detail by the Investigatory Powers Act 2016. It is beyond the scope of this book to go into detail here, but the main points to realise are that developing a relationship using a covert identity and/or repeated monitoring of a person involves strict controls, designed to protect both the investigative team and the suspect. For example, the CHIS must not be seen to be enticing the suspect into an illegal act.

There are also situations in a corporate environment where direct access to the disks that contain the evidence is impractical, such as email on large corporate mail servers. In cases like this, it may be acceptable to use a proxy technique, such as having the output from an authenticated mail client export details about mail traffic between individuals.

Figure 12.4 Wayback machine – BBC 1998.

It may also be necessary to directly monitor or record the live network traffic in an organisation. This is tricky on a number of levels. Firstly, most jurisdictions forbid spying on a public network without a warrant; in English law this is governed by the Investigatory Powers Act 2016. However, a business who owns their own internal network, have the right to intercept communications within it, as employees will usually have consented to this, explicitly in their contract, or implicitly through accepting the company's Acceptable Use Policy (often called an AUP). Even this kind of surveillance can be fraught with difficulties, as the investigator would only have the right (if granted by the company) to monitor internally destined traffic; monitoring traffic that was headed for the internet would still require a warrant.

12.4 Analysis

Once evidence has been properly acquired, the job of the investigator really begins. Even if a triage has already turned up a 'smoking gun', be it an incriminating word document, a series of indecent photographs or an email admitting murder, this is just the beginning. The presence of the evidence is nothing more than an indication that the device has been used in connection with the offence.

To prove the case the investigator has two key questions to answer:

(1) How did this artefact get there?
(2) Who put it there?

Although these are very straightforward questions, as you may imagine, answering them is far from straightforward. In this section, the process of answering these questions in a digital forensics' context, will be considered.

12.4.1 Investigative Basics

Before an effective investigation can take place, it is important that the intelligence case and objectives are known. It is quite common for a digital investigator to be asked for 'a full forensic' or to 'give me all evidence of fraud'. The answer to the first question might well be 'How much money do you have?' and the answer to the latter is, of course, 'What does fraud look like?' While the answers are tongue-in-cheek, these actual questions illustrate the gap between what lead investigators want *versus* the specific detail that is needed to retrieve the evidence they actually need.

The first objective of the digital investigator is to learn what the case is all about, but most importantly what the lead investigator is looking to prove or disprove. Quite often the latter will not know the capabilities of digital forensics, so it is important that the digital investigator is able to identify, from the lead investigator, the role that they believe the computers have played in their case. This can lead to the posing of questions such as:

(1) Can you prove or disprove this computer is associated with a particular eBay username?
(2) Has this computer downloaded indecent images of children?
(3) Has this computer been used for any form of communication between X & Y?

Clearly these are closed questions which can be answered yes or no. These are ideal types of question for the digital investigator and express a real-world situation. The detective may not necessarily know how this may be achieved by a computer, but the investigator will; that is where the two forms of investigation come together.

Usually, it is possible to distil the most complex case down to one or more simple questions. This is the key to setting out a focused and efficient investigative strategy. Once it is known what is to be established, then the areas that need to be searched or the tests that need to be performed, are generally self-evident.

Taking the example of an eBay fraud, the objective would be to prove or disprove that a computer had been used to defraud someone on eBay. Firstly, the eBay user account used in the fraud would need to be known or at least the item number which is specific to the offence. Times and dates of interactions and transactions are also useful for reasons that will be explained later in Section 12.4.3. Secondly, having established what is needed to be proven, the main tool would be a basic text search of the image of the computer's disk for the text fragments that tie that machine to the fraud. A good start would be for the username, the item ID number and any unique description connected to the case.

If the computer has been used in this offence, then a carefully constructed set of search terms should find some trace of it, even if the files have been deleted or if the computer has been reformatted and the operating system has been reinstalled. When a user deletes data on a computer it is not usually erased: the operating system simply marks the area as available for possible re-use. Fragments of files and webpages viewed years ago may be found.

If any of the search terms are found, then it is worthwhile progressing with the investigation. If not, it is likely that this computer cannot be proven to have been involved in the matter. The next step is to determine the significance of the evidence that has been found and this is discussed below.

12.4.2 Classic Dead Analysis

As described previously, the term 'dead analysis' refers to the analysis of an image rather than working on a running system. In this section, the types of artefacts left behind in the course of using an operating system and how they might be interpreted, will be discussed. Although this section will, where necessary, focus on Microsoft Windows operating systems by way of examples, all operating systems will have equivalent files that will perform the same function.

Before considering the specific types of artefacts on the disk, it should be recognised that at no time in the development of any operating system, did the developers give any consideration whatsoever to the needs of forensic examiners. What is being discussed here is the digital investigator's equivalent of Locard's Principle.

Log files from webservers for example, were intended for a specific purpose such as diagnosing problems with the software itself, yet they also reveal which computers have connected to them. It follows, therefore, that the skill in analysing findings, is based on an appreciation of what real-world user activities lead to these traces. Sometimes there is one obvious explanation, but if not, then the investigator will have to develop possible theories to explain the presence of a particular artefact. Occasionally, this will involve conducting experiments on another computer to determine if user actions are likely to produce the same artefact. The investigator's goal is to produce an explanation of the likely sequence of events performed by the user which have resulted in the creation of traces.

Computer operating systems and associated applications are complex pieces of software designed for speed and efficiency, a characteristic that turns out to be advantageous for the digital forensic investigator. One piece of commonly used software that implements a useful efficiency technique is the web browser (*e.g.*, Internet Explorer, Edge, Chrome or Firefox) which makes use of an 'internet cache'. Used primarily to facilitate faster loading of webpages that have previously been visited, this area of the system is a literal goldmine for the digital investigator. Under normal circumstances (leaving aside private browsing mode) an investigator should be able to locate most of the component parts of a webpage that the user has visited. This can include the webpage itself which will include all the operating code the internet browser uses to format the page before displaying it to the user. Any images that form part of the displayed webpage, can also usually be recovered. Specialist software can be used to rebuild the webpage to how it may have looked to the user. Clearly, evidence of this nature will be invaluable in presenting a case to a jury to show the webpage exactly as the suspect saw it.

Another important set of artefacts of web browsing are cookies, small text files left behind when users visit most websites. While they are not as indicative of the user's activity on the site, they show if a site has been visited and may store information about the current state of the user's web browser or data they have entered.

A further important aspect of the internet cache is the internet history. The location of this file will vary depending on the operating system and browser used but will contain a time stamped list of all internet activity that a user has undertaken. This can be particularly useful in providing a timeline of the internet activity on the machine. Its use and significance will be discussed below where timelines and attribution are considered.

As a convenience to users, most operating systems, and many applications, (particularly the MS Office suite) maintain a time ordered list of recently accessed files. These lists are composed of individual files that act as pointers to other files the user has been working on or viewing. These files contain a multitude of information that allows the operating system to both locate and open that file. This information can include the file path, where the target file is located, the size in bytes of that file, together with the time it was created, last modified and last accessed. All of these can go a long way to providing, at the very least, significant circumstantial evidence of the user's activity. The most important aspect of these files is how they are created, which is directly related to the first occasion when the target file is opened.

Most operating systems, including Windows, make use of small files called links (.lnk in Windows) or aliases which point at the actual location of the file. It can therefore be directly inferred from the presence of an alias or link file in specific locations, that a particular file has been accessed by a user at a specific time and date. Depending on the operating system and the type of file, it may also be possible to infer more about patterns of file access and use but a detailed discussion of this is beyond the scope of this introduction.

Although many applications create these alias files to record things like recently accessed files, often the main storage for data about how applications have been used, is in a special operating system database which is the other major gold mine of artefacts (the registry for Windows operating systems and plists in MacOS). This database maintains a multitude of settings related to both the operating system and individual applications—in many ways it can be thought of as the engine management system of a computer because it provides all the necessary information that a system needs to function correctly.

A full analysis of what information can be derived from a registry/plists is far beyond the scope of this section (indeed, it is the subject of many books in its own right), however some of the artefacts that can be found are:

(1) Details of the particular operating system, when it was installed, when it was last booted up and shut down and whether that was a normal shut down or due to a power loss.
(2) User ID of the last user to log in along with a timestamp.
(3) Any time corrections that might need to be applied relating to the user's location or for daylight savings time can be found here, as can details of any users who have created accounts on the system and when they did so.
(4) A list of the serial numbers of all storage drives that have been connected to the system together is held. This can be particularly useful when trying to associate a USB drive to the system as it is possible to identify not only, whether or not the drive was connected but the time and date that occurred and the drive letter it was allocated.

Central to the value of all these artefacts, is the concept of timestamps. Time is of particular significance in computing and of course, crucial in a digital forensics investigation. Despite the apparent simplicity of measuring time, it can actually be very complicated when discussing it in the context of digital artefacts. As noted earlier, each computer has a hardware system clock which maintains the time even when the computer is powered off. This time may be set by the user or may be configured to synchronize the time with an internet time server when the computer has a network connection; however it will also have a time zone set by the user. It is not unusual to find a computer time off by many minutes or to have the wrong time zone set. Therefore, it is important to check the BIOS time as well as the time zone set for the operating system, before examining any file timestamps so that they can be converted to actual local time if needed.

Each file and many individual settings will have a date and time stamp as part of the metadata which notes at least the following three values:

Created time – The time at which this file was created on the volume where it was located.
Modified time – The time a change was last saved to the file.
Accessed time – The time a file was last accessed (note this is any access and can be affected by events such as antiviruses or similar non-user interventions).

This collection is commonly referred to as MAC times and some operating systems record additional metadata timestamps about other events. The presence of these times allows much to be inferred from the activities that have occurred in relation to the file as shown later (Section 12.4.3).

Deleted files, and the file deletion processes, produce another rich range of artefacts for the digital investigator. In operating systems with a graphical user interface, like Windows and MacOS, the file deletion process involves moving a file from its original location to a special area called the 'recycle bin'. This allows the user to easily recover files if they are accidentally deleted. Each recycle bin (there can be several – one per user account) will maintain details of each file that is contained within it including the original location of the file before it was deleted and the time it was deleted. For the forensic investigator this is highly significant, as it can show precisely when a significant file was deleted which may or may not be incriminating. Only when the user asks for the recycle bin to be emptied is the file actually deleted.

As mentioned previously, operating systems are programmed for efficiency, and this can be used to the investigators advantage when considering deleted files which have been emptied from the recycle bin. All operating systems maintain a list of the files using a file allocation table of some type. These vary by name and type depending on the operating system however, commonly, they share the same characteristics, the main one being a list of files with various types of information about the file including its name, size, location on the drive and associated times. Each of these records will be of a fixed size. In the case of one such record (NTFS – New Technology File System) this is 1024 bytes, or separate characters.

When a file is deleted, its record in the file table is no longer relevant, however contrary to what you may think, the record is not deleted since that would be inefficient and take up too much system time.

Instead, a single character of the record is changed to indicate the file has been deleted leaving the rest of the record intact. While very handy for the digital investigator, the reason it is done is that it is 1023 times faster than overwriting the entire record (in the case of NTFS). Investigators can simply look for this deleted file indicator, which will remain intact until a new file is created and therefore a need exists to reuse that part of the database. Once located, either in full or in part, they are able to process the entire record and the deleted file can, under favourable circumstances, be recovered.

Figures 12.5 and 12.6 below show the differing views of a drive where files have been deleted. Figure 12.5 shows the view of the drive that is presented to the user by the Windows operating system (the logical view). As can be seen, no files are visible and by implication, are inaccessible by the user. Figure 12.6 on the other hand, is the view of the same drive using EnCase, a digital forensic software package, which reveals the files that remain on the drive. Careful examination of the file list will show that in the vast majority of cases, the first character of the file name has been changed to _ (underscore) indicating this file has been deleted.

There will of course come a time when the file table entry is itself overwritten when a new file is created. This, however, does not mean that all of the data from the file has been deleted. There may be many parts of the file still intact if the operating system has not needed to reuse the space; the challenge now is to identify and recover it.

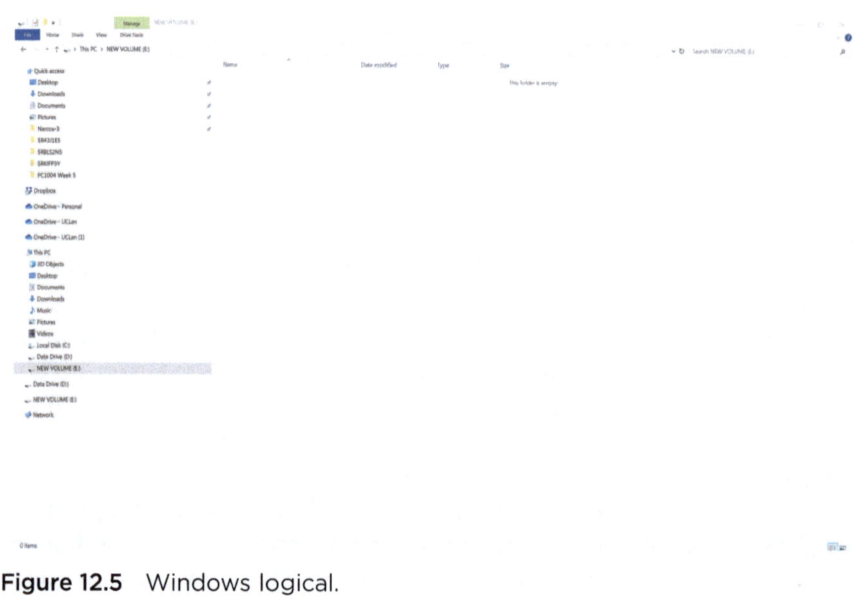

Figure 12.5 Windows logical.

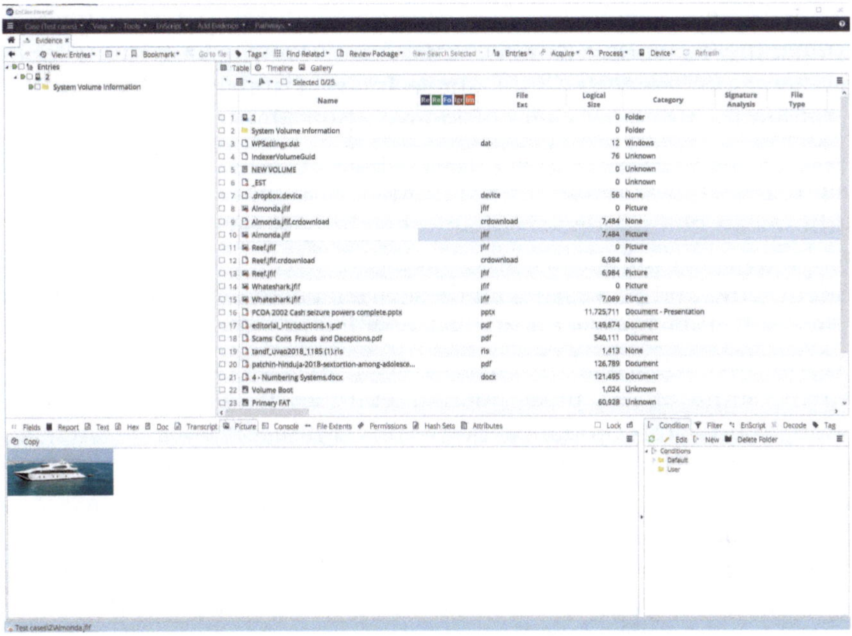

Figure 12.6 Forensic view.

Every file of the same type (*e.g.*, a JPEG image, a MS Word file, or a MS Excel Spreadsheet) will have an identical set of characters at the beginning which mark it as a certain type of file, known as a file signature. These signatures are unique and by appropriate searching of the areas of the drive marked as 'not allocated,' it is possible to recover either full, or fragments, of files. This technique is called data carving.

The development of newer, faster, solid-state drives has complicated this process. These drives both automatically purge deleted data and rearrange it. Consequently, data carving results in a significant number of false positives that need careful examination and elimination.

One final technique that can be useful, is called hash analysis. As discussed previously, the process of hashing generates a hash value of a drive which can be used to ensure the integrity of a file or disk image. This same process can be applied to a single file to generate a value that is unique for an individual file. Many applications create duplicate files, as part of their operations, or users may create duplicates of a file. By searching for files of the same hash value, it is often possible to see the evolution of a particular file as operations on it chart its movement through the drive by amending its MAC times.

Equally, where suspicious files are known from previous investigations, these can be easily searched for by their hash value. This is how law enforcement's Child Abuse Image Database (CAID) works, searching for the hash value of known child abuse images. Equally, there are many known good files that form part of the operating system and are found on every computer with that operating system, so these can be excluded from any searching processes on the same basis thus speeding up the investigation.

Identifying and processing all these artefacts manually on a drive is entirely possible, however it is a task best left to a computer which can perform the same task far faster and more accurately than any human. Some of the most popular commercial packages used in investigations are Guidance Software's EnCase, X-Ways forensic, Griffeye analyse and AccessData's FTK (Forensic ToolKit). At the other end of the scale, Autopsy 3 is a free, yet powerful, open-source tool which, broadly speaking, does the same job. All these packages can take a suspect disk image as input, automatically identify deleted and known types of files, reconstruct the internet history, and recover emails amongst other things.

In skilled hands, these tools can produce results quickly, however one of the most powerful tools is a simple sorting algorithm that allows files to be ordered in a variety of different ways. The significance of this will be discussed in the next section.

12.4.3 Timelines and Attributes

Having identified and located the key artefacts for an investigation, the next step is to provide some context to them. In many situations the actual artefact will speak for itself; for example, an incriminating photograph will always be an incriminating photograph. However, how it got there may well be the difference between proving a case or not. For these reasons, it is important to understand the provenance of a given file. One of the most effective ways of doing this is to produce a timeline of events.

Using any of the major tools, it is possible to sort all the files in time order, be that created time, modified time, or accessed time (as discussed above). Such a process will in many cases reveal interactions between the differing types of evidence discovered. This allows the investigator to explain, using a combination of their knowledge of which artefacts are produced and in what circumstances, the time those artefacts were created and acted upon. It can also indicate the type of concurrent activities occurring at the time, which can in turn give clues to the most crucial question: Who is responsible for the existence of a given file?

On the face of it, this is a very simple question. However, the reality of it is quite different. Without a direct witness as to whose fingers were on the machine at the time, it can be very problematic indeed. There are, however, several approaches that can help identify the user.

Most modern operating systems operate on a user account basis, and it is fairly easy to associate evidence with a specific user account. This does not, however, guarantee that the named user was in fact the individual using the account, although it is a good starting point. Many operating systems offer the ability to password protect user accounts. In most circumstances this is with a password of the user's choosing. These passwords are stored in encrypted format in an operating system database along with the date and time that they were changed.

There are several tools that can help crack the encrypted password. If the computer was seized while it was on, then the plaintext password might still be in one of the buffers. The users may have a short password, vulnerable to a brute force attack (trying every possible password of that length), to a dictionary attack (trying all the words in a dictionary) or by using an insecure method to encrypt the password that is vulnerable to a special table of hashed passwords, known as rainbow tables. Whatever means are used, by knowing the password and the time it was changed, the investigator can start to build a case about the user. Firstly, what is known about the password? What is its relationship to the user? In order to be memorable, most computer users set a password that has some resemblance to an element of their life: their pet's name, the type of car they drive; something that can easily be associated with them. Many others also have a single password that is incremented periodically (adding one to the number on the end, *e.g.*, Rover1, Rover2, Rover3, . . .).

The time that the password was changed can also be significant. It may be able to attribute the change directly to the user. Perhaps their whereabouts may be known at that time or there is some activity on the machine that can tie them specifically to its use. In fact, any occasion where conclusive known use of the machine can be established will be useful to connect a person to the use of the specific account. However, linking an account and password to a presumed user in no way guarantees that the account and password has not been shared. One way of dealing with this problem is timeline analysis.

Consider your own computer usage. How many times, when you use your computer, do you perform just a single task at a time? While writing this section, the author has responded to email, viewed webmail, logged into a couple of forums, making posts and accessed his online banking. All of these activities are attributable to the author, and when intertwined around the life of a specific individual can, depending on time gaps, paint a compelling picture that they were the individual responsible for the suspicious activity.

Even if there are no activities that can be specifically linked to the suspect, it may still be possible to build a case using this type of analysis. Consider your own routine when accessing your own computer. Do you have a set routine of the tasks you perform? The author certainly does and by routinely accessing certain applications and websites in a particular order he is leaving a footprint of his online behaviour that can be identified and traced. By cataloguing internet activity in a time ordered list, it is fairly easy to see defined browsing sessions as they start and end. By separating these out and comparing them with attributable actions, such as sending emails, then it may be possible to infer that the same activity is the result of the actions of the same individual.

Although none of these techniques on their own or taken together, can be said to guarantee who was at the keyboard at the relevant time (only a witness can really do that), they do create a scenario where a compelling case can be put forward questioning who else but the named user can be responsible. For example, consider this scenario: at 10.10 pm on Saturday 12th January 2022, a blackmail letter was sent from the user account 'Paula', using an anonymous webmail account. Analysis showed that the Paula account was password protected and the password was 'Rover'. Rover is the name of the dog owned by Paula, evidenced by vet records seized during her arrest.

The user account was accessed at 10:00 pm that day and there were no failed attempts to log in. Immediately after logging in, the user accessed several websites in an order that can be seen to be replicated on three occasions in the preceding week. On no occasion was any failed attempt to log in noted. On each occasion the user sent emails in the name of the user Paula, to what appear from the content, to be family members and wrote documents signed in the name of Paula that relate to business interests known to belong to her.

As a jury member what are the chances of all these things being undertaken by a person other than Paula? The more facts like this that can be strung together, the more compelling the case it was Paula and the harder it gets to answer the question 'Who else could it have been?'.

12.5 Reporting

12.5.1 Introduction

All the forensics disciplines have technical language and jargon that are unintelligible to a lay audience such as a jury, a judge or lawyer, or senior executives in a company. Digital forensics involves a mix of technical language and jargon, along with concepts and language used by ordinary people, as they use their computers and smartphones. Some of the technical language from a decade ago has seeped into everyday vernacular. When writing reports for some of the other forensics disciplines, it is entirely possible that few, if any, of the audience will have seen the inside of a laboratory or be familiar with any of the equipment or methods. With digital forensics reports, the investigator should not be tempted to use technical language even if they think it is commonly understood. This section provides a guide to where this might be appropriate, along with some of the rules that may apply for certain audiences: it then concludes with a look at some of the different ways an investigator can demonstrate their competence and expertise which nearly every report will require.

12.5.2 Digital Forensics Report

Irrespective of the purpose of the investigation that has been undertaken, the results will have to be conveyed back to the person who commissioned it. This is almost always done with the help of a written report. The nature of these reports, how they are constructed and comply with the rules, are now considered below.

Whatever type of report the investigator is commissioned to write, there is one principle that overrides all others: the report must be understandable. No matter how much evidence that may have been uncovered or what amazing technical feats have been performed to produce a cast iron case, it will be useless if the findings cannot be explained in a way that can be understood by an average jury member. For example, consider the concept of a link or alias file discussed in the previous section. How would an investigator explain the significance of a link file with the name 'Ransom Letter.doc.lnk' having being found in the recent documents folder, with an accessed timestamp on the same day as a kidnapping? Firstly, they would need to explain how the word processing application stores the link file in a special folder and secondly, this implies that the logged in user opened a file of that name at that time.

One thing that can detract from the readability of a forensics report, especially when the computing knowledge of the reader is highly variable, is the constant in-text explanation of every term used. The easiest and most effective way to address this is to provide a glossary which can provide as detailed explanation as is necessary, describing to the reader what the artefact is and how it has been generated.

One of the most effective ways of conveying complex technical concepts to non-technical audiences is through the use of metaphors and similes. For example, there is a clear link that can be drawn between a file allocation table and a telephone directory. This can be extended to explain the concept of file deletion where a line that is struck through the name and address in the phone book doesn't necessarily mean that the house it relates to has been knocked down. In computing, the ability to explain the complex concepts that underpin a case, using easy to understand language, is perhaps even more important than the ability to find artefacts in the first place.

A common confusion seen in reports is that the artefacts are produced and discussed in the order in which they are found, or perhaps worse still in no order at all. This is ineffective because it prevents the author from properly demonstrating how one artefact leads to another and, by doing so, diminishes the value of the evidence. A particularly effective way of presenting digital evidence is to identify the artefacts that lead to the incriminating piece of evidence and present them in order, with an appropriate level explanation of what is happening, to show how the evidence came to be in the place it was found.

For example, a file called bomb.jpg was found in the recycle bin of a computer. The following extract from an investigator's report shows how its provenance can be easily explained without the reference to overly technical discussion. (You can assume that supporting exhibits are produced to evidence the points made and all times are in GMT using the 24-hour clock.)

At 10:00:19 h on 22 December 2022, the user conducted a Google search for the term explosives. This was confirmed because the computer stores copies of webpages visited in a place called the 'internet cache' in order to help display pages faster.

A copy of the search page was found which had been written at 10:00:19 h on 22 December 2022.

At 10:00:46 h, a search results page from the Google search engine was returned which contained a link to the website www.bombmaking.com.

At 10:02:11 h, an entry in the user's internet history shows that the user accessed several pages on the www.bombmaking.com site in quick succession, including a page entitled bombs.htm.

A webpage entitled bombs.htm was found in the user's internet cache with a creation time of 10:03:01 h.

Examination of this page revealed that one of the images it displays was called bomb.jpg. Further searching of the internet cache showed that, at 10:03:02 h, a file of that name was created on this machine. A copy was found in the cache.

A file of the same name was found in the location c:\user\my documents\my pictures\bomb.jpg which was created at 10:04:13 h.

It was established through testing that such an entry can be generated by right clicking a picture on a webpage and using the 'save as' option.

At 10:15:02 h, a link file was created in the recent documents folder that references that file. This shows that the file had been opened on this machine. Further examination shows the user opened files of this type using Photoshop; an examination of the Photoshop history shows a file of the same type opened at that same time.

It was not possible to locate the file in the location identified which suggests the file has been deleted or renamed.

Examination of the recycle bin history showed that at 10:30:00 h, a file named bomb.jpg was deleted from c:\user\my documents\my pictures\.

The file was compared against the file found in the internet cache and found to have identical hash values. Hence, these two files are identical.

The investigator is then able to say that the user of this machine had searched for, accessed and saved the deleted file bomb.jpg.

It can be noted that there is very little discussion of what work had to be done to establish these facts: the focus is very clearly on the evidence and activities surrounding it. Although the methods used to determine the facts are important, they are not for this audience, they should appear in your contemporaneous notes and in any report for another expert, as it must be possible for someone else to reproduce your actions.

12.5.3 Report Format and Audiences

The use of digital forensics in a criminal or tribunal setting is perhaps an obvious application of the science, however digital forensics is regularly used in many other contexts. Civil proceedings make extensive use of digital forensics, as most business records are electronic and found in databases or email records. It is not unusual for digital forensics to be used in so-called e-discovery where hundreds of thousands of corporate email records need to be searched. In the case of e-discovery, there are often very technical reports going to the experts on both sides of the case. In these cases, where the audience is clear, then technical language is more appropriate.

By the time you read this, there will almost certainly have been a recent data breach of some large company reported in the news. Digital forensics plays a key role in diagnosing and defending against such incidents with reports needing to be written for CEO/CIO/Board level, as well as for technical staff. The public relations department of a company with a public data breach might also need general media friendly reports that could be issued as press releases. In addition, there is likely to be a significant law enforcement involvement because of the obvious Computer Misuse Act violation; any evidence gathering, and reports would need to be compliant with criminal evidence standards and appropriate for court use.

In all these cases, the science and techniques discussed will be the same, following appropriate standards such as the ACPO guidelines, making sense even if it is believed at this point that the findings are not destined for the courts—investigating an innocuous anomaly in some data may end up being the first part of a much larger puzzle.

The style of reports aimed at a private sector audience will be quite different from those aimed at a court or tribunal audience. As court/tribunal reports have a single purpose, to inform upon what has happened, the style should not include judgement and recommendation, as that is the function of the judge and jury. This is in contrast to a report on a data breach written for a CEO/CIO, where the purpose would be to answer questions posed by senior management. It is not unreasonable for them to want to know what has happened, but their interest goes further and would consider matters such as: How did it happen? What internal failings led to this? What do we need to do to prevent it happening again? While a court need not be concerned with these questions, they are vitally important in the context of recovering a business from a significant event.

When writing a report in such a circumstance, it is important to tailor the report so that it answers the questions the person commissioning it has posed. In doing so, it is important to understand what the managers want from the report. This can vary from an in-depth analysis of what had occurred to a recovery plan, for example. They may wish to learn lessons or value an insurance claim in respect of the incident. Nonetheless, the same standards of quality and truth as those found in a legal report, need to be maintained since such a report could appear in evidence in a civil claim, say, against an insurance company contesting a claim for cybersecurity insurance. It is therefore wise to ensure that it is written in a manner which is compliant with the requirements of the region's judicial system.

In the case of internal reports, consider that they may later become disclosable and potentially used against the organisation. One only has to look at recent judgements from the Information Commissioner's Office to see investigation reports, that detail the faults within their organisation's systems, later used as justification for large fines or even legal action by those affected. While all reports must be factual and contain relevant information, take guidance from those commissioning internal reports; they may wish some parts to be delivered orally or even through third-party lawyers to preserve legal privilege. Legal privilege is a complex concept out of scope for this chapter, so, if in doubt, seek professional legal advice.

Every jurisdiction will have a preferred format for reports; for English courts the Criminal Procedure Rules and Civil Procedure Rules set out what should be contained in such reports. These rules govern the limits and scope of the evidence an expert can give and the caveats that they must include. While dry reading, the format can be adapted for non-legal audiences which can make it easier to re-write for legal audiences if required later.

12.5.4 Competence and Expertise

One of the key elements of the procedure rules noted above, is establishing the competence of the report writer providing expert evidence. For many years, the lack of formal qualifications in digital forensics meant that an expert's own CV and claimed experience was sufficient; many digital forensics expert witnesses had multi-decade careers with no formal qualifications. One such person, Jim Bates, found his career ended abruptly in 2008 when he was established to have lied about having an electronics engineering degree despite many years of experience and being considered a 'founding father' of the UK digital forensics industry at the time.

Although vender certification on specific software is common, the availability of accredited degree programmes in security and forensics means that formal qualifications are now expected in many cases additional to experience. Professional bodies have now begun to accredit digital forensics programmes. In the UK, GCHQ certifies Digital

Forensics MSc programmes, and the Chartered Society of Forensic Sciences (CSFS) accredits both undergraduate and postgraduate digital forensics programmes. Digital forensics experts can also use membership in professional societies like the CSFS, the British Computer Society or the Association of Computing Machinery as part of their demonstration of competence.

One area that is only just becoming regulated, is the accreditation of digital forensics laboratories to ISO quality standards. The UK Forensic Science Regulator now has oversight of digital forensics, and this means that digital forensics practitioners will have to, at least, validate their processes and show that they comply with quality standards, as part of their demonstration of competence. This is a demanding process even for larger labs and smaller labs can find it disproportionately effects their ability to function.

Continuing professional development is important in all disciplines, but perhaps doubly so for digital forensics. Not only are the capabilities of the technology rapidly evolving but also the applications that may be developed and become popular. In addition, the ways in which people interact with and adopt technologies changes frequently, creating a dynamic socio-technical system. A competent digital forensics practitioner needs to be able to demonstrate knowledge and awareness of how a vast array of new and legacy systems are being used both by ordinary individuals and sophisticated criminals. This is quite a challenge but one that is fundamental to continued accreditation by the Forensics Regulator for both the lab and the individual.

12.6 Conclusion

Unlike most of the forensic science disciplines, digital forensics is a very young science which is still evolving. Part of the reason for this, is that the ways in which people interact with digital systems are evolving in parallel with the technology. In the last section, it was noted how this makes continuing professional development even more important in digital forensics than in other disciplines, but the problem is even worse than it appears. Not only do digital forensics practitioners need to keep up with rapid changes in both technology and how people are using it, but they also have to predict (to some degree) where technology advances are heading so they are prepared before they begin appearing on the street and in people's homes. These concluding comments consider the challenges facing the digital forensics discipline and highlight some of the areas that are likely to change in the coming years.

If you ask anyone involved in criminal digital forensics investigations what they see as the biggest threat, they will likely tell you it is sheer volume. In 2015, the BBC reported considerable delays in forensic examinations and a Channel Four report from 2022 showed a backlog of over 20 000 digital devices by UK Police Services (excluding Met and GMP). It is common for this backlog to generate a delay to an investigation of 12 months because of this. Throughout this chapter we have touched on the reasons for this: an increase in use of computing technology, an increase of data storage sizes as well as the many novel ways in which digital devices are used to commit crime. There are some tactics and techniques being employed to address the problem but they are clearly not enough. This backlog was exacerbated by the COVID pandemic which, despite seeing a drop in many physical crimes, saw a significant increase in all forms of cybercrime.

The 2022 strategic review of policing by Her Majesty's Inspectorate of Constabularies, Fire and Rescue Services identified cyber and digitally facilitated crime as one of the key threats to the UK over the next five years. They propose a number of action plans including training frontline officers to triage data. This brings with it clear risks, despite this already happening to a limited degree. As we have discussed, the rapid evolution of consumer technology can make much of this training obsolete before it is delivered. The increased use of encrypted disks and end-to-end encryption means that triage requires direct access to the device which, in turn, requires specialist skills, otherwise ACPO Principle 2 is violated and evidence could be tainted. Perhaps most salient, in a time of austerity post-COVID, is the funding of training. As written above, new recruits receive specific training on not just the practicality of digital investigation but the criminology and psychology behind it.

Some technology developments, such as the increased use of solid-state drives (SSD), will allow faster searches but perhaps the most effective approach to the problem is in limiting the extent to which digital examination is deployed.

Artificial intelligence techniques, such as machine learning, also show promise to quickly identify areas for triage. However, they suffer from requiring massive datasets to train the algorithms – this is partly mitigated by initiatives such as Digital Forensics as a Service from the Netherlands Forensic Institute which the Dutch National Police and the Dutch Fiscal Information and Investigation Service, amongst others, currently collaborate. In addition to the obvious advantages of cloud servers offering highly intensive computing resources and large amounts of storage to process forensic data rapidly, it also makes significant data available to researchers as training sets for machine learning algorithms. However, the adoption of centralised platforms is slow and fraught with data protection, privacy and security issues. A further barrier to the adoption of machine learning is that techniques such as neural networks are complex and it can be difficult to explain their output, causing obvious issues in legal cases. Despite the incredible capabilities of artificial intelligence, we are also discovering that since it is trained by humans, we can unconsciously train it in our biases and prejudices; Amazon attempted to use artificial intelligence to simplify recruitment and make unbiased hiring decisions. By training machine learning algorithms on the CVs of previously successful applicants, the majority of which were male and from a particular background, they inadvertently trained the algorithm to be biased against minorities and, in particular, women. Naturally, the project was quickly discontinued.

Other developments include the use of artificial immune systems based on danger theory. This is modelled on how the human immune system is thought to respond to necrosis or uncontrolled cell death through disease, injury or radiation, for example, to detect when something potentially malicious is happening to a live system. It will reserve relevant artefacts for later examination or, post-event, assist examiners prioritising their review. While work is already underway to make use of machine learning to automatically investigate computers, such approaches clearly require a paradigm shift by practitioners who have previously been focussed on the fullest possible examination.

Leaving aside the volume problem, there are two other major issues facing digital forensics practitioners. The first, and perhaps most problematic, is that of cloud forensics. Here, data is stored remotely, often with multiple versions, and in many cases in different jurisdictions. From a technical point of view, copying data from a cloud server is straightforward for the cloud provider but getting the authority to do so is far from easy. However, the legal picture is developing, and it is hopeful this may become a far more streamlined process in the near future. Currently, to access data in a foreign jurisdiction, preservation requests would have to be served and formal legal applications, under Mutual Legal Assistance Treaties, would need to be made. This process can take many months to complete, perhaps rendering the evidence eventually recovered of less value. This situation pre-supposes that a relevant treaty is in place; not all countries are part of this arrangement. Cybercrime is a global phenomenon, but despite the efforts of the Budapest Convention which took place over twenty years ago, cybercrime and data protection laws vary from country to country making it harder to show that an equivalent crime may have been committed in the jurisdiction where the server is located.

The second major issue is currency of knowledge. It was noted previously that continuing professional development is doubly important for digital forensics practitioners because both hardware and software are constantly changing.

Since the last edition of this book, WhatsApp has retained popularity despite some challenges from the likes of Telegram and Signal. In that time, WhatsApp has changed the way it stores messages locally on smartphones, invalidating previous technical knowledge and will no doubt continue to make such changes.

Not only having to keep up with the more obvious popular software, the investigator also must be alert to novel ways of communication. Over a decade ago, a popular covert communication method was to use a webmail client to compose draft messages but to not send them, which meant that email traffic was never seen because both parties simply logged into the same account and read the draft messages.

Today, a popular smartphone app is a game called Words with Friends which allows people to play the Scrabble word game remotely. It also has a built-in chat function which could be used for covert communication under the guise of playing a game. Another niche smartphone app is Glympse, which allows two or more people to track each other's location so an illicit meeting could be arranged without either party knowing the location until the last moment.

Early in this chapter, it was noted how small and dense storage media were becoming, in addition to the novelty items that USB storage devices may be disguised as, such as keyrings. However, the hardware issue goes beyond simple storage devices. The so-called Internet of Things (IoT) describes how ordinary everyday objects and appliances are being connected to the internet and provided with extra functionality. The classic example of a future IoT device has been the smart refrigerator that knows what it contains and when items expire, automatically ordering milk and other essentials when it detects items running low.

When writing the previous edition of this chapter, the authors speculated about digital forensics examiners querying smart home devices. In less than five years, this has become reality, with digital forensic labs regularly receiving devices such as Amazon Alexa, Smart TV's Ring doorbells, and even drones. One of the most common new devices is the car. Currently the level of cutting edge research on these novel devices is beyond anything the authors originally envisaged.

Disappointingly, but not unexpectedly, the means by which criminals use these devices is expanding at a similar rate, if not faster; drones flying controlled drugs and other contraband into prisons is already a common occurrence.

In the past, the digital forensics practitioner could rely on knowing the operating system and storage media for a computer at a crime scene and use a well-known, forensically safe method to recover the data. Now, even if the examiner can figure out which devices contain relevant data, they may not have an established method of extracting it, which can make each examination bespoke and time consuming.

The good news is, that regardless of the hardware, software or network connections, the general principles of computing technology and the fundamentals of digital forensics remain unchanged. Whether it is your refrigerator spying on your eating habits, your TV listening to your conversations or your tablet revealing your email, the underlying computing model remains the same. Skilled digital forensics practitioners need only learn how to find the data, then the task becomes familiar: analyse, look for connections and construct a timeline. Digital technology is a permanent feature of our lives, which means that few cases will come to court without a digital evidence element making one thing certain – whether they are investigating evidence or developing new skills, digital forensics practitioners will never be idle!

Acknowledgements

The authors thank Prof. Blaine Price, The Open University, for his kind assistance with previous versions of this chapter.

Further Reading

C. Atheide and H. Carvey, *Digital Forensics with Open Source Tools*, Syngress Publishing, 2011.

E. Casey, *Digital Evidence and Computer Crime: Forensic Science, Computers, and the Internet*, Academic Press, Cambridge, MA, 2011.

A. Gillespie, *Cybercrime: Key Issues and Debates*, Routledge, London, 2015.

G. Horsman, ACPO principles for digital evidence: Time for an update? *Forensic Sci. Int.: Rep.*, 2020, 2, 100076.

T. Holt, A. Bossler and K. Seigfried-Spellar, *Cybercrime and Digital Forensics*, Routledge, 3rd edn, 2022.

I. J. Lloyd, *Information Technology Law*, Oxford University Press, Oxford, 7th edn, 2014.
H.-M. Maras, *Computer Forensics: Cybercriminals, Laws, and Evidence*, Jones Bartlett Publishers, 2014.
ISO/IEC, *ISO/IEC 27037:2012 Information technology – Security techniques – Guidelines for identification, collection, acquisition and preservation of digital evidence*, ISO/IEC, Geneva, Switzerland, 2012.

13 Drugs of Abuse

Michael Cole*[a] and Mark Baron[b]

[a] Anglia Ruskin University, School of Life Sciences, Faculty of Science and Engineering, East Road, Cambridge CB1 1PT, UK; [b] University of Lincoln, School of Chemistry, Brayford Pool, Lincoln, LN6 7TS, UK
*Email: michael.cole@aru.ac.uk

13.1 Introduction

The problem of drug abuse continues to increase, and the drug analyst is always being presented with new and interesting challenges. This is particularly true of, for example, the synthetic cannabinoids, and new polydrug mixtures including Nyaope (a mixture of cannabis, heroin and hallucinogenic antiretrovirals), White pipe (a mixture of cannabis and methaqualone) and "Happy Water" (a mixture of caffeine, diazepam, ketamine, Ecstasy, methamphetamine and tramadol). Drug use is a dynamic problem which changes with, amongst other factors, preference amongst drug users, socio-economic climate and changes in legislation. Discussions with a number of constabularies and forensic science providers in the UK indicate that, at the time of writing, there is a move away from "designer drugs" with a return to the use of more "traditional" drugs including heroin and cocaine. Cannabis, of course, remains the most widely used drug in many parts of the world. The drug analyst must therefore be at least as versatile and dynamic as the drugs market. Each year the number of drugs coming under control grows

Crime Scene to Court: The Essentials of Forensic Science, 5th Edition
Edited by Niamh NicDaeid and Peter C. White

Published by the Royal Society of Chemistry, www.rsc.org

with 250 currently under UN control (2023 data), the majority of which are synthetic drugs with no natural origin or are chemically modified natural products. Further, there are changes in the use of language as exemplified by the changes in the relative uses of – MDA (methylenedioxyamphetamine) and MDMA (methylenedioxymethylamphetamine – 'Ecstasy'), with the decrease in the use of "Ecstasy" and the increase in the use of "MDA" and "MDMA". Drug analysis is one area of forensic science where a positive identification of a material must be made and as a consequence there is always a requirement for the development of novel and improved analytical methods, data interpretation, agreed language and reporting of the evidence to court either in written and/or verbal form. The challenge is further exacerbated in some legislative systems where the drug, and the isomers present, all need to be identified.

The legislation in the UK prescribes different charges and sentences for different drugs and for differing amounts of those drugs. In addition to the identification of any controlled substance that might be present, the analyst may also be required to quantify any drugs of abuse in the sample. Depending on the drugs thought to be involved, some techniques are more appropriate than others for these tasks and it is a knowledge of the drugs involved and which techniques to apply, and when, that form the major part of a drug analyst's expertise and activity. There is also continued interest, both nationally and internationally, in the comparison of drug samples with each other. This provides information concerning the manufacture, distribution and supply networks and movement of drugs at the street (user) level. These comparisons require different methodologies depending on the drug in question. If such data is to be shared and used in a meaningful way, however, then the same instruments and methods should be used in the participating laboratories who should all train their staff to a given set of standards and engage in the same quality control and management procedures and protocols. There is an increasing use of chemometrics and artificial intelligence in drug identification, comparison and profiling and it is imperative that the assumptions under the chemometrics used for such processes are upheld. This is becoming the subject of increasing research since many long-held assumptions are found not to be true. This being the case, the success of such endeavours depends on the drug class in question and the methods to be used. Understanding these processes is an important part of the drug chemist's work.

13.2 Drug Control Legislation in the UK

A number of principal pieces of legislation are employed to control the drugs of abuse, namely the Misuse of Drugs Act 1971, the Misuse of Drugs Act (Regulations) 1985, the Misuse of Drugs Act (Regulations) 2001 and the Psychoactive Substances Act (2016). There have also been a number of other legislative instruments amending these principal documents. The Misuse of Drugs Act is also often abbreviated as MDA. The 1971 Act lists the drugs in one of three classes as part of Schedule 2 to the Act. Examples of such drugs and penalties associated with them are given in Table 13.1.

A number of classes and individual drugs have subsequently been added and deleted, as detailed in the subsequent Amendment Orders, the Misuse of Drugs Regulations 1985 which controlled benzodiazepines and barbiturates for the first time, and the Misuse of Drugs Regulations 2001. Importation and exportation of controlled substances are prohibited by Section 3 of the Act, production, supply

Table 13.1 Examples of different drugs classified under Classes A, B, C and Temporary of the Misuse of Drugs Act 1971 and the Psychoactive Substances Act 2016 and the penalties imposed.

Classification under the Misuse of Drugs Act 1971	Penalty for possession	Penalty for supply	Examples
Class A	Up to 7 years in prison or unlimited fine or both	Up to life in prison or unlimited fine or both	Heroin, cocaine, LSD, methylamphetamine
Class B	Up to 5 years in prison or unlimited fine or both	Up to 14 years in prison or unlimited fine or both	Cannabis, barbiturates, amphetamines, cathinones (exceptions = Class C), cannabinoid agonists
Class C	Up to 2 years in prison or unlimited fine or both	Up to 14 years in prison or unlimited fine or both	Ketamine, benzodiazepines, piperazines, cathinone, pyrovalerone.
Temporary Class	Not an offence	Up to 14 years in prison or unlimited fine or both	There are currently no drugs in this class
Psychoactive Substance under the PSA 2016	Not an offence	Up to 7 years in prison or an unlimited fine or both	Nitrous oxide, some SCRAs *e.g.* 5F-MDA-19

or offer to supply a controlled substance are prohibited by Section 4. Possession is controlled by Section 5. Cultivation of any plant of the genus Cannabis is controlled by Section 6 of the Act, while Section 8 states that:

> A person commits an offence if, being the occupier or concerned in the management of any premises, he knowingly permits or suffers any of the following activities to take place on those premises . . .

and the list includes producing or attempting to produce a controlled drug, supplying or attempting or offering to supply a controlled drug, preparing opium for smoking or smoking cannabis, cannabis resin or prepared opium. Section 9 of the Misuse of Drugs Act is interesting in that it relates specifically to opium. The section makes provision for the control of smoking or other use of prepared opium, the frequenting of such a place where this occurs and also being in possession of pipes or other utensils made or adapted for use in connection with the smoking of opium or preparation of opium for smoking. New drugs continually appear on the market and so this legislative instrument is under constant revision.

The legislative materials also make provision for the control of any salt, ester or stereoisomer of the compounds listed in the legislation. A controlled substance is therefore defined as one which falls under the umbrella of being controlled by one or other, or both (the Act or the Regulations), of these pieces of legislation. In short, The Misuse of Drugs Act prescribes what cannot be done and the regulations prescribe what may be done, with the correct authority to do so.

Other legislative tools, *e.g.* the Traffic Offences Act 1986, make provision for the confiscation of the proceeds in trafficking in drugs of abuse. Recent Amendments to the Misuse of Drugs Act add specific compounds or small groups of compounds to the list of controlled substances and, rarely, remove a substance from the list. Sometimes drugs are reclassified and so it is important to ensure an awareness of all modifications.

In recent years there has been the global emergence of new drug substances that have the potential to cause similar effects to psychoactive substances which are under international control. Year on year there has been an increase in the number of new substances appearing on the market. Collectively these are referred to as New Psychoactive Substances (NPS).

NPS are defined by the United Nations Office on Drugs and Crime (UNODC) as;

> 'substances of abuse, either in pure form or a preparation, that are not controlled by the 1961 Single Convention on Narcotic Drugs or the 1971 Convention on Psychotropic Substances but which may pose a public health threat'.

'New' means that they have only recently appeared on the market and they are not included in the above conventions as opposed to having been only recently developed. NPS are primarily "designer" drugs that have been designed to evade international and national drug control legislation and so are often referred to as legal highs along with a host of other names such as herbal highs, spice products, bath salts, plant food and research chemicals. Most of the substances that have appeared during the last decade still remain outside of UN convention control and so fall under the NPS definition. However at a national level much activity has taken place to control supply and use of NPS through legislative control. As such the term "legal high" only correctly describes those substances that are not controlled by national legislation and this is country specific.

NPS cover a wide range of different drug groups including piperazines, benzodiazepines, arylamines, tryptamines, synthetic opioids, phenethylamines, synthetic cannabinoid receptor agonists (SCRAs) and synthetic cathinones with the SCRAs and cathinones forming the single largest group of substances and the synthetic opioids posing a particular serious risk to public health.

Initially, in the UK use was made of the MDA 1971 to control NPS. Prior to 2009 amendments to the MDA were not that common however, since 2009 there have been a number of amendments using statutory instruments that have brought under control several hundred NPS as identified in Table 13.2.

Some amendments have controlled named substances, however a large number of substances have been brought under control through the use of generic controls. For example, in 2009, seven groups of SCRAs were controlled with each group having its own generic definition. Other SCRAs falling outside of these controls (second generation cannabinoids) quickly appeared on the market and so in 2012 a further amendment modified the generic controls of 2009 and introduced additional groups. Since then, further amendments have been made in 2016 and 2019 to broaden the

Table 13.2 Compounds controlled by Temporary Class Drug Orders (TCDO) and other modifications of the Misuse of Drugs Act, 1971 since 2009.

Statutory instrument (year/number)	Enforcement date day/month/year	Order type	Novel psychoactive substance (NPS)
2009/3209	23/12/09	Amendment	Synthetic cannabinoids (Class B), γ-butyrolactone (GBL) (Class C), 1,4-butanediol (1,4-BD) (Class C), 1-benzylpiperazine (BZP) (Class C). Substituted 1-benzylpiperazines and 1-phenylpiperazines (Class C).
2010/1207	16/04/10	Amendment	4-Methylmethcathinone (mephedrone) (Class B). Substituted cathinones (Class B).
2010/1833	22/07/10	Amendment	Naphyrone and substituted cathinones (Class B).
2012/980	05/04/12	TCDO	Methoxetamine.
2012/1390	13/06/12	Amendment	Desoxypipradrol and pipradrol related compounds (Class B), phenazepam (Class C).
2013/239	26/02/13	Amendment	Synthetic cannabinoids (Class B), methoxetamine (Class B), ketamine and phencyclidine related compounds (Class B).
2013/1294	10/06/13	TCDO	25I-NBOMe, 25B-NBOMe, 25C-NBOMe, 25D-NBOMe, 5-APB, 6-APB, 5-APDB, 6-APDB, 5-IT, 6-IT.
2014/115	10/06/14	Amendment	Benzofuran compounds and others (Class B), NBOMe compounds (Class A).
2014/3271	07/01/15	Amendment	AH-7921 (Class A), various tryptamines (Class A), various LSD related compounds (Class A).
2015/215	11/03/15	Amendment	MT-45 (Class A), 4,4′-DMAR (Class A).
2015/1396	27/06/15	TCDO	3,4-Dichloromethylphenidate, 4-methylmethylphenidate, ethylnaphthidate, ethylphenidate, isopropylphenidate, methylnaphthidate, propylphenidate.
2015/1929	27/11/15	TCDO	Methiopropamine.
2016/650	27/06/16	TCDO	Phenidates (12 month extension of 2015/1396).
2016/1109	14/12/16	Amendment	Extends the generic control of SCRAs structurally related to JWH-018 (Class B).
2016/1126	27/11/16	TCDO	Methiopropamine (12 month extension of 2015/1929).
2017/634	31/05/17	Amendment	U-47700 (Class A); 12 phenidates including those under Temporary control (2015/1396) (Class B); 16 NPS benzodiazepines including diclazepam, etizolam and flubromazepam (Class A).
2017/114	27/11/17	Amendment	Methiopropamine (Class B).
2018/1356	01/04/19	Amendment	Gabapentin, pregabalin (Class C).
2019/1323	15/11/19	Amendment	Extends the generic control of SCRAs structurally related to JWH-018 (Class B).
2021/868	18/08/21	Amendment	Flualprazolam, flunitrazolam, norfludiazepam (all Class C)
2022/322	13/04/22	Amendment	1,4-Butanediol, γ-butyrolactone, 4-hydroxy-*n*-butyric acid (Class C to Class B).

scope of the generic definition covering SCRAs structurally related to JWH-018.

In 2011 a legislative change introduced a Temporary Class to the MDA which allows NPS to be controlled as a temporary class drug if it is thought to have the potential for misuse and that misuse will cause harmful effects. An NPS can only be classified as a temporary class drug for 12 months (by the issue of a Temporary Class Drug Order (TCDO)) during which time sufficient evidence needs to be gathered to bring the NPS under permanent control or the temporary class drug order is revoked. To date six TCDOs have been used to control specific substances although the two most recent orders issued in 2016 were 12 month extensions for two issued in 2015 for methiopropamine and a group of phenidates.

Despite the above measures, the use of TCDOs was only ever going to tackle those substances that were likely to be of most concern and a more general form of legislative control was needed to deal with the ever increasing number of NPS. This came in 2016 in the form of the Psychoactive Substances Act (PSA). The PSA covers psychoactive substances not controlled by the MDA. Penalties are less severe than the MDA with a maximum imprisonment of 7 years and the act does not include the offence of possession except in a custodial institution which has a maximum penalty of 2 years plus a fine. To obtain a prosecution under the PSA, the Forensic Science Provider (FSP), instructed in the case, provides an identification of the submitted psychoactive substance. Psychoactivity of new substances is determined using *in vitro* testing carried out by the Defence Science and Technology Laboratory (DSTL). If the substance identified by the FSP has been determined to be a psychoactive substance then DSTL provides this information in the form of an expert witness statement.

With the introduction of the PSA, it may appear that the Temporary Class of the MDA is no longer necessary however it remains a part of the drug control legislation as there may still be a need to deal rapidly with highly dangerous substances.

13.3 Drugs of Abuse and Their Sources

During the course of a drug analyst's career, they are likely to encounter a wide variety of drugs. What follows is a description of the principal classes of drugs likely to be encountered.

13.3.1 Cannabis and Products

The products of *Cannabis sativa* L. form the major part of the drug materials currently submitted to forensic science laboratories. The materials come in three major forms: herbal material, resin and oil. All contain the active constituent, Δ^9-tetrahydrocannabinol (Δ^9-THC), sometimes with a small amount of its isomer, Δ^8-tetrahydrocannabinol. The precursor of this is cannabidiol and in older samples in which decomposition of the sample has started, cannabinol will also be observed (Figure 13.1).

Herbal material of *C. sativa* comes in a variety of forms, from whole plants to small amounts of leaves, stems and flowering tops, and can also be encountered fresh or, more commonly, dried. The fresh material has a very characteristic smell. The leaves are palmate, with serrated edges. The stems are square and four-cornered. The flowering tops are recognised by the presence of buds, flowers and sometimes seeds. Three principal morphological characteristics allow the recognition of cannabis plant material. The first is single-celled trichomes, which are colourless and, unusually, point unidirectionally up the stem. The second feature is the cystolithic trichomes, with crystalline materials at their base. The third feature is the presence of glandular trichomes, which consist of a glandular structure on a thin

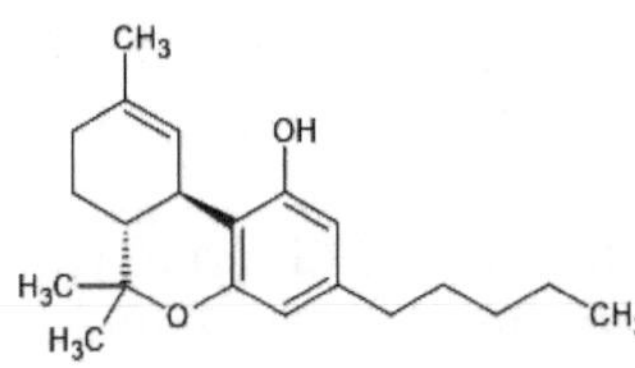

Δ^9-Tetrahydrocannabinol

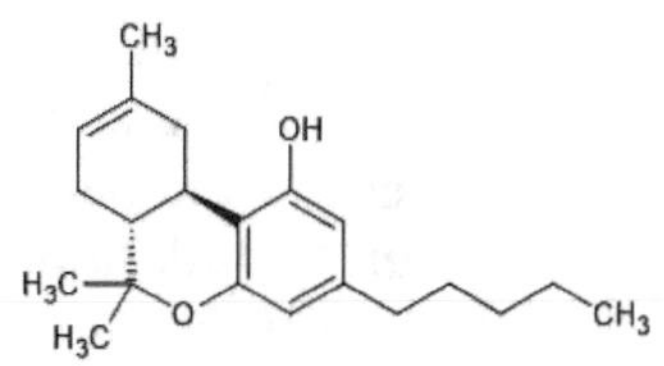

Δ^8-Tetrahydrocannabinol

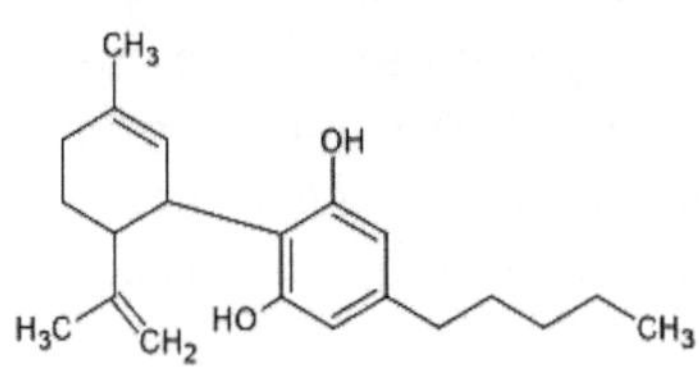

Cannabidiol

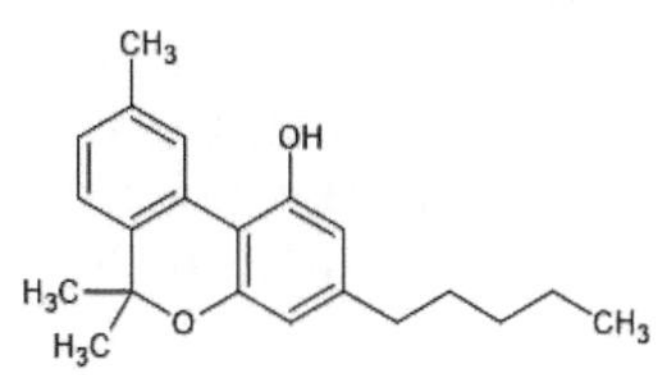

Cannabinol

Figure 13.1 Chemical structures of some cannabinoids present in *Cannabis sativa* L.

stalk of cells. It is within the glandular trichomes that the resin is stored. If all three of these morphological features are observed, the plant is almost certainly *C. sativa*. Even the closest relative of *C. sativa*, the hop (*Humulus lupulus*) does not have these morphological features, although it does have glandular trichomes at the base of the leaves which may, by the inexperienced eye, be confused with the trichomes of *C. sativa*. The herbal materials may be packaged in a number of different ways and contain differing amounts of leaf, stem, flowering tops and seed material. These materials contain a very wide range of concentrations of the active component Δ^9-THC. The form of the packaging and its contents may be used to identify the country or region of origin of the plant materials.

Resin comprises the material obtained when the glandular trichomes are collected, by a variety of means. These include passing large hessian nets through the crop, on to which the resin becomes stuck, as is done in North Africa, and rubbing of the plant materials between the palms of the hands, as occurs in Pakistan. Either way, the resin is compressed into blocks and shipped. The resin may contain other cellular debris and it is frequently possible to observe parts of both the glandular hairs and the cystolithic trichomes under microscopic examination. This material typically contains concentrations of Δ^9-THC.

The oil, commonly referred to as 'hash oil', represents the material that is extracted using solvent extraction of the plant materials. The concentration of the active constituents and other cannabinoids is much higher in this material.

In addition to bulk samples of the drug a number of different types of items might be encountered by the forensic analyst. These include wrapping materials which should be compared to determine whether there is a relationship between samples being analysed. In addition, there may be trace items, such as scales, knives, cutting implements and stamps, all of which may be the subject of investigation to determine whether they have been in contact with controlled substances and whether they can be used to determine any commonality between samples. Further, the remains of cigarettes used to smoke cannabis products, smoking utensils (*e.g.* roach stones) and even the contents of ashtrays can all be analysed for the presence of cannabis materials.

13.3.2 Cannabinoid Agonists

Developed initially for pain relief, synthetic cannabinoids (more correctly known as synthetic cannabinoid receptor agonists (SCRAs)) have recently found popularity in the illicit drug market since about 2006 although they may have been available earlier. These compounds, which occur as either solids or oils, are dissolved and then said to be sprayed onto dried vegetable matter of psychoactive plants which are then smoked. There is contradictory evidence as to whether these plants are actually part of these mixtures. The synthetic cannabinoids elicit similar responses to Δ^9-THC because they bind to the same cannabinoid receptors in the brain. They are sold under several brand names including "Spice", "Spice Gold" and variants. Often the packages are labeled as containing named plants but these are rarely if ever found in the mixtures when they are analysed. It is currently unclear where the drugs are synthesized although Asia has been implicated in media reports.

SCRAs fall into a number of different chemical classes including napthoylindoles (JWH-018, JWH-073, JWH-398), napthylmethylindoles, napthylpyrroles, napthylmethylindenes, phenylacetylindoles (JWH-250), cyclohexyphenols (CP47,497) and classical cannabinoids (HU-210). Some of their chemical structures, shown in Figure 13.2, are very similar to those of the traditional cannabinoids although some show little similarity hence the need to correctly refer to them as SCRAs. A generic description of the chemical structure of the synthetic substances has been given by EMCDDA and consists of a core group, a linker group, a linked group and a tail. This structural description is used in the SCRA generic definitions used in the MDA to control SCRAs to Class B. New SCRAs are continually appearing with modifications to the core group, linker group, linked group and tail, that take the SCRA outside of MDA control; hence the need for amendments to the MDA as shown in Table 13.2 since control of SCRAs was first established in 2009 and 2013. At the time of writing there are several SCRAs that fall outside of the MDA generic definitions, however if these are found in drug seizures they can be dealt with through a PSA prosecution.

These compounds often occur as mixtures within the drug materials. Synthesis of these drugs is often complex and involves enantiomeric separations so that the correct enantiomer may be isolated for use. Little is known about the toxicology, pharmacology and the side effects on ingestion of these drugs.

At the time of writing there are no presumptive tests for the whole group of these drugs which are conveniently analysed by GC–MS on a routine basis. However, presumptive tests may be used to achieve exclusions around the identity of these drugs. Initially there was limited availability of reference standards. However, in recent years, many reference standards for the known SCRAs have become available along

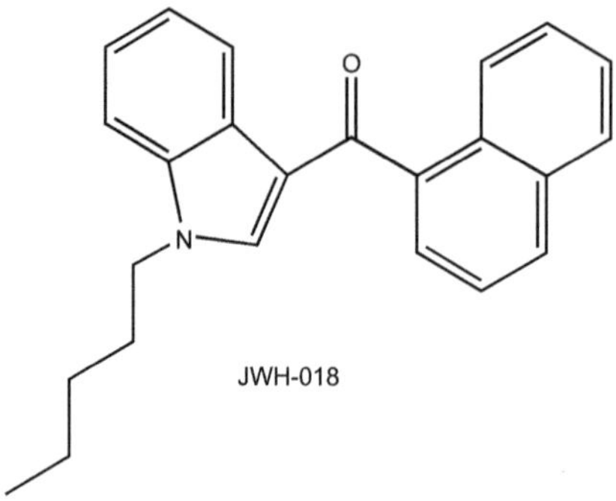

JWH-018

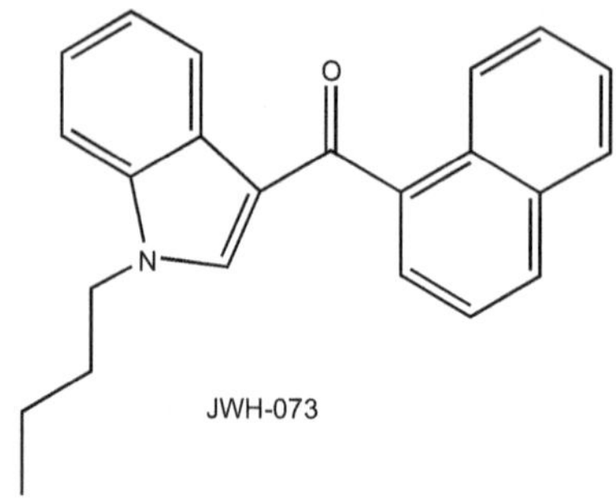

JWH-073

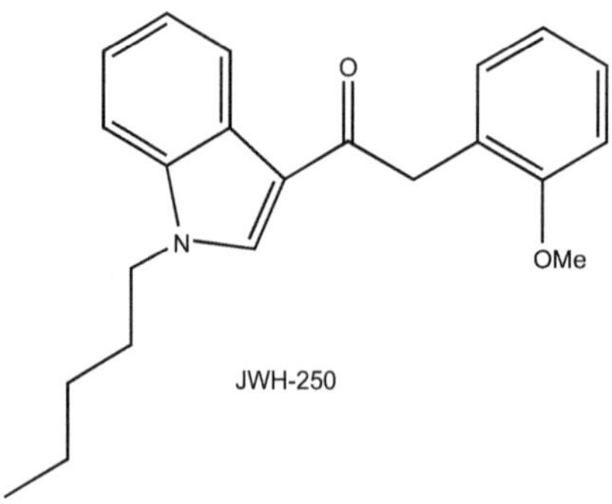

JWH-250

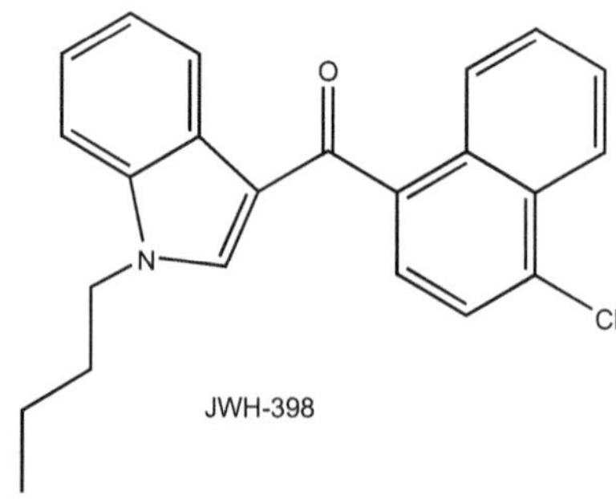

JWH-398

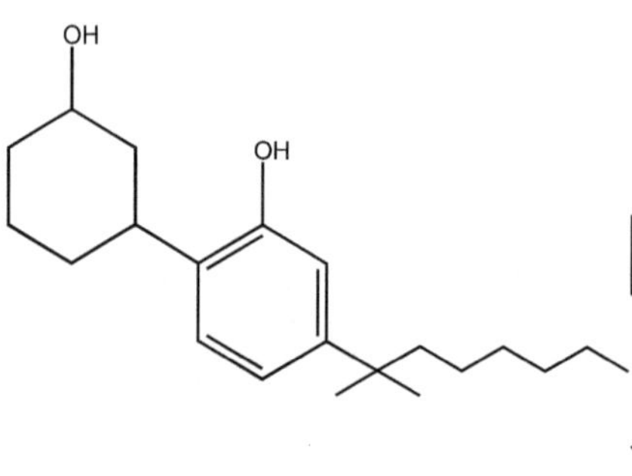

CP 47,497

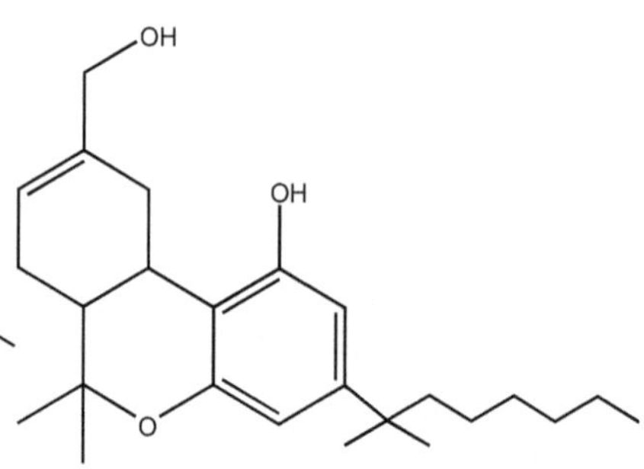

HU-210

Figure 13.2 Structures of exemplar SCRAs.

with analytical methods and reference data such as mass spectra. FSPs have developed in-house mass spectral libraries but also make use of third-party libraries such as that available from The Scientific Working Group for the Analysis of Seized Drugs (SWGDRUG). SWGDRUG has created a mass spectral library (70 eV positive electron ionization (EI^+)) of NPS that can be appended to the National Institute of Standards and Technology (NIST) database which is routinely used for searching and identifying unknown mass spectra obtained from GC–MS analysis. It is necessary on occasion, however, to use GC–MS in conjunction with one and two dimensional proton and carbon NMR to determine the structure and identity of novel SCRAs.

In terms of legislative status some of these drugs are controlled by name in a limited number of EU states and in the UK, for example, they are controlled as Class B drugs "cannabinoid agonists".

13.3.3 Heroin

Heroin is derived from the opiate alkaloids produced by the field poppy *Papaver somniferum* L. It is not a pure compound, but a mixture of the impurities, reaction intermediates and breakdown products manufactured during the synthesis of diamorphine; compounds carried through from extraction of the morphine from the plant and subsequent processing; excipients, *e.g.* sugars; and adulterants, *e.g.* other drugs, used to dilute, or cut, the drug. Diamorphine is derived from morphine, one of the alkaloids from opium resin. The seed pods of the plant are lacerated and latex exudes which then dries. From the dried latex, morphine is extracted using a series of solvent/solvent extractions and precipitations. The morphine is then acetylated, producing the 3- and 6-monoacetylmorphines and diamorphine, which are substances controlled under the Misuse of Drugs Act 1971. Since codeine is also extracted from the opium resin using these methods, acetylcodeine will be present in the final product. The chemical structures of alkaloids commonly found in heroin are shown in Figure 13.3.

Thebaine is also present because it is extracted from the opium resin in the same way as the morphine. Two other alkaloids of different classes are also co-extracted, namely noscapine and papaverine. The amounts of all these alkaloids present in the final products will depend upon a number of factors, including geographical origin of the drug, quality of the extraction chemistry, quality of the acetylation and quality of the purification processes. Since heroin is produced in a batchwise manner, no two samples will ever be the same. However, heroin samples from different geographical regions may exhibit certain chemical characteristics, as outlined in Table 13.3.

Interestingly it is possible that opiates produced by yeasts may appear on the drug market. It is now possible to genetically modify yeast to convert glucose to benzylisoquinolines of which the opiate drugs are part. This has been made possible by the successful genetic modification of yeasts which allows tyrosine to be converted to L-DOPA which in turn allows the whole biochemical pathway to be expressed. At the current time there is disagreement about the degree to which this would become a problem, but current legislative

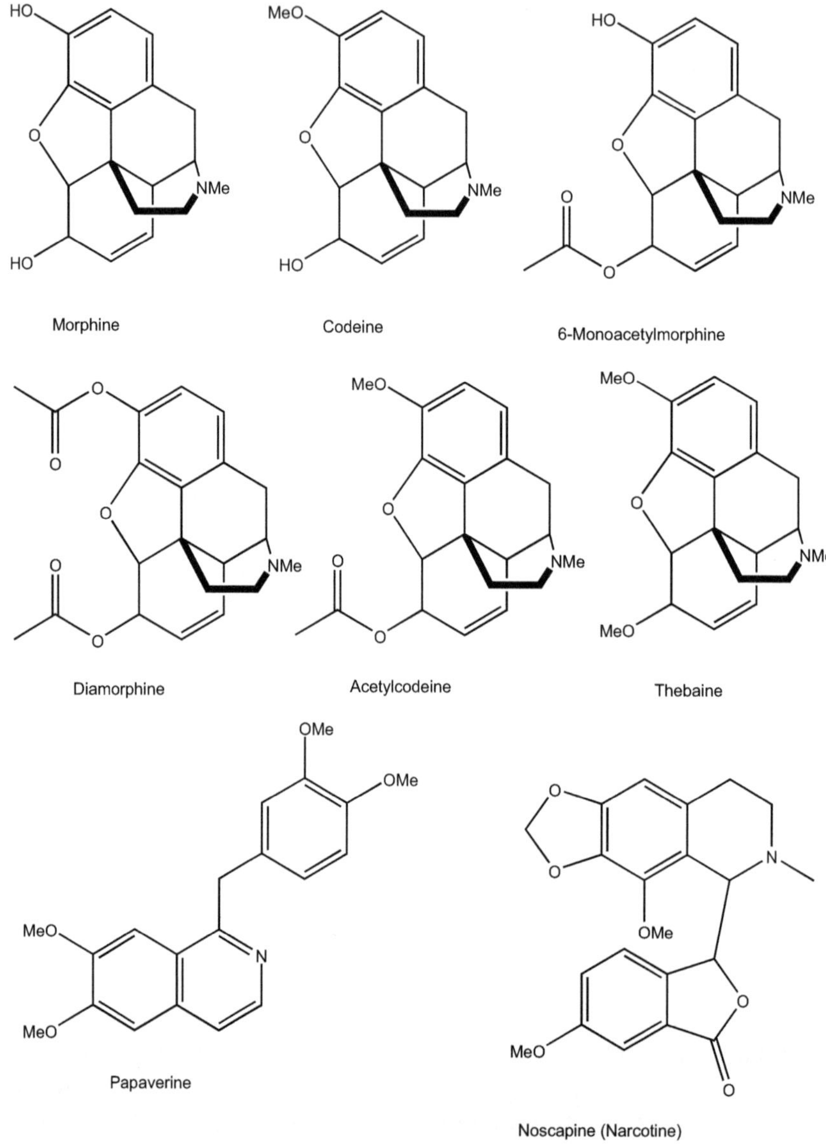

Figure 13.3 Chemical structures of alkaloids commonly found in heroin.

instruments would control such drug manufacture. It is due to this variation in appearance and chemical nature that the origins and relationships of drugs can be identified. This is examined in greater depth later in this chapter.

Cutting agents are used to dilute the drug and hide its impure nature and/or lack of active ingredient. Examples of common cutting

Table 13.3 Examples of how heroin samples can vary both chemically and in appearance, within and between regional sources.

	Drugs present			
Origin	Acetyl codeine	6-*O*-Monoacetyl morphine	Noscapine	Papaverine
S.W. Asia				
Type I Beige–brown, powder with small aggregates. 60% diamorphine, free base.	5%	3%	10%	4%
Type II White–off white 80% diamorphine, hydrochloride salt.	3%	2%	—[a]	—[a]
Middle East				
Type I Beige–light brown powder, aggregates rare 50–70% diamorphine, hydrochloride salt.	3%	2%	—[a]	—[a]
Type II White–off white, fine powder 70–80% diamorphine, hydrochloride salt.	2–3%	2%	Data not provided	

[a]—Drug not detected.

agents include sugars, caffeine, benzodiazepines and barbiturates. These too can be used to compare drug samples. Interestingly, the benzodiazepines are added as anxiolytics to control anxiety in heroin users and barbiturates are found in the heroin as a drug added to control pseudoepileptic fits which can occur as the diamorphine is cleared from the body.

Although bulk or trace samples of the powders may be encountered it is equally likely that utensils will be found. These include, for example, spoon surfaces. Spoons are often used to mix the heroin with a water soluble organic acid so that the diamorphine itself becomes water soluble. This solution can then be drawn through cotton wool, which acts as a crude filter to remove solid materials and injected. As a consequence, the cotton wool and syringes used may also be encountered. It should be noted that such items represent a considerable health and safety risk, and great care should be exercised in their handling to prevent 'needlestick injuries'. The spoons themselves are often interesting—the lower side will be black where heat from a flame has been applied, while the spoon itself may contain the residue from the extracted drug. The handle may be covered in small white flecks of rather purer heroin and breakdown products, which can serve to indicate that heroin was once heated in the spoon.

Several new synthetic opioids (NSOs) have emerged as a class of NPS. They include substances such as non-pharmaceutical fentanyls (*e.g.* butylfentanyl), isotonitazene and U-47700. These substances are highly potent as μ-opioid receptor agonists. For example, isotonitazene is reported to be 500 times more potent than morphine and of greater potency than fentanyl. In 2021 isotonitazene was added to Schedule I of the United Nations Single Convention on Narcotic Drugs of 1961. However, at the time of writing, isotonitazene is captured solely by the PSA in the UK. This is expected to change as the Advisory Council on the Misuse of Drugs (ACMD) has recently recommended control by the MDA as a Class A drug.

Heroin is also mixed with several drugs, both in the UK and overseas. For example, in the UK heroin and cocaine are mixed in a drug known as "speedball". In South Africa low grade heroin is mixed with cannabis and hallucinogenic antiretrovirals, for example efavirenz, in a highly addictive drug which is smoked, called "Nyaope". Other mixtures include "white pipe" in which heroin is mixed with methaqualone and tobacco and smoked through the neck of a broken glass bottle. Such mixtures present special analytical problems because the drugs within the mixture have different chemical properties and stabilities, all of which must be taken into account by the analytical chemist.

13.3.4 Cocaine

Cocaine is produced from the tropane alkaloids produced by the trees *Erythroxylon coca* Lam, *E. novogranatense* (Morris) Hieron and *E. novogranatense* var. *truxillense* (Rusby) Plowman. Its chemical structure is shown in Figure 13.4.

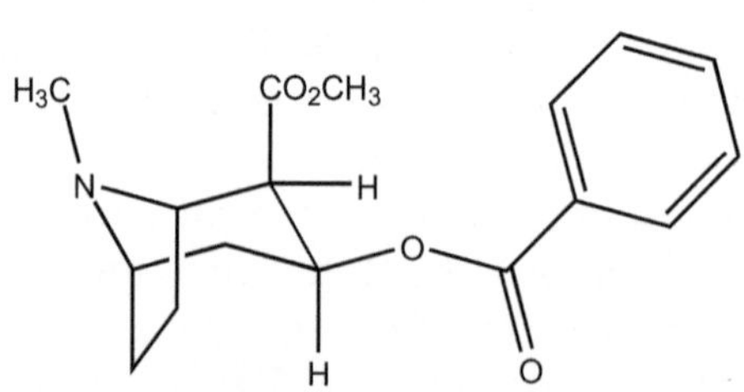

Figure 13.4 Chemical structure of cocaine.

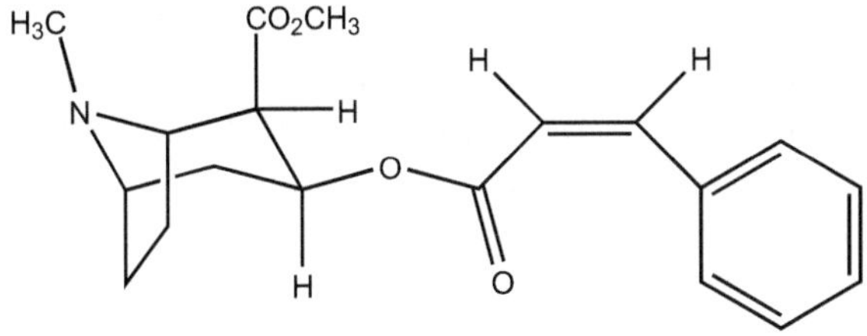

cis-Cinnamoylcocaine

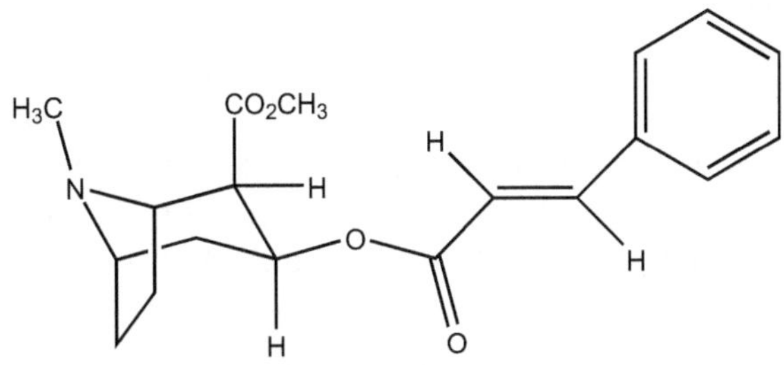

trans-Cinnamoylcocaine

Figure 13.5 Structure of cinnamoylcocaines found in cocaine samples.

The species examined depends on the region in South America from which the sample originates. As with heroin, the tropane alkaloids are extracted from the leaves by a series of solvent/solvent extractions and precipitations. A hydrolysis step follows, so that all of the ecgonine-based tropane alkaloids are hydrolysed to ecgonine. From this, ecgonine methyl ester is manufactured, from which the final product, cocaine, is synthesised. Since the material, like heroin, is produced in a batch process, no two samples are the same and it contains impurities from the natural products in the leaves, including other tropane alkaloids and esters with plant acids, *e.g.* the cinnamoylcocaines (Figure 13.5).

Again, it is the composition of the product that allows samples to be compared and geographical origin to be determined. Two principal forms of cocaine may be encountered. The first is cocaine hydrochloride. This is a white powder, which has a fluffy, snow-like appearance, hence one of its street names, 'snow'. If the free base is then re-prepared from the hydrochloride salt, hard, granular lumps of cocaine are formed. This form is known as 'crack'.

As with heroin, cocaine may also be cut. The cutting agent that is employed depends on the region in which the drug is encountered. In the United States, the drug is commonly adulterated with sugars, *e.g.* mannitol or sucrose, but in the UK local anaesthetics (*e.g.* lignocaine, procaine) are frequently used.

As with cannabis, trace samples may be encountered on materials used to administer the drug. This includes traces of drug on surfaces used to mix the drug. Any small pieces of rolled paper or banknotes should also be analysed for the presence of cocaine although it should be noted that bank notes in many currencies are highly contaminated with drugs, particularly cocaine.

13.3.5 Amphetamines

In the UK, under the Misuse of Drugs Regulations 1985, Schedule 1, Part (c), any compound structurally derived from phenethylamine is controlled. These compounds include amphetamine, methylamphetamine (MA), methylenedioxyamphetamine (MDA) and methylenedioxymethylamphetamine (MDMA); their chemical structures are shown in Figure 13.6.

These compounds can be produced *via* a number of different synthetic routes, using as starting materials either natural products derived from plants (such as ephedrine, safrole and isosafrole), or chemicals available from chemical companies (*e.g.* benzenemethylketone, BMK). It is unlikely that the natural products will ever come under control, but certain purified precursor chemicals are now only supplied to organisations who have the requisite license. Examples include BMK and ephedrine, both precursors of amphetamine.

These phenethylamines can be synthesised *via* several different routes, including, for amphetamine, the Leuckart synthesis, the nitrostyrene route and the reductive amination of BMK. The synthetic routes to *N*-alkylated and ring substituted amphetamines follow the

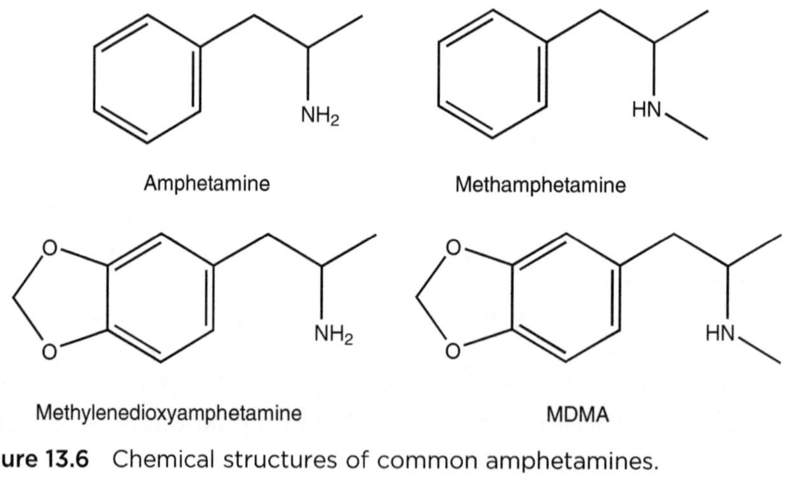

Figure 13.6 Chemical structures of common amphetamines.

same principle – just using the appropriate starting materials and alkylamines (for *N*-alkylated variants). The most popular routes depend on the country in which the drug is being synthesised. For example, in western Europe, the synthesis of amphetamine is principally *via* the Leuckart synthesis, but in parts of northern Europe and the United States the drug is manufactured principally *via* the reductive amination of BMK, and in Japan ephedrine is a favoured starting material. Since different synthetic routes are employed and the drugs are made in batches, the impurities and reaction by-products will also vary between batches. Some will be 'route specific', indicating which of the synthetic routes has been employed. Further, the overall composition of the drugs and impurities will vary between batches, allowing differentiation between samples from different batches and identifying those street samples that have come from the same parent batch. Very great care should be taken in the extraction and analysis of impurities in amphetamines because some (i) will react with each other and (ii) decompose during the extraction procedure. A further complication is that many of the impurities occur at 10^{-1}–10^{-3}% of the concentration of the amphetamine itself and so specialized extraction techniques are required.

In addition to traditional chemical methods of analysis and comparison, amphetamines are interesting because physical comparison can provide a wealth of information about the origin of the drug. The tablets that are manufactured are often coloured, and analysis of the colourant can occasionally lead to the identification of links between the samples. The logos which are stamped into the tablets can sometimes be used to link samples, particularly if the logo is rare or unusual. It should not, however, be assumed that because a tablet is the same colour and has the same logo that the drugs contained within the tablets are the same. This can only be determined through chemical analysis. One interesting facet of this drug type is the use of 'ballistics' to make comparisons between drug samples. The tablets are produced in large numbers in tableting machines. Should one of the dies in the instrument become flawed, the flaw is imparted to every tablet that is made in that die. This can be used to link samples together as having been pressed in the same machine even when the drug content of the tablet is different.

13.3.6 Cathinones and Related Compounds

Cathinones are a group of semi-synthetic and synthetic "designer drugs" which are structurally, and hence pharmacologically, similar

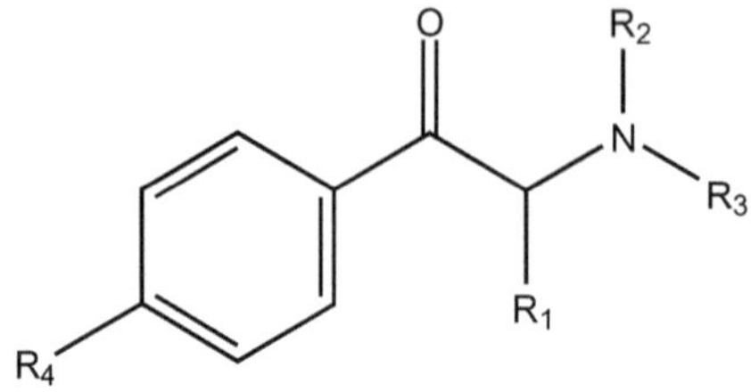

Figure 13.7 Chemical structures and classification of cathinones.

to the amphetamines. They are β-keto analogues of amphetamine. Some of the drug compounds occur naturally. For example, cathinone occurs in the leaves of Khat (*Catha edulis*). Chewing the leaves of Khat is used to achieve a stimulant effect and combat fatigue. The generic structure of cathinone is shown in Figure 13.7. Cathinone derivatives are either (i) *N*-alkylated at positions R_2 and R_3, (ii) ring substituted at position R_4 or (iii) formed by the variation of the α-carbon substituent at position R_1. These drugs are commonly encountered as powders which are then insufflated or injected as solutions or, less commonly, in tableted forms. These synthetic drugs are often sold under a variety of names as plant food, bath salts or "research chemicals" with a label "not for human consumption".

Synthetic cathinones were first manufactured in the 1920s but their popularity as a recreational drug only occurred in Europe from 2007 onwards. Certainly, in one police force in the UK, cathinone seizures rose sharply with the global shortage of safrole in 2008/9 and this mirrors a general trend across then UK. Safrole is the starting material for the manufacture of Ecstasy. As safrole became more freely available, from 2010 onwards, the popularity of the cathinones again declined. However synthetic cathinones have played a significant role in the increased popularity of NPS over the last decade with many structural variants appearing and their popularity continues to date. The European Drug Report 2021 reported that, along with the SCRAs, the cathinones represent 60% of NPS seizures reported in 2019 by EU member states.

13.3.7 *Psilocybe* Mushrooms

Occasionally, material from mushrooms with hallucinogenic properties may be encountered by the forensic chemist. The hallucinogens in the fungi are based on an indole–amine nucleus. In the genus *Psilocybe*, of the approximately 140 species, 80 are known to contain hallucinogens. The three most important species are *Psilocybe*

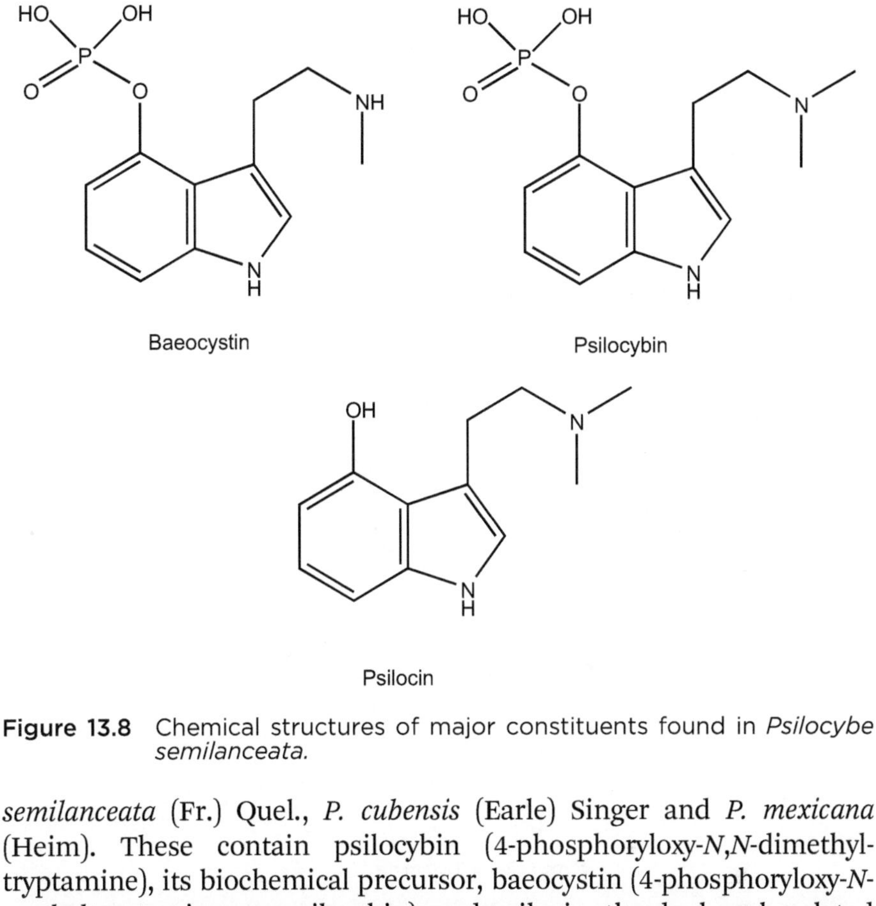

Figure 13.8 Chemical structures of major constituents found in *Psilocybe semilanceata.*

semilanceata (Fr.) Quel., *P. cubensis* (Earle) Singer and *P. mexicana* (Heim). These contain psilocybin (4-phosphoryloxy-*N*,*N*-dimethyltryptamine), its biochemical precursor, baeocystin (4-phosphoryloxy-*N*-methyltryptamine, norpsilocybin), and psilocin, the dephosphorylated compound, which is also a hallucinogenic metabolite. The chemical structures of the major constituents found in *Psilocybe semilanceata* are shown in Figure 13.8.

The effects of baeocystin are not known. The materials may be encountered either as dried mushrooms or as powdered materials. As with many mushrooms, identification of these species using morphology alone is a difficult area in which to work and ideally a qualified mycologist should be engaged to positively identify the species of mushroom. To assist in identification a number of techniques using DNA have now been established.

13.3.8 Mescal Buttons

The peyote cactus *Lophophora williamsii* (Lem. ex Salm-Dyck) Coult. is known to produce the hallucinogen mescaline (Figure 13.9). The

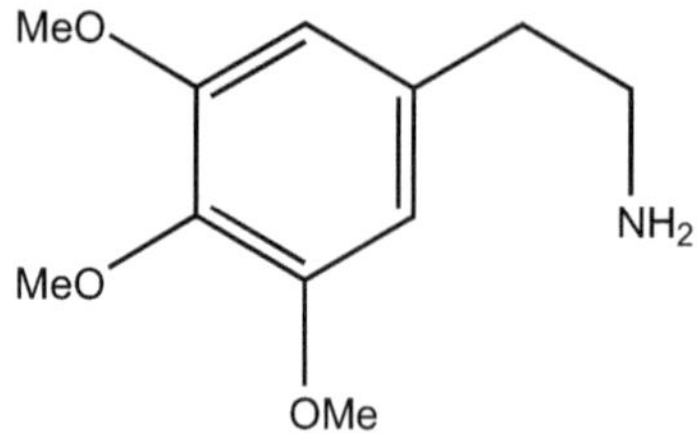

Mescaline

Figure 13.9 Structure of mescaline.

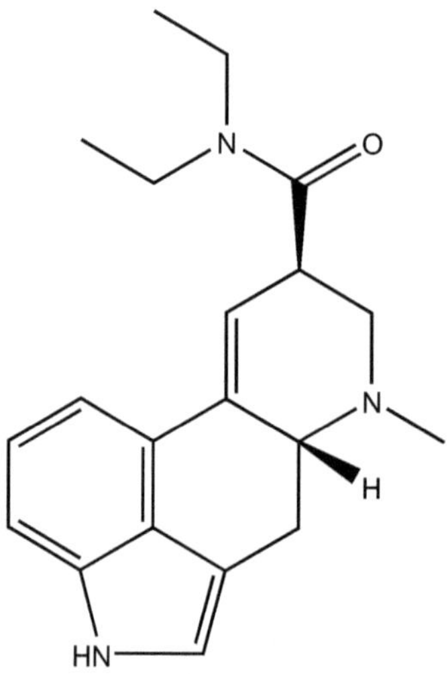

Lysergic acid diethylamide

Figure 13.10 Structure of lysergic acid diethylamide.

material is usually encountered as disc-like slices of the body of the cactus which have been dried. Occasionally (particularly in the USA), seizures are encountered where the active constituent has been synthesized from 3,4,5-trimethoxybenzaldehyde.

13.3.9 Lysergic Acid Diethylamide (LSD)

Lysergic acid diethylamide, also known as lysergide or LSD (Figure 13.10), is one of the most potent hallucinogens known and is usually encountered in one of four forms: blotter acids, gelatine blocks, small tablets and microdots. It can be synthesised following the isolation of lysergamide from morning glory, Hawaiian baby woodrose seeds, or the ergot alkaloids from *Claviceps purpurea* or *Aspergillus clavatus*. No single route dominates over the others.

LSD can be encountered in gelatin blocks, into which the LSD has been incorporated whilst the gelatin is liquid. It can also be encountered as small, highly coloured tablets, known as microdots. Both of these carrier media present problems in terms of the homogeneity of dosage units: one unit may contain a much larger or smaller dose than another. It is for this reason that the most frequently encountered dosage form is the blotter acid. Adsorbent paper is passed through a solution of LSD and then allowed to dry, leaving the drug impregnated in the paper. The papers are often decorated with emblems, patterns or symbols, that may pass over the whole sheet, or individual dose units. The dosing on LSD blotter acids is far more uniform, giving doses of 30–50 micrograms per dose unit.

13.3.10 Barbiturates and Benzodiazepines

In addition to synthetic drugs and drugs derived from plant products, some drugs are 'diverted' from licit sources. This is particularly true of the benzodiazepines and barbiturates. The principal dosage forms of both of these drug classes are tablets and capsules. However, a few of the drugs are prescribed as solutions for injection and the drug chemist ought to be aware of the possibility of encountering these drugs in this form.

13.3.11 Tryptamines

The tryptamines are a chemically diverse group of drugs based on the indole nucleus. The class includes LSD, psilocybin and psilocin, discussed above, and also the natural products ibogaine, bufotenin and harmaline in addition to the rarer dimethyltryptamine (DMT) and 5-methoxy-DMT. The method of ingestion of the drug depends upon the drug being administered. Doses and methods of administration are given in Table 13.4.

The drugs may be used as plant powders, extracts from the plants which are ingested or injectable preparations of the pure compounds. Consequently, it may be necessary to identify the plant material, the compounds in the extracts and compounds on paraphernalia associated with use of drugs of this type.

13.3.12 Piperazines

The piperazines, illustrated in Figure 13.11, are a group of drugs which increased in prevalence, but which are now not seen as frequently as in the past – reflecting the dynamic nature of the illicit drugs market.

Table 13.4 Common tryptamines with details of dose and administration route.

Drug	Dose (mg)	Administration route
5-Methoxy-*N*,*N*-dimethyl-tryptamine (5-MeO-DMT)	2–3	Intravenous
	6–20	Oral
Bufotenine	8–16	Intravenous
Harmaline	150–300	Oral
Ibogaine	Up to gram amounts	Oral
N,*N*-Dimethyltrypatamine (DMT)	4–30	Intravenous
	60–100	Smoked
	>350	Oral

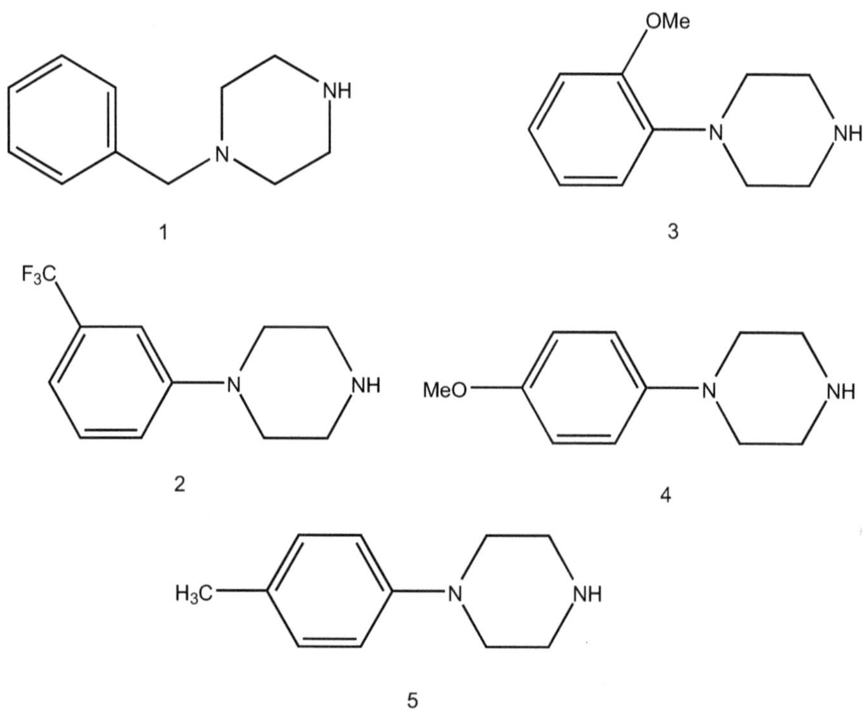

Figure 13.11 Structures of commonly encountered piperazines.

There are a number of piperazines likely to be found in drug samples, including *N*-benzylpiperazine (BZP), 1-(3-trifluoromethylphenyl) piperazine (TFMPP), *ortho*-methoxyphenylpiperazine (*o*MeOPP), *para*-methoxyphenylpiperazine (*p*MeOPP), and 1-(4-methylphenyl)piperazine (*p*MePP).

N-Benzylpiperazine (BZP) [**1**] was synthesised as an antiparasitic drug by Wellcome Research Laboratories in the UK in 1944 but found not to be effective. There was further interest in BZP when scientists discovered it might have antidepressant activity. Benzylpiperazine is a synthetic drug and is easily synthesised from benzyl chloride and piperazine hexahydrate. The related compound trifluoromethylphenylpiperazine (TFMPP) [2] is also found as a street drug. Introduction of a methoxy moiety is a common structural variation of designer drugs of the amphetamine type. The analogous piperazine derivatives are *ortho*- and *para*-methoxyphenylpiperazines (*o*MeOPP [3] and *p*MeOPP [4]). Methylphenylpiperazine [5] has also been encountered in seizures of drugs.

In general, piperazines are sold *via* the internet and non-internet-based herbal shops in tableted dose formats. BZP is commonly sold by itself as either 'BZP' or 'A2'. The typical dose of BZP is 75–150 mg taken orally, that of TFMPP is 25–100 mg. The most popular forms of pills are those that contain a blend of both BZP and TFMPP but there are 'blends' that contain the other substituted phenylpiperazines. The drugs are also frequently mixed with caffeine and ephedrine. BZP was first reported to the European Monitoring Centre for Drugs and Drug Addiction (EMCDDA) and Europol in 1999 through their early warning system on potential drugs. By March 2008, 15 EU member states and non-member Norway had reported seizures of BZP in tablet, capsule or powder form to either Europol or EMCDDA.

Piperazines are not controlled on a global basis but more so by individual countries. As mentioned above, piperazines came under control in the EU in 2008. BZP is now also controlled in other non-EU countries including New Zealand and Japan. For example, as of October 2008, BZP, TFMPP, 4-FPP (4-flurophenylpiperazine), MBZP (methylbenzylpiperazine) and MeOPP became Class C controlled drugs in New Zealand.

With reference to the UK, piperazine drugs became listed under the Misuse of Drugs Act 1971 (Amendment) Order 2009 in December 2009. BZP and the other psychoactive piperazine derivatives were listed in Part 3 of the Schedule as Class C drugs. The less frequently encountered piperazines (isomers of FPP for example) are far less frequently and strictly controlled across the world when compared to the frequently encountered BZP and isomers of TFMPP.

13.4 Identification of Drugs of Abuse

There are a number of materials which may be encountered during the course of a drug analyst's career. These include drugs in bulk form (anything that can be seen), trace samples, wrapping materials and paraphernalia that may be associated with drugs of abuse. Whichever of these is encountered, it is usual to use presumptive tests, sometimes thin layer chromatography (TLC) and then a chromatographic method linked to a spectrometer to identify and quantify drugs in a bulk sample. It should be remembered, however, that there are some drug groups for which presumptive tests and TLC will not provide a great deal of useful information (for example the SCRAs) and so the approach should be to move straight to instrumental methods. Trace samples are usually described, and the analysis then proceeds directly to a chromatographic method which is linked to a spectroscopic technique. In addition, other methods have been developed particularly for drug comparison and profiling including the analysis of the inorganic components of drugs, the use of stable isotopes and the use of DNA.

Although by no means perfect for eliminating drug classes, the visual appearance of a sample may well indicate the class of drug that may be expected in the sample. When examining an item, every precaution should be taken to prevent contamination between samples. Trace samples should not be examined in the same room as bulk samples. Trace samples can be materials in, or on, which drugs are likely to be found but cannot be seen. Another way to think of them is as samples which themselves could easily be contaminated. The potential for contamination is enormous and great care should be taken to eliminate it. This impinges on consideration of lab design, throughput of samples, siting of equipment, the movement of personnel and even the design of the fabric of the building, *e.g.* the air conditioning system.

In addition to drug samples themselves, there are a number of items of paraphernalia that may well be encountered on a regular basis by the drug chemist. These include mirrored or glass surfaces (*e.g.* oven doors) on which drug samples may be cut or mixed; scales or balances on which drugs are weighed and measured; knives and cutting implements that have been used to divide and apportion the drug materials; and items that have been used to administer or take the drug. These include spoons, syringes and items for smoking cannabis or methamphetamine-based products. Further, drug debris may also be present. For example, there may be residues from heating of a heroin sample on a spoon or smoked material from cannabis.

Provided that there is sufficient sample, the analysis should proceed through a logical sequence of events leading to drug identification, quantification and, on occasion, profiling and comparison. If only a tiny amount of sample is present, the non-specific non-instrumental methods are usually omitted. The analysis proceeds straight to the instrumental methods that provide definitive information concerning the identity of the drug. The analysis proceeds as shown in Figure 13.12.

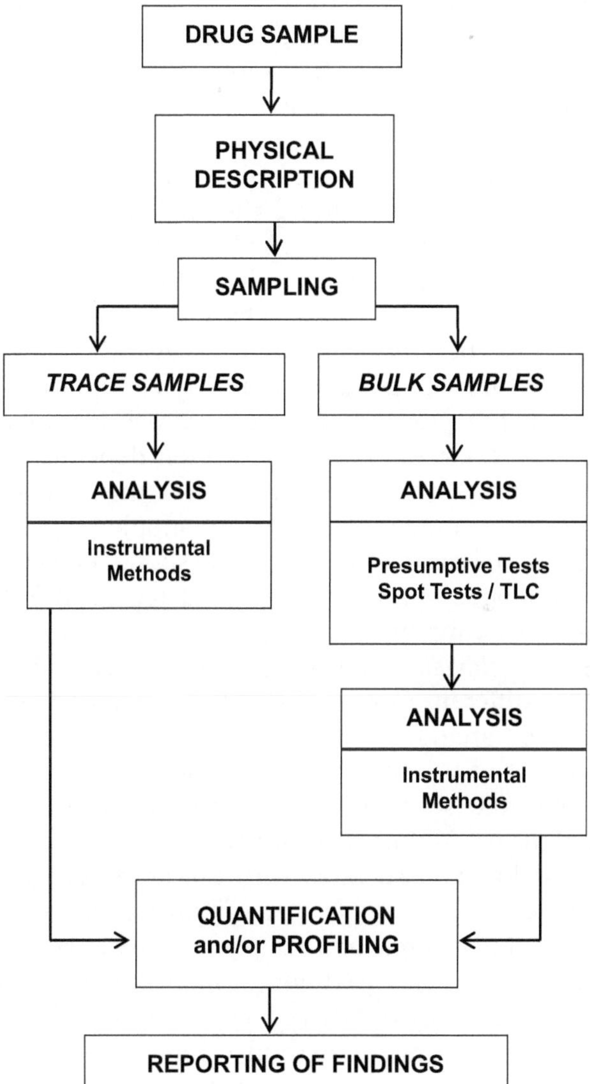

Figure 13.12 Scheme for the definitive identification and quantification of drugs of abuse.

13.4.1 Physical Descriptions and Sampling

Initially a full physical description of the item should be made, including a note of the integrity of the packaging. Physical comparison of the packaging can be useful in linking samples together, especially if one or more of the samples has changed chemically as happens for some drugs, *e.g.* cannabis products. In addition, with extruded polymers, *e.g.* cling film, it is possible to examine by polarized light the striations which are a result of the film manufacturing process. Under such conditions it is possible to observe striae not visible under normal daylight and sometimes it is possible to match striation patterns.

Tableted dose forms should be examined especially carefully. Such examinations are said to be around the "ballistics" of the sample. Where physical imperfections are found on the surface of a tableted dose, it can be compared to other tablets. Where the damage is the same, the tablets can be considered to have been pressed in the same die. However, just because the tablets have the same physical imperfections it cannot be assumed that they contain the same drug. Another physical characteristic of the drug which is important is the mass of the drug. Charging in offences often depends upon a consideration of the amount and concentration of the drug and whether this is likely to be for individual use or for supply.

Following physical examination, decisions should be made as to how the item(s) should be sampled. Each item that appears different should be considered to be a member of a separate group. Where there are a large number of packages in a group (*i.e.* many identical items) there are a number of different models that can be adopted (Table 13.5).

The items to be analysed should be chosen at random from the group. The easiest way to do this is to assign each member of the group a number and then choose the subsample using a random number generator or random number tables.

One of the problems associated with powdered drugs is that the various components separate because of particle size and density of the particles. It is for this reason that powders are only analysed after thorough homogenization. Methods available include placing the sample in a polythene bag or Waring blender to simply mix it up, or using the 'cone and quarter' method where the sample is mixed thoroughly and divided into four equal parts. The process is repeated for two opposite quarters. This process is repeated until a sample suitable for analysis (25–50 mg) is obtained. Where very large samples are to be analysed, *e.g.* a dustbin liner full of drug, a 'core' sample

Table 13.5 Numbers of items to be analysed from a group of materials thought to contain drugs.

Number to be sampled	Advantages	Disadvantages
All ($n = N$)	Certainty about composition	Excessive samples sizes
$N = 0.05N$, $0.1N$, *etc.*	Simple	Excessive sample sizes when N is large
$n = 20 + 10\%\ (N - 20)$ where $N > 20$	Different populations likely to be discovered in large sample if present	Excessive sample sizes when N is large
$N < x$ then $n = N$ $x \leq N \leq y$ then $n = z$ $N > y$ then $n = \sqrt{N}$	UNDCP methods ($x = 10, y = 100, z = 10$)	Excessive sample sizes when N is large
$n = 3$ or 4 regardless of N	Mathematically proven to work	Only used by a few law enforcement agencies

should be taken, using a glass tube to try to sample all of the different regions of the drug bulk. Small samples are easily removed from liquid samples. Stains and marks can be sampled using a swab soaked in an appropriate solvent. Rinsing is often used to remove drug samples from some items of paraphernalia, *e.g.* syringe barrels.

Choice of solvent is hugely important in drug analysis especially when drugs are to be compared. The solvents used should fulfil a number of criteria, including not reacting with the drug in question, not causing or catalysing the breakdown of the drug, being capable of dissolving the drug quantitatively and being compatible with subsequent analyses. For example, methanol is suitable for most very polar drug classes but would not be suitable for drugs containing opiates or tropane alkaloids because of the risk of hydrolysis of the drugs. Chloroform represents the best solvent for these drugs. However, chloroform is not suitable for cannabis samples because chloroform can break down in UV light releasing hydrochloric acid. Cannabinoids are acid labile and so this solvent is not suitable. Methanol is too polar and ethyl acetate does not dissolve the cannabinolic acids. For this reason ethanol is used.

Some drug groups, and drug mixtures are more problematic. For example, recent work on the benzyl- and phenyl-piperazines shows that some solvents do not dissolve the drugs quantitatively, some cause the breakdown of the drug and sometimes they react with the drugs under examination. In this case the tertiary alcohol, 2-methylpropan-2-ol. was used. However, this solvent presents its own challenges since it is a solid at temperatures below 25 °C. In an age where new drug classes are regularly encountered, dissolving the drug in preparation for analysis is an important consideration, especially if drug comparisons are to be made.

Whichever sampling technique is employed for a single item, sufficient should be left for a re-examination by any other interested party. This is particularly important in any adversarial legal system. This is referred to as "equal opportunity". If leaving a sample for re-analysis is not possible, the value of the data should be very carefully considered before the analysis is undertaken.

13.4.2 Presumptive Tests

If the sample is a trace sample, such as a swabbing, then there may only be enough to perform instrumental tests. If there is sufficient sample, however, the analyst should proceed to the next stage, the presumptive tests. These are cost effective, easy, simple tests, which provide accurate and reproducible colour changes on the addition of certain reagents to certain classes of drug. Importantly, they work even when there is a complex mixture of drugs and other material present. The principal importance of these tests is that they provide the analyst with an idea of the class or classes of the drug(s) contained within the seizure. Additionally, they can also provide exclusions around what is present or absent – for example in the analysis of SCRAs. Due to their simplicity, cost effectiveness and ease of interpretation, they are ideal for use by police officers and customs officers, who are not necessarily trained laboratory scientists, working at airports, seaports and other duty stations. It is normal to apply a battery of common tests to discriminate between drug classes and because there may be more than one drug class in the sample, as exemplified by the opiates, benzodiazepines and barbiturates that are all commonly found in heroin samples.

Presumptive tests have several disadvantages including the fact that they do not definitively identify which member of a particular class of drug is present. They are relatively insensitive, detecting down to 5–10 μg of drug. They also react with substances that are not under control, thus providing what are somewhat inaccurately described as 'false positives'. This term should be avoided if at all possible: the reaction is just as positive, only with a compound that is different from the one for which the analyst may be testing.

13.4.3 Thin Layer Chromatography (TLC)

Following presumptive tests, the next stage in the analysis of drugs is the use of TLC. The stationary phase, mobile phase and visualising reagent selected will have all been determined on the basis of the class of drug present, identified through the presumptive tests. Aliquots of the sample(s), together with both positive and negative controls, are spotted separately onto the plate which is then developed with the mobile phase. On completion of this stage the plate is removed from the developing tank and dried. The chromatographic plate is examined visually under ultraviolet light (long (360 nm) and short (254 nm) wavelengths) and any compounds observed are marked. The chromatogram is subsequently sprayed with the developing reagent. Again, the reagent used is determined by the drug class(es) present. More than one chromatographic analysis may be required so that the drugs are separated in different solvent systems depending upon the drug classes present. The distances moved by the drug relative to the distance moved by the mobile phase (R_f values) and colour reactions of any compounds observed are recorded in the laboratory notes. TLC is relatively rapid, enables several samples to be run on one plate and separates many of the major components of different classes of drugs of abuse. Compounds are identified based on two physico-chemical parameters: R_f value and colour reaction with a developing reagent, and comparison with the positive controls employed. The positive controls provide R_f and colour reaction data against which the sample can be compared, while the negative control demonstrates that the compounds observed are due to the sample applied and not to contaminants of the solvent. These controls are essential on every plate since the R_f values obtained will not be identical between operators or replicates of the same analysis, owing to several physico-chemical properties of the analytical system.

Due to the nature of the chromatographic system, three principal limitations are inherent to the technique. First, the separation between compounds (resolution) is limited, secondly, even with these data generated there is no definitive proof of the identity of the drug and thirdly the technique is not quantitative. The power of the technique lies in the ability to determine which members of a particular class of drug are present, and on some occasions, which adulterants are present too, in a short time interval. Further, TLC finds particular application in drug comparison where the concentrations of the materials being compared are relatively high, *e.g.* in cannabis products and in heroin. The power of the method lies in the rapid way it can be used to differentiate samples which are clearly different (both for cannabis and heroin samples) and those in which decomposition products, *e.g.* cannabinol in cannabis, are present and hence further profiling would not be valid.

13.4.4 Instrumental Techniques

Following the identification of the members of the class of drug present, instrumental methods are employed to identify the drug. The techniques used fall into three principal categories: chromatographic, spectroscopic and hyphenated techniques.

13.4.4.1 Chromatographic Techniques

The techniques that find common application are high-performance liquid chromatography (HPLC) and gas chromatography (GC). Other techniques, such as supercritical fluid chromatography (SFC), are not used routinely and hence not considered any further in this chapter although on occasion they do have certain advantages. HPLC separates analytes on the basis of chromatographic mobility in a stream of liquid flowing through a matrix of materials on which the compounds are retained to different extents and hence are separated. The time a compound is retained on the column (retention time) is recorded. The choice of stationary and mobile phase is again dictated by the drug class in question, illustrating once again the central importance of the correct interpretation of the presumptive tests. GC separates the analytes in the gas phase, again providing a retention time or retention index when the retention time is compared to that of an internal standard. The principal means by which compounds are identified through these methods is the comparison of the retention time of the components of the sample with those of a standard compound, chromatographed on the same instrument under identical conditions.

HPLC requires that the compound exhibits good liquid chromatographic properties and is readily detected *e.g.* through the absorbance of ultraviolet light or by fluorescence. It represents an improvement over TLC because of its greater sensitivity and resolving power of components in a mixture. However, alone, it does not definitively prove the identity of the drug. GC has similar advantages and disadvantages. The latter technique also presents problems, in that some drugs, *e.g.* certain benzodiazepines, are thermally labile (decompose when heated), or others, which are polar, acidic or basic, require modification (derivatisation) of their chemical structure before analysis. This requires the analyst to ensure that each compound is fully derivatised before analysis.

13.4.4.2 Spectroscopic Techniques

The principal spectroscopic technique that is employed is infrared (IR) spectroscopy. The IR spectrum is obtained by measuring the interaction of IR radiation with a drug. The spectrum can be interpreted in four ways. One method is to compare the spectrum of the sample with that obtained for a standard of the pure drug. Another method is by comparison of the fingerprint region of the spectrum (wavenumber range 450–1500 cm^{-1}) of the sample to the standard or to literature values. The fingerprint region is of particular value because it represents the low energy vibrations of the molecule due to absorbance of the IR radiation and is unique to any one analyte, and therefore provides definitive identification of the drug. Alternatively, the principal peaks (strongest absorptions) can be listed and the compound identified through examination of the literature values. In the final method the sample is compared with a library of data prepared in the laboratory, preferably by computer. This has a particular attraction because the spectra will have been prepared on the same instrument eliminating the differences observed due to any inherent variation between instruments. This should also be borne in mind when comparing data to literature values—operator-dependent and instrument-dependent differences will always be observed. In addition to the identification of the active drug, other drugs and excipients in a sample may also be identified by IR spectroscopy.

An HPLC system using a diode array multiwavelength ultraviolet/visible detector not only provides the retention time of a compound, but also its ultraviolet, or ultraviolet/visible absorption spectrum as it elutes from the end of the chromatographic column. This provides another means of compound identification and allows differentiation of different drugs. One such example is in the analysis of amphetamine, methamphetamine, MDA and MDMA. Amphetamine and MDA and methamphetamine and MDMA have the same retention times respectively, but the pairs can be separated on the basis of the wavelength of maximum absorption which is different due to the 3,4-methylenedioxy substitution on the benzene ring of MDA and MDMA. The spectrum should always be compared to a sample analysed in the same batch of solvent since variation in solvent strength and pH may lead to variation in the observed ultraviolet spectrum. HPLC coupled to diode array detection provides stronger evidence for the identity of the compound—the degree of proof that this represents is arguable and the discussion as to whether or not this technique provides definitive proof of the identity of a drug is beyond the scope of this chapter.

13.4.4.3 Hyphenated Techniques

Techniques which can provide definitive identification of drugs of abuse include the so-called hyphenated techniques, *i.e.* gas chromatography–mass spectroscopy (GC–MS), gas chromatography–Fourier transform infrared spectroscopy (GC–FTIR) and high performance liquid chromatography–mass spectroscopy (LC–MS) techniques. The chromatography system separates the analytes and the spectrometers provide unique signatures which can be used to identify a compound. The degree to which such techniques are successful depends on a number of factors. The more concentrated the sample, the better the signature for any given compound provided the chromatographic and detection systems are not overloaded. Often the drug chemist's problem is one of too much sample rather than not enough and better tuning of the instrument will provide a more accurate signature. Thus, in these cases, it is the unique nature, under a given set of conditions, of the mass spectrum (GC–MS and LC–MS) or IR spectrum (GC–FTIR) that provides data for identification of a compound. If the sample is very weak, or the detector is dirty, then interferences in the spectra will be observed and it will be difficult, if not impossible, to call a match between samples analysed under identical conditions to the standards, or between the samples and the literature values. Literature values should always be treated with caution, since the spectra obtained by both MS and IR are dependent on the conditions under which they were obtained. The instrument and operator in the laboratory will be different, in most cases, from those by which the literature values were obtained and this should always be borne in mind when data interpretation is undertaken.

For a hyphenated technique such as GC–MS to confirm the identification of an unknown then a reference standard needs to be run contemporaneously. This allows matching between the standard and unknown of both retention time and mass spectrum providing agreement between two independent techniques. If a reference standard is not available, then an identification can be made if reference mass spectra are available either through in-house MS libraries or third-party libraries for example the SWGDRUG NIST addendum mentioned in Section 13.3.2. However, under these circumstances, GC–MS is not a confirmation technique, and the result alone can only be considered an unconfirmed identification. Failure to match the unknown MS to a reference MS would be considered unidentified. However, the presence of an unidentified peak(s) in the chromatogram may warrant further investigation using techniques such as NMR (Section 13.4.4.4).

13.4.4.4 The Use of Nuclear Magnetic Resonance (NMR) in Drug Identification

NPS previously unseen and reported in the literature may appear in seized samples and so may be unidentified in routine laboratory analysis as described in the previous Section (13.4.4.3). NMR is capable of providing definitive structural information of the unknown substance although this requires skillful interpretation of the NMR spectrum. Often very simple extraction of the NPS into fully deuterated methanol is all that is required as sample preparation for proton NMR.

The structural information available from NMR also allows definitive identification to the level of structural and stereo isomers. For example, some isomers of drugs are psychoactive and some are not but hyphenated chromatographic techniques cannot easily be used to separate them. Additionally, the mass spectra of some isomers are too similar for definitive identification. One such example is that of 3- and 4-clephedrone. These structural isomers cannot be easily distinguished by GC–MS but NMR provides a method which enables discrimination and the two isomers can be identified as shown in Figure 13.13. Both 3- and 4-clephedrone are controlled by

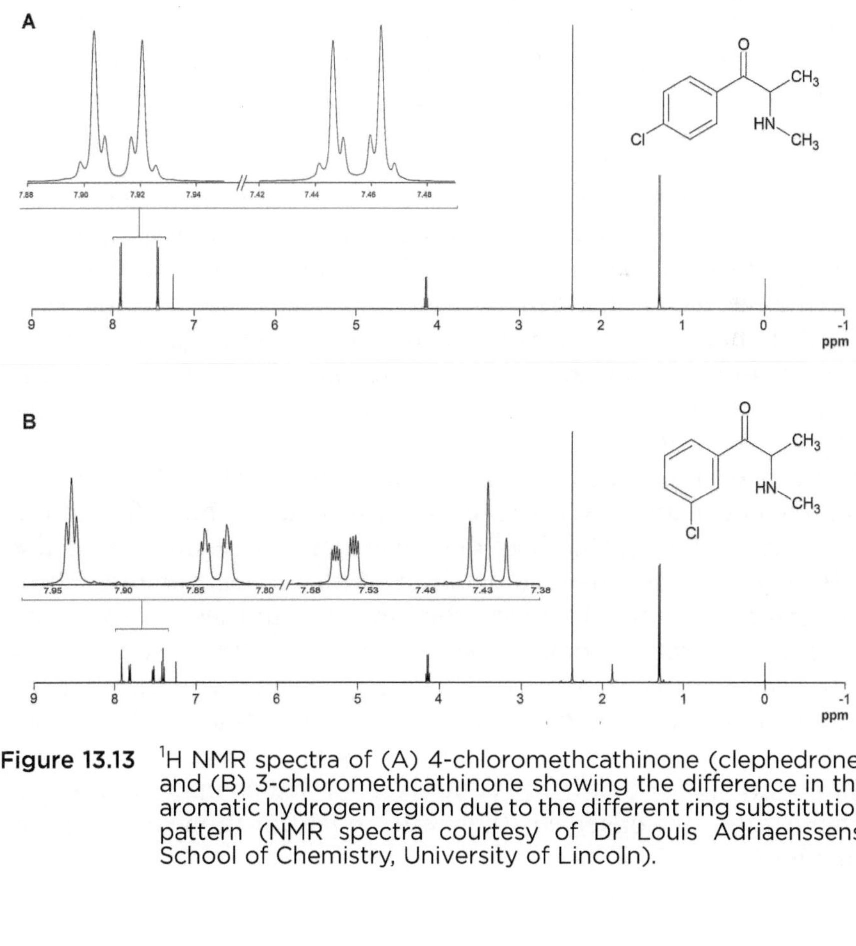

Figure 13.13 ^{1}H NMR spectra of (A) 4-chloromethcathinone (clephedrone) and (B) 3-chloromethcathinone showing the difference in the aromatic hydrogen region due to the different ring substitution pattern (NMR spectra courtesy of Dr Louis Adriaenssens, School of Chemistry, University of Lincoln).

the same generic definition in the MDA and so identifying the unknown as clephedrone is all that is required however this example demonstrates how NMR can be used where identification to a specific structure is required.

13.5 Quantification of Drugs of Abuse

Apart from forensic scientists wishing to identify drugs, diluents and other components in samples, a knowledge of the quality and quantity of the drug present is often required. Under the Misuse of Drugs Act 1971, there is also a requirement, for certain controlled drugs, to quantify the amount of drug in a sample, since the length of imprisonment or level of a fine is dependent on the quantity involved in a case. Furthermore, by determining the quantity of drugs and other components, comparisons between samples can be made which can aid drug profiling and intelligence studies.

To quantify the amount of drugs present in a sample, a number of techniques, including UV, HPLC, GC and GC–MS, may be employed. All are based on a measure of a change in a physico-chemical property measured by the detector in response to different amounts of analyte passing through the detector.

A few points are pertinent to drug quantification. Whether single-point, two-point or full linear regression methods are used, linearity of response must be established prior to use of the method as a quantitative technique. This is essential when the instrument is installed, but also whenever a component of the instrument is changed and is vital if the data generated is to be accepted in court. Further, data should be generated from the lowest to the highest expected concentration, thus reducing the risk of column priming and carry-over between samples. The samples should always be separated by blanks, *i.e.* the solution or solvent in which the sample has been prepared. This best demonstrates the cleanliness of the instrument, equipment, and operator. With the instrumental chromatographic systems described the output from these instruments is a graph (chromatogram) of detector response *versus* time. Peaks are detected for components as they elute from a column, but for quantification it is important that there is baseline resolution between the peaks of interest and any others, and the response for the compound being quantified is due solely to that compound. If these general rules are observed, then valid quantification data will be obtained.

With greater use of mass spectroscopy, hyphenated to chromatographic systems particularly LC–MS, it is possible to both identify and quantify drugs and their congeners in a single analysis. In addition to total ion counts (which provide the identity of the drug from the mass spectra) it is now possible use the mass spectrometer to measure the ion count for single ions. When ions are used which are unique to the analyte of interest it is possible to quantify a drug using selected ion monitoring and counting the numbers of particular ions. Greater accuracy and precision are achieved if a relative ion count is used where the count of the ions of interest from the drug are compared to the numbers of ions from an internal standard. Of course, the ions from the internal standard should also be unique to that compound and not found in any of the analytes being addressed. This is often achieved using deuterated internal standards where the standard is the same molecule as the drug of interest with a number of hydrogen atoms replaced by deuterium atoms. Deuterated internal standards experience very similar conditions within the analytical process (sample preparation and instrument) to the parent drug and are seen as the 'gold standard' for quantification with mass spectrometry based techniques. However deuterated internal standards are not available for all drugs particularly for the vast number of NPS that may be encountered.

13.6 Profiling of Drugs of Abuse

In addition to quantification of drugs of abuse, as indicated earlier it may be necessary to compare two or more samples of drugs to determine whether or not they came from the same parent batch. The term 'common origin' should be avoided since it is plainly obvious, for example, that all cannabis resin has a common origin: the plant *Cannabis sativa*. What is really meant is, 'Do the resin samples come from the same batch?' The same question can be asked of other plant-derived drugs, such as heroin and cocaine, or synthetic drugs, such as the amphetamines. It is now recognised by the European Network of Forensic Science Institutes Working Group on Drugs that there are four levels of drug comparison which can be used to determine whether the drug samples came from more than one batch:

(1) Drug identification
(2) Drug quantification
(3) Identification of the cutting agents
(4) Chemical impurity profiling

Each of these will yield different information about the drugs and their relationships. In addition, it is possible to determine any relationship between wrapping materials—the drugs may be different but the wrapper might be linked. Further, as described above for tableted drugs, the ballistics of any drug sample should also be considered.

Steps 1–3 of the comparison are carried out using all of the principles described above. The methodologies and controls are identical and represent the first stage in comparison. However, the information that such analyses reveal about sample relationships is variable. If a drug is identified and found to have been mixed with a very uncommon cutting agent then a link may be inferred, but it may also be that the same batch of drug was cut with two different agents and a match may be missed. It is vital, therefore, to examine the chemical impurity profile, the signature of impurities themselves, in the drug samples if meaningful data is to be obtained. This is because unless the cutting agent is itself one of these impurities, the proportions of these in the sample will remain unchanged on cutting the drug.

The chemical profile may take many forms. It may be a profile of the drugs themselves as in cannabis products or heroin. It may be the profile of the chemical impurities, *e.g.* those examined when amphetamines are compared. For synthetic drugs, to determine the whole profile of impurities, a concentration step for the impurities is usually involved. Either solid phase extraction or liquid–liquid extraction steps can be used. However, the essential point is that they must be rapid, cost effective, quantitative and extract all the compounds relevant to the profiling with as close to 100% efficiency as can be achieved, without decomposition or production of artefactual impurities. Other methods which have been suggested include the examination of inorganic ions found in drug samples, the analysis of stable isotopes in samples and the analysis of DNA in drug samples of plant or fungal origin. Having created the profile the analyst must then place an interpretation on the data. How the data is interpreted and reported depends upon the materials examined, the methods used, the question being answered, and the statistics used to assist in interpretation. Examples of each are given below.

13.6.1 Profiling of Cannabis Products

Cannabis resin and oil (herbal material does not give meaningful chemical profiling results but is especially useful as a source of DNA) contain tetrahydrocannabinolic acids. These are thermally labile and hence GC is not a good means of drug profiling because the drugs decompose on heating in the injection block of the instrument. Plants also exhibit phenotypic plasticity and cannabis profiles may not even be the same from the same plant. GC is used in some laboratories to quantify total Δ^9-tetrahydrocannabinol content, but HPLC is preferable since no decomposition occurs and both the acidic and neutral cannabinoids can be detected and quantified directly. It is also for these reasons that HPLC can be used or profiling these samples.

The profiling of cannabis samples is performed by extracting the herbal material, resin or oil with ethanol; an aliquot of the centrifuged solution is then analysed by HPLC. The chromatograms obtained are then either compared directly as shown in the example in Figure 13.14, where three cannabis samples have been analysed, or normalised (scaled with respect to a component in the mixture) and overlaid if computing facilities are available.

From these results the profiles for samples (a) and (b) cannot be discriminated but are different from the profile for sample (c). Hence this example illustrates how profiling can be used to determine if materials are related to each other. Such data does, however, illustrate the difficulty in drug profiling when HPLC is used because of its lack of resolution when compared to GC.

13.6.2 Profiling of Heroin

In recent times, with advances in instrumentation, both HPLC and GC have been applied to heroin profiling. The sample requires a simple pre-treatment to extract the opiate drugs noscapine and papaverine, which are then subjected to either HPLC or GC analysis. If GC is used, it is common to derivatise the sample prior to analysis to eliminate the problem of artefact formation. For example, it is known that transacetylation of diamorphine (the 3,6-diacetylated product of morphine) and morphine can occur if both are present in an injection block of a GC and certain solvents are used at the point of injection, with the production of both 3- and 6-monoacetylmorphine. Typical profiles obtained from two related heroin samples are shown in Figure 13.15.

A great deal of work has been carried out on data processing for heroin comparisons. This ranges from comparison of paper copies of chromatograms and examination of the ratios of different compounds, through to cluster analysis and the use of Euclidean distances for retrospective comparisons of large datasets. Such methods have been used in, for example, determination of country of origin of heroin samples and also to examine the relationships of samples at

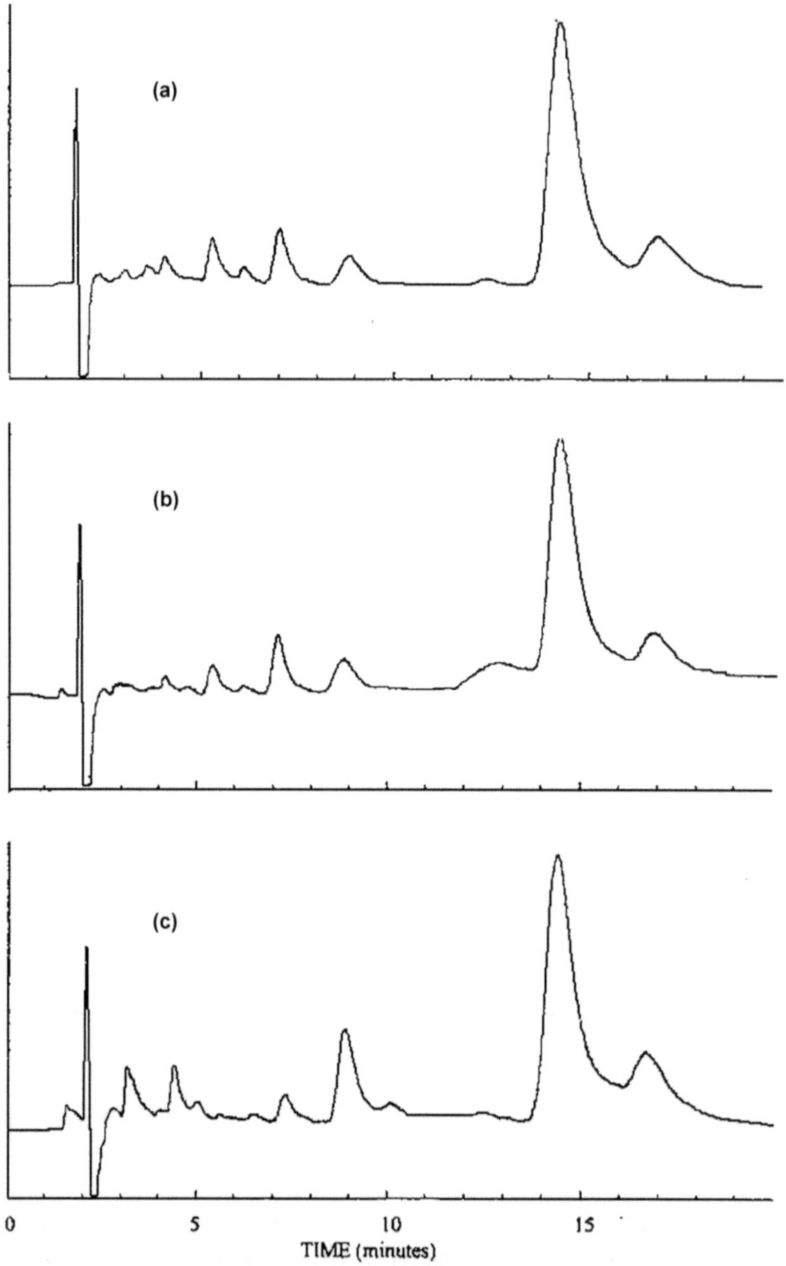

Figure 13.14 Profiling of cannabis samples by HPLC using UV detection. From these profiles it impossible to discriminate between samples (a) and (b) but these are different to sample (c).

the street level. However, there are still problems in terms of whether the assumptions in the underlying chemometrics are met and in data exchange, comparison and the setting up of large databases for

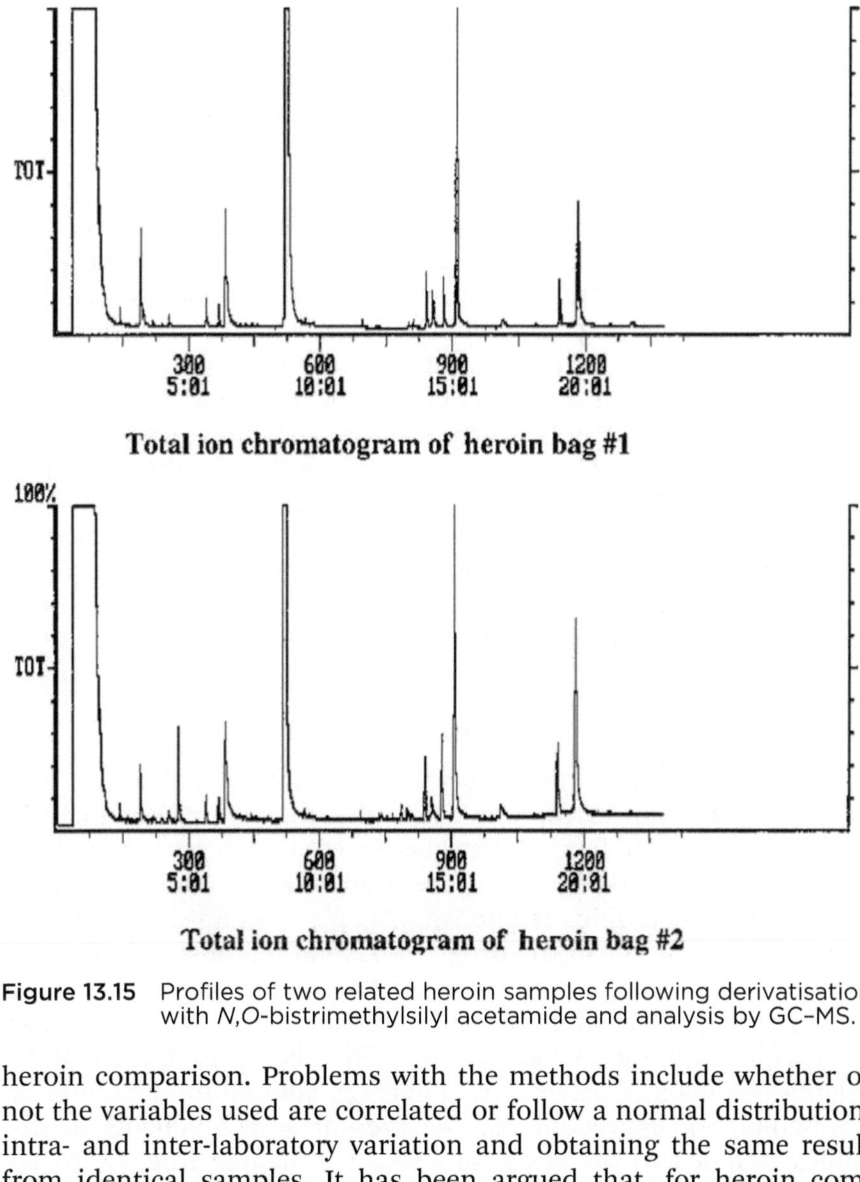

Figure 13.15 Profiles of two related heroin samples following derivatisation with *N*,*O*-bistrimethylsilyl acetamide and analysis by GC–MS.

heroin comparison. Problems with the methods include whether or not the variables used are correlated or follow a normal distribution, intra- and inter-laboratory variation and obtaining the same result from identical samples. It has been argued that, for heroin comparisons at least, there is some rationale for using a centralised laboratory.

13.6.3 Profiling of Amphetamine

Although there have been various studies of how to carry out amphetamine comparisons there are a number of problems associated with these methods because they have been formulated around amphetamine synthesised by the Leuckart route. A number of different synthetic routes are now used (including variations on the Leuckart, reductive amination and nitrostyrene routes). The route of preference depends on where the amphetamine is being manufactured as mentioned earlier, the Leuckart route is popular in north-east Europe, the reductive amination of BMK in parts of Scandinavia and the nitrostyrene route in parts of the Far East. The situation is further complicated by the fact that the impurities occur at extremely low concentrations, can potentially decompose in the solvent in which they are dissolved following extraction, require complex extraction and recovery steps, are relatively polar and reactive in nature; and profiles often contain a very large number of compounds including stereo and regioisomers. However, following extensive collaborative work it is now possible to reliably store amphetamine, extract the impurities, analyse them using GC–MS using a collaboratively tested and harmonized method and compare the data to that obtained in other laboratories. This work highlights the considerable amount of work that is required before such a position can be reached and it is only for very few drugs that such detail is known about the profiling process.

13.6.4 Drug Profiling Using Metal Ion Content

In addition to examination of the drug impurity profile obtained from the organic content of the drug, it is also possible to examine the inorganic content. Several studies have shown that, particularly where there is chemical modification of the drug, it is possible to use metal ion content to determine relationships between samples. The metals themselves can occur in the drugs for several reasons. They may be there because of the container in which the chemical reaction is being carried out, *e.g.* aluminium from a saucepan. They may be present as a consequence of the catalyst in the reaction being used, particularly when a reduction is being carried out, *e.g.* aluminium, platinum or tin. Sodium or potassium may be present as components of the additives added to the street sample after manufacture, *e.g.* if an additive is a sodium or potassium salt then the levels of these metals may be changed.

Several techniques are available to provide quantitative data for the analysis of metals in drug samples. These include flame atomic absorption spectroscopy (FAAS), electrothermal atomic absorption spectrometry (ET-AAS) and inductively coupled plasma mass spectrometry (ICP–MS). The latter has the capability of multielement analysis but is expensive to run and maintain.

There are several considerations to take into account before using such analysis methods. Appropriate blanks should be employed to ensure that the metals chosen for discrimination come from the drug samples and are not leached from the vessels used to prepare the samples or from the reagents employed to prepare samples. Appropriate techniques should be used for metal dissolution. Studies have shown that dissolving samples in nitric acid and microwaving may be more appropriate than vortexing or sonicating the same extract. The analytical method should be optimised for the metals under consideration and appropriate linearity of detector response demonstrated prior to any sample analysis. Under such circumstances the data so generated can be used in complex algorithms used to determine sample relationships.

The use of metal ions has a number of advantages over the traditional analysis of organic impurities. These latter may be volatile and may be subject to degradation processes. The consequence of this is that the organic content of the samples may not be stable. With metal ions, unless there is an addition to or deletion from the sample the metal content will remain the same. It is worth considering the way in which such data is reported. Good practice would suggest that the best way to report this data is for the metal content to be reported for each metal in its uncharged state. This is because the methods used determine the metal content in terms of concentration but do not, in general, provide information on the valency of the ions present. This too may be a function of storage conditions of the drugs.

13.6.5 Drug Profiling Using Stable Isotopes

In addition to examining the organic chemical impurities of the drug sample it is also possible to examine the composition of the drugs at the atomic level. Of particular significance are the stable isotopes of carbon and nitrogen although it is possible to use hydrogen, oxygen and sulphur in such work. Work has been undertaken on the profiling of cannabis products, heroin, cocaine, MDMA and amphetamine. The basic premise is that the isotopic ratios of carbon and nitrogen reflect either the area in which the drugs were grown and/or the chemicals which were used to prepare the drugs.

By way of example, considerable work has been undertaken on cannabis. It is possible to relate herbal material to the geographic area in which it was grown on the basis of the $^{12}C/^{13}C$ and $^{14}N/^{15}N$ ratios. It is known that the different ratios of the isotopes are particularly influenced by water availability and in studies where cannabis of known origins has come from areas of different rainfall it has been possible to discriminate between the samples. This is easily observed in a plot of the difference in the nitrogen ratios against the difference in carbon ratios. Groups of drugs from different areas can easily be differentiated using principal component analysis. This work has further been developed to the extent that once a number of samples have been analysed linear discriminant analysis can be used to make predictions about where a sample may have come from and under experimental conditions it has been found that such predictions can be accurate.

The use of stable isotopes is particularly advantageous because the drugs maintain, in general, their original elemental and isotopic profiles. This is in contrast to organic composition or occluded solvent content which can vary as time progresses as the drugs decompose or dry out.

13.6.6 Plant Drug Identification and Profiling Using DNA

On occasion it is necessary to identify drug materials from plant material but the necessary microscopic characteristics have been lost because of physical damage to the sample or the chemical characteristics required have been lost, *e.g.* because of the age of the sample or poor storage conditions. On such occasions the analyst may be able to use DNA analysis as a means of identifying a drug sample, or indeed making comparisons between drug samples. This is of particular relevance and importance where, for example, genetically identical plants have been found to exhibit phenotypic plasticity and hence different chemical profiles because they have been grown under different environmental conditions. One such drug where this occurs is cannabis.

Early work on DNA analysis of *Cannabis* focused on the use of random amplification of polymorphic DNA (RAPD) using non-specific primers. However, this has the disadvantage of problems with reproducibility. It is now possible to use RAPD with species-specific primers. This can be used to identify the species of plant—in drug chemistry this is generally cannabis—but cannot be used to relate two samples to each other. It is also possible to use the analysis of chloroplast DNA. Using optimized primers it is possible to amplify conserved DNA and then digest with a series of restriction endonucleases. Each of these will produce a series of DNA fragments and it is possible, using the correct suite of enzymes, to produce fragments that indicate that the sample is indeed cannabis. Again, profiling is not appropriate. However, using STR analysis it is possible to generate profiles, specific to cannabis, which can be used to profile and compare samples. Indeed, such approaches have already been used in casework for some time.

It is also possible to use DNA technologies to identify fungal materials, which are notoriously difficult to identify on the basis of morphology alone. Studies have also extended to the study of drug materials used in traditional and herbal remedies which have been used to perpetrate crime, including murder.

13.6.7 Numerical Methods

Detailed discussion of the methodologies used for comparison of drug samples using statistical methods is beyond the scope of this text. However, a few general principles should be outlined.

Before employing any numerical method, it is essential that the correct analytical methodologies have been followed. Additionally, for instrumental methods, it is important to determine that the instrument is functioning correctly and that a linear detector response is being obtained for the concentration range of the analyte in question. Appropriate blanks and check standards should be analysed between samples.

Once the numerical data is obtained, several different methods can be used to compare samples. How closely samples are related can be determined using Euclidean distances and illustrated using dendrograms. However, one assumption, often ignored, is that the analytes considered should be independent of each other in terms of their concentrations. This is true for some, but not for others, *e.g.* the concentrations of some opiates in heroin samples are not mutually independent. Under such circumstances the use of these techniques is not appropriate.

The quotient method, Canberra distances, principal component analysis (PCA), hierarchical cluster analysis (HCA), and Fisher's linear discriminant analysis are among the methods that have been used for the comparison of samples and the explanation of relatedness. However, at the end of the analysis the only truly reliable method for determination of whether two samples are related or not, is for the analyst to compare the chromatograms if chromatographic methods have been used, or some method of directly comparing numerical data for multivariate analysis of components of the sample.

13.7 Quality Assurance in Drug Analysis

In addition to simply analysing the drugs of abuse, it is important that quality assurance is considered. Every necessary control should be taken, and these have been considered above. Further, the instruments should be demonstrably functional, both qualitatively and quantitatively. This requires that standard test samples providing defined responses are analysed at the start and end of each analysis, and if the analysis is long, within the analytical sequence too.

Computerised data handling should be backed up by analogue output. In the UK this is an important requirement. This is because there is a myth among the lay population (and unfortunately among some scientists too) that if the answer comes out of a computer, it must be right. With numerical data handling and quantitative analyses, this is especially important.

Recent developments include the introduction of internationally accepted quality assurance practices, *e.g.* ISO 17025. These require all the above-mentioned precautions to be taken and, in addition, good laboratory note-taking. Everything that is done, right down to changing a pair of protective gloves, should be recorded in writing. Glove changes are necessary to prevent contamination between samples and it is unlikely that in a court room 6 or 12 months after the analysis was performed, you will remember whether or not you changed gloves between samples. If all the above points are considered, the drug analysis will be valid and successful.

In the UK, a Forensic Science Provider (FSP) instructed by law enforcement to analyse seized drug samples to evidential standards must be accredited to the standard ISO 17025 and include, within the scope of the accreditation, the codes of practice and conduct published by the Forensic Science Regulator. Accreditation is provided by the United Kingdom Accreditation Service (UKAS) and to achieve accreditation the FSP must demonstrate that all procedures are validated and documented and that analysts are competent to carry out the procedures. The scope of the accreditation of an FSP covers named drugs routinely encountered in casework and the techniques used for their analysis. Drugs such as NPS that are rarely encountered or not seen previously by the FSP will not be named within the scope of the accreditation. A flexible approach is therefore required and achieved by validating procedures for how the laboratory deals with such substances.

Acknowledgements

Many thanks to Catherine Watson and Fraser Johnstone of Key Forensic Services Ltd, Coventry, UK for valuable discussions on how forensic science providers in the UK have adapted to recent changes in legislation such as the introduction of the Psychoactive Substances Act in 2016.

Further Reading

M. Cole, *The Analysis of Controlled Substance—A Systematic Approach*, John Wiley & Sons, Chichester, 2003.

M. D. Cole and B. Caddy, The Analysis of Drugs of Abuse—*An Instruction Manual*, Ellis Horwood, New York, 1995.

T. A. Gough, *The Analysis of Drugs of Abuse*, John Wiley & Sons, Chichester, 1991.

Guidelines on representative drug sampling, European Network of Forensic Science Institutes Drugs Working Group, 2002, www.ENFSI.org.

L. A. King, *The Misuse of Drugs Act—A Guide for Forensic Scientists*, Royal Society of Chemistry, Cambridge, 2003.

L. A. King, *Forensic Chemistry of Substance Misuse: A Guide to Drug Control*, Royal Society of Chemistry, Cambridge, 2022.

E. Lock, L. Aalberg, K. Andersson, J. Dahlen, M. D. Cole, Y. Finnon, H. Huizer, K. Jalava, E. Kaa, A. Lopes, A. Poortman and E. Sippola, Development of a harmonised method for the profiling of amphetamine V. Determination of the variability of the optimised method, *Forensic Sci. Int.*, 2007, **169**, 77–85 (and references contained therein).

Rapid Testing Methods of Drugs of Abuse, United Nations Drug Control Programme Handbook ST/NAR/13/REV.1, United Nations, New York, 1994.

K. Wolff, *Detection of drug misuse: biomarkers, analytical advances and interpretation*, Royal Society of Chemistry, Cambridge, 2017.

14 Forensic Toxicology

Hazel Torrance

Post Mortem Toxicology Service, Scottish Police Authority Forensic Services, Glasgow, UK
Email: Hazel.Torrance@spa.police.uk

14.1 Introduction

14.1.1 What Is Toxicology?

Toxicology is the study of poisons (or toxins), including their origins and properties, and their effects on living organisms. The meaning of the term poison is dealt with later. A toxicologist is primarily concerned with the chemical analysis of specimens such as blood and urine for the presence of poisons and subsequently with interpreting the significance of the results obtained. In some countries outside the United Kingdom, the toxicologist is also directly involved in treatment of the poisoned patient. The remit of the toxicologist requires expertise not just in analytical chemistry but also in pharmacology and aspects of biology and medicine.

Appending the word 'forensic' to the name 'toxicology' simply means the use of toxicological investigations in court. Any type of toxicological work, in its widest sense, may appear in criminal or civil court proceedings and a list of examples is given below. However, because most court cases are concerned with the effects of poisons on humans, 'forensic toxicology' has traditionally been considered to be the study of the effects of drugs and poisons on human beings.

Crime Scene to Court: The Essentials of Forensic Science, 5th Edition
Edited by Niamh NicDaeid and Peter C. White

Published by the Royal Society of Chemistry, www.rsc.org

The investigations of these effects may be related to a fatality or to the change (or potential change) in behaviour of an individual for the purpose of some sort of medico-legal enquiry, the exact nature of which depends on the country involved.

In the case of deaths, these are typically caused by suicidal or accidental overdose of drugs. Alternatively, the death may have occurred under the influence of a drug or poison but was in fact a consequence of another event, such as a car accident or a fall from a height. Very occasionally homicidal poisonings occur involving drugs, such as insulin, but also other substances such as carbon monoxide gas, cyanide, or pesticides. In fatal investigations, the toxicologist works in close collaboration with the forensic pathologist, who carries out the medico-legal autopsy and furnishes samples of blood, urine and other tissues for analysis in the laboratory. The toxicologist is rarely required to attend the locus of a death or even the autopsy room, and specimens for analysis are normally delivered to the laboratory by an appropriate method, to preserve their integrity (preventing any tampering) and legal validity (preserving a chain of custody). In some jurisdictions, suppliers or traffickers of drugs which are subsequently implicated in a drug-related death are prosecuted on a charge of culpable homicide, and it is under these circumstances that the toxicologist is most commonly involved in homicide cases.

However, post-mortem forensic toxicology accounts for only a minor percentage of the field as a whole and most work involves living cases, such as performance testing of employees (workplace drug testing, when either legislation in force or else the employment contract prohibits employees from using or working under the influence of a proscribed panel of substances) or drivers of vehicles allegedly driving under the influence of alcohol and/or drugs. The forensic toxicologist is also employed in other types of living cases, notably the clients of drug abuse clinics and inmates of prisons, who are regularly tested to determine whether they are using controlled drugs, complainers in cases of alleged drug facilitated assault, including sexual assault ('date rape'), child abuse cases, in which a parent or guardian is accused of administering drugs to the child in their care, and child custody cases, in which a parent or guardian must demonstrate abstinence from drug use before the child is allowed back into their care.

Other types of toxicological investigation that may appear in court proceedings are:

(1) **Clinical toxicology.** The investigation of poisoning in a supervised medical setting, usually a hospital.

(2) **Environmental toxicology.** The investigation of the harmful effects of poisonous substances in the environment.
(3) **Regulatory/animal toxicology.** The investigation of the toxic properties of substances to fulfil the legal requirements for the licensing of commercial products.

The remainder of this chapter deals primarily with post-mortem forensic toxicology but includes sections on cases involving living subjects.

14.1.2 Forensic Toxicology in the United Kingdom

The systematic toxicological investigation of human organs for the presence of poisons is widely considered to have been initiated by Mathieu Orfila, a physician working in Paris whose treatise on toxicology was the first substantial work on the subject. Several students of Orfila returned to the United Kingdom in the 19th century and established the subject here in their home universities within the wider discipline of forensic medicine.

Toxicology services have therefore been provided traditionally by university forensic medicine departments. However, other organisations now largely provide these services including several private companies in England and Wales, the Scottish Police Authority Forensic Services in Scotland and Forensic Science Northern Ireland in Belfast. In other countries, laboratories are operated by law enforcement agencies, by government or by the medical services. Increasingly, commercial organisations have developed to provide scientific support including forensic toxicology for forensic investigations and this is almost universally true for high-throughput performance testing toxicology. Defence counsel has free access to these commercial suppliers, unlike the state-owned laboratories, either because these work only for the law enforcement agencies, or else, having been engaged on a case by the prosecution, they are unable to work simultaneously for the defence. Many laboratories are accredited by their national accreditation body (UKAS in the United Kingdom) for some or all of their scientific procedures, such as the analysis of blood samples, to the relevant quality standard ISO/IEC 17025:2017 (General requirements for the competence of testing and calibration laboratories).

Qualifications accepted by the courts for forensic toxicologists are degrees in medicine or science in an appropriate subject (chemistry, biochemistry or pharmacology) plus experience in a routine forensic toxicology laboratory. In common with other fields of forensic science, most toxicologists are trained in the laboratory in which they are employed. In the United Kingdom, there is no system for certifying individual scientists and their credibility is tested in court, but in other countries, notably the USA, systems are already in place for certification of individual scientists (Diplomate of the American Board of Forensic Toxicology (DABFT)).

14.2 Poisons and Poisoning

14.2.1 Definition

A poison is 'any substance which, taken into or formed in the body, destroys life or impairs health'. This definition does not restrict poisons to particular classes of chemical. Almost all substances are capable of causing harm if administered at a high enough dose, including such essentials for life as salt, water and even oxygen. As examples, consider the use of salt and water as an emetic to induce vomiting when a poison has been swallowed. If care is not taken to restrict the quantity of salt to a spoonful or two, a dangerously high amount of salt can be ingested and this can have its own toxicological consequences. Also there have been instances of individuals who recognise the dangers of dehydration when using ecstasy at a dance club, almost dying by drinking too much water (about 6 L) too quickly. Several factors are involved in determining the dose at which toxic effects are produced and these are considered below.

14.2.2 Factors Affecting the Toxic Dose of a Substance

Most substances can act as poisons at a high enough dose and the crucial question of whether or not a substance is a poison relates to the dose administered rather than to the substance involved. Additionally, prior to being poisonous the effects of a drug may have an effect on the ability to carry out complex tasks *e.g.* driving a car or operating machinery at work. Figure 14.1 may be useful in visualising the relationship between the size of the dose administered and the effects on the subject.

The graph indicates that low doses of a substance are consistent with life; *e.g.* if a drug is administered at the normal dose (the therapeutic dose), then life will continue and the organism or person may benefit from the substance. As the dose increases, however, mild

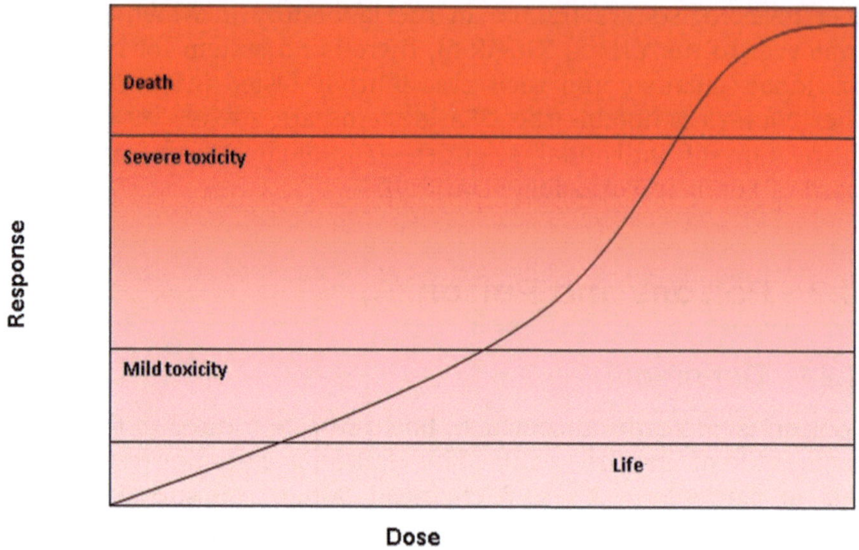

Figure 14.1 Relationship between the dose of a substance administered and the effects or response on the organism.

toxicity may occur, shown by the appearance of adverse or side effects. At this stage, the beneficial effects of a drug may be outweighed by the appearance of these undesirable effects. As the dose continues to rise, the adverse effects become more severe and may start to become life threatening. Severe toxicity occurs at too high a dose and is seen, *e.g.*, as depression of respiration: the breathing becomes more shallow and the rate slows down until it finally ceases altogether. At this point the threshold for fatal poisoning is surpassed. Note that there is no indication of timescale in the diagram. The dose may be administered at one time or in small amounts over a period of time, and similarly the response may appear in the short term (acute poisoning) or long term (chronic poisoning).

It should also be noted that, for some substances, the lower end of the dose curve starts to rise again as the dose approaches zero, giving a U-shaped curve. Such substances are often trace elements and vitamins which must be present in the diet. Both too much and too little is harmful: *e.g.* copper, iron and zinc are essential trace metals. If the dietary intake is too low, the term used is deficiency rather than toxicity. In the context of forensic toxicology, an analogous situation can arise in which a patient may take too low a dose of medication and suffer as a result. Common examples are underdosing on insulin (resulting in a diabetic coma), antiepileptic drugs (resulting in an epileptic fit, which can be fatal) or antipsychotic drugs (resulting in a psychotic episode in which the individual can be harmed or can hurt those in the vicinity).

What has been described qualitatively above is a dose–response curve and in an ideal situation the curve is predictable and the thresholds for each transition stage are clear-cut. This does not happen in the real world, and hence problems of interpretation arise, and these are the province of the toxicologist. Some of the factors that upset the simple dose–response curve are examined below. It is important to note here, as well as later, that there is no specific 'fatal dose' or 'fatal threshold' for a drug or other substance.

14.2.2.1 Carcinogenic and Mutagenic Substances

These include, *e.g.* asbestos and ionising radiation, which cause cancer or mutations, respectively. There is no simple dose–response relationship for them because there is no safe exposure level. Any exposure can cause a catastrophic injury to the living organism and, once initiated, it is essentially irreversible. In the case of ionising radiation this can cause damage to genetic material in living cells, *i.e.* a gene mutation—an alteration to the normal DNA sequence. This mutation, however, may not be apparent in the generation exposed to the radiation, but several generations later.

14.2.2.2 Age and Size

The doses of a substance required to cause mild toxicity, severe toxicity and death vary with the age and size of an individual; *e.g.*, children are often given a lower dose of a drug than adults because their body weight is less. Doses are normally calculated as quantity of substance per unit of body weight, *e.g.* milligrams per kilogram, to compensate for different body sizes. Children also differ from adults in the metabolism and disposition of substances which they ingest. The adult liver, for example, may behave differently from that of a child because it has already encountered substances during its lifetime. As individuals reach old age, their ability to cope with exogenous (foreign) substances is diminished and the threshold for toxicity may be expected to become lower.

14.2.2.3 State of Health

Ill-health can also affect the ability of the body to cope with a toxic insult due to a poison. This is obvious if the liver, the main metabolic organ of the body, is diseased, but many other organs can be involved, including the kidneys which excrete poisonous substances into the urine, and the heart, which may fail at a lower dose of substance than normal if there is already significant heart disease.

14.2.2.4 History of Exposure

The response to a particular dose of a substance is often affected by the previous exposure of the individual to the substance and can take two forms. The most common and better-known effect is the development of tolerance. This means that the individual becomes used to the effects of the drug such that a higher dose or more frequent administration of the same dose is necessary to obtain the original effects. For example, regular heroin users who inject heroin several times a day develop tolerance to the effects of the drug over a period of several weeks and may end up using doses that would have caused death if they had taken them at the beginning of their habit. Equally important in the forensic context is the loss of tolerance. Again, considering the case of a heroin abuser who has acquired tolerance to the drug over a period of weeks or months, this tolerance will be lost fairly rapidly during a period of abstinence from the drug, *e.g.* during a custodial prison sentence. If this individual returns to the same heroin habit after release from prison, there is a severe risk of accidental heroin overdose and death.

The second way in which previous exposure to a substance can affect the dose–response relationship is through the development of sensitisation. In effect, this means the development of an especially sensitive, idiosyncratic response at doses that may be lower than those normally required to elicit a toxic effect. One example of this type is volatile substance abuse (sometimes called 'glue-sniffing' or 'solvent abuse'). Regular sniffing of butane (the gas used in cigarette lighters) can cause sensitisation of the heart such that further exposure can cause cardiac arrythmia (irregular heart beat) and death.

14.2.2.5 Paradoxical Reactions

For some substances there exists the possibility of a paradoxical reaction by a small number of individuals. The dose–response curve described above does not apply to these individuals as the effects produced are quite different from the normal ones experienced by most of the population. An interesting example is a paradoxical excitement caused by some of the drugs in the benzodiazepine class.

These drugs normally cause sedation, relaxation and loss of anxiety, but in a very small number of cases the drugs can cause excitation and aggression. Although this effect is very rare, its occurrence is unpredictable, and it is sometimes invoked by the defence to explain bizarre or uncharacteristic behaviour in a client.

14.2.2.6 Genetics

Each individual has a unique set of genes inherited from their parents which provide the genetic codes for enzymes and other proteins used in the body, including those involved in the metabolism/biotransformation of drugs and in the response to drugs. The most important group of enzymes in drug metabolism is the family of cytochrome P450 enzymes, and for each member of the family several variations exist. Individuals may have different cytochrome P450 profiles depending on their genetic makeup and these differences lead to some individuals being 'fast' metabolisers or 'slow' metabolisers depending on their cytochrome profiles. The field dealing with the interface between pharmacology and genetics is known as pharmacogenetics and is of significant interest to the pharmaceutical industry because of the possibility of personalizing drug treatment regimes. The corresponding field in toxicology is toxicogenetics, and within forensic toxicology the intention will be to use the genetic profile in the interpretation of toxicology results.

14.2.3 Types and Examples of Poisons

Poisons may be grouped into different classes depending on their chemical composition and properties, and their effects on the human body. Physical classification of poisons might distinguish solids, liquids and gases whereas chemical classification would create acidic, basic and neutral groups of substances. Functional classification creates groups of substances such as drugs, pesticides, metals, *etc.* and pharmacological classification of drugs would subdivide them according to their effects, for example stimulants, analgesics, sedatives, *etc.* An early type of classification was made according to the mode of action.

14.2.3.1 Corrosive Poisons

This group includes acids, alkalis and other substances that cause physical corrosion of body tissues through direct contact. Corrosive poisons such as vitriol (sulphuric acid) and lysol (a mixture of substances similar to carbolic acid) were more commonly encountered in the early part of the 20th century, when there were fewer drugs to hand. This type of substance is usually taken orally and causes severe and very visible damage to the soft tissues of the mouth, oesophagus, stomach, and intestine.

14.2.3.2 Irritant Poisons

These include inorganic and metallic poisons such as lead, mercury and arsenic compounds. They are most often ingested orally and cause severe gastrointestinal irritation and upset, with nausea and vomiting, colic and diarrhoea or constipation depending on the substance involved. They are absorbed from the digestive system and are transported in the blood circulation to the organs of the body. These poisons are compounds of chemical elements which cannot be broken down or metabolised by the body to inactive substances: the body can only excrete them or deposit them where they will do the least damage, *e.g.* in the bones of the skeleton. The passage of metals such as lead through the kidneys into the urine causes severe tissue irritation, and kidney damage is a common result.

14.2.3.3 Toxins

Naturally occurring substances produced in living organisms are usually referred to as toxins rather than poisons. The best-known examples are venoms from poisonous snakes and insects, but the group also includes substances produced by bacteria such as salmonella and cholera and algae such as the blue-green algae which causes problems both in inland water reservoirs and in the sea in the neighbourhood of shellfish beds.

14.2.3.4 Systemic Poisons

This is the largest group of poisons and it includes those substances that act directly on biochemical processes within cells. The name refers to the blood circulation system, since these poisons must reach their site of action *via* the blood circulation after they are taken into the body by any one or more of a variety of routes, which are considered in more detail below. Within this group are the well-known alkaloid poisons such as strychnine; inorganic poisons such as cyanide and carbon monoxide; pesticides; and drugs.

The remainder of this chapter will be concerned primarily with poisons of this type.

14.2.4 Routes of Administration and Excretion

Administration and excretion are topics of particular importance in the interpretation of analytical toxicology results and are frequently the basis of questions asked in court. Drugs and other substances can be taken into the body by a number of different routes, which affect:

(1) The speed with which they act on the body.
(2) The fraction of the dose which is absorbed into the bloodstream (the bioavailability) and is able to reach the site of action of the substance.
(3) The ultimate concentration of the substance that is achieved in a particular body tissue.
(4) The effects experienced by the person receiving the substance.

Similarly, substances are cleared from the body by different routes, which affect:

(1) The length of time over which the substance remains active in the body.
(2) The methods by which the body inactivates or detoxifies biologically-active materials.
(3) The selection of samples for analysis.
(4) The length of time over which the substance can be detected.

A generalised diagram illustrating the absorption and excretion of a substance is shown in Figure 14.2. This relates the level of the substance in blood, to the time after administration. For the most part, substances must enter the blood circulation before they can exert an effect on the body; and often, but not always, the time when the strongest effects are observed can be related to the maximum blood concentration. The units of concentration currently favoured are milligrammes per litre ($mg\,L^{-1}$) of blood (or urine) and milligrams per kilogram ($mg\,kg^{-1}$) of tissue (such as liver). Alternative units are used if the former produce very small numbers, *e.g.* nanogrammes per millilitre ($ng\,mL^{-1}$), which is numerically equivalent to microgrammes per litre ($\mu g\,L^{-1}$), is 1000 times smaller than $mg\,L^{-1}$ and is typically used for drugs found in blood at low concentrations, *e.g.* tetrahydrocannabinol, the active constituent of cannabis.

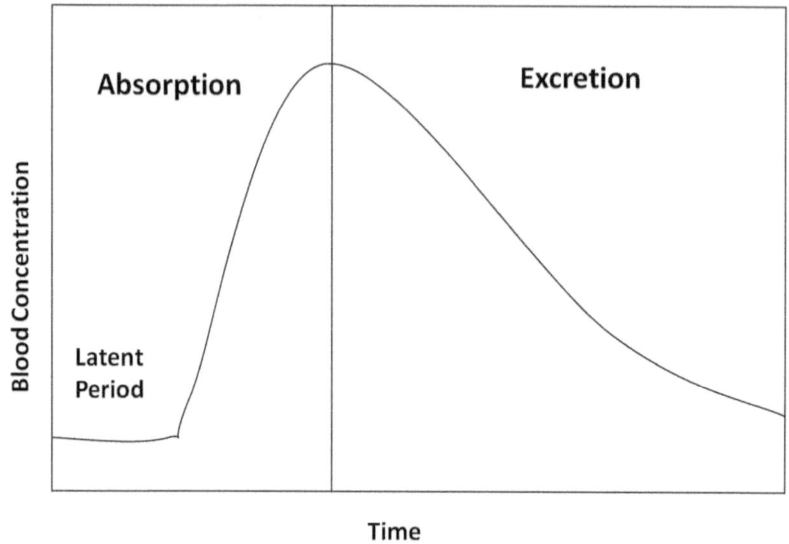

Figure 14.2 Absorption and excretion profile for a substance entering and leaving the human body.

14.2.4.1 Routes of Administration

14.2.4.1.1 Oral

Oral administration implies entry to the body through the mouth and digestive tract, whether taken deliberately, accidentally or unknowingly as a result of administration by someone else. To the toxicologist, entry to the body begins not when the substance is swallowed and reaches the stomach or digestive tract, but when it is absorbed from these organs into the bloodstream. In this respect the digestive tract from the mouth to the anus can be considered as a tunnel passing through the body: the space within it is not an integral part of the body itself. Some substances are not well absorbed when taken orally, including liquid mercury and some other toxic metals and their compounds. These mostly pass through the digestive system without entering the body tissues.

The most important questions usually raised in court concern, 'How long did it take for the substance to be absorbed into the body?' and, 'How much of it was actually absorbed?'

The first question depends on the substance involved: as mentioned above some substances are not absorbed at all. However, most drugs are formulated to be taken orally and are usually absorbed within a period of 30 min to 2 h. This also applies to many other poisons when taken orally. The pharmacologist can alter the rate of absorption by changing the physical form (formulation) of the substance. Slow-release capsules contain small granules with different rates of dissolution, or different coatings on the tablet. It is interesting to note that drug abusers have been known to make their own alteration to increase the rate of absorption of slow-release drugs, *e.g.* the OxyContin (oxycodone) tablet coating was purposefully removed by crushing the tablet or even licking the coating off prior to administration. Information on rate of absorption can be obtained from the medical literature for the drug in question, contained in pharmacopoeias and, *e.g.*, the Association of the British Pharmaceutical Industry (ABPI) Data Sheet Compendium.

A related question is, 'Where does the absorption take place?' Most substances are absorbed from the intestine rather than from the earlier part of the digestive system (mouth, oesophagus and stomach) or the end section (large intestine and rectum). There are exceptions, *e.g.* some substances are absorbed through the mucous membranes of the mouth. This is usually considered to be a different route of administration (buccal) and is similar to the intranasal route considered below. Similarly, drugs are absorbed through the rectum but this is also usually considered as a separate route when the substance is introduced as a suppository.

The second question, 'How much of it was actually absorbed?', refers to the pharmacological term bioavailability, which is important in determining the dose required for a desired effect. Bioavailability simply gives the percentage of a dose that reaches the blood circulation and is thereby available to exert a biological effect. If this is small, the dose must be increased proportionately. There are several factors that affect the bioavailability, two of which will be mentioned here.

- The physical and chemical form is important. Thus, cannabis resin is poorly absorbed from the digestive tract because it is an oily resin, and some salts of lead are poorly absorbed because they do not dissolve very well in water.
- The effect of the liver must be taken into account. When drugs are absorbed from the intestine, they enter the bloodstream through a network of blood vessels that surrounds the gut. The blood flows into the hepatic portal vein and thence to the liver, where the vein subdivides again into smaller blood vessels that pass through the liver (the portal blood vessels). From the liver, the blood is collected once more into the hepatic veins, which return blood to the heart *via* the inferior vena cava. In this way food absorbed from the gut is taken directly to the liver to be

metabolised for energy. However, medicinal drugs are also carried in the same direction and are immediately confronted by the metabolic capabilities of the liver. This is called the first-pass metabolism and explains why many drugs are rapidly inactivated after they are swallowed and an alternative route of administration is used. Heroin is one example: if it is taken orally, there is almost complete first-pass metabolism and the products formed, morphine and other metabolites (breakdown products) do not give the same euphoric effect as when heroin is administered by injection.

14.2.4.1.2 Intravenous

The intravenous route became available in the 19th century, when the hypodermic syringe as we know it was used to administer drugs. This device was almost immediately recognised as an efficient means of introducing abusable drugs such as cocaine and heroin directly into the bloodstream. The bioavailability of substances administered intravenously is almost 100% (some is left in the syringe) and, more importantly, the time taken to reach the blood circulation is essentially zero. This route of administration can induce a very rapid rise in the blood concentration of a drug such as heroin or cocaine, causing potentially lethal side effects on the body's physiological systems, notably on the heart rate and blood pressure. Drugs administered intravenously under medical care are usually diluted in a transfusion of glucose or saline (salt) solution or are pumped in slowly using a syringe pump.

14.2.4.1.3 Inhalation

Drugs and other substances can readily enter the blood circulation following absorption through the lungs, which are richly supplied with blood vessels. Common instances of this type of administration are the use of an inhalant (such as an antiasthmatic device) and smoking (tobacco and other drugs). Smoke particles and gases enter the lungs, and their constituents pass through the thin walls of the air sacs (alveoli) and the surrounding capillary blood vessels. Substances are absorbed rapidly by this route and effects are seen within seconds or minutes of administration. The intense 'high' obtained after smoking crack cocaine contributes greatly to its addictive properties. The bioavailability of inhaled drugs is about 40%.

14.2.4.1.4 Through the Mucous Membranes

The moist surfaces of mucous membranes are found in the nose, mouth, eyes, ears, vagina, rectum and lungs, amongst others, and all of these can and have been used to administer drugs. Perhaps the best known instances are the use of eye and ear drops, nitroglycerine tablets taken sublingually (*i.e.* under the tongue), chewing tobacco and snuff, and also the snorting (a more formal name is nasal insufflation) of drugs such as cocaine and amphetamine. The bioavailability of drugs by snorting is similar to smoking, about 40%.

14.2.4.2 Routes of Excretion

The excretion phase shown in Figure 14.2 consists of two main components, at least for many substances of interest here. The first consists of the metabolism or breakdown of substances such as drugs and the second involves voiding of the waste products from the body.

The role of the liver in metabolism has already been described. Most other tissues, including the blood itself, also have metabolic capabilities, which are designed to break down food and nutrients to provide foodstuffs energy and sustenance for the body. However, the biochemical reactions involved are not specific to but also act on most exogenous (foreign) substances. The metabolic process in this case often serves to inactivate or detoxify the exogenous materials prior to their excretion from the body. As part of the metabolic sequence (many biochemical reactions may take place one after the other), sugar molecules may be attached to the metabolites to render them more water soluble and easier to discharge into urine. These end products are called conjugates and, as illustrated in Figure 14.3, the metabolism of diamorphine (the controlled drug present in heroin) is *via* monoacetylmorphine, morphine and then finally ending up as the glucuronide conjugate of morphine.

The main routes for actual excretion of substances from the body are: (1) through the kidney into the urine, (2) through the liver into bile and faeces, (3) through the skin in sweat, and (4) through the lungs, exhaled in breath. An interesting consequence of excretion *via* bile into the digestive tract is that substances can be reabsorbed into the bloodstream further along the gut, giving rise to a loop called the enterohepatic circulation, *i.e.* from the gut to the liver, from the liver to the bile, discharge of the bile into the gut, reabsorption from the gut and then back to the liver. The net effect of this circulation can be to retain substances in the body longer than might otherwise be expected and in turn, this can lead to a prolongation of the time over which they can be detected by the toxicologist. A common example of this is the retention of cannabis metabolites in the body; these can be detected for up to several weeks after a single administration, in favourable instances.

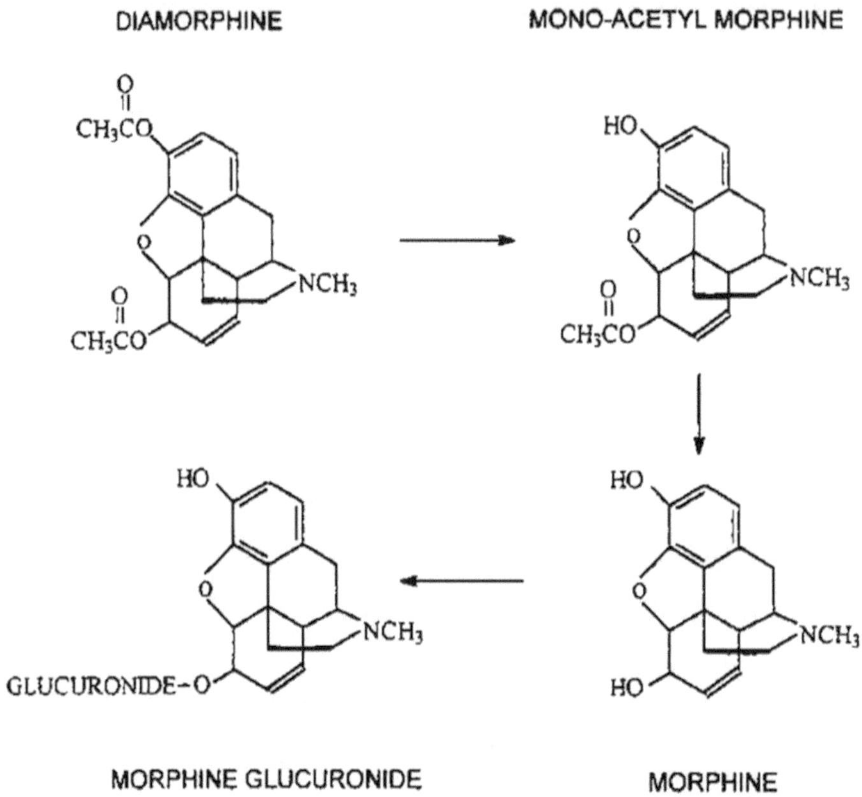

Figure 14.3 Metabolic sequence for the controlled drug diamorphine found in heroin. Monoacetylmorphine and morphine are intermediate metabolites. The final product, morphine-glucuronide, is a water-soluble product excreted in urine.

In one exceptional case, cannabis metabolites were detected for 4 weeks after a single cannabis cigarette was smoked.

Court cases often raise the question, 'How quickly was the substance cleared from the blood?' or, knowing the concentration of a substance in blood at the time of death, 'What was its concentration at a given time prior to death?' These questions can be answered if graphs such as those shown in Figure 14.2 are available for the substance in question, although extreme caution should be used and one must consider individual variation and circumstances.

The rate of clearance from blood is usually specified in terms of the half-life in blood. This is the time it takes for the concentration of the drug in the blood (blood level) to fall to half of its initial value. For example, the average half-life of morphine in blood is about 3 hours. This means that a blood concentration of, say, 1 mg of morphine per litre of blood will fall to 0.5 $mg\,L^{-1}$ in about 3 h, and this will fall to 0.25 $mg\,L^{-1}$ after a further 3 h period, and so on. Half-life data for drugs are available from reference tables in pharmacopoeias and other texts.

Unfortunately, the half-life of a substance can vary widely between individuals for reasons similar to those given earlier concerning dose and response. Data which are available from literature sources are usually given in terms of a range and mean (or average) value. For example, morphine has a half-life of 1–5 h with a mean or average value of 3 h. When the toxicologist is asked in court to calculate blood concentrations at a time prior to death, or prior to the time when a blood sample was taken, the calculation will give the average value and a range, based on data from the literature. For obvious reasons these calculations are referred to as back-calculations. The range can be very wide if a significant time interval is involved, in fact so wide as to be useless for the purposes of the court.

14.2.5 Patterns of Poisoning

14.2.5.1 Time and Place

As indicated earlier, poisoning incidents have tended to involve substances that are readily available in the locality or are familiar to those involved. These have changed during the course of history as the number and variety of chemicals available has increased, but today they also vary according to geographical location around the world. In Europe, the USA and other developed industrial regions, the most common substances involved in fatal poisoning are drugs, both licit and illicit, whereas in most developing countries agricultural chemicals (mostly pesticides) cause the highest proportion of deaths. Interesting regional variations occur even in the types of illicit drugs used, *e.g.* amphetamine is commonly abused in the UK and Europe but in Japan and South-East Asia methamphetamine has always been more popular. Similarly, in South and Central America and the Caribbean, cocaine is the major drug of abuse and heroin is uncommon whereas the pattern is reversed in South-East Asia. Indeed, even within a small country like the United Kingdom there can be large variation in drug trends over different areas. For example, in 2013/14 an outbreak of 19 deaths related to the use of ethylphenidate was concentrated within the Edinburgh area of Scotland.

Nevertheless, these patterns are subject to change, sometimes suddenly. When ethylphenidate was controlled under the Misuse of Drugs Act in the United Kingdom in April 2015, there was an overnight shift away from this drug. The former member countries of the USSR had a very small problem with illicit drugs but were faced with a rapid rise in the abuse of illicit drugs and the attendant rise in drug-related deaths after the collapse of the Soviet Union.

Finally, it should be noted that unexpected incidents can happen, as illustrated by mass poisoning incidents over the years. In Tokyo, the underground system was deliberately poisoned with the chemical warfare gas, sarin, in a politically motivated mass homicide attack. In Switzerland and France, mass suicides by ingestion of cyanide have been encountered in several locations among members of religious organisations.

14.2.5.2 Fatal Poisoning

14.2.5.2.1 Suicide

Most cases of fatal poisoning occur because of suicide or accident (~40% and ~15% of drug-related deaths, respectively) and most of these in the UK involve an overdose of medicinal drugs, consistent with the principle that people tend to use what is readily available. In the UK, suicidal poisoning with other substances such as cyanide usually involves a laboratory or factory worker who has access to the substance in their place of work. The important medico-legal question in these cases concerns the intent of the deceased, *i.e.* did they intend to commit suicide or not? Circumstantial evidence from the locus of the death can be helpful in this respect, including the presence of a note written by the deceased stating intent, the presence of medicine containers such as empty pill bottles and the absence of any indication of violence or struggle. The medical history of the deceased is also relevant, especially if suicide had been attempted previously. Suicide by deliberate overdose is not necessarily easy to perform: the quantity taken can be too little to cause death or be so large as to cause vomiting of the substance swallowed.

14.2.5.2.2 Accident

Accidental overdose cases also occur regularly, although it can be difficult to distinguish between accidental and deliberate overdose on the basis of toxicology alone. The toxicologist is often asked to estimate the quantity of substance taken, to assess if it could have been taken inadvertently. Two points should be taken into consideration when making this estimation:

- Estimation of the dose requires the analysis of each of the body organs, including the stomach contents and contents of the digestive tract. The dose cannot be calculated simply and accurately by multiplying the blood concentration of substance by the volume of blood in the average body. This is because the substance does not reside only in the blood but is distributed throughout the body tissues in varying amounts.

- The minimum dose taken can be estimated very approximately by comparing the concentration of the substance in autopsy blood with the concentrations obtained therapeutically when it is administered in a known amount. In many overdose cases, however, the blood concentration is 10 or more times higher than the normal therapeutic level and the likelihood of this being accidental is lower. Additionally, caution must be used in this approach due to the phenomenon of post-mortem redistribution, discussed later.

It is important to note that the blood concentration obtained after a given dose of substance varies widely between individuals and that, as for half-life calculations, a range and average value should be calculated. Also, the estimation of dose taken must consider the length of time the deceased survived before dying and therefore involves a back-calculation to the time when the substance was administered. As described above, these calculations have their own uncertainties which compound the problems of dose estimation.

14.2.5.2.3 Industrial Deaths

Deaths from poisoning in industry are unusual in the UK; deaths resulting from an accident caused by intoxication with alcohol or drugs are more common. Poisoning cases inevitably involve exposure to substances present in the workplace, and Health and Safety legislation such as COSHH (Control of Substances Hazardous to Health) carefully controls this type of exposure. Deaths are therefore the result of an accidental or unregulated exposure unless the premises involved are not conforming to legal requirements. Typical problems are exposure to carbon monoxide through faulty ventilation or to cyanide through splashing with cyanide solutions (rapidly absorbed through the skin) and asphyxiation due to entering a closed space in which the oxygen has been displaced by an inert gas such as carbon dioxide (*e.g.* in a brewing tank). If industrial cases of non-fatal poisoning are encountered, relevant data for interpretation purposes can be obtained from tables published by the Health and Safety Executive, among others.

14.2.5.2.4 Iatrogenic

These deaths are the result of medical treatment or intervention. They are infrequent and are often the result of errors in calculating the dose of a drug to be administered or else of the use of the wrong route of administration. Deaths due to a general anaesthetic are usually considered separately.

14.2.5.2.5 Homicide

Homicidal poisoning is unusual in the UK today but occurs more frequently in other parts of the world, again often involving agricultural chemicals. Mass deaths due to cyanide poisoning, which may have been at least partly homicidal, have also occurred in different countries. Such cases attract media attention because of their relative rarity. Examples which have occurred in the UK over the last few years include poisoning of spouses with arsenic, cyanide, paraquat, carbon monoxide, atropine and chloroform. The use of carbon monoxide can involve car exhaust but in one case involved a cylinder of the gas, which was obtained by a technician from his laboratory.

14.3 The Work of the Forensic Toxicologist

14.3.1 Role of the Forensic Toxicologist in Medico-legal Investigations

The relationship of the toxicologist to the legal authorities in charge of the case is normally that of a consultant with established expertise in the field of forensic toxicology. The work is usually carried out under contract, either on an individual case basis or as a service provider. In the post-mortem context, specimens for analysis are taken at autopsy by a pathologist authorised by the legal system to do so. Alternatively, specimens are taken by a forensic medical examiner (formerly known as a police surgeon) from an alleged offender, typically under the Road Traffic Acts (1988 and 1991), Sections 4 and 5. These specimens remain the property of the legal authorities and are delivered to the toxicologist who is instructed to undertake the analysis. In the wider context of forensic toxicology, urine, oral fluid and hair samples can be collected by any trained personnel.

Decisions on whether or not to have a toxicological analysis carried out, and what types of analysis are to be requested, are usually taken by the pathologist or by law enforcement personnel, in road traffic cases. The factors involved in these decisions are varied but include the facilities available in the toxicology laboratory, the time taken to obtain results and the cost of the analyses. It is recommended that the toxicologist is involved in making these decisions and that background information concerning the case is made available in order to direct the toxicological investigation and subsequent interpretation of results. Relevant information is contained in the pathologist's report, the police report, the medical history and (if involved) the fire service report. In cases where Novel Psychoactive Substances (NPS) or "legal highs" are suspected it may be useful to examine packaging found at the scene and identify the drug present before analysis of the biological samples.

The purpose of involving a toxicologist in an investigation is to establish if drugs or other substances played a significant role in the case, *e.g.* a death, and access to relevant case background can be of great assistance in indicating what substances to look for. The work can then proceed effectively and produce results in the shortest time and at the minimum cost. In most cases, once the immediate objective of the work has been obtained, *e.g.* a cause of death or a cause of impairment has been established, work can cease unless there is a specific need for further information. This might concern the presence of other drugs or substances available to the deceased or the accused. There is certainly no need for blanket coverage of all possible substances which the laboratory is capable of detecting and measuring, as this would take too long and cost too much.

14.3.2 The Forensic Toxicological Investigation

Different scenarios are possible for the toxicology investigation, depending on the nature of the case. The possibilities are as follows:

- **Specific substance**. The simplest situation is when a sample is submitted with a request for analysis of one or more specific substances. Typical examples are requests for analysis of alcohol in blood (almost all cases) or for analysis of alcohol and carbon monoxide in a fire death.
- **Partial unknown**. In this situation, the toxicologist is informed of a number of possible substances that were available to the deceased or the accused, all or none of which might be present in the samples. In this case, each of the possible substances is targeted specifically to determine whether or not it is present.
- **Complete unknown**. Usually, there is no information available concerning what was available to the deceased or accused. The pathologist may not have a cause of death but may suspect that drugs are involved. Alternatively, there may be an established cause

of death or impairment and the intention is to rule out the involvement of drugs and other substances in the case. In this situation, the toxicological investigation consists of a systematic application of all of the analytical methods available, beginning with the analysis of drugs and substances commonly found in the locality. The profile of the deceased or accused may be used as a guide: *e.g.* pensioners may be more likely to abuse a prescribed medication, although it is imperative that the toxicologist is not blinded by stereotypes. Also, as indicated earlier, if a significant concentration of a drug or other substance is detected, the toxicologist may stop further work after consultation with the pathologist or police.

Because cases often involve a fatality, there is not the same degree of urgency as in other contexts, with the exception that a homicide involving a poison may require the investigation to be carried out as rapidly as possible. Depending on the complexity of the investigation, the time required to produce a report may vary from less than 24 h (*e.g.* a simple blood alcohol measurement) to several weeks (for a full toxicological investigation of all types of drugs and other possible substances).

14.3.3 General Analytical Approach

Most of the cases handled by the forensic toxicologist involve some type of legal proceedings. In this context, ideally two different tests were usually required to provide a high standard of proof that the results are reliable, beyond reasonable doubt. However, with the advent of more sophisticated instrumentation including tandem mass spectrometry it is usually sufficient to carry out one test with multiple points of identification. If two tests are used, the first test is normally a screening test, which is used to establish if a drug or substance is present or potentially present. Ideally, this type of test should be quick and inexpensive, but also reliable at removing negative samples from the next test. If the first test is positive, a second analysis, called a confirmatory test (sometimes quantitative), is carried out. The primary characteristic of this test is that it should be unambiguous and give unequivocal proof of the presence and also the amount of a substance.

14.3.4 Different Types of Specimen

The different types of specimen and their value to the toxicologist are summarised in Table 14.1. Because of recent interest in oral

Table 14.1 Types of specimen and their value to the toxicologist.

Specimen	Attributes
Blood	Available in almost all cases and can be obtained at autopsy without opening the cadaver if it is known to present a risk of infection, although consideration whether this is a true femoral sample should be given. Blood levels of substances can be interpreted using literature data; note, however, that autopsy blood is usually partially clotted and haemolysed (*i.e.* the red blood cells have burst open) so it is not exactly equivalent to blood taken by a clinician from the living patient. Consideration should be given to the likelihood of post-mortem redistribution of drugs and post-mortem production of alcohol Window of detection of drugs is typically 24 hours
Urine	A non-invasive specimen which can be collected from living patients by any trained personnel. Available from only a percentage of autopsy cases. Contains most substances, including drugs and their metabolites, which have been taken into the body. The concentrations of drugs and metabolites in urine are often higher than in blood. Window of detection of drugs and their metabolites typically several days. Lower infection risk than blood; the major disadvantage of urine is that concentrations of substances in urine cannot be correlated with response in the same way as blood levels and therefore cannot be used to establish a cause of death or a cause of impairment, but only indicate a possible cause.
Liver	Available at autopsy from almost all cases especially when blood is not available due to decomposition. When available, the liver is a large organ and there is no problem in obtaining an adequate specimen for analysis. Drug concentrations are usually higher than in blood. Interpretation of drug levels in liver can be difficult because of lack of reference data.
Bile	Bile is obtained from the gallbladder in some autopsy cases. Drug concentrations can be higher than in blood. Bile contains many constituents that can cause interference in the analytical methods. Interpretation of drug levels in bile can be difficult because of lack of reference data.
Vitreous humour	Can be obtained from the eyes in most autopsy cases but this may be ethically unacceptable. Considered to be less prone to artefactual changes in the concentrations of alcohol and drugs. Few interfering substances present. Volume available is small. Difficult to interpret levels of drugs and other substances because of lack of reference data. Sometimes used for drowning cases, to measure potassium and sodium concentrations.

Table 14.1 (*Continued*)

Specimen	Attributes
Cerebrospinal fluid	Can be obtained from the brain in most autopsy cases. Difficult to avoid contamination with blood and brain tissue. Difficult to interpret levels of drugs and other substances because of lack of reference data.
Brain	Available in most autopsy cases. Difficult to analyse because of the presence of fatty material. Difficult to interpret levels of drugs and other substances because of lack of reference data. Useful for detection of volatile substance abuse as volatiles are retained in the fatty tissue of the brain for longer periods than in blood.
Lung	Used for detection of gases and volatile substances. Only qualitative results are obtained, as concentrations are not meaningful.
Hair	Useful for detecting both drugs and other substances such as metals and arsenic. Provides a long window of detection of substances compared to blood and urine. Contains a 'profile' of the sample donor's exposure during the lifetime of the hair (about 1 month per cm). Only qualitative results are obtained, as concentrations are not meaningful.
Nail clippings	Used mainly for assessment of chronic exposure to environmental materials.
Oral fluid	Available from living subjects only. Easily collected by expectoration or with a swab by any trained personnel. Drug concentrations may be related to those in blood plasma and so may provide information on impairment. Of interest for roadside testing of drug drivers.

fluid and hair analysis, more detailed accounts are given for these in Section 14.5.

14.3.5 Tools of the Trade—Methods of Analysis

There are generally 3 steps to analysis of biological samples:

(1) Sample pretreatment
(2) Extraction
(3) End-step analytical procedure

14.3.5.1 Sample Pretreatment

This involves physical and chemical manipulation of specimens to render them suitable for further work, *e.g.* filtration of urine, centrifugation of blood, homogenisation of tissue, adjustment of pH (degree of acidity).

14.3.5.2 Extraction and Concentration

Drugs and other substances are often present at very low concentrations in specimens, typically comprising less than one part per million of specimen (*i.e.* 1 mg per litre of blood or per kilo of tissue). To detect and quantify these substances, the bulk of the matrix containing the analyte (substance being analysed) must be removed and discarded. This entails a process of extraction and concentration of the analyte(s) using either an organic solvent such as ether and dichloromethane or a solid absorbent, often based on silica.

14.3.5.3 End-step Analytical Procedure

The purified extract is analysed by one or more techniques based on spectrometry, chromatography or immunoassay.

Spectrometric techniques regularly used in toxicology include UV spectroscopy and mass spectrometry. These depend on the isolation of a pure substance from the specimen and the information obtained from the technique is a spectrum. This is valuable information for the identification of unknown substances, and/or confirmation of the identity of a substance thought to be present. Spectrometric methods become more powerful when combined with another technique, especially a chromatographic separation method.

Chromatographic methods in this context should be considered simply as methods of separating out the constituents of extracts of biological specimens. The type of information obtained is a chromatogram or chart, characterised by the presence of peaks, each representing one or more constituents, along a timescale generated by the detector. The chromatogram gives two pieces of data: the time taken for each constituent to pass through the chromatographic system (the retention time) and the size (height or area) of the peak on the chromatogram. The former is characteristic of the substance concerned and the latter is proportional to the amount. A chromatographic method can therefore be used for identification (based on the retention time) by using a reference table of retention times of drugs and other substances, and also to quantify the amount of substance present (based on peak size). Taken alone, a single chromatographic method cannot give unequivocal identification, as more than one substance may have the same retention time. The reliability of identification can be improved by using more than one system or as indicated above by combining the chromatographic method with a spectrometric method. The most common examples of the latter are gas chromatography–mass spectrometry (GC–MS), high performance liquid chromatography-diode array spectrometry (HPLC–DAD) and liquid chromatography-tandem mass spectrometry (LC–MS/MS). GC–MS and LC–MS/MS are considered to be the most reliable methods for the identification of substances. Taken together, they are applicable to all materials of interest although alternatives can also be used.

Immunoassay tests are based on the use of antibodies, the protein molecules formed in living organisms to control infection or invasion by foreign organisms such as viruses. Each different type of foreign material (antigen) entering the blood causes the formation of an antibody which will 'recognise' and bind to that material forming an antibody–antigen complex. Using a variety of methods, antibodies can be obtained which recognise drug molecules. These can be used to test for the presence of drugs in a specimen if there is some way of detecting the formation of an antibody–drug complex. Many ways of detecting the formation of these complexes have now been developed and these form the basis of different immunoassays, including radioimmunoassay (RIA), enzyme multiplied immunoassay technique (EMIT), fluorescence polarisation immunoassay (FPIA, often known by the commercial trade name TDx) and enzyme-linked immunosorbent assay (ELISA), which is one of the most widely used techniques. These tests are used as screening methods as they are relatively simple to perform and large numbers of specimens can be processed. They are, however, expensive to buy as commercial kits and are not available for all types of drugs—the most commonly available kits are for drugs of abuse. The nature of immunoassays is such that they are usually not specific for a single substance, but for a group of similar substances. This can be advantageous, as a single kit can be used to screen for the presence of several substances, *e.g.* for the opiates. The disadvantage is that the lack of specificity renders the immunoassay test unsuitable for unequivocal identification of a particular substance or for quantification of an analyte in a specimen. They are also not suitable for picking up changes in drug trends when new drugs with different chemical structures enter the market.

14.3.6 Chemical Classification of Drugs

An area of confusion that sometimes arises, especially during an examination of a toxicology report in court, concerns the difference between the pharmacological and chemical classification of drugs.

The former is familiar to most people. Drugs are classified according to their effects on the body, *e.g.* narcotic analgesics, stimulants, hallucinogens, or antidepressants. The chemical classification, however, deals only with the chemical structures and the resulting chemical properties of the drugs, in particular, whether a drug is acidic, basic or neutral. For example, acidic drugs containing carboxylic acid groups include anti-inflammatory drugs (such as ibuprofen and salicylic acid) and the barbiturates. Basic drugs contain an amine structure and constitute the largest group of drugs, including the narcotic analgesics such as morphine, stimulants such as amphetamine and cocaine, major antipsychotics such as chlorpromazine, antidepressants such as amitriptyline, and most other prescribed drugs. Neutral drugs are fewer in number and this group contains *e.g.* anabolic steroids. To the analyst, the chemical properties of drugs can be used to subdivide them into three fractions (acidic, basic and neutral drugs) and toxicology reports may refer to these chemical classes, due to their testing methods, as well as to particular drugs (*e.g.* alcohol, paracetamol) and groups of drugs (*e.g.* benzodiazepines).

14.3.7 The Toxicology Report

A toxicology report is issued only by an authorised member of the toxicology personnel after the results of the analyses have been reviewed and checked and, preferably, following consultation with the requester to ensure no further work is required. In practice, the analytical work can be carried out entirely by the reporting scientist or under the supervision of the scientist with assistance from technicians, or by a combination of these two. Of course, much of the work is actually performed by laboratory instruments under the supervision of the analyst and results are processed by the software. The degree of personal involvement is steadily diminishing with time. In some contexts, the laboratory can be fully automated and all that is required of the personnel is to load the specimens in a rack. The reliability of the results produced must be safeguarded by the instigation of a quality system in the laboratory. More than just quality control, a quality system specifies the procedures used for all aspects of the work of the laboratory, including the final report and its contents.

At its simplest, a toxicology report contains the following information:

(1) Date and reference number of the report.
(2) The case name (the deceased or specimen donor).

(3) The instructing authority *e.g.* the police, coroner or procurator fiscal and who provided the specimens.
(4) The date on which the specimen(s) was received.
(5) The tests that were carried out.
(6) The results of the tests.
(7) The signature of the analyst.

In addition, according to the practice of the laboratory and legal system concerned, the report may also contain the written interpretation of results and opinion of the toxicologist who signed the report. More recently, additional information regarding the accreditation status of the laboratory and the specific tests is expected to be displayed. In the end the toxicology report should contain all the information regarding the samples and results pertaining to the case, but also explains the context in which the results were determined, the accreditation of the laboratory and any deviations from documented procedures *e.g.* lower sample volume tested resulting in a less sensitive method.

Presentation of a toxicology report in court may consist of attesting that the report and its contents are an accurate record of what was done and of the results obtained. This much is the province of the 'professional witness'. If, in addition, the toxicologist is asked to give an opinion on the results, expert testimony is involved. Court appearances by a toxicologist on simple matters of fact are now relatively rare, as legislation in the UK allows a report to be accepted by the court without the verbal testimony of the signatory, if neither prosecution nor defence has any problems with it. It follows that court appearances now occur only in complex or difficult cases in which the opinion of the toxicologist will be carefully examined and cross-examined. In future, the toxicologist will be subject to more searching cross-examination with respect to the analytical methods used. Method validation procedures are largely standardised in toxicology and will give an estimation of the statistical uncertainty associated with analytical results.

14.4 Interpretation

As indicated previously, toxicology results are primarily either qualitative, indicating whether a substance is present or absent, or quantitative, indicating the concentration of a substance in a specimen. Given the purpose of the investigation, the interpretation of results concerns their significance with respect to a cause of death or a cause of intoxication.

14.4.1 Qualitative Results

These are results which can give an indication of the presence of a drug or drug group. Often these are derived from either an analytical method that cannot produce accurate quantitative results, including many immunoassays, or from analyses carried out on specimens in which the concentration is not of diagnostic significance. The most important examples of the latter are tests carried out on stomach contents or on urine. It is perhaps obvious that the concentration of material in the stomach is less important than what is present or perhaps the total quantity present.

14.4.2 Quantitative Results

These are normally concentrations of substances in blood, urine and other tissues. Their interpretation is intended to relate the concentrations to effects on the sample donors. However, concentrations of substances in urine are more open to misinterpretation and misunderstanding. The problem arises because urinary concentrations vary widely between individuals, even more than in blood, and this in turn is because the volume of urine excreted changes according to the state of hydration of the body. The volume is also affected by foodstuffs such as coffee and drugs such as diuretics, which influence the working of the kidneys. Analytical methods used for urine specimens nevertheless often yield quantitative results and caution must be exercised if asked to interpret these concentrations with respect to incapacitation.

Although some experts are willing to provide an opinion of this type, it is generally accepted and recommended that such interpretations should not be made except under exceptional circumstances, usually where additional information is available. In these cases, the concentration of drugs and other materials in urine do not prove incapacitation, and it is better to base the case on an examination of the accused by a forensic medical examiner or other physician at the time of the offence. Toxicology can then be used to support this examination.

One final point concerning urine concentrations. In the USA, threshold values (also called 'cut-off' values) have been specified for drugs of abuse in urine and if a drug is detected but its concentration is below the cut-off, then the sample is considered to be negative. These cut-off values are 'administrative' thresholds rather than limits imposed by the sensitivity of the method and were established to deal with problems of inadvertent exposure and analytical methodology. There are no statutory cut-off values in the UK but there are relevant guidelines available depending on the purpose of the test, *e.g.* guidelines published by the European Workplace Drug Testing Society.

The medical literature indicates that no concentration–effect relationship exists for many drugs, because the individual response is very variable, as described earlier. This means that, in practice, the interpretation process is limited and may be, at best, an estimate based on the information available. Perhaps the best-known example of this is the wide variation observed in the effects of alcohol on individuals: some people may function quite well at blood alcohol concentrations that would cause others to be extremely unwell. As a guide, the toxicologist can indicate what would happen to the average person at a given concentration and point out the possibility of individual variation.

In all branches of forensic science, interpretation is based on a comparative approach. Toxicology results are compared with a database of concentrations found in previous cases. This reference information comes from a variety of sources:

(1) The records of the toxicologist's own department.
(2) Reference textbooks and pharmacopoeias.
(3) The medical literature, consisting of research papers published in peer-reviewed journals.

The information in this database is segregated according to the type of case, including normal-therapy cases, non-fatal poisoning cases and fatal poisonings. The toxicologist can then decide if the results are consistent with therapeutic administration, an overdose or a fatal overdose. What it is often not possible to do is to give an opinion with certainty that a measured concentration of a substance in blood would have produced a particular effect.

14.4.3 Specific Problems of Interpretation

A salient problem of interpretation is caused by polydrug use among drug abusers. The effects of drugs administered individually are difficult to predict, but when two or more are taken together then the net combined effect is highly unpredictable. Two sedative drugs such as alcohol and heroin would have a combined effect which is greater than that of the individual drugs, but the combined effect of a stimulant and a sedative such as cocaine and heroin is very uncertain. Additionally, some drugs (including nicotine) can induce or inhibit metabolism of other drugs resulting in unexpected increased or decreased drug concentrations.

Care must also be taken in the interpretation of toxicology results since there could be alternative explanations for the presence of drugs in a sample. Several instances have been recorded in the literature in which drugs were found in test specimens (particularly urine samples) which were derived from legitimate sources rather than from illicit materials. For example, poppy seeds used for baking have sometimes been found to contain traces of opium acquired from the poppy capsule. If enough of these seeds are consumed, morphine can be detected in urine. Similarly, some cases were encountered in the USA in which herbal tea contained coca leaves ('Health Inca Tea'), giving rise to positive cocaine tests in urine. Thirdly, some drugs which can be prescribed or even purchased are metabolised in the body to controlled substances. A wide range of medicines give either amphetamine or methamphetamine, and codeine is metabolised to morphine. Finally, the analyst must take care in testing specimens for amphetamine to avoid false positives due to the presence of similar substances such as ephedrine, a constituent of cold remedies and ginseng tea and phenylethylamine, present in decomposed samples.

The effects of passive smoking cannot be neglected. Cannabis and crack cocaine are usually smoked, and those present in the same room or car as the smoker will absorb some of the drug due to passive smoking. Recent work has also indicated that rooms used for smoking crack become contaminated by cocaine 'fall-out'—minute crystals of the drug which settle on the furnishings of the room. People, including children, coming in contact with these can give (false) positive results in a drug screening test. For this reason, cut-off values for urine tests make allowance for the maximum level which could possibly be obtained by passive exposure.

The fact that some drugs are very rapidly metabolised in the body, and cannot be detected a short time after dosage, must not be overlooked. The best-known example is diamorphine, which is present in heroin and is converted rapidly to morphine. Heroin abuse is usually inferred from the presence of 6-monoacetylmorphine and morphine in the blood or urine and from other circumstantial evidence. Synthetic cannabinoid receptor agonist drugs are particularly known for their extensive rapid metabolism and more often it is essential to test for the metabolites to increase the chance of detection.

In addition to the problems of interpretation already mentioned, it is also known that drug concentrations can vary according to the source of the blood specimen (*e.g.* from the heart or from a major blood vessel) and can change after death due to passive physical processes of diffusion in the organs. It is generally not possible to predict or allow for these post-mortem changes and redistribution when reporting drug concentrations, and the part of the body from which blood samples are taken at an autopsy should be chosen with care, to minimise these effects. Guidelines for forensic pathologists recommend that blood samples should be taken from a peripheral vein such as the femoral vein rather than from the heart or aorta. In any event, the source of the blood should be indicated on the submission form which accompanies the blood sample to allow for appropriate interpretation.

14.5 Specific Areas of Interest and Case Studies

The purpose of this section is to provide some additional detail about areas of particular interest, with examples of how toxicology cases are handled and how the results can be used in subsequent legal proceedings.

14.5.1 Alcohol (Ethanol)

14.5.1.1 Chemistry

Ethanol is the most common member of the group of substances known as alcohols, characterized by the presence of a hydroxy group in their chemical structures, and is colloquially referred to simply as ‘alcohol’. Ethanol is a colourless flammable liquid, boiling point 78.4 °C, which is used widely in the laboratory as a solvent for organic materials.

14.5.1.2 Alcohol Concentrations and Amounts

Ethanol is completely miscible with water and is a constituent of alcoholic beverages which contain a wide range of alcohol concentrations, usually between 0.5% and 50% by volume. All of these beverages originate in the fermentation of sugars from fruit or cereals, yielding a maximum alcohol concentration of about 15% by volume, after which yeast is deactivated by the alcohol produced. Higher concentrations of alcohol are usually obtained by distillation, up to 94.8% by volume, which is unpalatable, and in beverages such as whisky is usually diluted to approximately 40% by volume or less.

The alcohol content of a beverage is usually expressed as '% by volume', *e.g.* a beer of strength 5% by volume contains 5 mL of alcohol per 100 mL of beer. Alcohol has a density of 0.789 $g\,mL^{-1}$ and so 5 mL alcohol weigh 5×0.789 g = 3.945 g. The weight of alcohol in an alcoholic drink is calculated as

$$\text{volume of drink} \times \text{alcohol content } (\% \text{ by volume}/100) \times 0.789$$

For example, the weight of alcohol in a 125 mL glass of wine containing 13% alcohol by volume is $125 \times 13/100 \times 0.789 = 12.8$ g.

By contrast, alcohol concentrations in blood and urine are usually expressed as weight per volume, *e.g.* in many countries (including Scotland since Dec 2014) the prescribed limit for alcohol in blood when driving is 50 mg per 100 mL of blood (abbreviated to 50 mg%) which is numerically equivalent to 0.5 $g\,L^{-1}$. Lower concentration units are used for convenience for alcohol in breath and the corresponding limit for alcohol in breath is 22 mg per 100 mL breath (22 mg%).

14.5.1.3 Pharmacology

14.5.1.3.1 Effects of Alcohol

Alcohol is a central nervous system (CNS) depressant which acts by (1) augmenting the effects of the inhibitory neurotransmitter gamma-aminobutyric acid (GABA) and (2) decreasing the effects of the excitatory neurotransmitter glutamate. CNS depression can manifest itself in different ways, including sedation, amnesia, unconsciousness and ultimately death. Paradoxical disinhibition occurs when those parts of the brain that govern self control and restraint are depressed by alcohol, which occurs at an early stage of alcohol intoxication.

It can be seen from the calculations given above that even a single drink may contain a substantial dose of alcohol, greatly in excess of any other drug consumed, and as mentioned earlier, even low concentrations of alcohol can enhance the effects of other drugs such as sedatives.

Like many drugs, the effects of alcohol are dose-related and, because blood alcohol concentrations are also dose related, the effects can be correlated with blood alcohol level, as summarised in Table 14.2. However, there is a great deal of variation in the response

Table 14.2 Effects of alcohol *versus* blood alcohol concentration.

Blood alcohol concentration (mg%)	Anticipated effects
0–50	Little effect, except perhaps more talkative.
50–100	Tendency for loss of fine motor control, causing perhaps slurred speech. Also, inhibition and self-critical faculty may begin to be lost and there may be mild euphoria.
100–200	Generally, both physical and mental functions become significantly impaired as the blood alcohol concentration increases through this range, perhaps resulting in poor balance, staggering when walking, poor driving performance, poor judgement, sedation, emotional instability and nausea. Many fatal cases in forensic medicine occur with blood alcohol concentrations in this range, when the subject performs badly but is not so intoxicated as to be incapable of getting into trouble.
200–300	Subject displays typical signs of being 'drunk and incapable': increased impairment of physical and mental abilities, including impaired vision and speech; difficulty in standing; confusion; sedation, possibly coma.
300–450	Severe intoxication resulting in unconsciousness, impaired respiration, coma and unresponsiveness in all but very tolerant subjects. Possibly fatal.
>450	Probably fatal, due to respiratory paralysis.

of the individual subject to alcohol due to the factors mentioned earlier plus the effects of mood, expectation, and experience with alcohol of the user and the environment in which alcohol is consumed—alone or in a social setting. Tolerance to the effects of alcohol develops with repeated consumption but impairment still occurs, *e.g.* in tasks requiring fine control such as driving a vehicle. Physiologists refer to 'fine motor control', meaning motor control of the small muscles that carry out fine movements, which are also involved in control of a vehicle. Numerous studies have shown that the risk of an accident when driving increases exponentially with blood alcohol concentration. Questions about the effects of alcohol on individuals occur regularly in court, but only a general indication can be given of the range of effects that might be expected at a given blood alcohol level.

14.5.1.3.2 Absorption

Absorption of alcohol into the bloodstream starts in the stomach but mostly takes place in the small intestine. Absorption is usually rapid and is complete within 0.5–1.5 h after the last consumption of alcohol. Food in the stomach can delay absorption by slowing down gastric emptying into the intestine.

14.5.1.3.3 Distribution

After absorption has occurred, alcohol is distributed uniformly throughout the body water, including all fluids, organs and tissues, and the fetus if the subject is pregnant. It follows that knowledge of the water content of a tissue or fluid allows alcohol concentrations in different body compartments to be compared and this is the basis of the Widmark equation, which relates the blood alcohol concentration to the amount ingested.

14.5.1.3.4 Metabolism

Metabolism of alcohol also begins in the stomach but mostly takes place in the liver, where it is oxidized by two different systems. Alcohol dehydrogenase (ADH) is the main metabolic enzyme, producing acetaldehyde which is subsequently oxidized by aldehyde dehydrogenase to acetic acid. This can be assimilated into the normal metabolism of the body and so alcohol can act as a nutrient, at least for some time. The second oxidizing system is the microsomal enzyme oxidizing system (MEOS), which is important in the metabolism of alcohol at high concentrations. The rate of alcohol metabolism is influenced by the genetic profile of the subject, due to polymorphism of the enzymes involved, and varies widely between individuals as well as having a geographical variation between populations.

Alcohol concentrations in the body are high compared to those of most other drugs and saturate the metabolic systems, such that they are working at maximum capacity until the alcohol concentration falls to a low level. Technically this is described as 'zero-order kinetics', in which the rate of alcohol metabolism is independent of the alcohol concentration, *i.e.* a constant amount is lost per hour regardless of how high the alcohol level is in blood. The average rate is 19 mg% per hour within a range of 9–29 mg% per hour, although these figures vary depending on the source of the data. The rates do not, however, vary according to the size of the person as an increase in body weight also means that the liver is proportionately larger. Body size does affect the total amount of water in the body, which in turn affects the alcohol concentration reached for a given dose of alcohol, *i.e.* a small person will reach a higher blood alcohol level than a large person if they both consume equal amounts of alcohol.

14.5.1.3.5 Elimination

Alcohol is eliminated from the body mostly by metabolism in the liver but a small percentage (2%) is excreted in urine and breath plus additional small amounts in other secretions such as sweat. Concentrations of alcohol in urine are higher than in blood, reflecting the higher water content of urine compared to blood, during the elimination phase. However, the reverse is true for the absorption phase since there is a lag time between alcohol reaching the blood and then distributing into urine. Alcohol is present in breath at concentrations which are much lower than in blood: the ratio of blood to breath alcohol concentrations is variable but on average it is 2300 : 1, the figure used in UK road traffic legislation to define the prescribed limit for alcohol in breath.

14.5.1.4 Forensic Toxicology

14.5.1.4.1 Prevalence

Alcohol is one of the most commonly encountered drugs in forensic toxicology. During the 10 year period 2012–2021 more than 32 000 post-mortem cases were examined by Forensic Medicine and Science, University of Glasgow. Alcohol was present in 50% of these cases.

14.5.1.4.2 Alcohol Toxicity

Alcohol causes both acute and chronic effects. Acute intoxication occurs while alcohol is still in the blood with a range of dose-related effects, as described in Table 14.2. Chronic effects of alcohol abuse involve a number of body systems but especially the gastrointestinal, neurological and cardiovascular systems. Alcohol is associated with ulcers in the oesophagus and stomach, and in the liver, alcohol causes fat deposition leading to hepatitis, cirrhosis and an increased incidence of cancer. Alcohol also causes pancreatitis, damaging the islets of Langerhans which produce insulin, leading to chronic pain and type 1 diabetes. Alcohol is associated with encephalopathy (disease of the brain) including dementia and with cardiomyopathy (disease of the heart muscle).

14.5.1.4.3 Acute Abstinence Syndrome

Sudden withdrawal from alcohol ingestion, especially after a prolonged period of excessive alcohol abuse, can lead to life-threatening withdrawal symptoms, including seizures and delirium tremens. Ethanol is a CNS depressant and prolonged abuse leads to tolerance, in which the brain adapts to the depressant effects of alcohol to restore as much normal activity as possible. If alcohol is removed, the sedative effect disappears and the brain moves into a hyper-excited state in which neurological damage (excitotoxicity) occurs. Management of this condition, *e.g.* in a detainee in a jail cell, involves administration of a sedative.

14.5.1.4.4 Artefactual Production of Ethanol

Ethanol can be produced in blood and tissues after death by processes similar to fermentation, especially if the body has been damaged, *e.g.* in an accident, allowing ingress of microbes and the environmental temperature is high enough to permit enzymatic reactions to take place. Concentrations of alcohol produced in this way can be significant, up to 150 mg% or more, although lower concentrations are more likely to be due to post-mortem production. Ethanol can also be produced in stored blood samples but this can largely be prevented by addition of fluoride to inhibit biochemical changes.

One approach to this problem involves the analysis of other samples including urine and vitreous humor, if available, which are considered to be less prone to post-mortem change than blood. The presence of alcohol in blood but not in urine or vitreous may indicate post-mortem production. However, in cases of undiagnosed diabetes, sugar in the urine can lead to production of ethanol in these samples. An alternative approach is to analyse the suspect sample for the presence of metabolites produced during life following ingestion of alcohol or else to look for markers of post-mortem production of alcohol. For the former, alcohol produces glucuronide and sulphate metabolites only in living subjects and so these serve as markers that alcohol was ingested before death. On the other hand, post-mortem formation of alcohol can be indicated by the presence of other volatiles in blood, and these have been used to interpret blood alcohol levels in aircraft accident deaths.

14.5.1.5 Methods of Analysis

Two methods of analysis of alcohol are of interest in forensic toxicology. One is the laboratory procedure used for samples of blood and urine and the other is the evidential breath test device. Hospital laboratories may also use a clinical biochemistry procedure but the methods described below are more appropriate.

14.5.1.5.1 Laboratory Method

Blood and urine samples are analysed routinely by gas chromatography with flame ionisation detection. The method is rapid, has good accuracy and precision, better than 2% relative standard deviation, and is selective, since the analysis is carried out in parallel using two gas chromatography columns with different selectivities. It is also a robust and clean procedure as the samples are prepared for headspace analysis rather than being analysed directly, avoiding column contamination with blood proteins and other nonvolatile components. In the headspace technique a measured aliquot of blood or urine for analysis is sealed in a partially filled glass vial with a cap containing a rubber septum. A series of calibration standards containing known concentrations of alcohol is prepared at the same time under identical conditions. The accuracy and precision of the measurements are improved by adding a fixed amount of a reference material to each sample as an 'internal standard'. The vials are then placed in a warm oven before being analysed.

Alcohol in the samples equilibrates with the air in the vials, giving concentrations in air which are directly proportional to those in the samples. A small portion of the air (headspace) is removed with a hypodermic syringe by an automatic sampling device and injected into the gas chromatograph, producing a chromatogram for each sample with peaks for alcohol and the internal standard. The ratio of the area of the alcohol peak to that of the internal standard peak is calculated for each sample. A calibration curve is prepared by plotting the peak area ratios for the standards against their known alcohol concentrations and this curve is used to obtain the alcohol concentrations in the case samples being tested.

14.5.1.5.2 Evidential Breath Test

Breath alcohol measurements are made with living subjects under a variety of circumstances including road traffic safety, employment screening and doping control in sport. Instrumental breath test devices are used both at the roadside as a preliminary screening test and indoors, typically in a police station, as an evidential breath test. The portable roadside devices use an electrochemical (fuel) cell whereas the evidential versions use an electrochemical (EC) cell or infrared (IR) spectrometry or both (EC/IR). Current EC/IR instruments are sensitive, specific and automated, requiring little operator involvement in the measurement process. They must be maintained regularly and be in a stable temperature environment.

The automated test procedure involves a calibration check at the prescribed limit using a gas standard of ethanol in air from a cylinder, analysis of two sequential breath samples from the subject, and a second calibration check. The calibration checks must be within approximately 10% of the target value and the two breath test results must agree within 15%. Additional checks are usually made to detect mouth alcohol (residues from recently ingested materials containing alcohol) and insufficient breath samples. The instrument prints out a record of the test and its results plus any error messages or warnings.

The reliability of these instruments has increased compared to previous versions, and problems caused by interfering substances in breath, such as acetone in the breath of diabetics, have largely been overcome.

14.5.2 Road Traffic Safety

14.5.2.1 Legislation in the United Kingdom and Europe

The relevant legislation includes the Road Traffic Act 1988 as amended by the Road Traffic Act 1991, The Road Traffic Act 1988 (Prescribed Limit) (Scotland) Regulations 2014, the Railways and Transport Safety Act 2003, The Drug Driving (Specified Limits) (England and Wales) Regulations 2014 and the Drug Driving (Specified Limits) (Scotland) Regulations 2019. Originally the 1988 Act Section 4 deals with driving, or being in charge, when under influence of drink or drugs, this is more commonly referred to as the "impairment" charge with the identity and concentration of alcohol or drugs less important, with more emphasis on the individuals impairment to drive. Section 5 deals with driving or being in charge of a motor vehicle with alcohol concentration above prescribed limit. The 2003 Act legitimized the use of oral fluid samples in addition to blood and urine. Samples submitted under Section 4 are usually analysed for the presence of drugs since alcohol is covered by Section 5 and is mostly analysed in breath.

In December 2014 a change to the drink driving legislation, applicable to Scotland only, came into effect, this saw a reduction in the drink driving limit from 80 mg% to 50 mg% in blood. This lower alcohol limit is viewed as a "zero tolerance" approach to drinking alcohol and driving. Whilst the number of recorded offences relating to drink driving did not change significantly, some would suggest this shows a change in attitudes and has therefore had a positive impact on society. There have also been calls for the rest of the UK to follow suit in this reduction and bring the whole of the UK under the same limit which has also been used across many other European countries for several years.

The most recent significant change to driving legislation was the introduction of prescribed limits for drugs whilst driving in England

and Wales coming into force in March 2015 (The Drug Driving (Specified Limits) (England and Wales) (Regulations 2014)) and subsequently the same limits coming into force in Scotland in October 2019 (The Drug Driving (Specified Limits) (Scotland) (Regulations 2019)).

14.5.2.2 Driving Under the Influence of Drugs (DUID)

The roadside test device for alcohol described in Section 14.5.1.5 has proved to be an effective and successful tool for law enforcement, creating a demand for a similar device for the detection of DUID. Commercial development of test methods initially evaluated urine 'dipstick' tests in which a urine sample is collected from a driver and tested for drug presence using a disposable device resembling the home pregnancy test and based on an immunoassay technique (ELISA) described earlier. Similar tests were also developed for oral fluid and sweat, along with handheld electronic readers to provide an objective indication of the test result. These systems are now in use throughout the world and have been found to be effective in deterring drug driving, at least in the initial period after their introduction when public awareness was at its highest.

It is important to note that Section 4 of the UK Road Traffic Act is primarily concerned with the prevention of driving while the driver is impaired due to drink or drugs rather than with driving with drugs present in the blood circulation (a 'zero tolerance' approach). Section 4 requires an assessment to be made of whether or not a driver is impaired, which is subjective and difficult to do at the roadside. The preliminary impairment test (also known as the field impairment test) was introduced in the Railways and Transport Safety Act 2003 to provide the police with a way of assessing impairment, based on familiar tests of physical and mental ability such as balancing on one foot, walking in a straight line and understanding and following instructions. If the preliminary test indicates impairment, a medical examination follows in the police station. Outside the UK, the problem of assessing impairment has been circumvented by the introduction of zero-tolerance legislation, which creates offences based only on the presence of specified controlled drugs in the driver and does not attempt to create limits as for alcohol.

The new Drug Driving legislation across Scotland, England and Wales provides quantitative concentrations of prescribed and non-prescribed drugs in blood as shown in Table 14.3 below, for which a statutory offence will be considered. Drugs either not routinely

Table 14.3 The limits in blood for drugs routinely prescribed and those not routinely prescribed. Levels exceeding these will be considered as a statutory offence under the Drug Driving legislation now introduced in both Scotland and, England and Wales.

Drug not routinely prescribed	Drug limit in blood (ng mL^{-1})	Routinely prescribed drug	Drug limit in blood (ng mL^{-1})
Benzoylecgonine	50	Amphetamine	250
Cocaine	10	Clonazepam	50
Δ^9-Tetrahydrocannabinol	2	Diazepam	550
Ketamine	20	Flunitrazepam	300
Lysergic acid diethylamide	1	Lorazepam	100
Methylamphetamine	10	Methadone	500
Methylenedioxymethamphetamine	10	Morphine	80
6-Monoacetylmorphine	5	Oxazepam	300
		Temazepam	1000

prescribed or illicit have relatively low concentrations giving effectively zero tolerance to their use whilst driving. However, clearly many drugs are prescribed regularly and therefore it is important to have a balanced approach to assigning prescribed limits, this is evident by the much higher concentrations stated for these drugs.

14.5.3 Alternative Specimens

14.5.3.1 Hair

Hair has gained popularity as a matrix for drug testing because it can be used to distinguish between chronic and naive (first time) use. Cranial (head) hair is most common, but axillary or pubic hair can also be used. Drugs circulating in blood diffuse into the hair root and are trapped in the keratinized shaft of the hair as it grows. This leads to the parent drug being more prevalent than its metabolites. The mechanism by which drugs are bound to hair is still not fully understood, but it is related to the melanin (pigment) content. Hair with a lower melanin content, *i.e.* blonde hair, has less potential to bind drugs than hair with a higher melanin content, *i.e.* dark hair. Also, basic drugs (a large group which includes most drugs) are positively charged in the blood and bind more strongly to hair than acidic drugs (a smaller group which includes some sedatives, anti-inflammatory drugs and drug metabolites) which are negatively charged. In this way hair incorporates a record of drugs that have been in circulation during its growth period.

A huge advantage of hair is the long time window of detection for drugs. Hair grows at approximately 1 cm per month, therefore a 12 cm length of hair will represent about 12 months of hair growth. Segmental analysis can be performed on the hair; this involves cutting the hair into smaller lengths and analysing each separately. The length of the segments will depend on the reason for the analysis but for longer lengths of hair 3 cm segments are usually used and 1 cm segments for shorter lengths. The number of segments is also limited by the amount of hair available, since approximately 20 mg of hair are required for an analysis.

External contamination of the hair is a major consideration when carrying out hair analysis. Sources of contamination can range from drugs smoked in close proximity to the sample donor to sweat from a parent touching a child's hair. Hair samples must be 'washed' before any analysis is carried out to try to eliminate the possibility of contamination. The samples are typically immersed sequentially in a solvent such as dichloromethane and sodium dodecyl sulphate (SDS) buffer, and ultrasonicated for a short time. The wash liquid is then removed and the process repeated. The washes can be analysed to see if any drugs are present, but if the washes are negative no further washing is required. A compromise needs to be reached between removing as much external contamination as possible without losing any drugs from the inside of the hair. Usually, the presence of metabolites in hair confirms that drugs detected have been ingested rather than deposited from outside. However, some drugs readily degrade to their metabolites, *e.g.* cocaine hydrolyses to benzoylecgonine, which can make it hard to distinguish between ingestion and external contamination. After the wash procedure, drugs are extracted from the hair by prolonged immersion in a suitable solvent and analysed using conventional methods.

14.5.3.2 Oral Fluid

Testing oral fluid for drugs has increased exponentially over recent years. It has many advantages over other matrices, especially in its ease of collection and representation of recent drug use. Oral fluid can be described as the mixture of fluids and other components sampled from the oral cavity. It is made up mostly of saliva, with other fluids such as oral mucosal transudate and cell and food debris. Drugs in the circulation diffuse through lipid membranes in the salivary glands into saliva. The chemical nature of the drug and the pH of the blood and saliva affect how well the drug moves from blood into saliva. Some drugs are highly protein bound in blood, *e.g.* diazepam, resulting in very low concentrations being found in the saliva.

As in hair analysis, oral contamination is also an issue in oral fluid analysis. In this instance the difficulty lies in distinguishing between drugs which have diffused into saliva from blood and those which have come from contamination of the oral cavity through smoking or ingestion. It is well accepted that the active constituent of cannabis, Δ^9-THC, in oral fluid comes purely from contamination of the oral cavity when cannabis is smoked as opposed to diffusion from the blood.

Expectorate (spitting) samples are unpleasant to collect for the donor and the collector and therefore are not often used. Several commercial companies have developed different collection devices, most of them centred on a sterile fibre pad which is placed in the mouth and left for a period of time to absorb oral fluid. This pad is then placed in a tube, usually with some buffer to preserve the sample before the analysis is done. Oral fluid analysis is comparable to blood analysis with respect to sample preparation; however, the concentrations found in oral fluid tend to be much lower than in blood.

Because of its ease of collection and the possibility of detecting recent drug use, oral fluid is now widely used at the roadside to test drivers suspected of being under the influence of drugs. In the UK the Railways and Transport Safety Act 2003 made testing oral fluid for drugs at the roadside legally possible. Several commercial kits for roadside testing of oral fluid for drugs of abuse are available, consisting of a collection device and an immunoassay-based presumptive test which gives a positive or negative reading for one or more drug groups within a few minutes. These are only screening tests and any results need to be confirmed by the use of laboratory techniques based on mass spectrometry.

14.5.4 Drug Overdose Case Studies

Recent substance-related deaths in the UK have often involved illicit drugs including diamorphine (heroin), methadone, cocaine and amphetamine derivatives such as methylenedioxymethamphetamine (MDMA, 'ecstasy'). In these cases, the role of the toxicologist is central to the investigation of the case but often problems arise in court in providing the clear-cut interpretations needed by the legal process. This problem is compounded by polydrug use, in which the effects of simultaneous administration of more than one drug are difficult to predict. This problem is compounded further by the use of Novel Psychoactive Substances (NPS) or "legal highs" where there is usually very limited information on the effect on the human body.

The pharmacology of the drugs involved in the following case studies can be obtained from other sources.

14.5.4.1 Case Study: A Heroin-related Death

In this case, a known intravenous drug user was found dead at home with a syringe beside him. Some heroin powder was also found nearby. Interviews with his friends suggested that he had not injected for some time. At autopsy, death was ascribed to the inhalation of gastric contents, due to drug intoxication (confirmed by subsequent toxicology). A blood sample taken at the autopsy was found to contain 0.05 mg 6-monoacetylmorphine per litre of blood, 0.6 mg morphine per litre of blood and 0.15 mg morphine-glucuronides per litre of blood. In addition, a low concentration of codeine was detected, 0.07 mg per litre of blood. The heroin was analysed and found to contain 40% diamorphine by weight (diamorphine is the pharmacological name for the active constituent of heroin: the normal diamorphine content of 'street' heroin in the UK is 30–50%, but in the locality of the case, the average purity was 10–15%).

In this investigation, there is a clear indication that drugs of abuse might be involved, and the toxicologist was specifically directed to analyse the blood sample for the common street drugs. Heroin is rapidly metabolised in blood to 6-monoacetylmorphine (6-MAM), morphine and morphine-glucuronides (Figure 14.3) and these are the usual target substances for the analyst. Codeine is a member of the group of opiates but is not a metabolite of heroin or morphine but is derived from codeine originally present in opium from which heroin is produced. The illicit manufacture of heroin does not purify the product and the codeine remains in the heroin as an impurity. The presence of codeine indicates, but does not prove, that heroin was administered rather than, say, morphine tablets. The best indication that heroin was used comes from the presence of the heroin metabolite 6-MAM in blood or urine.

The blood sample was screened using an immunoassay method, which gave a positive result for opiates but not for any other illicit drug groups. This test did not indicate which opiate drug was present since it lacked specificity. As a result, attention was focused on determining which type of opiate was present, and the sample was analysed by LC-MS following the preparation of a basic drug extract. Morphine which is unconjugated is represented as 'free' morphine which the morphine-glucuronides represent further metabolism. These analyses produced data for the positive identification of heroin metabolites and codeine and for the quantities present.

Interpretation of the results requires a knowledge of morphine concentrations found in clinical patients treated with morphine or diamorphine (the mean and range for normal therapeutic concentrations) and also morphine concentrations found in overdose cases. Consultation of a pharmacology text, pharmacopoeia or specialised textbooks indicates that blood concentrations of morphine found in normal therapy are usually less than 0.1 $mg\,L^{-1}$, although can be very high in cancer patients treated long-term with morphine, due to development of tolerance. Blood levels found in heroin-related deaths are usually above 0.2 $mg\,L^{-1}$ although the range is from zero to several milligrams. Many published studies of heroin-related deaths report cases in which no morphine was detected in blood. The reasons are not known but may include failure of the heroin to be uniformly distributed in the circulation, because of death occurring very suddenly. On the other hand, very high concentrations have been found in regular injectors who are tolerant to the drug.

In the present case, the concentration of 0.6 mg morphine per litre of blood was clearly above the normal therapeutic range and provided a plausible cause of death. Regular users of heroin might live at such a concentration, but the deceased in this case had not injected for some time and had presumably lost his acquired tolerance. He might also have overdosed because the heroin used was of higher purity than normal in his area. The cause of death was attributed to inhalation of gastric contents: this is a common finding in drug-related deaths, as the normal swallowing reflex is lost as a result of depression of the nervous system by the drug. Also, nausea is a side effect of heroin use, especially in new users.

Additional information comes from the presence of 6-MAM in the blood sample and from the morphine-glucuronides result. 6-MAM is the initial metabolite of heroin and is rapidly converted to morphine. As a result, it can only be detected for a short time, if at all, after administration of heroin. Its presence in blood indicates that the deceased died a short time after administration of the drug. Morphine-glucuronides can give an indication of survival time after administration of heroin. In this case morphine was present mostly in the free form and only 20% had been converted to the conjugate form, indicating that the deceased has died less than 3 h (one half-life of morphine) after administration of heroin.

During examination of this case in court, the toxicologist concerned was asked if the morphine concentration found in blood was a fatal concentration. It should be clear that a simple yes/no answer is inappropriate and that some qualifications need to be added. However, the toxicologist was able to confirm that the morphine concentration was in the range found in previous heroin-related deaths and could provide a cause of death, when taken into account with the other information available.

14.5.4.2 Case Study 1: Polydrug Fatality

A man in his twenties died following administration of heroin supplied by a friend, who shared half of the heroin 'deal', a small bag of heroin. The deceased was on a methadone maintenance programme at the time and also used diazepam and alcohol. The friend was charged with culpable homicide on the basis that he supplied a drug which caused death. Autopsy findings included a recent injection mark on the inner aspect of the left arm and the presence of significant pulmonary congestion and oedema, typical of respiratory depression.

The autopsy blood sample was found to contain morphine 0.15 $mg L^{-1}$, diazepam 0.6 $mg L^{-1}$, methadone 0.3 $mg L^{-1}$ and alcohol 120 mg%. Morphine concentrations were discussed in the previous example. Methadone concentrations in maintenance patients average around 0.3–0.35 $mg L^{-1}$ blood but can be much higher depending on the dose. In 189 male fatalities attributed to methadone alone, the methadone concentrations in post-mortem blood ranged from 0.08 to 4.6 $mg L^{-1}$ with a mean of 0.82 $mg L^{-1}$ and a median of 0.63 $mg L^{-1}$. Diazepam concentrations in blood following regular high dose administration (30 mg per day) are in the range 0.7–1.5 $mg L^{-1}$.

None of the substances detected in this case were present at a high concentration which might reasonably have caused death on its own, bearing in mind the wide range of morphine concentrations found in heroin and methadone-related deaths, which overlap substantially with therapeutic concentrations. The cause of death in polydrug cases is usually attributed to the combined effects of several CNS depressants—all four drugs in this case. However, the prosecution contention was that it was the heroin that caused the death and that the friend was culpable. Two questions need to be answered. Would the deceased have died if heroin had not been administered? Would the deceased have died in the absence of the heroin? Different outcomes have been obtained for cases of this type depending on the background circumstances and on how the court decides on these two questions.

14.5.4.3 Case Study 2: A Paracetamol Overdose

A middle-aged woman with a history of suicide attempts was admitted to hospital at 9.30 pm after taking an overdose of paracetamol and an antidepressant (prochlorperazine) earlier that evening, at 6 pm. A sample of blood was taken for analysis in the hospital laboratory while normal supportive treatment (gastric lavage) was carried out in the ward. At 1 am, the lab result indicated a level of 270 mg paracetamol per litre of blood. Medical intervention was indicated by this result and the drug acetylcysteine was administered. Unfortunately, a 10-fold overdose of this drug was administered despite the dose being queried by the pharmacist and duty nurse. A booster dose of the drug was subsequently administered, also in a 10-fold overdose, at 4 am. The patient died at 6 am of a cardiac arrest.

The toxicological investigation of the case following an autopsy covered a full screen for drugs in the autopsy blood sample, including paracetamol and basic drugs (which include antidepressants). The results were 102 mg paracetamol per litre of blood and 0.07 mg prochlorperazine per litre of blood. All other analyses were negative.

Interpretation of these results must first deal with the drugs individually. Paracetamol causes liver failure and death (after 2–3 days) if taken in a high dosage. There are clear guidelines available to physicians which relate the blood level of paracetamol to the time that has elapsed since the drug was taken and to the probability of liver failure. In the present case, a blood level of 270 $mg\,L^{-1}$, 4 h after the drug was ingested, indicated a high probability of fatal liver damage. The treatment to be followed is to administer a drug such as acetylcysteine which protects the liver from such severe damage. The level of paracetamol measured in the autopsy blood specimen was 102 $mg\,L^{-1}$, indicating a drop of 170 $mg\,L^{-1}$ in 8 h approximately. If the deceased's liver had been functioning normally, the level should have fallen to below 70 $mg\,L^{-1}$ (the half-life of paracetamol is 4 h). Incipient liver damage is indicated by this result.

The inadvertent overdose of the antidote, acetylcysteine, resulted from poor labelling of the drug formulation and from subsequent miscalculation of the dose. Several similar cases have been published in the medical literature, which led the manufacturer to design new packaging. There was no analytical procedure available for this drug and it was not measured in the autopsy blood specimen. The small amount of toxicity information available related to tests on dogs, although this indicated that the drug was safe even at high doses.

Prochlorperazine taken on prescription produces a therapeutic range of 0.01–0.04 mg per litre of blood. In the present case, the deceased had survived about 12 h after taking the drug. The half-life is approximately 6 h so the drug at its highest level might have been 2–4 times higher (*i.e.* up to 0.28 mg L^{-1}), which places it in the range associated with toxicity.

Conclusions reached on the basis of toxicology alone are unclear, especially concerning the role of the drug acetylcysteine in the death, in the absence of toxicity data. Paracetamol and prochlorperazine were at concentrations likely to cause toxicity. A death certificate was issued ascribing the death to drug overdose and in the subsequent inquiry, the court accepted this as the cause of death.

14.5.5 DUID Case

A young woman was accused of causing death while driving under the influence of drugs. She failed to negotiate a bend on the road and her car left the road and hit a pedestrian on the pavement. Approximately 2.5 h after the accident, a blood sample was obtained from the accused by a forensic medical examiner. On analysis this was found to contain a trace of alcohol, not less than 2 mg per 100 mL blood; MDMA (ecstasy), 0.26 mg L^{-1} of blood; diazepam, 0.12 mg L^{-1} blood; desmethyldiazepam (a metabolite of diazepam), 0.33 mg L^{-1} of blood; Δ^9-tetrahydrocannabinol (Δ^9-THC, active constituent of cannabis), trace amount detected; and 11-nor-Δ^9-tetrahydrocannabinol-9-carboxylic acid (metabolite of Δ^9-THC), 0.13 mg L^{-1} blood.

This is obviously another case of polydrug use. In this case the important question to consider was what effects the drug combination was having on the driver at the time of the accident. The elapse of 2.5 h before a blood sample was obtained and the effect of this delay on blood concentrations also had to be considered. Since the sample was taken from a living subject and the timescale was known, it was possible to carry out back-calculations to estimate the concentrations of some of the substances at the time of the accident:

- Alcohol. The average rate of clearance of alcohol from blood is 19 mg% per hour. Over a period of 2.5 h, the blood alcohol concentration would on average have fallen by 45 mg 100 mL^{-1}. The measured alcohol concentration in her blood specimen was not less than 2 mg 100 mL^{-1} blood and so at the time of the accident it is estimated as having been $2 + 45\frac{1}{4} = 47$ mg 100 mL^{-1} blood.

A concentration of 47 mg alcohol per 100 mL blood is below the prescribed limit of 50 mg 100 mL^{-1} blood in Scotland. This calculation assumes (1) that the driver was in the elimination phase of alcohol metabolism, *i.e.* any alcohol consumed before the accident had been completely absorbed into her system and was in the process of being cleared from her system; (2) that the average rate of alcohol elimination applies: in practice, alcohol clearance slows down as the alcohol concentration in blood approached background levels, as in this case.

- MDMA. The average rate of clearance from blood is reflected in the plasma half-life, the time taken for the concentration in plasma to decrease by 50%. In the case of MDMA, the half life is 6–9 h. Back-calculation gives an estimated blood concentration at the time of the accident of 0.32–0.35 mg L^{-1} blood, depending on the half-life value used.
- Diazepam and its metabolite, desmethyldiazepam. Diazepam clears slowly from blood, with a half-life in the range 21–37 h. Back-calculation gives an estimated blood concentration at the time of the accident of 0.13 mg L^{-1} blood. Desmethyldiazepam is a metabolite of diazepam and is an active sedative/tranquiliser drug in its own right. This also clears slowly from blood with a half-life of 50–99 h. Back-calculation gives an estimated blood concentration at the time of the accident of 0.34 mg L^{-1} blood.
- Δ^9-THC. This is the active drug substance in cannabis. It clears rapidly from blood and is often not detected in road traffic cases. A trace of this substance was detected in the blood sample, which indicates that there would have been a higher concentration of this active cannabinoid in the driver's blood at the time of the accident. It was therefore likely that the accused was driving whilst under the influence of the active drug constituent of cannabis. It is usually difficult to detect Δ^9-THC in whole blood samples after about 3 h have passed from when cannabis was used. Under favourable conditions, *e.g.* when controlled administration of cannabis takes place under laboratory conditions and fresh plasma is collected, the detection time may be longer, up to 6 h or so, but this is not necessarily the case when whole blood is analysed.
- Δ^9-THC metabolite. This has no drug activity. Its usual significance is as an indicator that cannabis has been administered at some previous time. This substance persists in the blood for many hours after cannabis has been administered and can be detected for up to 24 h, possibly longer.

Alcohol, MDMA, diazepam and cannabis all have the potential to cause driver impairment.

14.5.5.1 Conclusions

The combined effects of this polydrug mixture would include aspects of each of the individual drugs, making it difficult to predict the net effect. None of the drugs detected was present at the high concentration associated with overdose and toxic side effects, but they were present at concentrations typical of therapeutic use (diazepam) or recreational drug use (alcohol, MDMA and cannabinoids). Alcohol (even at low concentrations) is sedative in nature, as are diazepam, desmethyldiazepam and Δ^9-THC. Diazepam and desmethyldiazepam can cause sedation, especially in new users of the drug. Even single doses of diazepam have been shown to cause significant impairment due to CNS depression. Δ^9-THC causes intoxication resulting in physical and mental impairment which have been shown to increase the likelihood of accidents when driving, especially in combination with other drugs such as alcohol and benzodiazepines. A single dose of cannabis can cause impairment when driving for up to 3 h. By contrast, MDMA has the effects of a stimulant (feeling of energy, alertness, *etc.*) as well as being a mild hallucinogen (affecting perception of the environment, mood, *etc.*). The main effect manifests as euphoria. Stimulants can adversely affect driving performance in different ways: (1) by diminishing the skills needed by the driver for accurate control of the vehicle, (2) by increasing risk-taking behaviour and (3) by causing fatigue after the initial stimulation wears off.

It is difficult to predict the effects of drugs on the individual from their blood concentrations alone, especially when several drugs are taken at the same time. However, the drugs found in the present case have the potential to cause impairment of a driver, especially because of the simultaneous presence of four drugs (alcohol, diazepam, desmethyldiazepam and Δ^9-THC) which depress the CNS. Taken together, the blood analysis results would support the opinion of a forensic medical examiner or police officer made at the time that there was a condition present in the driver which could be due to drugs.

14.5.6 Hair Analysis Case

A man in his mid twenties was found dead at home. He had been released from prison 6 weeks before his death. Analysis of an autopsy blood sample found diazepam 0.4 mg L^{-1} blood; desmethyldiazepam 1.7 mg L^{-1} blood; methadone, 0.3 mg L^{-1} blood and methadone metabolite (EDDP) 0.3 mg L^{-1} blood. The cause of death was given as methadone and diazepam intoxication.

A 4 cm hair sample taken at autopsy was segmented into four 1 cm portions, each of which related to a 1 month period prior to death. However, the hair did not cover the 1–2 week period prior to death as the roots were not included. Analysis of the hair revealed the pattern of drug use by the deceased (Figure 14.4): this included the presence of heroin metabolites in all four hair segments but methadone only in the segment closest to the scalp. The presence of drug metabolites indicates that the drugs were consumed and were not due to superficial contamination. The results should not be over interpreted, but they do indicate that the deceased was a regular heroin user (from the presence of 6-MAM) and that he had not used methadone before his release from prison.

14.5.7 Fires and Explosions

In the investigation of a fatal fire, the role of the toxicologist is often to establish the presence of carbon monoxide in the blood of the fatality. This has an important medico-legal significance as, taken in conjunction with the pathologist's observations concerning the presence or absence of soot in the air passages, it establishes whether or not the deceased was alive at the start of the fire. If death occurred before the fire, the possibility of homicide must be considered.

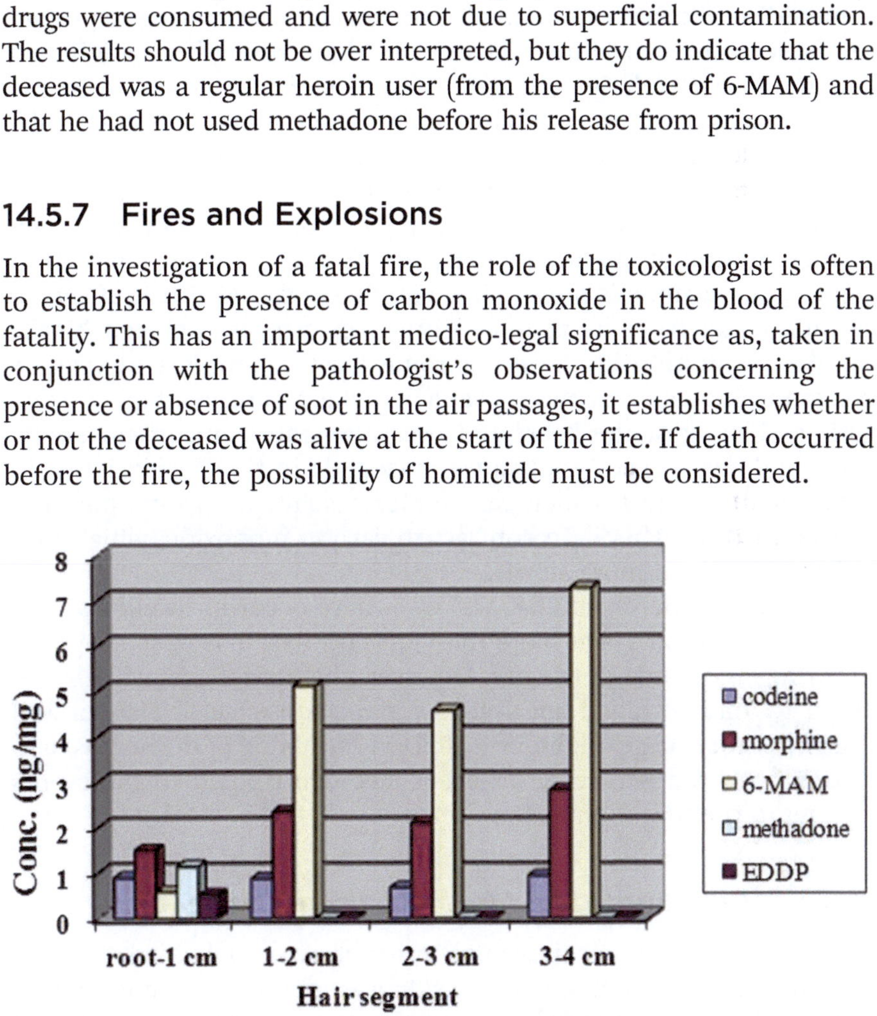

Figure 14.4 Analysis of a hair sample from a fatality.

Carbon monoxide is a colourless, tasteless, odourless gas produced by partial combustion of carbon-containing materials. It is produced almost immediately during the course of a fire and those present in the fire locus are exposed essentially from the beginning of the fire. The gas is inhaled and absorbed through the lungs into the blood circulation system. The significance of carbon monoxide in fires is that casualties can be overcome by the effects of the gas while asleep or at rest, perhaps unaware that a fire has started. If they subsequently become aware of the fire and start to exert themselves, the sudden demand for more oxygen can cause fainting and collapse, resulting in failure to escape.

In addition to carbon monoxide, the toxicologist may be requested to measure other toxic materials in the blood which have originated in the fire, particularly hydrogen cyanide, which is produced by the effects of heat on nitrogenous materials such as polyurethane, polyacrylonitrile, wool and silk. Cyanide levels are usually measured in whole blood specimens, even if plasma or serum is available. Cyanide is present at low levels (up to 0.25 $mg\,L^{-1}$ of blood) in the normal population and at slightly higher levels (up to 0.5 $mg\,L^{-1}$ blood) in smokers, as cyanide is present in cigarette smoke. The toxic effects of cyanide result from its action within body tissues, where it blocks cellular respiration at the cytochrome level. Symptoms of cyanide poisoning include dizziness, headache, chest pain, confusion, staggering, slowing of the heart and breathing rates, unconsciousness, coma and death. The significance of cyanide in fires lies in its rapid incapacitating effect, which can render casualties unconscious in a very short time. This is in contrast to carbon monoxide, which may build up in blood more slowly.

Most fire deaths in the UK and elsewhere occur in dwellings, and research has shown that more than 80% of these have been exposed to hydrogen cyanide in the fire. However, the toxicologist will also be asked to analyse blood specimens for the presence of alcohol and drugs, because these might have affected the ability of the deceased to escape from the fire scene or even have contributed to the events leading to the initiation of the fire.

14.5.7.1 Case Study 3: A Fire in a Leisure Centre

A fire occurred in a leisure centre in which six people died. The fire was small and did not spread outside a single room. The casualties occurred because of the spread of smoke, heat and toxic gases along a corridor, blocking the escape of the occupants who had been inside

Table 14.4 Carbon monoxide and cyanide levels in fatalities resulting from a fire in a leisure centre.

Casualty	% Combined haemoglobin–carbon monoxide in blood (%HbCO)	Cyanide level (mg L^{-1})
1	48	0.9
2	48	0.8
3	41	3.1
4	34	0.5
5	29	2.0
6	38	2.3

rooms. The results of the toxicological analysis of blood samples are given in Table 14.4 and show that the casualties had breathed in significant amounts of both carbon monoxide and cyanide, leading to their collapse and death.

The toxicologist can also contribute to the investigation of deaths resulting from explosions, as many of these fatalities are actually caused by a fire accompanying the explosion. Also, analysis of blood from fatalities in a gas explosion may assist the explosion investigator in identifying the nature of the gas involved. Usually this is either consumed in the explosion or dissipated after the event and cannot be detected by the investigator at the locus. However, it may still be detected in the blood of the fatalities who were alive before the explosion and inhaled the explosive mixture of gas and air involved.

14.5.7.2 Case Study 4: A Gas Explosion in a Dwelling House

Early one winter morning, a gas explosion partially demolished a building containing four flats in a city suburb. Five people were killed, including the four members of one family in a ground-floor flat (fatalities 1–4) and the resident of an upper flat (fatality 5). The causes of death were established at autopsy as blast injury, crush asphyxia (due to collapse of the building) and inhalation of smoke and fire gases from the subsequent fire.

The initial investigation produced three possible sources of flammable gas: the mains gas supply (methane), old coal-mine workings under the flats (firedamp, a mixture of methane and ethane), and bottled gas (butane). Subsequent investigation revealed that the actual source was the mains gas supply. The type of gas was confirmed by the toxicological analysis of the blood specimens taken at autopsy, which showed the presence of methane only; see Table 14.5. Also, the levels of methane were in inverse relationship to the levels of carbon

Table 14.5 Carbon monoxide and methane levels in fatalities resulting from an explosion and fire in a dwelling house.

Fatality	Combined haemoglobin–carbon monoxide in blood (%HbCO)	Methane (mL L^{-1})
1	2	1.14
2	8	Trace
3	11	0.28
4	32	0.15
5	71	0.04

monoxide, indicating which of the fatalities had survived longest after the explosion. The methane in the building had been vented by the explosion and survivors would have breathed out any gas in their blood, but simultaneously would have inhaled carbon monoxide from the fire which was ignited by the explosion.

The father of the family appeared to have initiated the explosion with his activities in the kitchen, and was found to have suffered blast injuries but not to have inhaled any carbon monoxide. By contrast, the upstairs neighbour (fatality 5) did not have any methane in her blood sample but had the highest level of carboxyhaemoglobin.

Further Reading

R. C. Baselt, *Disposition of Toxic Drugs and Chemicals in Man*, Biomedical Publications, Seal Beach, CA90740, 12th edn, 2020.

Goodman and Gilman's The Pharmacological Basis of Therapeutics, ed. L. Brunton, B. A. Chabner and B. Knollmann, McGraw-Hill, New York, 13th edn, 2018.

Clarke's Analysis of Drugs and Poisons, Pharmaceutical Press, London, 4th edn, 2011.

O. H. Drummer, *The Forensic Pharmacology of Drugs of Abuse*, Arnold, London, 2001.

S. B. Karch, *Karch's Pathology of Drug Abuse*, CRC Press, Boca Raton, FL, 5th edn, 2016.

P. Kintz, *Analytical and Practical Aspects of Drug Testing in Hair*, CRC Press, Boca Raton, FL, 2006.

B. Levine and S. Kerrigan, *Principles of Forensic Toxicology*, Springer Nature Switzerland AG, 4th edn, 2020.

S. Davies, A. Johnston and D. Holt, *Forensic Toxicology: Drug Use and Misuse*, RSC, 2016.

15 Explosions

Sharon Broome,* Clifford Todd, Linda Jones and Maurice Marshall

Defence Science and Technology Laboratory, Porton Down, Salisbury, Wiltshire, SP4 0JQ, UK
*Email: smbroome@dstl.gov.uk

15.1 Introduction

The investigation of accidental or illegal explosions, and the scientific analysis of their causes, has a long history which in the UK started formally in 1871. An explosion at a factory in Stowmarket making guncotton led the then Home Secretary to instigate an inquiry. A Royal Engineer, Captain Vivian Majendie, who was an expert on explosives, led the investigation. He recruited a chemist, Dr August Dupre, to assist. The arrangement proved so successful that the Home Office decided to continue it, leading to the present-day Forensic Explosives Laboratory (FEL). One of the first fruits of their collaboration was the 1875 Explosives Act which embraced various aspects of explosives including their manufacture and storage. Subsequent bombing outrages led Parliament to enact the 1883 Explosive Substances Act which deals with the criminal use of explosives and devices; this Act was intended specifically to deal only with the most serious offences, hence the special provision made requiring the fiat of the Attorney General for any prosecution under it. More recently, these Acts were amended by the Manufacture and Storage of Explosives Regulations 2005 and subsequently replaced by the Explosives Regulations 2014.

Crime Scene to Court: The Essentials of Forensic Science, 5th Edition
Edited by Niamh NicDaeid and Peter C. White

Published by the Royal Society of Chemistry, www.rsc.org

Many of the questions which Captain Majendie and Dr Dupre were asked to address are still relevant to today's forensic explosives investigators, including the following:

(1) Was it an explosion?
(2) Was it an accident or a bomb?
(3) Is this an explosive?
(4) Was this a viable device, or a hoax?
(5) Are these items or materials intended for making explosives or bombs?
(6) Could these instructions be used to manufacture explosive materials or devices?
(7) Has this person been in contact with explosives?
(8) Have explosives been stored in this place?
(9) Are there similarities between these items or incidents that link them together?
(10) Could the items have an innocent use?

Actual cases involve a variety of circumstances and the questions which are appropriate will vary accordingly.

In order to address the above points, the forensic explosives scientist requires a sound grounding in the requirements of the judicial system and the ethical principles underlying it, together with a detailed knowledge of the science and technology of explosives, practical experience in the construction of all types of explosive devices, and a clear understanding of their effects.

15.2 Explosives Technology

15.2.1 What Is an Explosion?

A convenient working definition is 'a sudden and violent release of physical or chemical energy, often accompanied by the emission of light, heat and sound'.

To the human observer an explosion seems instantaneous; however, the chemical reaction actually proceeds at a finite, albeit high, speed, progressing through the material as a definite 'front' or 'wave'. Two types of event may be defined: a 'deflagration' is an event where the decomposition within the explosive occurs at a speed equal to or less than the velocity of sound within the material, and a 'detonation' is an event where decomposition occurs at a speed greater than the velocity of sound within the material. Sometimes deflagrations are referred to as 'low-order explosions', and there is a mistaken tendency to assume that they are less serious than detonations. This is an error: many of the most devastating accidents with explosives have involved deflagrations rather than detonations.

Typically, deflagrations occur with velocities of less than 2000 m s^{-1} while detonations may reach velocities of 6000–8000 m s^{-1} for certain high-performance military explosives.

15.2.2 Types of Explosion

Explosions may be characterised by the source of their energy, *i.e.* physical or chemical, and also by their locus, *i.e.* whether dispersed- or condensed-phase.

Examples of the various classes are:

(1) **Physical.** An exploding pressure vessel, *e.g.* an overheated gas cylinder.
(2) **Chemical.** Explosion of a mass of black powder, or gunpowder as it is more commonly known.
(3) **Dispersed.** Detonation of a cloud of flour in air.
(4) **Condensed.** Detonation of a block of plastic explosive.

Each of these types can be further classified according to the speed and duration of the explosive event, and this can be linked to the type of practical effect observed, enabling the skilled investigator to draw some meaningful conclusions from examination of the types of macroscopic and microscopic damage found at the scene of an explosion. Thus, high velocity detonations cause shattering and cutting of metal, whereas low velocity events result in tearing and heaving.

15.2.3 Types of Explosives

Although dispersed-phase explosions can be of great power, as can physical explosions, the vast majority of practical explosives used for either commercial or military purposes are condensed-phase chemical explosives. These energetic materials are best considered according to their function; thus we have:

(1) **Pyrotechnics.** Used for the production of heat, light, sound or smoke, *e.g.* in fireworks or signalling flares.

(2) **Blasting explosives.** Used, *e.g.*, to break up rock in quarrying operations.
(3) **Initiatories.** Used to transform a small mechanical or thermal impetus into a violent shock wave capable of causing detonation of less sensitive energetic materials such as blasting explosives. Initiatories are also referred to as primary explosives, indicating their role at the start of a chain of explosive events (the explosives train). Analogously, the less sensitive explosives used for the main charge are referred to as secondary explosives.
(4) **Propellants.** Energetic materials which deflagrate in a controlled fashion to allow their energy to be used *e.g.* in propelling rockets or projectiles from guns.

Apart from the above scientific definitions, the forensic scientist needs also to be familiar with the legal definitions; *e.g.* Section 3 of the UK's Explosives Act 1875 states:

The term 'explosive' in this Act—

(1) Means gunpowder, nitro-glycerine, dynamite, guncotton, blasting powders, fulminate of mercury or of other metals, coloured fires, and every other substance, whether similar to those above mentioned or not, used or manufactured with a view to produce a practical effect by explosion or a pyrotechnic effect; and
(2) Includes fog-signals, fireworks, fuses, rockets, percussion caps, detonators, cartridges, ammunition of all descriptions, and every adaptation or preparation of an explosive as above defined.

This Act is updated and further expanded upon, under the Explosives Regulations 2014. Other countries have their own legal definitions. Some examples of plastic explosives (a type of secondary explosives) are shown in Figure 15.1.

15.2.4 Chemistry of Explosives

The requirements are simple: the ingredients of the explosive need to undergo some very rapid chemical reaction liberating large amounts of energy. In practice the most difficult part is achieving control of this process so that the explosive only reacts when required, and not otherwise.

The earliest known explosive was gunpowder, more correctly known as black powder, first developed by the Chinese over 1000 years ago. This is an intimate mixture of charcoal and sulphur (the fuel) and

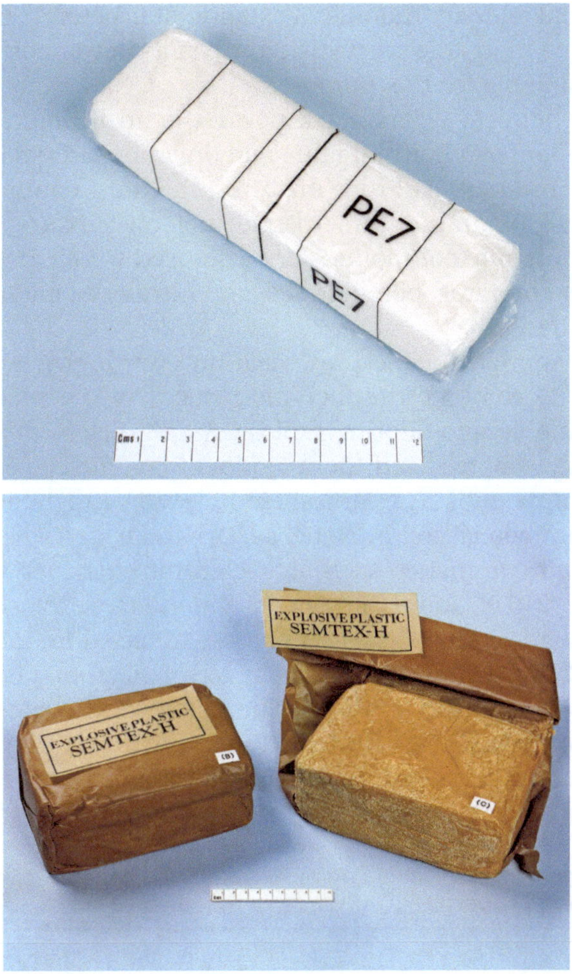

Figure 15.1 Examples of plastic explosives.

potassium nitrate (the oxidant). The chemical reactions which occur are in fact highly complex, resulting in the formation of oxides of carbon, sulphur and potassium, together with potassium–sulphur compounds, all in a range of oxidation states. Black powder illustrates the way in which the behaviour of explosives can be tailored to a particular application. If spread out in a thin layer in the open air it will merely burn violently, but if confined in a pressure-tight container it can be made to detonate if ignited. Under normal conditions black powder deflagrates and can be usefully employed as a propellant, *e.g.* in the guns of Lord Nelson's *HMS Victory* at Trafalgar, or in modern-day display rockets on Guy Fawkes' night. This versatile material is also very widely used in current military munitions as an igniter powder; the fiery particles of molten inorganic slag which form on its decomposition make it particularly suitable for this purpose.

In broad chemical terms black powder is an intimate mixture of fuels and oxidants. Other inorganic and organic materials can also be combined to produce explosive mixtures, the most common example being the use of ammonium nitrate–fuel oil mixtures (ANFO) in quarrying. Inorganic compounds are also used widely in pyrotechnic mixtures, *e.g.* chlorates, perchlorates, and nitrates of alkali metals are common oxidants.

The other approach, which can yield an even higher explosive performance, is to combine the fuel and oxidant in the same molecule. Although some organo-fluorine explosives are known, these have not been generally adopted, and the majority of organic explosives are organic nitro compounds. The earliest example was nitroglycerine (glyceryl trinitrate), commercialised by Nobel as 'Dynamite' or 'Gelignite'; others include the nitroaromatics such as 2,4,6-trinitrotoluene (TNT), or the nitramines, *e.g.* RDX. Some commercial and military explosives together with their chemical names and properties, are listed in Table 15.1.

The use of improvised explosives by terrorist groups is long established and still widespread today. The nature of the explosives used has varied somewhat with time, and with the availability of the particular precursor chemical. It was on this basis that the UK made amendments to the Poisons Act 1972 in order to restrict sales of

Table 15.1 The common and chemical names of organic explosives with their properties.

Common name	Chemical name	Properties
TNT	2,4,6-Trinitrotoluene	Low-melting solid; can be conveniently cast into bombs and shells; dissolves in organic solvents
Nitroglycerine	Glyceryl trinitrate	Viscous liquid; toxic vapour; the neat liquid is sensitive and unstable
PETN	Pentaerythritol tetranitrate	White crystalline solid
RDX	Cyclotrimethylene-trinitramine	White crystalline solid; high density; high detonation velocity; moderately soluble in acetone
HMX	Cyclotetramethylene-tetranitramine	White crystalline solid; high density; high detonation velocity; high melting point (275 °C); very stable; moderately soluble in acetone
Tetryl	2,4,6-Trinitrophenyl-methylnitramine	Yellow solid

certain materials that can be used to manufacture home-made explosives; as a consequence of which, certain specified chemicals may now only be held by members of the public after obtaining a license. Suspicious transactions of other named materials are required to be reported to the police. Examples of some improvised explosive materials are shown in Figure 15.2, with the properties of some commonly encountered materials listed in Table 15.2.

Figure 15.2 Some improvised explosives.

Table 15.2 Some improvised explosives with their properties.

Common name	Chemical name	Properties
TATP, acetone peroxide (AP)	Triacetone triperoxide	White crystalline material; very sensitive primary explosive.
HMTD	Hexamethylenetriperoxide diamine	Fine dry white powder; very sensitive primary explosive, particularly as regards impact.
DDNP	Diazodinitrophenol	Yellow–brown coloured powder; high impact and friction sensitivity.
ETN	Erythritol tetranitrate	White or colourless crystalline material; chemically similar to PETN.
Urea nitrate	Urea nitrate	Off-white–colourless crystalline powder; relatively insensitive and of comparable performance to ammonium nitrate-based explosives.

15.2.5 Initiation and Detonation of Explosives

As mentioned above, the most important practical problem in explosives technology is the safe control of initiation. This has led to the widespread use of the concept of the 'explosives train'; a small quantity of a very sensitive explosive is used to receive an initial stimulus in the form of thermal, mechanical or electrical energy and amplify this so as to start a reaction in a larger mass of less sensitive material.

This leads to the concept of the 'detonator' invented by Nobel. Commercially, three main types are produced: 'plain detonators', 'electric detonators' and 'shock-tube detonators'. Figure 15.3 shows examples of the variety of detonators available, as well as a casework example of an improvised detonator.

In electric detonators, a sealed pencil-shaped metal tube contains a metal wire filament embedded in a heat-sensitive match composition; when an electric current is passed through the wire the resulting heat from the filament ignites the match composition (the 'fuse head'). This in turn ignites the next element in the detonator. In practice this may be a delay composition (the 'delay element') used to provide greater control in the sequencing of industrial blasting operations, or the ignition may pass directly to the small charge of primary or initiatory explosive (the 'priming charge') which once ignited burns to detonation. The detonation of the primary explosive in turn causes

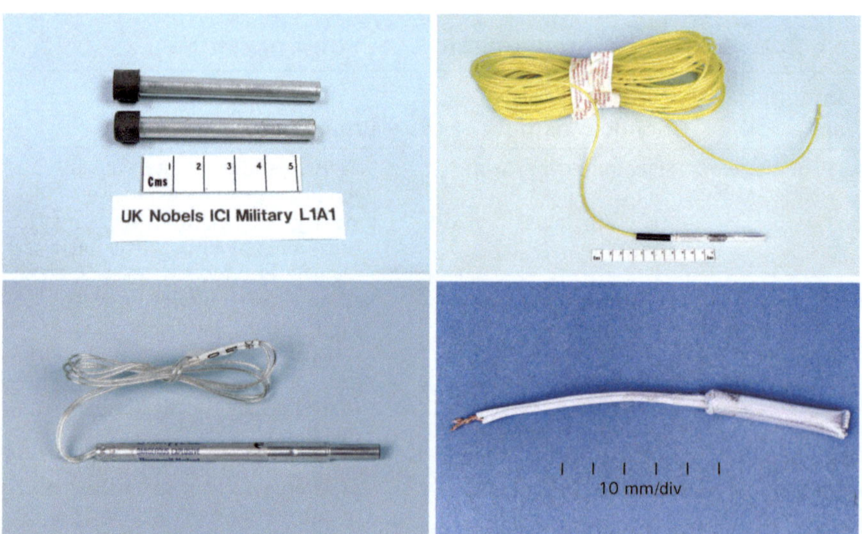

Figure 15.3 Detonators.

Table 15.3 Some explosive materials used in detonators.

Common name	Function
Mercury fulminate	Primary explosive
Lead azide	Primary explosive
Lead styphnate	Primary explosive
Potassium chlorate/lead thiocyanate	Match composition
Barium peroxide/selenium	Delay element
PETN	Secondary explosive
Tetryl	Secondary explosive

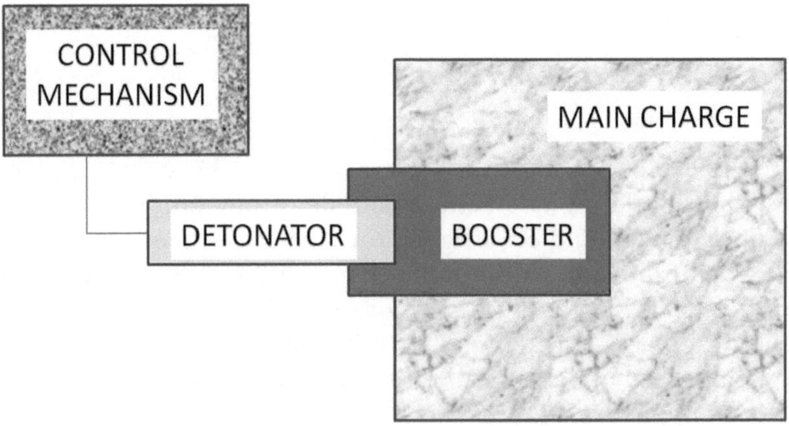

Figure 15.4 Elements of an improvised explosive device.

detonation of a slightly larger quantity of a secondary explosive (the 'base charge') which acts as a booster, amplifying the effect several-fold. In a plain detonator the wire filament is replaced by an igniferous fuse, and in shock-tube detonators a reactive explosive mixture is used to coat the inner walls of a narrow bore plastics tube, which when initiated, propagates a low energy signal at a speed of approximately 2000 m s^{-1} along the length of the tubing.

Table 15.3 lists some examples of the energetic materials used in detonators for each of the functions shown above.

15.2.6 Essential Elements of an Improvised Explosive Device (IED)

Great ingenuity has been displayed in the creation of improvised explosive devices, but common elements can be identified in terms of basic functions. Figure 15.4 shows the essential elements in schematic form.

15.3 Facilities Required for Forensic Explosives Examinations

In the UK, strict regulations apply to the keeping and use of explosives, and subject to a few exceptions, which are specified in the Explosives Regulations 2014, no laboratory or individual may possess explosives without appropriate licences. Similar regulations exist in many other countries. Apart from these legal aspects, safety considerations also need to be taken into account, and explosives should not be handled without adequate training and facilities for the purpose.

15.3.1 Safety

Explosives explode. This simple rule needs to be borne in mind at all times when handling energetic materials of any kind; all processes and equipment should be designed on the basis that at some time an unintended explosion will in fact occur and therefore the safety procedures must be such as to minimise the effect. Most nations have detailed explosives regulations covering the safe storage, transport and use of commercial and military explosives. However, these often cannot be applied to the case of illegal materials, and scientists investigating such matters must be careful to minimise hazards both to themselves and to others.

A particular hazard, in recent years, has been a significant increase in the number of incidents involving the use of home-made explosives, such as the very sensitive improvised primary explosives TATP (triacetone triperoxide) and HMTD (hexamethylene triperoxide diamine). These have been produced both for serious criminal purposes and by amateur experimenters, but in all cases are extremely hazardous, to the people making them and all who come within their immediate vicinity. Death and serious injuries have resulted from the manufacture of these materials in the UK, irrespective of the intended reason for making them. In addition, the possibility that improvised explosive devices may incorporate concealed anti-handling features should always be addressed.

Explosive devices should always be examined at the scene by properly qualified bomb disposal personnel before being disturbed in any way whatsoever; the bomb disposal personnel will also be able to provide the necessary advice on safe packaging and transport arrangements. In some cases, only a small amount of sample may be safe enough to submit to the laboratory, and possibly the entire item may have to be destroyed; recovering the debris from the controlled explosion of the material will enable subsequent residue analysis. Practical experience shows that the most dangerous devices are often those made by amateur experimenters, *e.g.* schoolchildren.

Specialist facilities must allow for four activities: receipt, storage, examination and disposal.

15.3.2 Receipt

The arrangements for receipt of items have some features in common with those general in forensic work, *e.g.* the possibility of cross-contamination between submitted items must be rigorously and demonstrably precluded. The general requirements for preserving the integrity of trace and contact evidence are discussed in previous chapters. Furthermore, the system must always preserve the chain of custody of evidence. In addition, explosives laboratories need pre-planned procedures for dealing with devices which may deteriorate into an unsafe condition or are suspected of incorporating anti-handling features. Thus immediate access to bomb disposal equipment and personnel is essential, together with radiographic facilities designed to allow safe examination of explosive devices and provide details of the internal construction of such items before disassembly.

The mechanics of receipt have to cater for a range of items which can cover anything from a small sealed package of hand swabs for trace examination to a blown-up motor vehicle, fully laden vehicle borne IED or several tonnes of bomb scene debris. In practice it is convenient to have different physical arrangements for the various streams of evidence types whether bulk or trace, explosive or non-explosive, clean or dirty.

15.3.3 Storage

Detailed safety guidelines and regulations cover the design of explosives storehouses (magazines). These require *e.g.* separation of detonators and bulk explosives; they also include limits on the mass of explosives which may be kept in particular designs of store, and also specify security features to prevent unauthorised access. Given the often uncertain nature of some of the material submitted to forensic explosives laboratories, it is useful to maintain an isolation magazine for any suspect materials.

A typical laboratory magazine is strongly constructed of reinforced concrete, provided with spark-proof electrical lighting and lightning protection, and fitted with high-security doors and locks. The whole is enclosed within a substantial barricade intended (1) to protect the magazine from an explosion in any other adjacent building, and (2) to deflect upwards the blast wave that would occur if the contents of the magazine itself were to explode.

The laboratory will also need to store non-explosive exhibits ranging from items for explosives trace analysis to bulk debris in a way that meets all the requirements for preservation of the chain of custody and protection of exhibit integrity, including cross-contamination awareness.

15.3.4 Examination

The items to be examined can be categorised as bulk explosives, invisible traces of explosives and non-explosive materials. Given the requirements to consider other evidence types, laboratory facilities may be expected to include areas designed to allow examination of exhibits that are DNA (Deoxyribonucleic Acid) sensitive, with all the associated environmental monitoring, Staff Elimination Databases *etc.* that entails.

15.3.4.1 Examination of Bulk Explosives

The design of facilities for bulk explosives is governed by both the type and mass of explosive being handled. In general, different work areas need to be provided for examination of detonators and primary explosives, pyrotechnics and incendiaries, and secondary explosives. Where practicable it is also wise to segregate activities involving large and small quantities. For example, it is often safer and more convenient to break down devices containing large explosive masses in a separate building, take small samples and then examine these in more detail in the main laboratory. Buildings for explosives work require appropriate electrical fittings to prevent spark and dust hazards, together with lightning protection and antistatic precautions; care needs to be taken when designing a facility to ensure that electro-explosive devices are examined away from areas of radio-frequency transmission activity and particular attention needs to be paid to fire prevention and escape in the building design.

In a laboratory for examining detonators and primary explosives the floor and benches are normally covered in conducting rubber sheet which is earthed to reduce hazards from sparks due to static electricity, as is all the laboratory equipment. All personnel entering the laboratory have to check themselves on a special meter to ensure that they are properly earthed. In dry climates it is necessary to humidify the laboratory atmosphere to reduce hazards from static electricity. The detonators and primary explosives are manipulated behind stout armoured screens (or possibly within an armoured cabinet) which need to be of a design that has been type-tested to prove that it really does withstand the effects of an explosion of the quantity likely to be involved. It is not sufficient to rely on normal laboratory safety screens, which are most unlikely to contain the explosive effects of a typical commercial detonator. Similarly, fume cupboards used for chemical treatment of primary explosives (which are generally toxic) should be suitably armoured with arrangements to vent blast pressure waves due to any detonation.

Special machinery is essential if explosive articles are to be opened to allow examination: lathes, cutters, manipulators and screens must all be type-tested to prove their suitability and safety before use. The superficial simplicity of the process conceals many potential hazards and it most definitely should not be attempted without proper facilities and training, *e.g.* the use of remote disassembly workshops.

15.3.4.2 Detection of Invisible Traces of Explosives

This can include swabs or other samples taken from people's hands or property, or samples obtained from items that have been involved in an explosion. As with all other contact trace evidence where the objective is to detect very small quantities of the analytical species, it is essential to carry out the examination in a controlled environment where it can be demonstrated by objective measurement that the results obtained are meaningful and not the consequence of contamination either from the general environment or other items at the laboratory.

In general, this will require precautions such as the use of clean disposable overclothing, separate work areas for different trace operations, regular environmental monitoring and cleaning, and control of access to ensure that people and items which enter the trace area are free from explosives. It should also be borne in mind that such contamination can occur at the crime scene from which the samples or items are recovered, and similar precautions must be taken there, as at the laboratory. In particular, samples for trace analysis should not be taken by people who are likely to be contaminated with explosives in their normal work environment, such as bomb disposal operators or other military personnel, and should also be taken at the earliest opportunity during the scene examination process.

15.3.4.3 Non-explosive Items

Live explosives will often be accompanied by non-explosive items, either as part of a device or in a collection of suspected bomb making equipment. These may include instructions, precursor chemicals, containers, timing devices, power supplies, arming mechanisms and switchgear. Facilities for mechanical examination and disassembly, measurement, photography and microscopy will be needed.

15.3.4.3.1 Post-explosion Debris

This could include the remains of an exploded firework or more generally an exploded device, a bomb-damaged motor car or tonnes of debris from a bombed building. Each of these evidence types requires a different approach and different facilities. Conventional well-lit, clean laboratory benches covered with fresh disposable paper will suffice for examination of the remains of a small exploded device.

Handling bomb-damaged motor vehicles requires special trolleys and lifting gear, particularly since such vehicles are likely to be structurally unsafe and may easily collapse on people nearby if lifted without properly designed equipment operated by trained personnel. Examination of bulk debris from bombed buildings requires a system for the mechanical handling of large quantities of material, together with facilities for drying, sieving and searching, a component of which is shown in Figure 15.5.

Debris from bomb scenes can present a range of hazards: there may be materials such as asbestos present in building debris, which is also liable to be contaminated with sewage; damaged vehicles may contain flammable liquids; and all are liable to be contaminated with bio-hazard materials, particularly from events where people have been injured or killed. Thus, laboratory facilities for this type of work need appropriate mechanical ventilation for dust and vapour extraction, supplemented as necessary by respiratory protective equipment. Specialist contractors may be required to assist with the examination process, whose involvement must be captured within the laboratory's Quality Management System. Vaccination of staff against a range of diseases is also advisable.

15.3.4.3.2 Instructional Material

Increasingly the police in the UK are intercepting terrorist plots at the early stages of the attack planning; this means that the forensic explosives expert will be asked to make comment on the explosives significance of a limited amount of recovered material. Often

Figure 15.5 Debris sifter.

instructional videos and/or documentation will be submitted and the expert will be required to use their experience to comment on the credibility of the information available, in order to assist the court in determining the potential harm that could have been caused had the plans come to fruition.

15.3.5 Disposal

A major consideration in explosives work of any kind is the need to provide for the safe and environmentally acceptable disposal of all waste and unwanted material. It is not permissible simply to mix explosives waste with normal waste, and most particularly not with waste chemicals. Waste explosives and explosives removed from improvised devices are likely to be less stable than newly manufactured material as a result of rough handling, inadequate storage and contamination and should therefore be treated with special care. Material may be destroyed chemically or more commonly by burning at an appropriately licensed facility.

15.3.6 Reference Collections and Databases

Reference collections are commonly used in forensic science; they are of great value in explosives work. Thus samples of compositions, and examples of packaging and labelling are all most useful in the identification of individual commercial explosives. Although the identification of an intact cartridge of explosive bearing the manufacturer's label may appear to be a trivial task, identification of the same item from a few small, damaged fragments of wrapper and a milligram of undetonated composition is a more challenging problem. Likewise, specimens of items such as detonators and blasting accessories are useful for physical comparison. Such collections of physical hardware can advantageously be supplemented with a library of commercial literature which can prove especially helpful in identifying unusual or old items.

A set of well-characterised samples of known explosives are essential as comparison and calibration standards for chemical analysis techniques. Collections of reference spectra are also helpful, but the possibility of errors in commercially published spectral libraries must be borne in mind, and it is wise to establish the authenticity of spectra used in critical work.

A carefully indexed database of devices should be maintained to facilitate correlation between incidents, yielding pointers to individuals and organisations involved in a series of outrages.

15.4 Forensic Science Questions

15.4.1 Was It an Explosion?

This may at first sight appear to be a fairly straightforward question to answer. Generally, it will involve examination of the damage at the scene, either at first hand or, less satisfactorily, by viewing photographs, reviewing witnesses' accounts of the event and laboratory examination of debris from the scene. The occurrence of sudden loud reports, flashes, violent projection of debris, smashing, tearing, or rupturing of structural materials and the formation of craters at the seat of a supposed explosion are all useful indicators. However, one instance where it can be a very difficult question to answer is in the case of air crashes, where there will be a huge amount of damaged material, only a very small proportion of which is likely to show any characteristic explosive damage. This situation is dealt with in more detail in the next section.

15.4.2 Was It an Accident or a Bomb?

This has to be a combined process both of elimination of possible accidental causes, such as leaks of flammable gases or the presence of flammable dust clouds; and wherever possible, the positive identification of physical or chemical evidence resulting from the bomb.

Physical evidence can involve recovering fragments of the bomb from the debris: contrary to popular misconception, explosions do not always vaporise everything in the immediate vicinity. Rather it is more accurate to say that 'bombs shatter and scatter'. Thus teams of trained and diligent searchers have been able to recover evidentially significant fragments of the device in many of the cases that have occurred in the UK. Such fragments may include pieces of the bomb container, the waterproof plug from the detonator, parts of any timing mechanism such as clockwork, mobile telephone, electronic circuitry, batteries, and wires.

Fragments of the device are likely to be found lodged in comparatively soft objects near the seat of the explosion, *e.g.* in vehicle tyres and seats in the case of under-car booby traps. The bodies of victims are also potentially good receptors for bomb fragments; medical staff should routinely be asked to take CAT scans (computerised axial tomography scans) or X-ray victims and pass on any fragments recovered for scientific examination, either as part of the Disaster Victim Identification (DVI) process or as part of the normal police investigation.

A small victim-actuated bomb in a book is shown in Figure 15.6, together with the accompanying bag in which it was transported. Such an item would be intended to explode when the book was opened.

Although containing only some 25 g of plastic explosive, such a device would be likely to maim or kill anyone opening it. The fragments recovered after explosion can be seen in Figure 15.7; information of considerable investigational and evidential value might potentially be obtained from such material. The metal fragments, for example, might yield chemical residues of the explosive, the battery fragments type and batch numbers; the watch might have tool marks or manufacturer's marks, and it might be possible to link the wires with similar wires from a suspect's premises. Likewise, the book and the bag might provide links to a suspect.

Incendiary devices that have partially functioned often produce a fused mass of charred and burnt material, particularly if the device

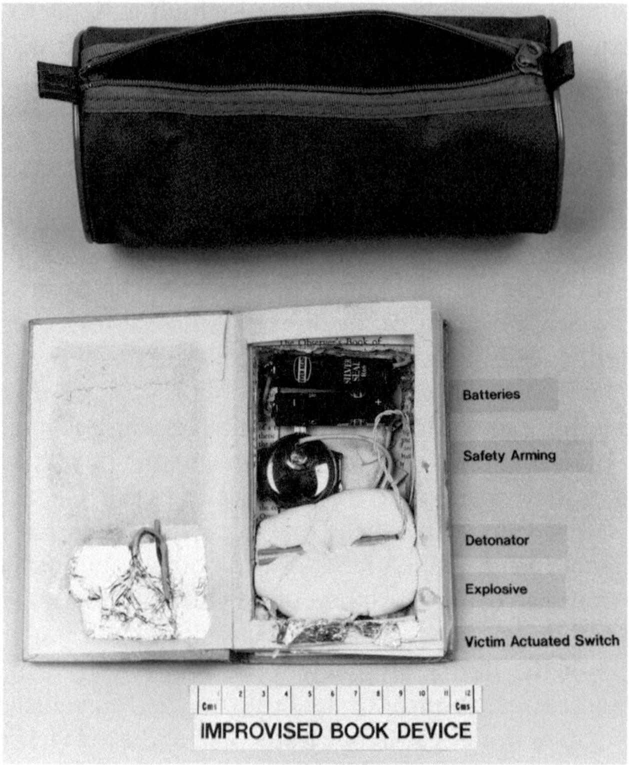

Figure 15.6 A small improvised bomb in a book.

was in a plastic container. In such cases radiography can often be used to reveal details of the construction of the device from otherwise intractable evidence.

In addition, explosions cause characteristic damage to many receptor surfaces as a result of the unique combination of high-velocity shock and transient high temperatures. Surfaces near to the seat of an explosion, at a range of a few centimetres to some metres depending on the size of the explosion, are likely to be exposed to the effects of very hot gases, and tiny particles, ejected at high velocity. Hot gases can give rise to the phenomenon referred to as 'gas-wash', which gives rise to a mottled irregular surface effect due to the partial melting and erosion of the target material. Fast moving particles can result in 'microcratering' on nearby surfaces, particularly metal. These tiny craters, characteristically with a rolled lip, range in size from fractions of a millimetre to 1–2 mm diameter. An example of a microcrater on a piece of aluminium is shown in Figure 15.8.

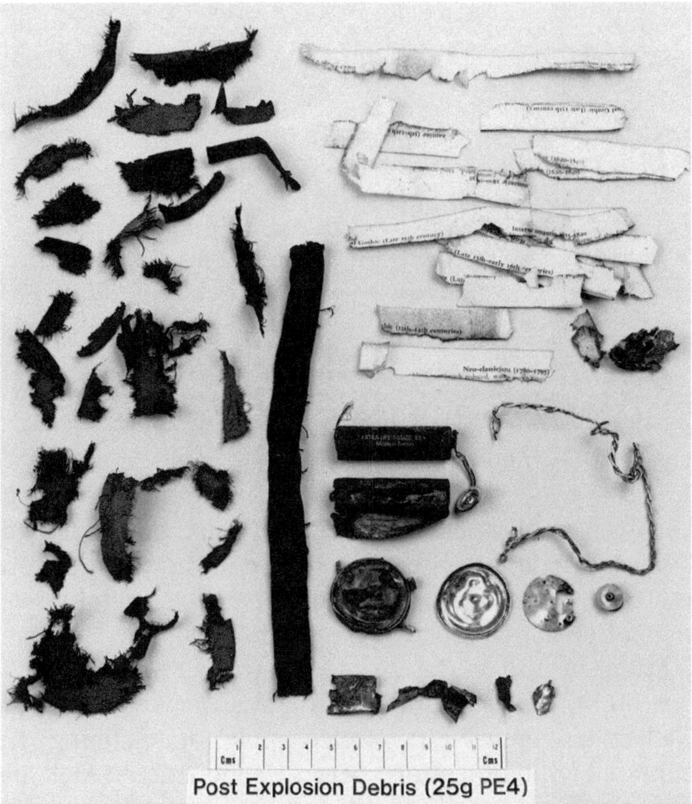

Figure 15.7 Fragments from a small bomb in a book.

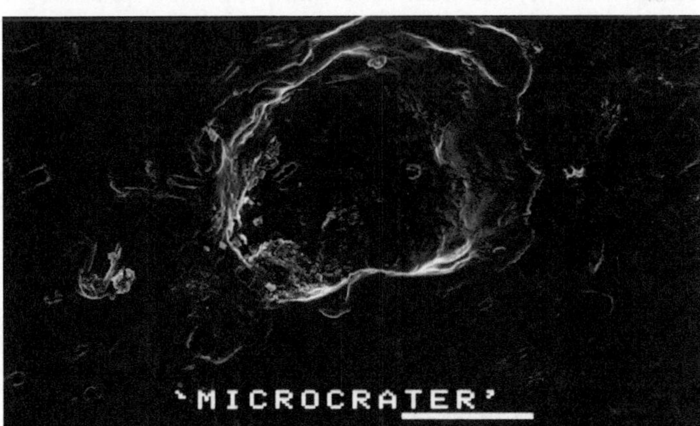

Figure 15.8 A microcrater in aluminium revealed by SEM.

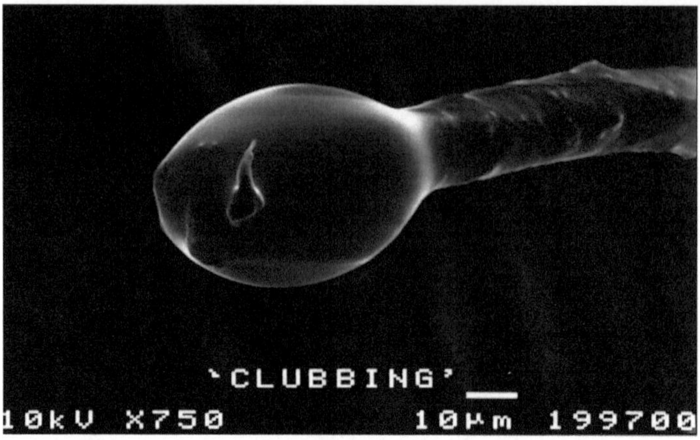

Figure 15.9 Clubbing on a nylon fibre as seen by SEM.

Many explosives also leave deposits of soot on cooler adjacent surfaces where partially combusted material and reaction products condense. Large metal fittings and items such as metal street signs and railings often act as receptors for soot in this way.

Examination of the surface damage with a Scanning Electron Microscope (SEM) can be highly informative. For example, nylon textiles may display characteristic 'toffee apple' or 'clubbing' damage; the momentary high temperature causes the polymer to melt and then re-solidify to produce a small spheroid on the fibre end which is readily visible in the SEM and an example of this is shown in Figure 15.9. Other characteristic effects also occur on metals; ferrous alloys may display changes in crystal structure, and explosive cladding of one metal upon another can also occur.

Chemical residues of explosives or their decomposition products may also be sought and if found provide useful evidence. In the case of uncommon chemicals with no non-explosive uses (*e.g.* tetryl or RDX), the detection of residual traces is strong evidence. This is especially true in the case of improvised explosives such as TATP and HMTD, since they are not used or produced commercially at all, owing to their sensitivity and poor storage qualities.

The gross physical damage at the scene of a suspected bombing is also informative, since careful examination will enable the scientist to distinguish between a dispersed-phase explosion and a condensed-phase explosion and in the latter case to locate the seat of the explosion. Structural damage, crater depth, and breakage of window glazing are all useful in assessing the mass of explosive involved; however, it must be realised that despite developments in modelling technology, calculations of the size of bombs from observations of scene damage may be subject to large margins of error, particularly when dealing with novel (*i.e.* uncharacterised) material. For this reason, an estimate in terms of relative scale of charge may be used, *e.g.* less than a hundred grams, less than a kilogram, in the order of tens of kilograms *etc.*

Particular difficulty can be experienced in determining whether an aircraft crash is due to the explosion of a bomb, or some other cause. In this type of case it is necessary to distinguish between what may be very extensive mechanical damage due to the crash and damage that could only be due to an explosion, and so a joint examination between experts from both areas of expertise is crucial. Another important factor is the need for the use of shared best practice between police specialists in managing potential terrorist crime scenes and accident investigators, so as to ensure evidence is handled to the necessary standard for the judicial system from the beginning of the process. Where reconstruction of the crashed aircraft is feasible, this can be extremely helpful in locating the position of an explosion. Tracing the path of shrapnel from a bomb can sometimes be accomplished by examination of the aircraft seats, *e.g.* by threading stiff wires through the fragment penetration holes, thereby enabling location of the site of the bomb. Where a bomb has been placed in hold baggage, careful examination of the recovered baggage can sometimes enable identification of which items of baggage were adjacent to the bomb and which specific item of baggage contained the bomb. This can provide exceptionally valuable assistance to the investigator, enabling possible links back to the individual responsible for checking the baggage onto the aircraft.

15.4.3 Is This an Explosive?

The forensic scientist is likely to encounter explosives samples in five main ways:

(1) As a bulk of an unknown substance, whether in isolation or together with components suspected of being related to bomb-making activities (including precursor chemicals).
(2) From a suspect device that has been rendered safe by bomb disposal personnel. Apart from suspect explosive, other components of the device such as batteries, electrical circuitry and igniters are also likely to be submitted.
(3) As traces on items of post-explosion debris.
(4) As traces on items suspected of having been in contact with explosives.
(5) As traces on people suspected of having handled explosives.

15.4.3.1 Examination of Bulk Explosives

Although some organic compounds are well-known explosives, *e.g.* TNT, there are many mixtures of ingredients which may be explosives depending on the precise formulation and circumstances. Thus, it is necessary to make use of both chemical analysis to identify suspect materials and also a variety of physical tests to demonstrate both the energetic nature and detonability of samples.

The very first step is of course to look at the sample: if it happens to bear a manufacturer's label identifying it, this can be quite helpful! However, it is still necessary to confirm the true identity of labelled items since the label may not be accurate. Occasional submissions to the laboratory in the UK have included those items which purport to be one thing, but which when disassembled, were shown to have very different compositions due to modifications carried out by the experimenter. Visual examination with a low power microscope is of considerable value, particularly for recognising materials such as home-made black powder.

The simplest test is to observe the behaviour of a small (tens of milligrammes) quantity of the material when it is ignited. This test must be done behind a substantial armoured glass screen, preferably within an armoured fume cupboard since some explosives produce toxic vapours on decomposition, *e.g.* lead salts.

Points to note are:

(1) Does it burn? Smoulder? Melt? Char?
(2) What colour is the flame? Are there flashes or sparks?
(3) Does a residue remain, and if so what is its appearance?
(4) Did the burning material emit a characteristic odour?

A more elaborate test is the Lidstone cartridge test, which assesses the detonability of small quantities of suspect materials. A 2 g sample of the suspect material is packed in a 0.303 brass cartridge case, a standard detonator inserted in the open neck of the case and the assembly fired in a tank filled with wood pellets. The wood pellets are sieved and the remnants of the cartridge case recovered. Comparison with firings of known explosives allows a rough assessment of the sample's explosive performance to be made. Examples of the different categories of damage are shown in Figure 15.10.

Since some explosives, particularly mixtures, react differently when initiated in very small quantities compared with how they behave in larger quantities, a larger-scale version of this test, known as a gap test, is sometimes used. This test uses up to 200 g of the questioned material, in standard-size steel tubes, initiated with a standard-size

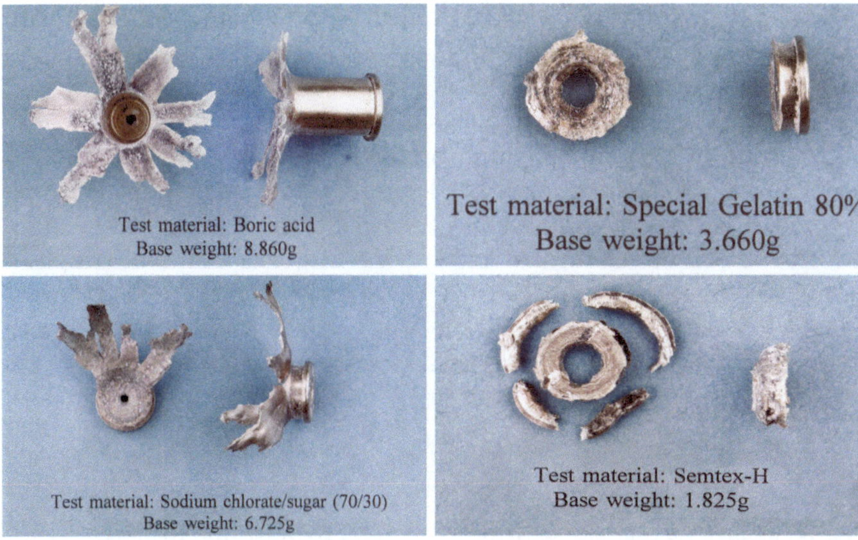

Figure 15.10 Categories of damage resulting from the Lidstone cartridge test.

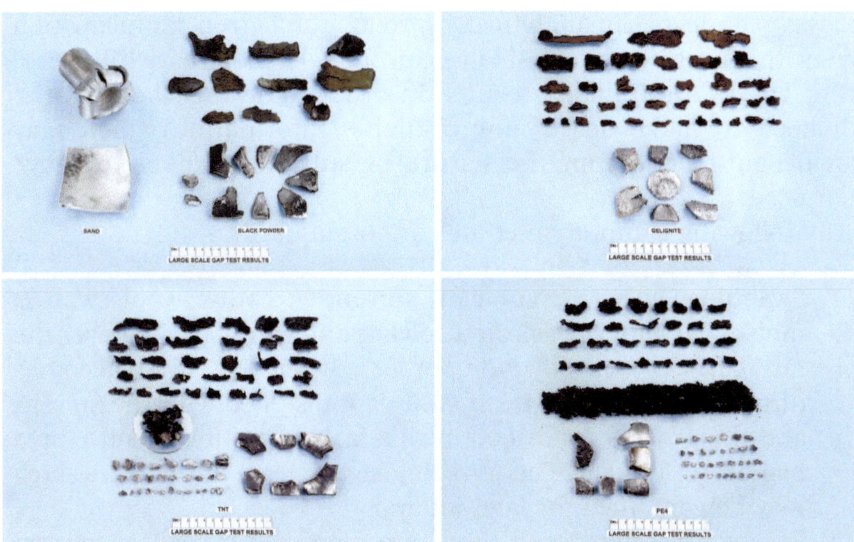

Figure 15.11 Degrees of damage caused by different explosives used within a gap test.

pellet of commercial high explosive. This, however, requires a properly licensed bomb chamber in which to conduct the test. Examples of the different degrees of damage seen for different explosive materials are shown in Figure 15.11.

Thereafter a whole range of chemical analyses can be performed, including wet chemical techniques such as colour-change tests, Thin-Layer Chromatography (TLC), or solvent extraction and gravimetric analysis, and instrumental methods such as Fourier Transform Infra-Red Spectroscopy (FTIR), Raman Spectroscopy, Ion Chromatography–Mass Spectrometry (IC–MS), Scanning Electron Microscopy with Energy-Dispersive Spectroscopy (SEM–EDS), X-Ray Diffraction (XRD), Ultra High Performance Liquid Chromatography (UHPLC), Gas Chromatography–Thermal Energy Analyser (GC–TEA), Gas Chromatography–Mass Spectrometry (GC–MS), High Resolution Mass Spectrometry (HRMS), and Liquid Chromatography tandem Mass Spectrometry (LC–MS–MS).

15.4.3.2 Examination of Items for Traces of Explosives – Recovery and Sampling

It is necessary to recover samples for detection of invisible traces of explosives from a wide variety of types of evidence, and sampling techniques need to be chosen accordingly. Because of the enormous range of possible explosives that may be encountered it is necessary to devise an analytical approach for a given sample which covers the range of most likely candidate explosives whilst providing timely and reliable results. In order to do this the scientist will need to make use of any available information which may provide guidance about the nature of substances likely to have been used.

The following sampling methods are often used:

(1) **Swabbing, either dry or with solvent.** A cotton wool swab is moistened with the selected solvent and gently rubbed over the test surface (typically non-porous); when the swab becomes dirty or dry a further fresh swab is moistened with the solvent and the process continued in this fashion until the entire area has been sampled. The resulting swabs from any one test area may be combined for later analysis.
(2) **Solvent washing.** Small items are placed in a beaker, covered with the chosen solvent, and then the whole agitated in an ultrasonic bath to aid recovery of soluble materials. Larger items may be held over a beaker whilst a few millilitres of solvent is run repeatedly over the surface.
(3) **Headspace sampling.** Items for trace analysis are received sealed within nylon bags. The headspace procedure collects the vapours inside nylon bags (known as the headspace) for volatile substances

using adsorbent tubes. The procedure is non-destructive and can be used in conjunction with other sampling techniques.

(4) **Vacuum sampling to trap particulates and vapours.** A glass syringe barrel is fitted with a filter support and a microporous filter and then a length of disposable tubing connected between the needle fitting and a vacuum pump to apply suction, *via* a glass tube containing an adsorbent. The open end of the syringe barrel is rubbed over the sample surface (typically porous), so that any mobile particles are trapped on the filter for subsequent examination. Any vapours, *e.g.* TATP, that may pass through the filter, are trapped by the adsorbent tube.

Solvents for swabbing or washing need to be chosen to suit the type of explosive being sought and the analytical method which is to be used. Table 15.4 lists some commonly used solvents and Table 15.5 gives some examples of the use of these different methods in the collection of samples for trace explosives analysis from a motor vehicle.

Table 15.4 Solvents used for recovering traces of explosive.

Solvent	Type of explosive	Comments
Water	Inorganic, *e.g.* nitrates, chlorates, perchlorates	Slow to evaporate
Methanol	Organic, *e.g.* nitrate esters, nitroaromatics, nitramines	Toxic, poor volatility
Ethanol	Organic, *e.g.* nitrate esters, nitroaromatics, nitramines	Poor volatility
Acetone	Good solvent for organics, especially nitrocellulose	Good volatility, but leaves aqueous residue unless dried thoroughly
Iso-propanol	Organic	Moderate volatility
Ethyl acetate	Good solvent for a wide range of explosives	Available in high purity
Methyl *tert*-butyl ether	Organic	Good volatility
Diethyl ether	Organic	Good volatility, but serious fire hazard
Acetonitrile	Peroxides, *e.g.* TATP, HMTD	Toxic, but good for use with LC–MS techniques

Table 15.5 Sampling of the surfaces and materials in a motor vehicle.

Substrate	Method
Glass and painted metal	Solvent swabbing
Impermeable plastic surfaces	Solvent swabbing or headspace sampling
Porous surfaces, woven materials	Vacuum or headspace sampling
Small loose items, *e.g.* keys, coins	Solvent washing

15.4.3.2.1 Controls

With any of the methods outlined above, no results can be considered to have forensic significance unless proper environmental control measures are taken before the work of trace analysis starts. These need to be designed to take account of the sensitivity of the analytical method used, and the potential exposure to contamination of the exhibits and sample solutions, throughout the process of sampling and analysis. As more sensitive detection methods are used, more stringent environmental control measures are needed. When nanogram quantities are being sought, operators should, as a minimum, wear fresh disposable overgarments and gloves to preclude transfer of suspect materials either between casework items or between samples and the general laboratory environment or the operator. Control swabs should always be taken from the actual work surface and also the operator's gloves and oversuit, and these control swabs should be prepared and analysed in the same way and at the same time as the actual trace analysis samples from the case under examination.

15.4.3.2.2 Sampling Kits

A number of forensic laboratories prepare kits of materials ready for use in sampling at crime scenes. Kits for the collection of explosives traces should contain items such as swabs and solvent which, by means of a rigorous quality control process, have been shown to be free of explosives traces. Three official kits are issued to police in Great Britain for recovery of explosives traces. The general swabbing kit is shown in Figure 15.12; a similar but smaller kit, with fewer components, is used for swabbing suspects' hands, and a portable vacuum sampling kit is supplied for porous or woven surfaces.

15.4.3.2.3 Sample Pretreatment and Concentration

This will depend again on both the likely candidate explosives and the analytical method chosen. For example, particles of sugar and sodium chlorate mixture recovered by vacuum sampling can be dissolved in water and analysed by IC based separation techniques to identify anions, cations and sugars. The same basic technique can be applied to aqueous extracts from vacuum sampling tubes, water wetted swabs or washings. The extracts can be concentrated by partial evaporation to increase analytical sensitivity.

Liquid chromatography and gas chromatography are used in conjunction with different selective detectors for determination of organic explosives traces. For organic explosives of moderate to high volatility

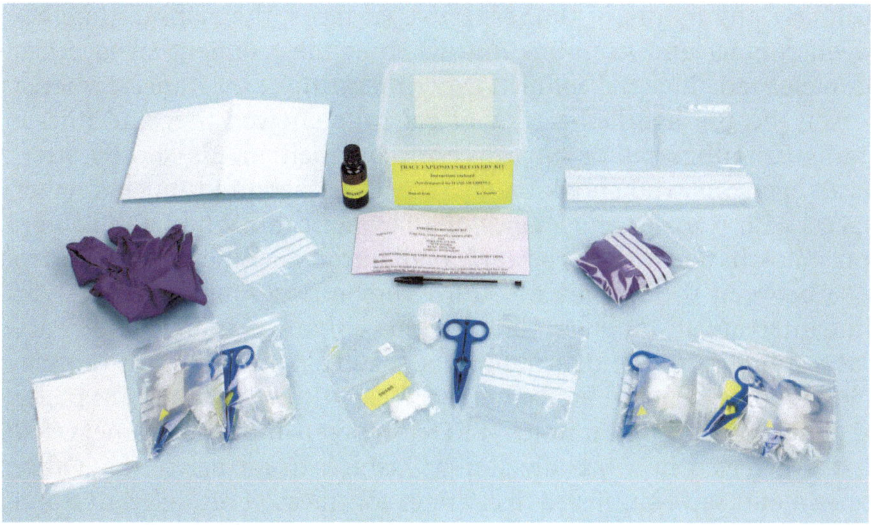

Figure 15.12 Trace explosives recovery kit.

such as nitroglycerine or RDX, capillary gas chromatography with a selective chemiluminescence (thermal energy analyser) detector or a mass spectrometer is suitable. For involatile explosives, or for less thermally stable explosives such as TATP and HMTD, liquid chromatography, coupled with single or tandem mass spectrometry *e.g.* HRMS, provides an alternative approach. Although these techniques are extremely sensitive, being able to detect picogram amounts of some explosives in pure solutions, they are generally liable to be seriously affected by the large amounts of non-explosive material present in real samples such as oils, fats, soot and soil. A good technique is to elute the initial sample extract through a simple solid-phase chromatographic column which adsorbs the explosives selectively for subsequent elution with a different solvent. The Forensic Explosives Laboratory has taken this approach with a sample pre-treatment method that allows a single sample, be it swab, wash or vacuum, to be separated into different fractions for analysis by all of the above techniques.

15.4.3.2.4 Interpretation of Explosives Trace Evidence

A surface such as the seat of a car or a person's hand may acquire traces of explosives either by direct contact with explosive, or by secondary transfer from another item contaminated with explosive. The quantity of explosives traces found will depend on a whole range of factors—the type and amount of explosive present on the donor, the duration and intimacy of contact between the donor and recipient, the nature of the respective surfaces, the elapsed time between contact and sampling and whether the recipient surface has been cleaned since the contact. Experiments done on transfer when a person directly handled a stick of gelignite showed they had micrograms to milligrams of nitroglycerine on their hands shortly afterwards. Persons who shook hands with them acquired much smaller traces, with transfer efficiencies of a few per cent. When a third person shook hands with the second, higher transfer efficiencies of up to 35% between the second and third person were observed for short time intervals. It was suggested that this enhanced transfer efficiency was a function of the distribution of the explosive traces on the second person's hand. Some very limited experiments with the transfer of pentaerythritol tetranitrate (PETN) between individuals suggested that it behaved in a broadly similar fashion to nitroglycerine. Other experiments have suggested decay rates for traces of nitroglycerine and RDX on hands, of three to five orders of magnitude in the first day, for normal handwashing regimes. Much lower decay rates might be expected for traces on inanimate surfaces subject to infrequent cleaning or otherwise undisturbed. Decay rates for solid explosives such as RDX and PETN, having relatively low vapour pressures, would also generally be lower than, for instance, the relatively volatile nitroglycerine.

Consideration must also be given to the likelihood that an explosives trace might have been acquired in the course of legitimate activity, *e.g.* quarry workers or military personnel might come into contact with explosives in the course of their normal duties. Nonetheless, the finding of traces of organic high explosives can be quite significant. Studies of casework samples received in our laboratory show that only a few per cent of samples taken from suspects' hands, clothing or premises exhibit detectable explosives traces, with a somewhat higher proportion of suspects' vehicles yielding positive results. In addition, surveys of traces of organic high explosives in the general public environment in Great Britain have shown that such traces are rare. However, higher levels of inorganic species of possible explosives significance, such as sodium, ammonium or nitrate ions are likely to be widely encountered in the general environment, so this needs to be taken into account when interpreting the significance of inorganic trace findings. Hence, 50 ng of RDX, for example, is likely to be significant evidence, but a similar finding of nitrate ions would be entirely insignificant. The quantity of nitrate ions would certainly need to be one, and maybe two, orders of magnitude higher before any kind of evidential significance might be attached, depending on the circumstances of the case.

In drawing conclusions regarding any aspects of forensic casework, it is important to maintain the impartiality of the forensic expert. Although the issue of cognitive bias is dealt with elsewhere in this textbook, and is most relevant to areas of forensic science where the expert opinion involves conducting a comparison between materials, it is still essential that a case assessment process is undertaken at the start of the case and that this is peer reviewed. Also, that an independent critical findings check is undertaken at the conclusion of the work, to ensure any alternative hypothesis has been considered.

15.5 Photography

Because of the hazardous nature of explosives and bombs, it is not possible to take most explosives related exhibits to court. As a consequence, it is essential to produce a high-quality photographic record of exhibits, particularly when devices are being taken apart, in order to show all the relevant details to the court. Previously, large negatives were used to give greater detail in the final print, which were more suitable than 35 mm film for production of courtroom-quality photographs. Nowadays, this has been superseded by digital photography, but this similarly requires specialist high-resolution equipment. It also requires dedicated recording equipment and significant archiving capability, to demonstrate to an evidential standard that the original image has not been altered or tampered with in any way. The increased use of technology within the UK judicial system means that good quality imagery can be shared more easily with those within the court.

15.6 Links with Other Forensic Science Disciplines

Explosives examinations generally take precedence over other forensic tests for two fundamental reasons: (1) safety, and (2) because unless there are items of explosives significance then it is unlikely that a case can be established under explosives-related legislation.

As in all complex forensic cases, discussion of requirements between scientists from different disciplines and the investigating officer is essential if the best evidence is to be protected and realised. Practical compromises between the requirements of explosives trace analysis, DNA, fingerprints, fibres and tool marks need to be agreed and the work planned accordingly since each aspect can make a vital contribution in many cases. Although specialists in forensic explosives examinations, it is essential that the expert also has an appreciation of the need to preserve a range of evidence types. Conducting joint examinations between different experts will ensure evidence recovery is maximized. Increased regulation may also require a degree of expertise in other trace evidence types, such as DNA sampling of explosive items which cannot leave the safety of the licensed facility.

15.7 Case Study

As in all forensic science, correct interpretation of the results is crucial. A delicate balance has to be struck, ensuring that all the relevant facts are made known, that the appropriate weight is given to each aspect of the evidence, and that the work is recorded in a manner which allows meaningful review by others skilled in the art so that, for example, other possible interpretations of the results can be explored by different experts assisting the various parties in a case. That said, how are the various scientific and technical threads pulled together? Consider the following hypothetical case.

15.7.1 The Scenario

A violent sharp report is heard just after 3 am, apparently from the grounds of a large country house, whose alarmed inhabitants summon the local police. Upon arrival the police make a search of the grounds and find that the base of one of the large (and valuable) statues in the ornamental gardens has been damaged; they decide to secure the scene and wait until daylight before making a detailed search.

Meanwhile other officers have been called to an accident just a few miles away involving a motor car and a small lorry. The behaviour of the lorry driver and his passenger causes them to make further inquiries and their suspicions are increased when the passenger is reported to have been convicted of burglary some years before.

Further inquiries are made, and a link to the attempted theft of the statue suspected. By now daylight has broken and examination of the scene around the statue has yielded a bootprint from a patch of soft ground, and revealed a pattern of damage on the stonework of the statue's base. Further searching results in the discovery of a pair of long thin wires, various fragments of broken stonework, and an assortment of general litter and debris of uncertain relevance; in addition, a tyre print is found in mud alongside the driveway. Scene photographs are taken, the recovered items carefully bagged and identified, and the damaged area of the statue's base is swabbed using a trace explosives recovery kit; the whole is then submitted to the laboratory for examination.

Chemical analysis of the swabs rapidly indicates the presence of PETN residues, and this, coupled with examination of the damage to the statue, suggests an unsuccessful attempt to cut the statue from its base using detonating cord. After discussion with the scientist the police decide to have the suspects' lorry, their clothing and their hands tested for explosives traces. It is agreed that in this instance the scientist will give priority to the trace explosives analysis and leave examination of the various items from the scene of the explosion until later; this simplifies matters since it neatly allows clearly separate examination of the 'clean' and 'dirty' exhibits and avoids any possible suggestion of contamination between them.

Most of the results from the trace analysis are negative. However, positive results for PETN are found for a pair of rubber gloves found in the cab of the lorry, from a penknife found in the pocket of the passenger's jacket, from the area of the pocket that held the penknife and from the passenger's trousers. The swab samples taken from both men's hands are negative.

A search of both men's homes reveals nothing relevant; however, a lock-up garage rented by the passenger yields a bag containing what initially seems to be a length of plastic clothes line; however, it has the words 'Danger—Explosive' printed along its length. This sample is therefore examined by a bomb disposal operator who packages it safely and advises forensic examination to confirm if it is in fact detonating cord. The rubbish bin from outside the garage is also searched; some discarded packaging appears significant and is also submitted to the laboratory.

These new items are identified at the laboratory as a particular brand of commercial detonating cord containing PETN, and the remnants of a box which once contained a particular type of commercial electric detonator widely used in quarrying.

Meanwhile the inhabitants of the house where the crime occurred have found another pair of rubber gloves discarded in their garden at some distance from the statue; unfortunately there have been several nights of heavy rain before the gloves were found. Examination of the gloves for explosives traces does indeed prove negative.

Searching of the debris from the explosion scene reveals a small burnt and deformed rubber plug, which the scientist is nonetheless able to identify as similar to the plug from detonators made by a particular company; there are also the wires, which prove to be a long length of a standard twin-core cable attached to two short lengths of plastic insulated single-strand wire whose colour and diameter match that from a particular range of commercial detonators. In fact the packaging from the lock-up garage also matches and the three items: wires, plug and packaging all belong to the same product range from the same company.

Further examination of the lorry reveals a partially discharged battery lying in a corner of the loading area and a torn packet matching the brand of rubber gloves found both in the lorry's cab and at the scene. One of the tyres on the lorry is found to match the tyre print in the garden.

15.7.2 The Prosecution Case

It was suggested that the facts were consistent with a bungled attempt to steal a valuable statue; detonating cord had been wrapped around the base in an effort to cut the statue loose so that it could be removed. In fact, all that had resulted was noise and damage. The frustrated thieves had then fled, leaving behind the rubber gloves worn by one of them; the second pair of gloves was found in the cab. No fingerprints were found at the scene, and the accuseds' hands were free of explosive traces because they had worn gloves. However, the penknife that was found in the passenger's jacket pocket had been used to cut the detonating cord and hence had contaminated the pocket with the traces of PETN found. Likewise, the gloves found in the cab had been worn when handling the detonating cord, hence the traces of PETN on them. The absence of traces on the gloves found later in the garden could be due to their having been washed away by the heavy rain.

The detonating cord found in the garage contained PETN; the packaging found in the rubbish bin at the garage matched the detonator plug and wires found at the scene.

15.7.3 The Passenger's Defence

The lock-up garage had been rented by a group of five men, including himself, each of whom had a key. The bag found in the garage belonged to one of the others.

While working in the garage on the previous day he had needed some rope to secure a bundle of wood; when he looked in the bag he saw what seemed to be a coil of clothes line and so he cut a length off with his penknife. However, a white powder then fell out of the line; he didn't know what it was, but realised it wasn't suitable for his purpose so he put it back in the bag, wiped the knife blade on his trousers and put the knife back in his pocket. He did all this while kneeling on the garage floor.

The reason he was in the lorry was that he had been out poaching and had been walking home when his friend drove by in the lorry and gave him a lift.

15.7.4 The Lorry Driver's Defence

He had been employed recently to remove heavy rubbish from the gardens of the house and had legitimately been near the statue in the course of his work, hence the bootprint. Likewise he had parked his lorry in the driveway, and this must be the explanation for the tyre print.

15.7.5 What Really Happened?

The passenger was telling the truth.

The lorry driver was in league with the owner of the bag, who was one of the group renting the garage. The bag-owner had left the crime scene on foot when he realised that the noise of the explosion had alerted the inhabitants of the house. In his haste he had discarded the gloves found later in the garden. The lorry driver had no choice but to leave *via* the driveway unless he wished to abandon his vehicle at the scene. This case study illustrates the fact that there are often a number of explanations for a given set of circumstances. The forensic scientist's task is to conduct impartial examinations and formulate unbiased conclusions within the remit of their specific discipline, in order to help the court reach an informed judgement. In most cases the scientific findings will form only a part of the jigsaw, the other pieces of which may radically affect the interpretation of the science.

15.8 Conclusion

In conclusion, the forensic explosives expert requires access to a wide range of facilities to enable them to examine the type of exhibit that they might encounter – from detecting nanogram quantities of improvised explosives on a person's clothing to the ability to examine vehicle borne explosive devices within a controlled forensic environment. Increased access to information on explosives on the internet has the potential to make available credible information to the resourceful enthusiast, as well as those with more devastating intentions. The expert must ensure they remain up to date with the latest material to allow them to provide an objective assessment of its viability to assist the court. At the same time, the increased regulation of precursor chemicals must also keep apace.

Acknowledgements

Further Reading

J. Akhavan, *The Chemistry of Explosives*, The Royal Society of Chemistry, Cambridge, 4th edn, 2022.

Forensic Investigation of Explosions, ed. A. Beveridge, Taylor & Francis, London, 2nd edn, 2011.

T. L. Davis, *The Chemistry of Powder and Explosives*, Angriff Press, Hollywood, California, 1943, vol. 2.

T. Urbanski, *Chemistry and Technology of Explosives*, Pergamon Press, Oxford, 1964, vol. 3.

H. J. Yallop, *Explosion Investigation*, The Forensic Science Society, 1980.

16 Firearms

Mark Mastaglio*[a] and Angela Shaw[b]

[a] Forensic Firearms Consultancy (FFC) Ltd., 16 Upper Woburn Place, London, WC1H 0BS, UK; [b] United Kingdom Accreditation Service (UKAS), 2 Pine Trees Chertsey Lane, Staines-upon-Thames TW18 3HR, UK
*Email: enquiries@forensicfirearmsconsultancy.com

16.1 Introduction

The practitioner in the discipline of forensic firearms examination, sometimes called forensic ballistics, can be required to attend the scene of a shooting and the post-mortem examination, carry out subsequent examination of items at the laboratory and, ultimately, to present expert testimony at court.

The earliest references to what we would recognise as a firearm date from the 14th century. In simple terms a firearm incorporates a mechanism, mounted in an ergonomically designed frame, which discharges a projectile through a tube. This is where the phrase 'lock, stock and barrel' comes from. Dictionary definitions of the noun 'firearm' usually refer 'to a portable weapon that discharges a projectile by means of an explosion; a rifle, gun, pistol, *etc.*' The UN define a firearm as any portable barrelled weapon that expels, is designed to expel or may be readily converted to expel a shot, bullet, or projectile by the action of an explosive, excluding antique firearms or their replicas.

Crime Scene to Court: The Essentials of Forensic Science, 5th Edition
Edited by Niamh NicDaeid and Peter C. White

Published by the Royal Society of Chemistry, www.rsc.org

In the UK a 'firearm', as defined by Section 57(1) the Firearms Act 1968, means:

(a) a lethal[†] barrelled weapon;
(b) a prohibited weapon;
(c) a relevant component part[‡] in relation to a lethal barreled weapon or a prohibited weapon;
(d) an accessory to any such weapon designed or adapted to diminish the noise or flash caused by firing the weapon…

This definition obviously applies to military-style weapons, *i.e.* those designed to kill; target rifles and pistols, which clearly can be used as weapons; air guns (including those powered by compressed carbon dioxide); and shotguns, together with less obvious items such as lachrymator canisters and electronic stun guns. Certain items, so-called relevant component parts, when they are capable of being used as part of a lethal barrelled weapon or prohibited weapon, such as barrels, revolver cylinders, frames or receivers and pressure bearing mechanisms, also constitute firearms for the purposes of the legislation, as do so-called silencers, more correctly called sound moderators, and flash suppressors.

Ammunition in the 1968 Act means 'ammunition for any firearm and includes grenades, bombs and other like missiles, whether capable of use with a firearm or not, and also includes prohibited ammunition.' This definition clearly applies to conventional bulleted cartridges and shotgun cartridges but can also be applied to airgun pellets, blank cartridges (be they adapted with a projectile or not) and cartridges that contain a lachrymator or irritant such as CS (*ortho*-chlorobenzalmalononitrile) or a pepper related substance (capsaicin or pelargonic acid vanillylamide).

The forensic firearms examiner must be competent in a wide range of examinations. Frequently asked questions include:

(1) Is this item a firearm or a relevant component part of one?
(2) What part of the firearms legislation is applicable?

[†] A quantitative threshold defining a '*lethal barrelled weapon*' was introduced by the Policing and Crime Act 2017 (PCA 2017) which created section 1B of 1968 Act. This latter section now further defines a firearm as '*a barrelled weapon of any description from which a shot, bullet or other missile, with kinetic energy of more than one joule at the muzzle of the weapon, can be discharged*'.

[‡] The PCA 2017 also introduced section 1D into the 1968 Act defining what is meant by '*a relevant component part*'.

(3) Did this gun discharge this cartridge case or bullet?
(4) How many guns and what type were used in the incident?
(5) Could the gun have gone off accidentally?
(6) Is this a bullet hole?
(7) Where was the shooter in relation to the damage or victim?
(8) What was the order of the shots?
(9) Could a gunshot wound be self-inflicted?

The application of what can be referred to as firearms chemistry, a branch of analytical chemistry, is also an important tool in the forensic investigation of shooting incidents. This includes the study of gunshot residue (GSR), its origin, formation, identification and persistence.

The aim of this chapter is to provide the reader with a basic grounding and overview of the subject and stimulate further interest.

16.2 Historical Development

The beginnings of forensic firearms examination can be traced back to the 19th century; however, it was during the first 30 or so years of the 20th century that a few notable pioneers took the discipline further, entrenching it as one of the essential tools of forensic science.

In 1900 Dr Albert L. Hall of Buffalo, USA, published the article 'The missile and the weapon', which detailed his experience as a clinician dealing with gunshot injuries and the identification of bullets. Work continued in Europe and the USA, and the subject progressed through the work of Professor Balthazar in France who published papers in 1912 and 1922 and Major Calvin Goddard who worked in the Bureau of Forensic Ballistics in New York. During the 1920s and 1930s a plethora of important papers were published, further establishing the discipline. In the UK, pioneers included the gunsmith Robert Churchill, and Major Gerald Burrard who went on to write 'The Identification of Firearms and Forensic Ballistics' in 1934. During this period expert testimony was given in several high-profile cases including the St Valentine's Day massacre in the USA and in the trial that followed the murder of an Essex policeman, William Gutteridge, who was shot dead on 27 September 1927. In the latter case the application of the relatively new discipline of forensic firearms examination directly led to the conviction of the two accused.

16.3 Firearms and Ammunition Terminology

As in all specialist subjects, technical terms and jargon abound. Firearms terms used by the layman can be confusing, and sometimes are just plain wrong. Arcane units are also encountered. At this point it is best to define some of the more frequently used terms and units and explain their significance.

16.3.1 Firearms Terminology

16.3.1.1 Action

This refers to the firearm's working mechanism.

16.3.1.2 Barrel

The barrel of a firearm is a tube which directs the projectile, or projectiles, under the influence of the propelling force, which may arise from propellant powder generated gases or from compressed gas. In the UK the length of the barrel is taken from the point of ignition to the barrel's end, that is its muzzle.

16.3.1.3 Breech

This is the part of the firearm at the end of the barrel where the cartridge or propellant is loaded. The breech face is that part of the breech bolt or breech block that abuts against the head of the loaded cartridge.

16.3.1.4 Rifling

This consists of a series of helical lands and grooves machined in the surface of the inside of the barrel, the bore. James Bond movie fans will recall that during the opening credits, the silhouette of 007 is pictured with the viewer looking down the barrel of a gun; the helical lands and grooves are visible in the animation. The lands are the raised portions, and the grooves are the valleys that run between them. Rifling imparts rotary motion to the projectile, stabilising the projectile in flight. The ballistic stability, known as gyroscopic stability, imparted to a spinning bullet is the same as that given to an American or rugby football when it is correctly thrown, the ball is spun around its long axis when it is launched, ensuring that the ball tends to fly nose first. The ball does not tumble and so can travel a prodigious distance. A ball thrown with no spin would not travel as far or be as accurate as one launched with spin and with the same force. When the relatively soft bullet travels down the hard steel bore of a rifled firearm, the rifling engraves marks on to the bullet which can identify the type and the specific firearm from which the bullet was discharged.

16.3.1.5 Calibre

One of the most misunderstood and potentially confusing terms is calibre. The calibre of a firearm can be defined as the diameter of the barrel's bore. In rifled firearms this is interpreted as the diameter of a circle whose circumference would be tangential to the tops of the lands. The calibre of the firearm's ammunition is a nominal description, which includes a numeric reference to the bore diameter. It can be alphanumeric, such as 9 mm Parabellum, the 9 mm referring to the nominal bore diameter and the word 'Parabellum' being a descriptor derived from the Latin phrase '*si vis pacem, para bellum*', which translates as 'if you wish for peace, prepare for war'. The calibre designation can also include the dimensions of the cartridge, for instance, the current NATO rifle cartridge is 5.56 × 45 mm, where the 5.56 mm refers to the bore diameter and the 45 mm to the cartridge case length. The calibre designation 30–06, a rifle cartridge, includes a reference to the nominal 0.30 inch bullet diameter and the 06 refers to 1906, when the calibre was introduced into US military service. The amount of propellant in the cartridge can also be referenced *e.g.* 45–70, the 45 refers to the nominal bullet diameter and the 70 to the powder weight in grains.

As can be seen, calibre can be expressed in millimetres or inches: 9 mm Short is also known as 0.380 Auto. Each calibre can have synonyms, *e.g.* 9 mm Parabellum is also known as 9 × 19 mm or 9 mm Luger. The last synonym illustrates that a calibre designation can also include the name of a firearms designer or manufacturer, or the type of gun the calibre was designed for, such as a 0.40 Smith & Wesson. To further complicate the issue some calibres can be discharged in firearms of a different calibre, *e.g.* the 9 mm Short/0.380 Auto cartridge can be discharged in a 9 mm Parabellum pistol.

Shotgun calibre designation is even more arcane and dates from the days of muzzle loading, when the calibre was described by the number of lead balls of the same diameter as the gun's bore which would weigh 1 lb. This means, in the case of 12-bore (also known as 12-gauge), the most common shotgun calibre, 12 spheres of the same diameter as the shotgun's bore would weigh 1 lb. All this may seem confusing and a little daunting to the student firearms examiner, especially when one realises that there are thousands of different calibres. Fortunately, for the forensic firearms examiner, a relatively small number of calibres are encountered on a day-to-day basis. In UK casework the commonest calibres are 0.22 Long Rifle, 0.25 Auto (6.35 mm Auto), 0.32 Auto (7.65 mm Browning), 0.38 Special, 0.357 Magnum, 9 mm Parabellum, 9 mm Short, 0.45 Automatic Colt Pistol, 0.455 Revolver, 5.56 × 45 mm, 7.62 × 51 mm, 7.62 × 39 mm, 0.410 and 12-bore shotgun cartridges. Modified blank calibre cartridges (adapted with the insertion of a projectile such as a ball-bearing, airgun pellet or piece of shaped lead), principally in 8 mm K, 9 mm PAK and 9 mm Rimmed are also encountered.

16.3.2 Cartridge Terminology

A cartridge is a whole round of ammunition. In modern terms it consists of a cartridge case, loaded with a primer, propellant and a projectile, or projectiles. Figure 16.1 illustrates the different kinds of cartridge that are most frequently encountered.

A cartridge is sometimes wrongly called a bullet; the term bullet refers specifically to the projectile. Shotgun cartridges usually contain multiple projectiles together with cartridge wadding. When a cartridge is discharged, marks can be transferred from the firearm on to the relatively soft cartridge case and its primer. These marks, as identified in Figure 16.2, can enable the make and specific firearm to be identified and can originate from any part of the firearm action that contacts the cartridge such as the breech face, firing pin, extractor, ejector and chamber.

16.3.2.1 The Cartridge Case

Bulleted rounds normally use cartridge cases made from brass, although materials such as aluminium and steel are also used. Shotgun cartridge cases are usually made from plastic or paper, with a brass or steel base.

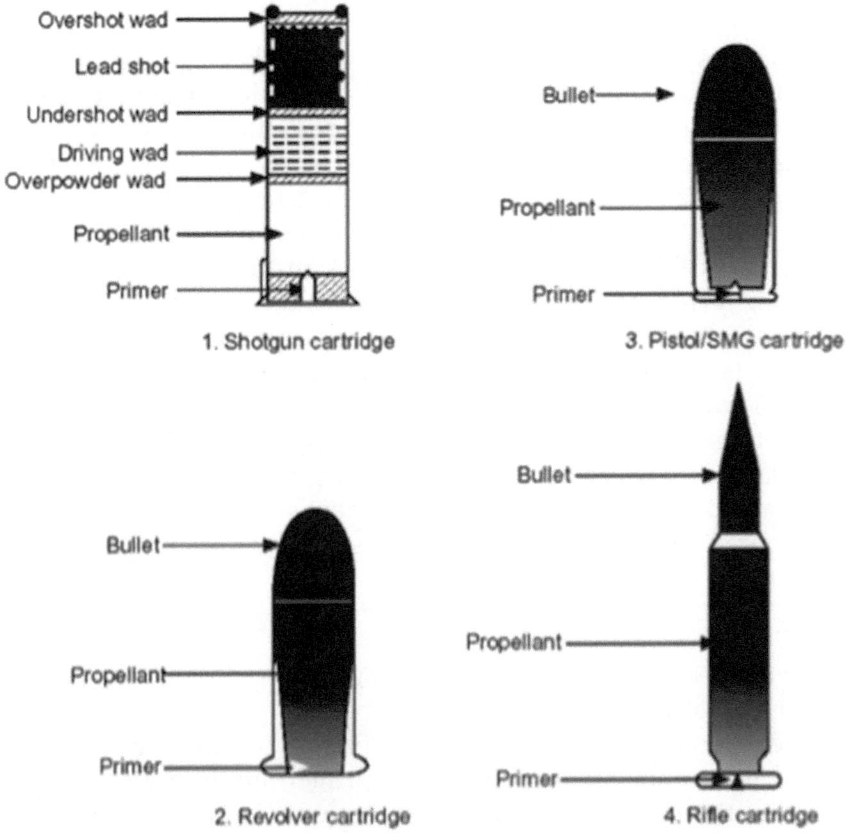

Figure 16.1 Schematic representations of commonly encountered cartridge types.

Figure 16.2 The base of a spent cartridge case showing the position and shape of the firing pin impression in the centre of the primer, the extractor mark at 3 o'clock and ejector mark at between 6 and 7 o'clock.

16.3.2.2 Propellant

Propellant is a mixture of chemicals which when ignited burns in a predictable manner producing large quantities of hot expanding gases which can be used to propel a projectile. The fundamental requirement is to transform a portion of the propellant's stored chemical energy into the kinetic energy of the projectile. The first type of propellant that was developed, gunpowder, also known as black powder, is still used today for vintage design firearms, it consists of an intimate mechanical mixture of potassium nitrate, sulphur and charcoal. Modern so-called 'smokeless' propellants contain nitrocellulose (single-based propellant), a substance that has the oxidiser and fuel components in the same molecule, and sometimes mixed with nitroglycerine (double-based propellant). The discharge of a propellant in a firearm can result in trace evidence, *i.e.* propellant grains and combustion products, being deposited on substrates, such as a victim's body or their clothing, close to the discharge.

16.3.2.3 Primer

This is the component of the cartridge that contains an impact-sensitive explosive mixture which detonates when impacted with sufficient force by the firing pin or striker and then ignites the propellant. For centre-fire cartridges the primer sits in a recess in the centre of the cartridge base. This explosive mixture is housed in a brass, nickel or gilding metal cup.

There are two main types of centre-fire primer design, Berdan and Boxer. The former consists of the explosive mixture covered with a foil or paper disk; the primer being detonated by the impact of the firing pin forcing the explosive mixture against an anvil integral to the cartridge case. In the case of the Boxer primer the anvil is integral to the primer. The latter design is more commonly encountered in modern day casework.

In rimfire cartridges the primer composition is present inside the cartridge rim. The compositions and analysis of both propellant and primer trace evidence are discussed in Section 16.10.

16.3.2.4 Bullets

A bullet is the projectile loaded into the cartridge case. Bullets for revolvers are normally made of lead, but those for self-loading pistols and rifles are encased in a hard metal covering, known as a bullet jacket. Bullet jackets are normally made from cupronickel, brass or steel. The jacket is present to prevent the lead being stripped off in the gun's bore and to prevent misfeeds in self-loading guns. The lead used for bullets and shotgun pellets is usually hardened by the addition of small quantities of a hardening agent such as antimony. Other metals and alloys, such as brass, bronze, aluminium and steel are also used to make bullets.

The most commonly encountered bullet shape is the round-nosed bullet. However, there are a range of shapes, the design being suited to their purpose. For example, some bullets have a flat nose; they are essentially cylindrical in shape and when they strike a card target, they punch a circular hole in it. This style of bullet is known as a 'wadcutter' and is designed to be used by target shooters. Some bullets have a hollow nose or a partially exposed core; their function is to expand on entering a quarry, be it animal or human, and thus increase the wounding potential. Military-purpose rifle bullets can incorporate steel penetrators, be designed to pierce armour, have an incendiary capability or have a pyrotechnic component so that the discharged bullet's trajectory is visible (the latter are known as 'tracer' bullets).

16.3.2.5 Shotgun Cartridges

These too are loaded with propellant and a primer, but usually contain multiple shot pellets, together with card, fibre or plastic wadding. The number of pellets that can be loaded into a shotgun cartridge depends on the size of the pellets and the calibre of the cartridge. For example, a 12-bore cartridge loaded with a 32 g load of number 6 size shot will contain approximately 320 pellets. The largest shot encountered in a British loaded cartridge has a diameter of approximately 9.1 mm. Some shotgun cartridges are loaded with a single projectile, known as a slug, but these are rarely used in crime in the UK. In recent years concern over environmental pollution has increased and there has been a move away from traditional lead-based shot. Use of other types of shot, such as tungsten or bismuth–tin alloy shot, has become more common.

16.3.3 Ballistics

Ballistics is the science of the propulsion, flight and impact of projectiles, and the word is derived from '*ballista*', a Roman military weapon in the form of a large, mounted crossbow. There are two main areas of study, internal and external ballistics.

16.3.3.1 Internal Ballistics

This is the study of the motion/behaviour of the projectile while it is still in the firearm, from the point of initial discharge to when the projectile leaves the barrel. The time taken from pulling the trigger of a firearm to when the projectile leaves the barrel can be as short as a few milliseconds, but what happens during this short period profoundly influences the design of both the firearm and the ammunition. What happens to the cartridge case and projectile during this short period of time is of vital importance to the forensic firearms examiner.

16.3.3.2 Exterior Ballistics

This is the study of the motion of the projectile from when it leaves the gun's muzzle to when it comes to rest. The forensic firearms examiner is sometimes asked to determine various parameters concerned with the exterior ballistics of a projectile, such as extreme range and its residual wounding potential. Depending on the type of firearm and the range over which it is fired, the time of the projectile's flight can be a fraction of a second to many seconds. A rifle bullet discharged vertically upwards could take a minute or so to return to earth. What happens to the projectile when it is close to and emerging from the gun's muzzle influences the projectile's stability and is also of interest; this is sometimes referred to as intermediate ballistics. The terminal ballistics, *i.e.* how the projectile interacts with a target, is also of interest when it comes to interpreting damage or wounds.

16.3.4 Units

In the UK and USA the grain is the unit of mass which is most often used to measure propellant and bullet weights. In the rest of Europe SI units are more frequently used. There are 7000 grains to the pound and 1 g is equivalent to 15.432 grains. The foot pound is a unit of energy frequently encountered in the UK and USA when referring to the kinetic energy of a projectile. In Europe the equivalent SI unit, the joule, is used: 1 ft lb is equivalent to 1.356 J.

16.4 Firearms

Firearms can be categorised in several ways according to their form and function:

16.4.1 Long Arms

These are intended to be fired from the shoulder. They have a butt stock which can be drawn into the shoulder to accommodate the recoil experienced when a cartridge is discharged. Rifles are so called because of the presence of rifling in the bore. Some types of short-barrelled rifle may be referred to as a carbine, a word derived from 'carabineer', a reference to French cavalry troops who carried a short-barrelled musket.

The design of a rifle depends on whether it is intended to be used in a sporting context, such as target shooting or hunting, or whether it has a military purpose. They have two things in common; they are designed to be accurate over several hundred metres and to be fired from the shoulder.

Most rifles incorporate a magazine capable of holding multiple cartridges, although single-shot rifles are also infrequently encountered in casework. The firing mechanism can be bolt-action, like the old British military 0.303 Short Magazine Lee Enfield rifle (SMLE); self-loading; lever action; pump action; or have a full-automatic capability such as modern military so-called assault rifles which can fire in the self-loading or automatic modes.

Another important type of long arm which is frequently encountered in firearms-related crime is the shotgun. These have smooth-bore barrels and are designed to discharge pellets also known as 'shot'. Typically, they are designed for game/pest/vermin or clay pigeon shooting. When the cartridge pellets and wadding emerge from the barrel the pellets disperse as a function of distance. Most shotguns have a constriction of a few thousandths of an inch at the muzzle, known as choke; this serves the function of controlling the pellet dispersion. It is unusual to encounter full-length shotguns in crime, other than in domestic incidents; the shortened, so-called sawn-off, shotgun is more common.

The usual barrel length of a full-length shotgun is some 26–30 inches (66–76 cm), whereas a typical barrel length for a sawn-off shotgun is 12 inches (30 cm). In extreme cases even shorter barrels are seen. In addition to shortening the barrel the butt stock is usually also cut off, leaving a pistol-like grip. The resultant shortened weapon is easier to conceal, but a significant amount of the gun's mass has been lost, so when a cartridge is discharged the recoil is fiercer. Shotguns can be single-barrelled, double-barrelled with the barrels side by side or over and under, pump action, self-loading or bolt-action.

16.4.2 Handguns

These are usually much smaller than long arms, in general use less powerful cartridges, are designed to be used over shorter ranges and can be fired with one hand. In common with long arms, early examples were smooth-bored which are infrequently encountered in casework. In general handguns can be split into a further two categories, revolvers and pistols.

16.4.2.1 Revolvers

Revolvers were developed in the early part of the 19th century and became increasingly common from the 1850s onwards. These are instantly recognisable to the layman from their depiction in cowboy movies. They all have one thing in common: they use a rotating cylinder, a kind of magazine, which holds the cartridges. On pulling the trigger the cylinder rotates to bring the next chamber in line with the barrel. The most common cylinder capacity is six cartridges, although some can hold twelve or so cartridges.

16.4.2.2 Pistols

The type of handgun most frequently encountered in case work is the self-loading pistol, also known as the semi-automatic pistol. Like the revolver, this is a repeating firearm. It is not uncommon for this type of gun to be referred to as an automatic; this term is incorrect and its use in this context should be avoided. Single shot pistols are also occasionally encountered in casework. Self-loading pistols use a detachable magazine which is usually inserted in a recess in the pistol grip. Self-loading pistols are so called because the firing action automatically removes the spent cartridge case from the pistol's chamber and ejects it from the gun; a new unfired cartridge is then stripped from the magazine and loaded into the chamber. A subsequent pull on the trigger will discharge the newly loaded cartridge and the cycle begins afresh. It is worth noting that since the introduction of the UK ban on small firearms in the wake of the Dunblane tragedy of 1996, many novel designs have emerged increasing the size of traditional handguns so that they are not caught by the ban.

In the last decade or so the number of handguns used in the commission of crime in the UK has increased. A significant proportion of these handguns have been converted from guns which were designed and legally sold as blank cartridge firing guns. These types of guns would originally have had a fully or partially obstructed dummy barrel. The level of conversion sophistication varies hugely: some have a piece of plumber's copper pipe for a barrel, others a highly engineered rifled barrel. Most of the converted guns encountered in the UK are capable of causing lethal injury at close range.

16.4.3 Automatic Weapons

This category of firearm is usually intended for military and, in certain countries, police use. They include machine guns, submachine guns, assault rifles and some handguns. They all have one thing in common: they continue to fire so long as the trigger is pulled and there is ammunition in the magazine.

Machine guns are large and tend to use rifle calibre ammunition. They are very rarely encountered in crime in the UK, and the same is true of military assault rifles. Most of these types of weapons tend to find their way into criminal hands from war zones or through leakage from military sources.

Sub-machine guns are smaller than machine guns and use pistol calibre ammunition. Over the last couple of decades their frequency of use in UK crime has increased, their small size and high rate of fire, potentially 1200 rounds per minute, make them particularly attractive to criminals. However, the competent use of such guns by criminals is unusual as a high rate of fire requires the shooter to know how to control the gun's recoil.

16.4.4 Air Guns

Most air guns can not only cause a fatal injury but are also capable of discharging metallic pellets with a kinetic energy in excess of 1 J; in the UK they are classified as firearms. They can be handguns or rifles; they all employ a mechanism whereby compressed air (or carbon dioxide) propels a small pellet or dart from the gun. The commonest type uses a spring and piston to compress a column of air which is pushed through a transfer port. Other types include pre-charged mechanisms, gas piston mechanisms, pump-up so-called pneumatic mechanisms and those powered by compressed carbon dioxide containing cylinders similar to those used in soda-syphons. Although millions of such guns are in circulation they are rarely used in fatal shootings. When they do occur, such incidents frequently involve juveniles and claims of accidental discharge. A more common crime where the firearms examiner may be asked to help is in the investigation of criminal damage cases involving air guns.

In the last couple of decades there has been a large increase in the availability of so-called 'Airsoft' guns which have been designed to look like conventional firearms but are usually mostly made from plastic and discharge small plastic pellets at low velocities. These types of guns, provided they are of relatively low power, are not classified as firearms in the UK.

Conventional firearms manufacturers, such as Beretta and Walther, have also allowed other manufacturers to make pellet-firing copies of their conventional self-loading pistols.

Air guns using self-contained compressed air cartridges, which look like conventional rifles, revolvers and self-loading pistols, have also been developed and successfully marketed. However, such guns are now banned in the UK after a rise was noted in their conversion into propellant/bullet discharging weapons.

16.4.5 Miscellaneous Firearms

The firearms examiner will also need to recognise and examine a range of weapons that do not readily fall into the above categories, *e.g.* electric shock devices such as stun guns, lachrymator devices, and disguised firearms like pen guns, walking stick guns, key fob guns and guns disguised as mobile phones. All of these are classified as firearms and prohibited under UK firearms legislation.

The forensic firearms examiner may also be asked to examine and classify old designs of firearm that may be viewed as being antiques.[§] It is therefore essential that the examiner has a thorough grounding in the history and development of firearms and the legislation that governs their use and possession.

16.5 Firing Mechanisms

There are a relatively small number of firing mechanisms or action designs. The most frequently encountered actions are described below.

[§] Prior to 22 March 2021 there was no statutory definition of what constituted an '*antique firearm*'. Following the enactment of the Antique Firearms Regulations 2021 this changed; there is now a legally enshrined criteria based on the gun's date of production, type of propulsion system and calibre that sets out whether it can be regarded as an antique firearm for the purposes of the Firearms Act 1968.

16.5.1 Revolver

This is usually a handgun which uses a cylinder with multiple chambers which rotates around an axis, bringing the chambers in line with the bore when the firing mechanism is activated, allowing successive cartridges to be discharged. Most revolvers can be operated in one of two modes, either single-action or double-action. Single-action involves cocking the mechanism by manually pulling back the exposed hammer until it engages on an internal sear, pulling the trigger allows the hammer to fall and the cartridge is discharged. The double-action mode allows the cartridge to be discharged by a single pull on the trigger, the hammer is retracted and the cylinder is rotated by the action of pulling the trigger. Most double-action revolvers can also operate in the single-action mode. In both cases the cylinder revolves and brings the next cartridge in the adjacent chamber in line with the barrel. The spent cartridge cases remain in the gun and must be manually removed from the cylinder's chambers – they are therefore not usually found at a scene of a shooting.

16.5.2 Self-loading

This is used in repeating weapons, allowing cartridges to be successively discharged by using some of the energy of the discharge to activate the removal of the fired cartridge cases and the loading of unfired cartridges. A separate pull on the trigger is required for each discharge.

16.5.3 Automatic

This allows successive cartridges to be discharged with a single pull of the trigger; the gun will fire until the cartridge supply is exhausted or until the trigger is released. Spent cartridge cases are extracted and ejected automatically from the gun.

16.5.4 Bolt Action

This uses a breech bolt. Unlocking the breech and pulling the bolt to the rear and then pushing it forwards will cock the weapon and remove any spent cartridge from the chamber and load the subsequent cartridge ready for firing. A bolt action is used in rifles and some relatively low-quality shotguns and large calibre pistols.

16.5.5 Pump Action

This utilises a manually activated slide mechanism situated underneath and parallel to the barrel. When pulled to the rear the slide opens the breech, extracting any cartridge or cartridge case; pushing the sliding mechanism forwards loads a cartridge from the gun's magazine into the chamber and closes the breech. The gun, typically a rifle or shotgun, is then cocked and ready to fire.

16.5.6 Lever Action

This design uses a manually activated lever, usually situated on the underside of the gun, which when pulled down opens the breech, extracting and ejecting any cartridge case or cartridge. Returning the lever loads a cartridge and closes the breech with the gun cocked and ready to fire.

16.5.7 Side Lock

This action describes the position of the firing mechanism, which is situated in removable side-plates as opposed to being integral to the gun. This action is mostly seen in good-quality side-by-side shotguns.

16.5.8 Box Lock

This action is generally found in double-barrelled shotguns; the firing mechanism is situated inside the gun's frame.

16.5.9 Hinged Barrel

The barrel or barrels are pivoted to the gun's frame. Opening of the action is normally facilitated by a catch. The manual act of opening and closing the barrel(s) cocks the firing mechanism.

16.5.10 Repeating Actions

16.5.10.1 Blow Back

This is the simplest of these designs. It relies on the recoil of the gun, the force of which is transmitted to the breech face from the base of the cartridge case. The breech block will move to the rear once its inertia has been overcome, and this results in the extraction and ejection from the gun of the spent cartridge. At the rearmost point of travel, the gun's recoil spring will force the breech block forward, a cartridge will be stripped from the magazine and loaded into the breech. The gun is now cocked and ready to be fired; the cycle will then begin again.

16.5.10.2 Locked Breech System

For more powerful cartridges used in self-loading pistols, simple blow back designs do not keep the breech closed long enough for the pressure to reduce to an acceptable and safe level. In these cases, the breech must be locked to the barrel until the pressure diminishes. Various solutions have been designed to achieve this, such as short recoil, rotating barrel and roller locking systems.

16.5.10.3 Gas Operated

Many high-velocity rifles such as military assault rifles, together with some large-calibre pistols and self-loading shotguns, use a gas-operated system to cycle the gun's action and unlock the breech. In this case some of the gas resulting from the discharge of the propellant charge is tapped off to cycle the action.

16.6 Manufacturing Processes

One of the most frequent tasks the forensic firearms examiner is asked to perform is to establish whether a given gun can be associated with specific recovered bullets or cartridge cases. In order to do this the examiner must know what kind of marks a firearm can transfer to the ammunition. Additionally, an assessment of the significance of these marks is required. Firearms are tools made up of a number of machined and finished components. Each one of these components can act as a tool transferring marks to the cartridge, the cartridge case, bullet, air gun pellet or plastic shotgun wad. Knowledge of how these components are manufactured is needed to enable the examiner to assess the evidential significance of the transferred marks; whether they are so-called 'individual characteristics' or whether they arise from design features which could be common to a group of firearms, so-called 'class marks'. The examiner also needs to be able to differentiate between individual marks and 'sub-class' marks which are unintentional manufacturing marks that can be present on a batch of firearms.

Individual characteristics arise from the irregularities or imperfections of the tool surface, and the profile of the tool may also change during the manufacturing process due to wear or damage. The latter point also applies to the firearm's internal components, which as has already been noted are themselves tools, and hence also subject to wear and damage. These individual characteristics can result in tool marks being transferred to the ammunition component enabling the likelihood that a specific firearm was responsible to be assessed.

16.6.1 Rifling Process

In general, there are two ways in which the barrel's bore can be rifled: by either removing metal or displacing it. The principal metal-removing technique is by cutting, either by using a hook, broach (both single and gang) or scrape method. All involve a tool with one or more cutting surfaces. Metal displacement techniques use pressure to impart the lands and grooves into the bore, the principal methods being button rifling and hammer forging.

After the rifling process the bore is crowned, the muzzle is chamfered to protect the rifling, and the bore is lapped or burnished. Some firearms, such as military rifles, also have a thin chromium layer added by an electrical process to protect the bore from erosion. These processes can also result in individual, class and sub-class characteristics being transferred to the firearm.

16.6.2 Firearms Components

Firearms components such as the breech block/bolt, firing pin, extractor, ejector and magazine are made by a variety of engineering processes. These processes, such as milling, casting, turning, drilling, reaming, grinding, stamping together with the finishing processes, can also lead to the components exhibiting individual, class and sub-class characteristics.

16.6.3 3D-printing

The last few years has seen the rapid development of increasingly sophisticated and affordable technology associated with 3D-printing (additive manufacturing) together with ever more available online open-source access to engineering drawings of firearms and their components. This has led to increasing numbers of such manufactured firearms and their components being encountered in casework. It is therefore essential that the forensic firearms examiner keeps abreast of such novel manufacturing techniques.

16.7 Attendance at Crime Scene and Post-mortem Examination

The forensic firearms examiner is frequently required to attend the scene of a shooting incident and, in the case of a fatal shooting, the post-mortem examination. This is a crucial part of the investigation where vital information can be determined and passed on to the investigating officer.

16.7.1 Crime Scene

Only highly experienced and demonstrably competent personnel are authorised to attend scenes in the lead forensic firearms examiner role. As in all scene attendance, precautions must be taken to minimise the risk of contaminating or compromising any trace evidence. This usually involves taking full DNA and fingerprint precautions by wearing a bodysuit, double gloves and face mask. Special consideration must be given to the risk of GSR contamination.

Safety is paramount when dealing with firearms, and the forensic firearms examiner may be asked to 'make safe' a gun. In any event it is important that the location and status of the gun is recorded before it is handled. Specific notes should be taken of whether the gun is cocked, the position of any safety catch or mode of fire selector, the position of spent and unspent cartridges in the cylinder of a revolver and whether the gun is loaded.

The most obvious questions asked at the scene involve the assessment of possible firearms-related damage; what type of projectile might have caused it; the possible location of cartridge cases, bullets, *etc.* and their identification; and the possible position of the shooter. In cases where the evidence allows a full scene reconstruction, the bullet's flight path can be determined. The scene, including all damage sites, firearms and spent ammunition components, should be photographed with a scale placed next to each item when photographed. The use of digital imaging and when required, computer-aided three-dimensional scanning techniques can greatly assist subsequent interpretation.

16.7.1.1 Possible Bullet Damage

The question of whether a bullet caused some given damage is easy to answer if a bullet can be recovered from the damage site. If no bullet is present, then a close examination of the damage and the use of two simple chemical spot tests can help.

The position and size of the damage site should be recorded. The appearance of the bullet damage will depend on the substrate material, and the shape, composition, velocity and orientation of the bullet relative to the substrate. Since Locard's principle implies that there will be a reciprocal exchange of material when two objects come into contact, it is highly likely that the bullet will leave material, such as lead, copper, or GSR at the impact site and the substrate will transfer material to the bullet. Spot tests for lead and copper/bullet jacket material can be used to assist in the determination of whether any bullet material is present at an impact site. In the last few years portable X-ray fluorescence (XRF) spectroscopy scanners have become increasingly available and can assist in determining whether bullet trace debris is present.

Microscopic lead particle deposition arising from GSR and vaporised bullet lead can also be detected. An assessment of the possibility of ricochet damage can also be made by careful examination and application of chemical spot tests/XRF spectroscopy.

16.7.1.2 Projectile Flight Path

The shape of the bullet damage can indicate the angle of impact. For example, a circular bullet hole and wipe may indicate an orthogonal or near orthogonal impact, whereas an elliptical hole with or without an associated bullet wipe can indicate an angled impact. The shape of the soot or powder deposition can also assist in angle of impact determination. The extent of any such deposition should be measured as this can assist in determining the range of fire.

If the substrate is reasonably thick, such as a wooden fence post, it may be possible to determine the bullet's path by the angle of the bullet track. Also, the appearance of the damage on both sides of the damaged object can assist in determining the projectile's entry and exit sites. If the material is glass, a larger crater is often seen on the opposite side to impact side. Over short ranges bullets tend to travel in straight lines. Consequently, if there is more than one damage site attributable to a single bullet, a straight line joining up these damage sites will give the bullet's flight path. The use of string or thin

Figure 16.3 A car windscreen struck by three bullets. One of the bullet's path is visualised with the use of a trajectory rod and laser allowing the position of the shooter to be estimated.

cylindrical rods and/or small lasers can be very helpful when extrapolating or interpolating a bullet's flight path. See Figure 16.3. This type of reconstruction often leads to an accurate assessment of the shooter's position.

16.7.1.3 Cartridge Case Location

The position of one or more spent cartridge cases can help to determine the position of the shooter; their locations should be noted so that they can be plotted on a scene plan and/or graph paper. Several cartridge cases grouped together can help to indicate the position of a self-loading firearm when it was discharged. Caution is required in such situations as ejected cartridge cases, being almost cylindrical in shape, can bounce and roll several metres on a hard surface such as a road/pavement, less so on a yielding surface such as a field. Nevertheless, the extent of any cartridge case groupings should be measured.

16.7.1.4 Shotgun Discharge Damage

Shotgun cartridges usually contain multiple pellets, so when discharged the pellets can cause multiple damage sites. At close range, within 1 m or so, the pellets can act as a single projectile, they then start to disperse/diverge from the initial pellet column. Consequently, the resultant damage to the target can be in the form of a single hole, possibly with scalloped edges or a pellet pattern with multiple pellet impacts. The size of this pattern should be measured as this will assist in the estimation of the range of fire. Large quantities of pellets can be present at the scene, together with shotgun cartridge wadding. Plastic buffer material, which is used in shotgun cartridges containing large pellets, may also be present. At close range this buffer material and cartridge wadding can also cause damage. The shape of the damage caused by wadding impact can also give insight into the potential range of fire.

16.7.2 Post-mortem Examination

The forensic firearms examiner can play a vital role in assisting the pathologist during the post-mortem examination. The principal questions asked are similar to those asked at the scene of a shooting: what was the gun used and how far away was it from the injured party when it was discharged?

The scientist works with the pathologist to identify the entry and exit wounds. This is not always straightforward, as it is a myth that the exit site is always larger than the entry site. From establishing the bullet's path through the deceased, it is sometimes possible to indicate where the shooter may have been relative to the victim. The use of forensic radiology techniques in recent years has greatly assisted in the interpretation of bullet paths through the body.

If projectiles, be they bullets, shotgun wadding or pellets, are recovered from the deceased, the calibre of firearm can be determined. The shape and extent of soot or powder deposition around an entry wound can indicate the orientation and distance the gun's muzzle was from the deceased. Similarly, the size of a shotgun pellet pattern together with the presence of cartridge wadding injuries can give an insight into the range of fire.

Where there is doubt as to whether the death was accidental, suicide or homicide the determination of the range at which the firearm was discharged is of crucial importance. When assessing a potential hypothesis concerning the mode of death, such as whether it was suicide or homicide, the evidence from the scene examination is brought together with the pathological findings.

Clothing from the deceased and any firearms-related items recovered at the scene and post-mortem examination should be submitted to the laboratory for examination.

16.8 Examination at the Laboratory

16.8.1 Firearms

Safety is paramount and all firearms must be checked to see if they are loaded when they are submitted to the laboratory. If required, full trace evidence anti-contamination precautions should be taken. During this safety check the condition of the gun should be noted. Protocols should be in place to ensure that live or spent ammunition is not submitted in the firearm or in the same exhibit packaging. Sometimes guns are submitted in a condition where the safety has not been assessed, such as when the breech is jammed closed, *e.g.* through corrosion. This can happen if a gun has been buried or has been recovered from a river. The use of an X-ray scanner can be useful in such situations to see if there is a cartridge in the gun's chamber.

Once the gun has been made safe, the full examination will usually take place after all other trace evidence work has been carried out. Trace evidence recovery may occur during a joint examination with scientists from other disciplines, or the forensic firearms examiner may be cross-trained.

When the firearm is first examined it should be photographed. The nature and positions of any foreign material adhering to it should be noted. Notes including a full description of make, model and type of firearm should be taken. The overall length, barrel length and magazine capacity should be recorded, as these parameters may have a bearing on the legal classification of the firearm.

A low-power stereo microscope can be used to examine the condition of the breech and bore. An endoscope is also a useful tool to assist in bore examination. If the bore is rifled, the direction of twist and the number of lands and grooves should be noted. A bore wipe, using a cleaning patch, filter paper or swab, should be taken as this will remove some of the GSR particles present which can be analysed. The position and nature of any markings stamped on to the firearm should also be noted; these can include the serial number, proof marks and date codes. Proof marks are stamped on to the firearm when the firearm is sent by the manufacturer to a proof house. At this testing facility the firearm will have been test fired so that it can be authenticated as being safe to be released into the marketplace.

The serial number may have been partially or totally removed to conceal the firearm's provenance. It is sometimes possible to discern the serial number by use of a stereo microscope, otherwise the examiner can use several techniques, such as chemical and electrochemical etching, ultrasonic cavitation and magnaflux, to recover it.

One of the propositions frequently put forward is that the firearm discharged accidentally or inadvertently. To address this, it is essential to determine if the safety catches and mechanisms are working. It is also important to determine the firearm's trigger pull by carrying out a quantitative assessment of the force required to pull the trigger. To see whether the firearm is prone to accidental discharge, a series of drop and impact tests are performed.

Once these tests and observations have been made an assessment should be carried out to see whether the firearm is safe to test fire. This is particularly important with modified, converted, reactivated or improvised (so-called craft guns) firearms. If the firearm is deemed unsafe, it may be appropriate to discharge a primed cartridge case, *i.e.* one that does not have a bullet or propellant loaded into it. A remote test firing device should be used with suspect firearms.

Firearms should be test fired using a range of ammunition with different bullet and primer materials. Varying the nature of the bullet/primer material will help to maximise the range of characteristic markings transferred to the ammunition from the gun. Undistorted bullets can be recovered by firing into a water tank or box containing cotton wool or bespoke material. The recovered test fired bullets and cartridge cases should then be placed into an appropriately labelled bag.

In order to assess the lethal potential of a discharged projectile it may be necessary to measure its velocity using a chronograph. Once the velocity has been determined the kinetic energy of the projectile can be calculated. A chronograph can also be used to enable other ballistic calculations to be made, such as the extreme range of the projectile(s).

In order to classify air guns with respect to the UK firearms legislation it is necessary to determine the muzzle kinetic energy of the discharged pellets. Air pistols have a limit of 6 ft lbs, other types of air gun 12 ft lbs; if the gun is capable of discharging pellets in excess of these limits further legal restrictions apply to their possession. Additionally, following the change in UK legislation in 2021 a suitably calibrated chronograph can be used to determine whether a specific gun can discharge a projectile with a kinetic energy in excess of the 1 J threshold defining a firearm.

Comparative penetration tests using ballistic soap, gelatine or other homogeneous media can be carried out when assessing the wounding potential of a projectile. Figure 16.4 shows the effect of a high-velocity

Figure 16.4 The effect of a 7.62 NATO bullet on ballistic gelatine. Note the back-spatter and temporary cavitation.

bullet passing through ballistic gelatine. The use of soft tissue and bone simulants can also be used to assess bullet behaviour on impact.

The firearm can also be test fired at various ranges in order to duplicate the soot, propellant or shotgun pellet pattern observed at the scene, at the post-mortem examination or on recovered clothing. To obtain precise range estimations the recovered gun and ammunition should be used. Sometimes this is not possible, and approximations must be made. Figure 16.5 shows the soot and powder deposition for four different types of 0.357 Magnum calibre ammunition discharged in a Smith & Wesson revolver as a function of distance.

The use of high-speed digital cameras in the last few years has greatly assisted in the assessment and visualisation of firearm and projectile behaviour and are now routinely used in forensic firearms units.

16.8.2 Ammunition

Unfired and spent ammunition should be examined to determine its calibre and make. The calibre is usually readily identifiable by the experienced examiner; however, if it is an unusual cartridge then its dimensions should be determined, and reference made to specialist validated books and databases. The viability of unfired ammunition should be determined, either by dismantling the cartridge and examining the primer and propellant or by test firing the cartridge. Spent bullets are weighed and any rifling characteristics present determined by identifying the number of lands and grooves and their

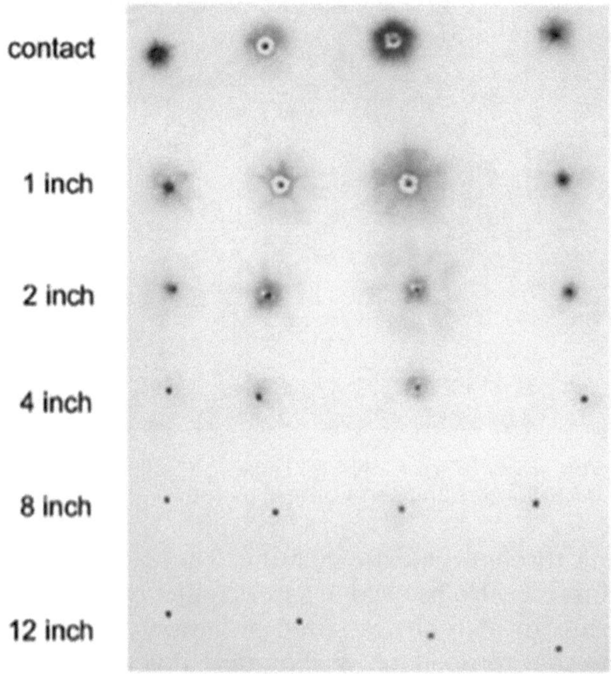

Figure 16.5 Soot and powder deposition for four different brands of 0.357 Magnum calibre cartridges as a function of distance of the gun from the target.

direction. The dimensions of the land and groove widths can also be measured. These rifling characteristics can be used with suitable databases, such as those created by the Federal Bureau of Investigation (FBI) and the Bundeskriminalamt (BKA), to determine what make and model of firearm may have been used.

The presence of any trace evidence, such as powdered glass or paint, can help to determine what the bullet may have struck: damage to the surfaces of the bullet other than the nose can indicate a ricochet, and the angle of impact can also be assessed by the angle of the plane of damage to the bullet. Similarly, any gun-dependent class marks present on spent and unfired cartridge cases or cartridges are assessed. The principal marks as identified previously in Figure 16.2 are the shape, position of the ejector and extractor, firing pin impression, breech face and chamber marks. These can all help to identify the type, make and sometimes model of gun used.

Any shotgun pellets submitted should be examined to determine if they are conventional lead pellets, and a representative sample should be weighed, and their diameters measured, so that their nominal shot size can be determined. The size of shotgun cartridge wadding will give the calibre of the parent cartridge and may also have pellet impressions pressed into its upper surface. The size of these impressions will also enable the shot size to be estimated. Plastic shotgun wadding should be closely examined, since the shotgun's barrel, especially if sawn-off, can transfer individual characteristics on to the wad.

16.8.3 Clothing

Any garment with suspected gunshot damage should be photographed and the positions of any damage noted, and the extent of the damage determined. The use of low-power stereo microscopy is helpful in looking at the nature of the damage sites.

The spot tests for lead and copper, previously discussed, should also be used to assess entry and exit holes. The extent of soot and propellant deposition can assist in range of fire determination.

Chemical tests for propellant pyrolysis products can also assist in the visualisation of any propellant distribution and may give insight into range of fire as can the use of an infra-red (IR) enabled digital camera when attempting to visualise soot on dark coloured fabric.

The use of a mannequin can be helpful in visualising the bullet's path though the body and clothing.

16.9 Comparison Microscopy

Individual characteristic gun marks present on cartridge cases and projectiles, such as bullets, air gun pellets and plastic shotgun cartridge wads can be compared to one another by using a comparison microscope. This piece of equipment is in effect two microscopes joined by an optical bridge, and uses a prism to enable two objects to be viewed simultaneously, side by side, at the same magnification.

The rifling or bore marks present on the bullet, air gun pellet or plastic shotgun cartridge wad which have arisen from the passage into and down the barrel are compared to other recovered samples or test firings using the comparison microscope. Similarly firing pin, extractor, ejector, chamber, or breech face marks transferred by the firearm to the cartridge case or cartridge can also be compared to other submitted samples and test fires. Figures 16.6 and 16.7 show a bullet and a cartridge case comparison, respectively.

Figure 16.6 Use of a comparison microscope to match individual characteristics on the land impressions of two bullets.

Figure 16.7 Use of a comparison microscope to match individual characteristics on the primers of two cartridge cases.

Live cartridges can also be examined to see if there are any gun-dependent marks on them which have resulted from being worked through the action or magazine of a self-loading firearm. Currently the principal skill required by the examiner, together with years of experience, is pattern recognition. In order to positively assess the likelihood that any gun-dependent tool mark present on a cartridge component was derived from a specific gun the examiner must be satisfied that the level of tool mark agreement exceeds what may have originated by chance. This process is subjective in nature and based on the examiner's training and experience.

Any critical comparison microscopy conclusion should be subject to peer review.

Recent research has built on work carried out in the 1950s to give a statistical foundation to the theory of tool-mark identification. This work has resulted in the formulation of the Consecutive Matching Striae (CMS) method for firearms identification using comparison microscopy. In CMS the examiner applies a conservative criterion of runs of aligned striae to establish a possible single source. Despite introducing a degree of quantification within the methodology, the CMS technique is also subjective – as it relies on the examiner's observations and interpretation.

The subjective nature of both the pattern recognition and CMS techniques means that any interpretation and evaluation of a comparison may be affected by some form of unintended bias. In the interpretation process the results are based on the individual's relevant experience, depth of knowledge and skill. Every effort must be made to make it logical, transparent, balanced and robust. Usually, the opinions are formed in the context of supplied case information, introducing the possibility of contextual bias. Limiting the supplied information to both the examiner and peer reviewer can reduce the risk of introducing bias. In the latter case the reviewer should be unaware of the examiners conclusions to reduce confirmation bias.

Current research and technological advances in cartridge case and bullet comparisons centre upon the measurement and/or image capture of surface topographies in either 2 or 3 dimensions. These outputs can then be compared using computer-based algorithms. The aim being to develop more objective methods to assess the likelihood that discharged bullets and cartridge cases could have originated from a specific gun. The development of large databases to calculate evidential strength will also assist in providing a more robust foundation to the attribution of a specific gun to a given discharged bullet and/or cartridge case.

Most government or police forensic firearms laboratories have what is known as an open case file (OCF). This consists of a collection of ammunition components, such as bullets, cartridge cases and plastic shotgun wads which have been recovered from the scenes of unsolved crimes. The collection is subdivided into calibre, and in the case of bullets, rifling characteristics. When appropriate spent or unfired ammunition and its components are submitted to the laboratory, they are screened through the OCF to see if there is any previous use of the same gun. Similarly, when a firearm is submitted, its test-fired bullets and cartridge cases are screened through the OCF. Depending on the size of the OCF, this can be a long process. In recent years there have been huge developments in the automation of the comparison microscopy process using hardware and software to digitally capture the individual characteristic marks and search them against the uploaded OCF database. Recent advances in digital file formats (particularly the .x3p format) may assist in the sharing of uploaded digital ballistic sample images between institutions using different makes of automated comparison technologies. Epoxy resin casting techniques have relatively recently been optimised so that replicas of spent bullets and cartridge cases can be shared between laboratories for comparison microscopy purposes without the need to send the actual crime samples.

However, it is still the case that any potential match must be manually looked at by an experienced examiner.

16.10 Firearms Chemistry

16.10.1 Gunshot Residue (GSR)

This section illustrates the value of GSR as trace evidence and the factors affecting its detection and identification in crime. When a gun is fired, thousands of microscopic particles, called gunshot residue, are produced by the ammunition. They are emitted together with hot gases from the muzzle of a gun as shown in Figure 16.8 and from any gaps or openings in the gun's action and can be deposited on the firer, any persons or surfaces sufficiently close to the firer, and the gun itself.

The recovery and identification of GSR particles can help address questions such as, 'Has the suspect recently fired a gun?' That the particles originate from a firearm is rarely contested. Of far more interest to the court is how these particles came to be on a suspect's clothing or hands. GSR evidence is one of the most heavily scrutinised types of trace evidence in criminal investigations.

Figure 16.8 High-speed image showing the emission of combustion gases from the muzzle and cylinder gaps of a revolver. Reproduced from https://kuulapaa.com with permission from Herra Kuulapaa, Copyright 2013.

16.10.2 Origin and Chemistry of GSR

Modern percussion primers are the products of more than 100 years of research. The earliest primers contained potassium chlorate as the main ingredient, but mercuric fulminate mixtures were introduced in 1831 and used widely thereafter. Mercury caused cartridge embrittlement and barrel corrosion, so the first non-mercuric, non-corrosive primer was introduced in 1928. This 'Sinoxyde' primer was composed of lead styphnate (an initiator), barium nitrate (an oxidant) and antimony sulphide (a fuel). This primer composition continues to be used alongside lead-free primers which were introduced in the 1980s.

When the primer is detonated by the gun's firing pin, the primer composition decomposes exothermically and drives hot gases and molten material into the cartridge case, igniting the propellant. The burning propellant decomposes, producing large volumes of hot gaseous products which dramatically increases the pressure within the cartridge. The cartridge case expands tightly against the chamber walls (this is known as obturation and prevents the rearward escape of combustion gases), while the high pressure forces the bullet out of the cartridge, into the barrel and out of the gun.

The high temperature inside the cartridge creates conditions in which the individual components of the primer can fuse together, in a matter of ten thousandths of a second. This very short time frame accounts for the variability and inhomogeneity that is observed in the composition and numbers of GSR particles produced. After the material has cooled, discrete particulate residue remains, containing combinations of the elements lead, barium and antimony and any other components present in the primer. The number of variations in the components is limited and therefore the number of different types of residue is small. Other variations can arise from components used in the construction of the primer cap, the firearm and the bullet, *e.g.* the use of tin foil to seal the primer cup. Nevertheless, different types of GSR particle can be identified and used to eliminate a particular gun or spent cartridge case as the source of residue found in a shooting incident.

In any population of residue particles there will be a variation of the proportions and combinations of the chemical components. The compositions in most cases will not be in fixed or consistent proportions and can even vary from shot to shot. This is one of the reasons that experimental reconstruction to investigate the numbers and composition of GSR produced during a specific crime scenario is particularly difficult. This can be understood by considering the very short time available for the components of the primer to mix and fuse together during the detonation and subsequent ignition of the propellant. Particles containing all three elements *i.e.* lead, barium and antimony were previously considered unique to the discharge of a firearm, however they have been subsequently observed elsewhere as will be discussed later (Section 16.11.4). For this reason, GSR is now widely referred to as being characteristic of originating from a firearm. Particles without all three components are referred to as 'indicative' or 'consistent' with a firearms origin.

The particles of GSR may also contain contributing elements from the cartridge case, bullet, gun barrel and propellant. Firearms that are not maintained and exposed to climatic conditions such as rain can rust with the formation of a layer of hydrated iron(III) oxide on the surface or interior of a steel barrel. Iron, sometimes at a major level, can be observed in the GSR produced from discharging ammunition from such guns. Particles from the bullet can also often be observed alongside particles originating from the primer. High levels of small spherical lead particles often with a trace of antimony, used to harden bullets, are produced by unjacketed leaded ammunition and lead shot.

16.10.3 Detection of GSR

The earliest methods for the detection of residue on the hands of a suspect were based on simple colour tests. These tests enabled the detection of lead, antimony and barium from primers and other elements, such as copper and zinc, from the cartridge case. These tests suffered from a distinct disadvantage in that they were not source specific and many sources of these elements other than firearms exist. Today these tests can only be used for screening or as presumptive tests, but they can be very useful in demonstrating the presence of lead or copper around a bullet entry hole.

In the early 1970s work began independently at the Metropolitan Police Forensic Science Laboratory in London and the Aerospace Corporation in California on a new method of identifying individual particles of gunshot residue using the scanning electron microscope equipped with an X-ray analyser (SEM-EDX). Importantly, this work established that discrete particles composed of lead, barium and antimony appeared to occur only in percussion primer residue. There was no known source other than firearms, and hence the first conclusive method for identifying GSR was established. This particle analysis method is still employed today.

16.11 Analysis and Identification of GSR Using SEM-EDX

Briefly, the SEM functions by scanning a finely focused beam of electrons over a sample. The impinging electrons interact with the sample and produce an image of a particle. Furthermore, the electron beam results in a displacement of electrons and emission of characteristic X-rays from the particle. These X-rays are then used to identify the elemental composition of the particle. Overall, the SEM locates particles of interest, generates a characteristic X-ray spectrum and software classifies the particle.

In general, GSR particles will be rounded because they have been generated during a fusion process, although an irregular appearance does not necessarily result in a particle being ruled out. The final decision on whether a particle constitutes GSR will depend on the morphology, size, chemical composition, homogeneity and the nature of the sample as a whole and must be carried out manually by an experienced analyst. In terms of size, the largest particles will generally be lost first and so most particles found on clothing are of the order of 1–10 μm in size (1 μm is one thousandth of a millimetre). Figure 16.9 shows a typical SEM image of a GSR particle.

Over the past 20 years or so the original acceptance criteria for the chemical components constituting a particle of gunshot

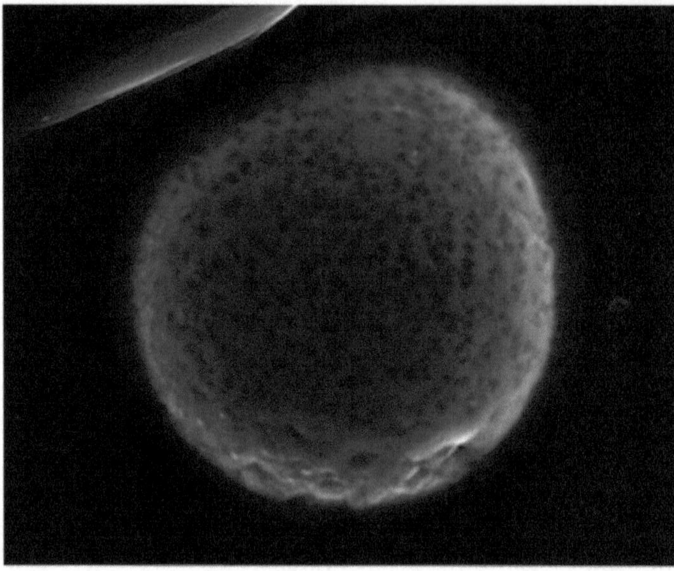

Figure 16.9 A scanning electron micrograph of a GSR particle.

residue have been refined and widened. As a general rule, the following five compositions of GSR particles are seen in casework and can be used to distinguish between different sources of residue:

(1) Lead–barium–antimony (characteristic).
(2) Lead–barium–antimony–aluminium (characteristic) and barium–aluminium (indicative).
(3) Lead–barium–antimony–tin (characteristic) and barium–antimony–tin (characteristic).
(4) Lead–barium–calcium–silicon (characteristic) and barium–calcium–silicon (indicative).
(5) Lead–barium–calcium–silicon–tin (characteristic) and barium–calcium–silicon–tin (characteristic).

In the absence of characteristic GSR particles it cannot be said that indicative particles originate from a firearms source as they can be commonly encountered in the environment. This is particularly the case with particles containing lead–antimony, barium–antimony and lead–barium (two-component particles) or lead, or barium or antimony (one-component particles). There are several relatively common, non-firearm related sources of these particles and it therefore cannot be stated in the absence of characteristic particles that these have a firearms origin. For these reasons in most cases in the United Kingdom one- and two-component particles are not reported.

It has been observed in casework in the United Kingdom that the discharge of ammunition containing lead–barium–antimony–aluminium primers, particularly shotgun ammunition, generates characteristic GSR particles in combination with a profusion of barium–aluminium particles. Often the barium–aluminium particles outnumber the characteristic particles. The reason for this could be the thermodynamic fusion of these two elements is preferentially more favourable than combinations of the other elements present.

Particles containing barium–lead where the barium peak is larger than the peak for lead are unusual in the general environment and are indicative of GSR originating from 0.22 calibre rim fire ammunition. The particles can be characteristic if the spent cartridge case or gun from the crime is available for comparison of the residue type.

Ammunition containing the older primer formulations based on mercury is still available, particularly in criminal circles. Such ammunition can produce residue particles where the mercury itself is not always present or detectable since it is so volatile, leaving, for example, only antimony–tin particles. This can mean detection and conclusive identification of the particles as originating from a firearms source is not always straightforward.

More recently, lead-free primers have been developed. The first and most common is Sintox, which is based on the chemical elements titanium and zinc. Newer lead-free formulations now exist based on potassium and boron, elements that are more difficult to detect in the SEM. Ammunition containing these components has so far rarely been seen in crime in the UK.

16.11.1 Deposition and Transfer of GSR

As indicated earlier, GSR is emitted from the end of the muzzle of a gun and from any openings such as the breech face or cylinder gaps in a revolver on to the firer and anyone standing next to the firer. The majority of GSR particles are deposited within approximately 3 m from the end of the barrel and approximately 1 m on either side of a gun. The amount of residue produced will depend on the type and condition of the gun, the type of ammunition and calibre of the firearm. Greater quantities of residue may be deposited if a shot is fired inside rather than outside, since external climatic conditions such as wind, rain and temperature can influence the amount deposited.

After the bullet emerges from the gun's muzzle and strikes a substrate it can transfer gunshot residue from its surfaces, in particular the base, to the perimeter of the entrance hole in the target. This appears to occur irrespective of the range of fire. On passing through the target surface most of the residue will be removed from the bullet's surface, such that little or none will be present on the exit hole. The presence or absence of gunshot residue therefore allows differentiation between the entry and exit holes. Examination for gunshot residue on the clothing of a shot victim, or *e.g.* a car window which has been shot at, can help establish whether the shot was fired close to or from afar and also the composition of the residue.

Handling a recently fired gun with residue on its surfaces can also result in a transfer of residue to the person's hands. This is termed secondary transfer. It is generally accepted that only a small proportion of the residue will be transferred at each contact. Consequently, it is often not usually possible to distinguish between residue that has been deposited as a result of firing a gun from that which has originated from handling a gun with residue on its surfaces or from standing next to someone firing a gun.

16.11.2 Persistence of GSR

Although thousands of particles can initially be produced upon firing a gun, in reality only a fraction of these ultimately remain on a suspect's clothing by the time they have been arrested and their clothing seized. Primer residue is chemically stable and does not evaporate or dissolve. It is lost by physical disturbance, with washing and wiping considerably accelerating the loss. Up to approximately 90% of GSR particles can be lost in the first hour following deposition. On the hands of a firer, residue may only remain for a few hours at most and may be lost much more quickly. The face and hair of a suspect usually suffer from less disturbance and persistence can be a few hours longer. On clothing, GSR can remain for up to a day on the outer surfaces if the same clothing is continuously worn. Walking, running or brushing the surfaces of the clothing will increase the rate of loss. However, residue may be transferred to a pocket from the hand or from a gun and can remain there indefinitely if not subject to any disturbance.

16.11.3 Recovery of GSR

Several methods of collecting primer and organic residue are employed in forensic science laboratories. The optimum recovery of primer residue is by a taping method. Small adhesive strips or carbon tape mounted on small metal stubs are pressed repeatedly over the surface of a suspect's hands, face, hair or clothing. Any particles of gunshot residue will adhere to the tape which can then be analysed in the SEM.

The forensically significant components of organic residue, namely nitroglycerine and 2,4-dinitrotoluene, are best collected using either a swabbing or vacuuming technique. Any primer or organic residue can be extracted from the resultant swabs or filters and the organic residue analysed using chromatographic techniques such as GC-TEA or HPLC. These methods are not employed routinely in most laboratories since examination for primer residue alone is more discriminating and sufficient to establish exposure to a firearms source. The chromatographic methods can also be used to detect the components of explosives, and such organic analysis is still undertaken in Northern Ireland.

Generally, it is outer items of clothing that can be expected to offer the best opportunity of detecting GSR following a firearms offence since residue will remain for longer on clothing than on skin and hair. Other items which are also routinely examined include gloves, balaclavas, bags which may have been used to convey firearms before or after an incident and vehicles from or at which shots may have been fired.

16.11.4 GSR-similar Particles

A variety of materials have been reported in the scientific literature as potential alternative sources of particles that may be mistaken for GSR. In general terms the particles are generated during a hot process, thus conferring a spherical often molten appearance.

Fireworks are often mentioned as being a possible source of particles similar to GSR. The main ingredient is black powder and to this is added an oxidiser, metal fuel and binders. To produce the colour observed when a firework burns, several different chemical compounds are employed such as potassium chlorate, strontium nitrate, barium nitrate and aluminium or magnesium powder. Antimony is often used to produce a glittering effect and lead produces an audible crackling sound. The potential can be seen for the production of pyrotechnic residue containing similar combinations of elements to GSR. One report of finding isolated lead–barium–antimony particles in the population of firework residue has been published in the scientific literature, but this has not been reproduced in other studies.

A profusion of barium–aluminium–magnesium particles has been observed in casework samples, particularly around the month of November. However, the presence of other particles commonly containing elements such as potassium, chlorine, copper and most notably magnesium would alert an experienced scientist that they were not observing GSR.

Lead–barium–antimony particles have been observed in samples taken from certain car brake linings and hence could possibly be present on people employed in occupations such as car mechanics. However, most of these particles also contained high levels of iron, magnesium, sulphur, silicon and titanium. The particle's appearance is usually one of a compacted mass which lacks the features of particles produced at high temperature and therefore not typical of GSR.

Cartridge-operated tools such as nail guns use blank cartridges as the energy source for providing the driving power. Blank cartridges contain all of the components of a round of ammunition except the projectile. In the UK most cartridge-operated tools employ rim fire ammunition, and the resulting residue can, from the coating on the nails, contain higher levels of zinc than GSR.

16.11.5 Contamination Avoidance

The largest single factor that must be considered when interpreting GSR findings is the possibility of contamination. In general, small amounts of GSR particles are anticipated so the opportunities for cross-contamination must be kept to a minimum and the risk must be assessed within the overall context of the case.

At a crime scene, as with any evidence type, appropriate protective clothing must be worn by anyone entering the scene. Clothing and samples from suspects' skin and hair surfaces need to be recovered as soon as possible after the incident. This will depend on the circumstances of the case and the time interval between the incident and the apprehension of a suspect. Generally, the suspect will be taken back to a police station where the samples will be taken from them along with their clothing. It is very important that any suspect is not conveyed in a vehicle used by armed police officers or taken to a police station that has a firing range, or sampled by someone involved with firearms, as this may compromise the evidence. At the scene it is critical that the investigator does not handle any guns or spent cartridge cases and then come into contact with a suspect or other item that is to be examined for GSR. Separate individuals should be involved in the collection of evidence that may be highly contaminated with GSR and that which may only contain low levels. Any other items recovered at the scene, such as abandoned items of clothing, bags, gloves or masks, should as with most forensic evidence be packaged, sealed and exhibited as soon as is practical to ensure its integrity.

At incidents involving firearms it is common practice for armed police officers to attend to secure the safety of the general public. In the most severe incidents this can mean incapacitating a suspect to prevent them from discharging a firearm that may injure themselves or others. The officers will also be involved in searching suspects thought to be in possession of a firearm and addresses where firearms may be located. There is a distinct possibility that firearms officers will have GSR on their hands and operational clothing, and that this may be transferred through physical contact to a suspect. The officers may also deploy a distraction device such as a flash-bang either in training or when operational. These devices produce smoke and a loud bang and produce distinctive particles containing zirconium, barium, aluminium, tungsten and chromium among other chemical elements. If found on a suspect's clothing, they provide further information in relation to the arrest of the suspect. All details of arrests of suspects involving firearms officers need to be considered when interpreting the GSR findings in a case.

Stringent procedures should be in place to ensure that the risk of any cross-contamination at the laboratory is kept to an absolute minimum. Quality control samples must be taken at the time of examination of an item and are generally taken from the examination bench, the outside of the packaging of an exhibit and possibly also the gloves and protective disposable coveralls from the sampler. Full personal protective equipment should be worn when conducting examinations and gloves should be changed frequently. Most laboratories undertake routine environmental monitoring by taking samples from the areas where items are examined and, in the area, where the SEMs are utilised. This ensures that the environment does not present a contamination risk to the integrity of the exhibits being examined and in the absence of these measures the integrity of an exhibit can be challenged.

16.11.6 Interpretation of GSR Findings

As can be seen, deposition of GSR is a highly variable event and subject to a large number of factors that determine what is eventually found.

This represents the challenge in interpreting the findings in a case to determine if someone may have been in possession of or fired a gun.

A brief outline of the circumstances will normally be available, but the practitioner must establish the full case circumstances as far as is practicable if they are to come to a reliable interpretation. Key factors include the timings between the incident and the recovery of the items, the recovery of a weapon or spent cartridge cases, the number of any shots fired, whether armed police officers were involved and what the suspect has said in response to the allegations. This last point is important in establishing if the suspect has any other explanation for the presence of GSR on their clothing, *e.g.* being a legitimate holder of a firearms licence or a member of a gun club. However, this tends not to be the case in most incidents involving firearms and the suspect may admit to being present but not to discharging or handling the weapon. It is then for the firearms chemist to consider the relative likelihood of the GSR evidence in terms of the questions asked by the investigator compared to any reasons given by the suspect.

In order to mitigate against cognitive bias in reporting GSR cases there are a number of strategies that are variously used in current practice and include independent checking of the classification of GSR particles by two competent and experienced scientists (the analyst and one other) independently of each other; utilising the case assessment and interpretation approach to document expected outcomes before commencing work and subjecting this to independent review; full independent review of the interpretation of GSR findings; scientists being able to demonstrate initial and ongoing competence and participate in internal and external quality assurance trials. The latter ideally would involve submission of a mock case unbeknown to the scientists which would be reported in the same manner as a normal case thereby testing the whole process from examination for the presence of GSR through to reporting findings to the customer.

It is often alleged that a suspect used a particular hand or drew a gun from a pocket or the waistband of their trousers. Finding residue in pockets suggests that a gun or a hand with residue on it has been placed inside the pocket at some time. Residue in pockets is less likely to be the subject of accidental contamination since it requires an additional step in the transfer process. This is important in cases where it is alleged that a suspect may have been in possession of a firearm. It is a common misconception that the finding of residue on a particular hand of a firer proves that this was the firing hand. Residue is redistributed and lost almost immediately after it is deposited. However, finding residue on the hands, face or hair of a suspect supports recent deposition and can be very helpful in an investigation. Finding residue on the back of a garment rather than on the front or sleeves can indicate a proximity to firing rather than being the firer. Significant amounts of residue are rarely found inside the waistband as this area is subject to much disturbance.

It must always be considered that if the time interval between an incident and recovery of a garment that was being worn is longer than a few days then the finding of GSR is more likely to be attributable to a more recent event. Conversely, the absence of GSR does not exclude a firearms involvement as residue may have been lost in the intervening time.

Although the presence of GSR in the environment is rare, in recent years there has been an increase in the criminal use of firearms and a rise in the number and use of armed police officers, so the risk of accidental contamination has risen. People who associate with firearms users might unknowingly get a small amount of residue transferred to them. For these reasons, very little can be said about the finding of single particles.

Residue is also important in helping to establish range of fire. It is by no means as precise an indication as that given by the firearms examiner following laboratory tests, but it is useful in instances where no visible blackening or powder patterns have been left or where the crime weapon has not been recovered. It can also be used when differentiating between entry and exit holes.

Further Reading

M. G. Haag and L. C. Haag, *Shooting Incident Reconstruction*, Elsevier Academic Press, San Diego, 3rd edn, 2021.

B. J. Heard, *Forensic Ballistics in Court: Interpretation and Presentation of Firearms Evidence*, Wiley-Blackwell, 1st edn, 2013.

T. A. Warlow, *Firearms, The Law, and Forensic Ballistics*, CRC Press, Boca Raton, 3rd edn, 2012.

J. Wallace, *Chemical Analysis of Firearms, Ammunition, And Gunshot Residue*, CRC Press, Boca Raton, 2nd edn, 2018.

R. Nichols, *Firearm and Toolmark Identification: The Scientific Reliability of the Forensic Science Discipline*, Academic Press, 1st edn, 2018.

17 Forensic Entomology

Dorothy Gennard

University of Lincoln, School of Life Sciences, Joseph Banks Laboratories, Lincoln, LN6 7DL, UK
Email: dgennard@lincoln.ac.uk

17.1 Introduction

Insects and arthropods, such as mites, spiders and indeed barnacles, are often the unexpected providers of evidence of a criminal or civil misdemeanour. These organisms have an evidential role in legal cases comparable to that of other biological evidence. Insects respond to opportunities to gain food, a reproductive site, a place to lay eggs and a place in which to develop through the stages of their life cycle. Their presence at crime scenes is interpreted by forensic entomologists in the light of knowledge about insect life cycles and patterns of succession, combined with how environmental conditions influence their growth, in order to provide an estimate of the time since colonisation of a body, dead or alive, or other information of interest to a court.

Forensic entomologists work on cases of suspicious death, alleged abuse and neglect, or infestations and nuisance in urban contexts such as food preparation areas, building sites or hospitals. The materials examined by them can range from food sold for human consumption, plant material including wood, to animals and humans living and dead. In each instance, forensic entomology can contribute

Crime Scene to Court: The Essentials of Forensic Science, 5th Edition
Edited by Niamh NicDaeid and Peter C. White

Published by the Royal Society of Chemistry, www.rsc.org

to the range of evidence presented to assist courts in reaching their conclusions. Their evidence is particularly valuable subsequent to the 48–80 hour period after death when the body's physiological changes are no longer readily interpretable by forensic pathologists.

Assistance in the context of civil cases rely more on knowledge of a specific order of insects or other arthropod and its ecology than on insect succession, and could be the result of an individual seeking the help of a forensic entomologist to bring a claim for reimbursement for personal loss or inconvenience. For example, infestation of buildings by insects may not only be an issue for environmental health officers but can also, on occasion, involve forensic entomologists where the matter forms part of a court case.

Such a civil case might be in the context of pest control in rented properties. For example, a person who has become ill because of eating tainted food stored in a cupboard may file a case in the civil court alleging that there is a cockroach infestation and that a pest control company should have been employed by the property owner to remove the pests. A forensic entomologist may be employed by the person to investigate the nature and size of the cockroach infestation at the flat and whether control measures allegedly previously put in place by a company employed by the landlord had been adequate.

Fundamentally, any insect or indeed arthropod can provide evidential information of importance in forensic investigations. However, greatest attention has been paid to those insects recorded in association with a corpse and these species have become the subject of ecological investigation by forensic and other entomologists.

The insect species of value in a particular case may vary geographically. In general, in Europe and the USA, they include necrophagous insects (*i.e.*, those feeding on dead bodies) such as the blowflies *Calliphora vomitoria* Linnaeus (Figure 17.1a), *Calliphora vicina* Robineau-Desvoidy (Figure 17.1b), greenbottles (Figure 17.1c) and flesh flies (Sarchophagids) (Figure 17.1d). Other species more frequently associated with organic matter, compost and manure, such as the black soldier fly *Hermetia illucens* (Linnaeus) have also been used to determine post-mortem intervals. All that is needed is scientific literature on its ecology and experimental data relating to the length of time that species takes to develop through individual life stages from egg to adult emergence at constant temperatures. Determination of size (length) of larvae in each larval stage and duration at set temperatures in particular stages may also give important information that can be translated into a measure of time since death.

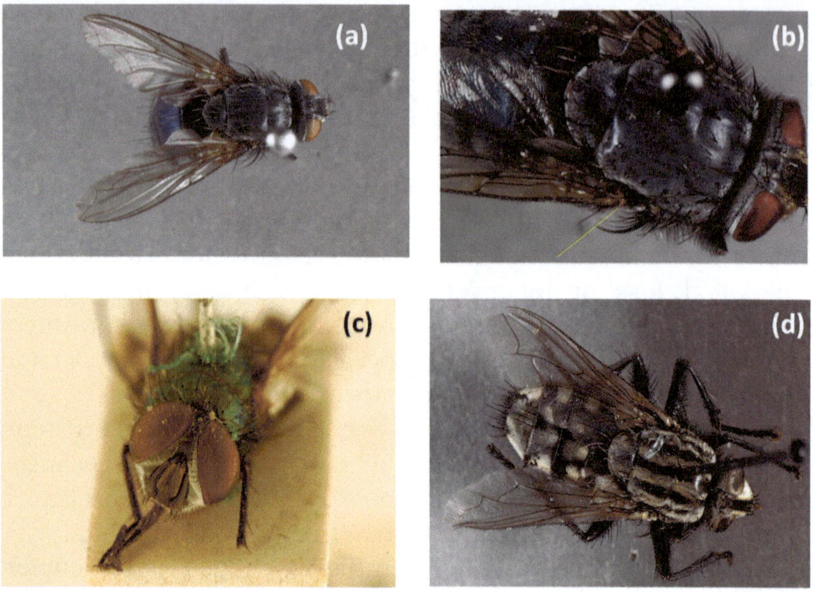

Figure 17.1 Species which are initial colonisers on bodies (a) *Calliphora vomitoria* (Linnaeus) a species more frequently considered a rural coloniser. (b) *Calliphora vicina* Robineau-Desvoidy a member of the Calliphoridae which tends to frequent urban environments. The marker indicates a feature used to identify this species. (c) An example of *Lucilia* species (greenbottles). (d) A member of the Sarcophagidae (flesh flies). Reproduced from ref. 1 with permission from John Wiley & Sons, Copyright 2007.

17.2 Insect Succession on a Corpse

As it decomposes, a predictable succession of insect species colonises an unburied corpse. The stages of decomposition are not discretely discernable; speed of decomposition of a particular body is environmentally dependent and strongly influenced by temperature and the rapidity of insect colonization. Worldwide a number of necrophagous species are recorded on corpses. Cosmopolitan species such as *Calliphora vicina* and *Lucilia sericata* (Meigen) are found in many countries but particularly in temperate zones. They are early colonisers of bodies. Other blowfly (Calliphoridae) colonisers include: *Calliphora livida* Hall; *Calliphora vicina*; *Cochliomyia macellaria* (Fabricius); *Lucilia sericata* (also known as *Phaenicia sericata*) and the black blowfly *Phormia regina* (Meigen), species of fleshfly (Sarcophagidae), sepsid flies, and dung flies (Scathophagidae). The initial coloniser species may vary with geography. In Dubai, for example, flesh flies colonized a body before blowfly species.

Beetles found on bodies include clerid beetles such as *Necrobia rufipes* De Geer; silphid beetles such as *Nicrophorus marginatus* Fabricius; the dermestid beetle species *Dermestes frischii* Kugelann, *Dermestes marmoratus* Say, and *Dermestes caninus* Germar; histerid beetles such as *Saprinus* sp.; members of the Nitidulidae for example *Omosita colon* (Linnaeus); predatory Staphylinidae such as *Creophilus maxillosus* (Linnaeus). The latter species is a secondary coloniser feeding on fly eggs and larvae, rather than the corpse.

Succession on buried corpses is slower, not least because soil impedes the exit of decomposition odours from the body, restricts the numbers of insect species that can gain access to the corpse and also because the grave temperature fluctuates much less and tends to be lower than the ambient air temperature. Decomposition is restricted as a result. The decomposition rate also varies dependent on grave conditions, including soil type and moisture level. Even after 10 years, a buried body can be preserved with recognisable skin and organs.

Much of the early knowledge about the assemblage of insect species colonising bodies, buried and unburied, arises from the work of a Frenchman called Jean Pierre Mégnin (1828–1905). He recorded four species of flies on buried bodies: *Calliphora vomitoria*, *Muscina stabulans* (Fallén), *Anthomyia* sp., and *Phora atra* (Meigen) a member of the Phoridae. He also noted the beetle species, *Rhizophagus parallelocollis* Gyllenhal, and springtails (Collembola) such as *Achorutes armatus* Lameere (*Ceratophysella armata* Börner) and *Heteromurus nitidus* (Templeton). Mégnin's records of insect colonisers of dead bodies has been confirmed and added to over time.

On an exhumed body scuttle flies (Phoridae) are most likely to be the dominant, or indeed sole colonisers to be recovered, often represented by a single species such as: *Conicera tibialis* Schmitz. *Triphleba hyalinata* Meigen, *Megaselia rufipes* (Meigen) or *Megaselia scalaris* (Loew). Scuttle flies can appear on a buried body around 1 year after interment. The often-termed coffin beetle *Rhizophagus parallelocollis* (Rhizophagidae) and the rove beetle (Staphylinidae) *Philonthus ebeninus* (Gravenhorst) most frequently colonise interred bodies in the second year.

Some fly species show preference for a subterranean lifestyle and are indicative of particular past treatments of the corpse. The Phorid *Conicera tibialis* and the Muscid *Hydrotaea* (*Ophyra*) sp. have been associated with bodies previously retained indoors prior to burial, as has the more cosmopolitan Sphaerocerid species, *Leptocera caenosa* (Rondani). This latter species is linked with human faeces, toilets, cracked soil pipes and caves as well as from graves. Whilst bodies in wooden coffins have been linked to the presence of the Phorid fly *Triphleba hyalinata*.

The degree of insect infestation on a corpse is dependent upon the depth of soil covering the body. *Calliphora vomitoria*, *Calliphora vicina* and *Lucilia sericata* have a very limited ability to colonise bodies beyond the first 5–10 cm of even loose soil. Muscid flies (*Muscina stabulans* (Fallén) and *Muscina prolapsa* (Harris)) can colonise remains more deeply buried in uncompacted soil at depth of 40 cm. Whilst adult female *Conicera tibialis* flies have been recovered from corpses buried 200 cm (2 m) deep. This species is thought to have at least two generations per year and that the second generation can recolonise the buried corpse apparently without coming to the soil surface to mate.

Other arthropods, besides insects, are found on buried bodies. Springtail species, microscopic organisms ranging from 2 to 12 mm long are commonly found in soil, at a density of 7500 m^{-2}. They are widely distributed but can be specific indicators of significant features related to buried bodies. Besides being in leaf litter, beneath tree bark and in decaying logs, springtails are found beneath snow and in caves. Some species are carnivorous, others consume algae, fungi, plant detritus, and also faeces, material commonly associated with graves.

17.3 Seasonal Influences on the Colonisation of a Corpse

Insects are cold-blooded and therefore respond to seasonal conditions. They require a level of solar radiation to provide a minimum external temperature to which the body is warmed before the insects are able to utilise energy for activities such as feeding, flying, mating and egg laying. In Europe, the majority of insects relevant to forensic science casework are active in three of the four seasons. With rare exceptions such as *Calliphora vicina*, in southern counties of the UK, the majority of necrophagous insects, apart from scuttle flies and winter gnats, are absent in the winter months. In the future the effects of climate change may influence this situation.

In other countries, a greater seasonal and geographic variation result in the same species being considered inactive in one particular season in some specific geographic regions of a country but not others. The black blowfly *Phormia regina* is inactive in the southern state of Texas (USA) in the summer months. Whilst further north it is present throughout the summer.

Daily weather conditions play a large role in determining insect activity. Rain prevents insects such as flies from flying although flesh flies can fly in the rain. Wind speed also has a strong influence on whether flies are able to fly and higher wind speeds tend to limit many necrophagous insects' ability to colonise a body, unless they are able to gain access by walking (an optimum wind speed for Calliphorid flight being 0.7 m s^{-1}). Season and weather conditions, including the presence of rain and high winds prior to finding the body, have to be taken into consideration in order to provide a post-mortem interval (PMI) figure as near the actual time of death as possible.

Within a particular temperature range, insect growth is directly linked to temperature. This fact enables the forensic entomologist to relate the insect's larval growth and development to the temperatures in the environment (crime scene) concerned. At higher temperatures, above this optimum range, the speed of growth and development reduces until the point of death. In temperate countries lower temperatures are the more significant. Such base temperatures are the minimum temperatures at which the insect can survive but will not grow. These are known to vary, and insects are genetically acclimated to particular locations. For example, in the southern counties of England a base temperature of 1 °C is considered appropriate for *Calliphora vicina*. In the northern counties of England, the base temperature for the same species is 3.5 °C. The species base temperature for development must be subtracted from the ambient temperature. Base temperature is therefore a significant component of the equation to determine time since death, using accumulated degree hours (ADH) or accumulated degree days (ADD) to obtain a minimum post-mortem interval (PMImin).

17.4 Arthropod Distribution and Interpreting a Crime Scene

Whether an insect (and thus a body on which it is feeding or ovipositing) has originated from a place other than the location in which it was recovered, is a question often asked in relation to a crime scene. Insect species distribution is known to be determined by season, altitude and habitat. Knowledge of entomology can place or confirm a suspect or victim at a particular location based upon the insect species present. The occurrence of species characteristic of a particular type of habitat, distinctly different from the one in which the body was found, may indicate that the body has been moved. For example, to find live larvae on a body on the Shetland Islands of a species such as *Lucilia richardsi* Collin (Figure 17.1c) which is common in central and southern England would also be cause for consideration of the possibility of movement of the body.

The altitude at which a person dies is also of significance in terms of the body's potential for insect colonisation and the determination of PMI of the victim. At high altitudes such as in mountain ranges, there is both a temperature effect and little food available to sustain large populations of insects which could potentially colonise a corpse. Therefore, the greater the height above sea level, the less abundant are the species of insects present. Whilst an altitudinal effect causes depletion of insect numbers, higher altitudes do not necessarily discourage the presence of specimens of a species common at lower levels. Hodecek (2022) examined 160 forensic cases from Switzerland. *Calliphora vicina* was the dominant species recorded in the Lausanne area in 69% of cases whilst the remaining three dominant species exhibited an equivalent but lower level of occurrence: *Lucilia sericata* (30%), *Calliphora vomitoria* (29%); *Lucila caesar* (Linnaeus) (27%). The strong influence of altitude, season and habitat type on species presence was again demonstrated even though 61% of cases were from deaths within habitation. *Calliphora vomitoria* was highly correlated with increasing altitude and was recorded for the majority of the year. Only three species (*Calliphora vicina*, *Calliphora vomitoria* and *Lucilia caesar*) were recorded from corpses located above 1500 m. *Lucilia sericata* was confirmed as a warmth-loving species and not recorded above 1200 m and frequently associated with cases of death in or near human habitation. In contrast *Calliphora vomitoria* and *Lucilia caesar* were predominantly associated with rural cases and shady locations. There are differences in species assemblage composition and also the presence of particular species in different geographical locations even on the same landmass. It is important that crime scene investigators are aware of the insect species previously recorded in forensic cases, the species ecology, and an inventory of the local necrophagous species.

Sometimes the behaviour of insects and flies in particular, produce unhelpful and variously shaped artifacts at a scene through regurgitation and defaecation, generating stains of varying shape and quantity which have been mistaken for blood spatter. Insects can alter the shape of wet blood stains by walking through them, thereby altering the scene. However, insect artifacts may also have their uses, particularly where the insects have fed on blood from a drug user. Toxicological experiments using LC-MS/MS techniques demonstrate that comparable levels of codeine/morphine, cocaine/benzoylecgonine, and 6-monoacetylmorphine/morphine can be recovered from the blood of a cocaine or heroin user sourced from fly artifacts on both absorbent and non-absorbent control surfaces. So, fly artefacts may potentially have forensic science value, rather than just being sources of contamination, confusion or nuisance.

Flying insects are not the only source of a body's past history. Human ectoparasites such as fleas and lice on a corpse are significant. Members of both families require the body to be alive, or be maintained at a temperature little different from that of a live body, for them to remain on the corpse. Fleas leave a body very soon after its host (person or animal) has died. Their presence on a body is significant in cases of drowning, as fleas are likely to survive no more than a day submerged in water. Fleas provide an indicator of time since submergence of a body that was likely to be still at or near normal living body temperature when the person went into the water. On the other hand, lice such as the human ectoparasite *Pediculus humanus* var. *corporis* DeGeer (Figure 17.2) are thought to survive from 3 to 6 days on a submerged corpse. Consequently, if all the lice recovered from a body are dead the conclusion is that the person could have died at least 3 days before.

Lice are not the only ectoparasites of forensic significance. Mites (Acari), members of the soil fauna, can be found on, around and below dead bodies throughout the stages of decay. Human bodies have been shown to support over 75 species of mite from 20 families including the Astigmata, Endeostigmata, Mesostigmata, Oribatida, Prostigmata and Ixodida (hard ticks). Where a body was concealed at

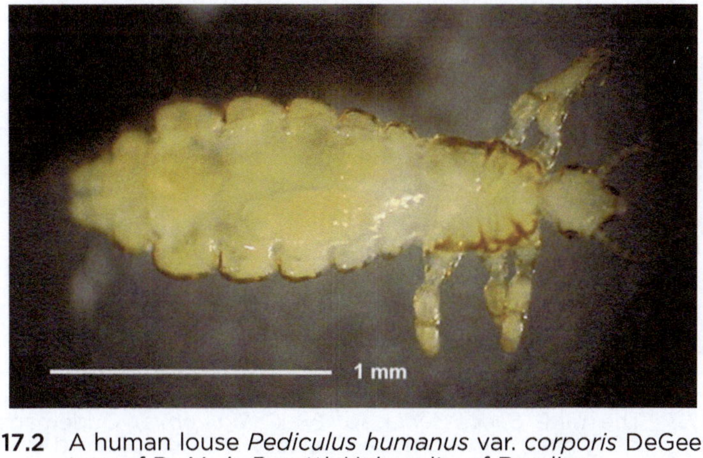

Figure 17.2 A human louse *Pediculus humanus* var. *corporis* DeGeer. Courtesy of Dr M. A. Perotti, University of Reading.

low temperatures or became mummified, mites may be the only arthropods present. They are attracted by the odours of butyric fermentation and of advanced decomposition. The mites on a body can come from a number of sources including being carried onto the corpse on the bodies of colonising flies: these are termed phoretic mites or phorents. Flies lay their eggs on a body whilst mites feed on the bacteria, fungi or protozoa *etc.*, present on the corpse or on the fly eggs themselves. Because mites have a more rapid life cycle than Diptera they can provide valuable confirmatory evidence about the timing of initial colonisation of the body by flies and hence the post-mortem interval. Mites arising from an entirely different habitat, can also transfer onto a corpse *via* the perpetrator. The specificity of mites to particular environments makes them valuable as trace evidence.

The role of mites as a means of determining post-mortem interval was recognised in the late nineteenth century. Mégnin used a mite species he termed *Tyroglyphus mycophagus* (now *Sancassania mycophagus* (Berlese)), along with details of the other insects present on a body (with information from the pathologist's investigation) to determine the time since death of the desiccated corpse of a female newborn found in Paris between 1877 and 1878. This was the first time that mites had been used to determine how long ago someone had died and only the second instance of the use of forensic entomology in France (Figure 17.3).

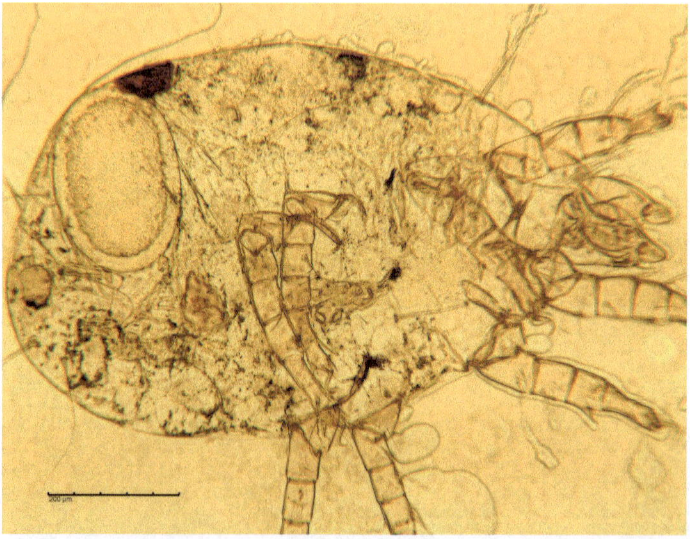

Figure 17.3 The mite *Sancassania* sp. (= *Caloglyphus*) Oudemans 1916, Astigmata, Fam Acaridae. Courtesy of Dr M. A. Perotti, University of Reading.

17.5 Insect Access and Corpse Decomposition

How accessible a corpse is to insects determines speed of colonisation. If a corpse is kept in a closed, insect-proof environment such as a cupboard; sealed beneath floorboards; or tightly wrapped in a bag, blankets or clothing, colonisation is restricted and insect succession influenced. All of these places are well known for concealing corpses. Bodies can also be hidden in a freezer or large refrigerator, or in a car from which insects have been excluded to conceal them. This will influence the estimate of time since death and so could provide the perpetrator with an alibi. (Delay in oviposition of up to 3 days might be expected where a body was stored in the boot of a car.) In each of these circumstances the initial colonisation will be delayed. When a concealed body is found it may not have decayed as rapidly as it would if it had been left exposed on the soil surface. The history of the body since death, in terms of how it was kept and where it is found, is an important aspect to consider when calculating the time since death.

In addition to environmental data and the condition of the corpse, the possible causes of death may be information provided to an entomologist by a forensic pathologist. Sometimes cause of death will be obvious from the condition or location of the body. In other circumstances the insect evidence may require further examination using toxicological or molecular analysis to provide greater detail for the court. Crime scene managers will make the decision regarding additional analysis in discussion with the forensic entomologist.

17.6 Effects of Chemicals on Insect Development and PMI

That insects can survive on tissue from a corpse where death was due to poisoning has been known for at least 110 years. Larvae of flies from the Piophilidae, Psychodidae and Fanniidae families feeding on bodies poisoned with arsenic have been recovered from some crime scenes. These larvae are able to bioaccumulate the element. Such evidence can be passed to a forensic toxicologist for further study. However, arsenic is not the only element that has been recovered from insect larvae. Mercury has been extracted from larvae, puparia and adult blowflies in circumstances in which they have been feeding on fish containing methylated mercury. The mercury passes along the food chain. Rove beetles (Staphylinids), feeding on the fly larvae feeding on such contaminated fish, also bioaccumulate any mercury retained by the maggots. Analysing such secondary predators at the scene will provide a means of confirmation of the identity of any drugs previously consumed by the corpse.

Insects can indicate the presence of drugs in individuals who have taken such substances for medical or recreational purposes or from victims in cases of suspected sexual assault such as date rape. These drugs can potentially be extracted from the entire larva (*i.e.*, the maggots consuming the bodies of individuals), from the larval foregut, or from the puparial stage. Examples of such drugs include opiates (codeine), tropane alkaloids (cocaine, its metabolite benzoylecognine), benzodiazepines (triazolam and oxazepam), barbiturate (phenobarbital) and tricyclic antidepressants (alimemazine and clomipramine). All of these types of drugs have been recovered from blowfly larvae that previously fed on tissue intoxicated with these drugs. Using the larvae as a means of investigating drug presence can be a more sensitive method of determining the identity of the drug than using tissue from the corpse.

Attempts to relate the quantity of drug consumed by the victim to the amount of drug recovered from the insect however, have met with varying degrees of success. The particular drug being investigated appears to be an important factor in whether or not there is a successful outcome. Even using gas chromatography (GC) and gas chromatography–mass spectrometry (GC–MS) to investigate drug content, for example, the presence of cocaine and its dominant metabolite benzoylecgonine, in both the larvae and decomposing skeletal muscle, may not allow accurate determination of drug concentration in the corpse. In other cases, correlating the levels of the opiate, codeine in *Lucilia sericata* larvae with those in the corpse, has been successful.

Drugs may have an effect on larval life cycle durations. This has to be taken into account when estimating time since death, a consideration that has been shown to be particularly true of flesh flies. Larvae of *Sarcophaga peregrina* (Robineau-Desvoidy) when fed on tissue drug-intoxicated with cocaine and its metabolite at levels of the median lethal dose and double the median lethal dose, were shown to have accelerated development. The drug reduces the time in the larval stages, causing early pupation and adult eclosion (emergence). Similarly, consumption of heroin increases speed of larval growth by up to 29 hours. In contrast, a different species of flesh fly, *Sarcophaga tibialis* Macquart, when fed on heroin-laced food spends considerably longer in each life stage than do larvae fed on food without heroin. Similarly, morphine, slows the rate of development to the adult stage in the blowfly *Lucilia sericata*. This is because *Lucilia sericata* is capable of excreting the drug during the post-feeding stage. The effect of a particular drug or its metabolites (or both), on the speed of development therefore appears to be species specific, at least in the flesh flies, because of variation in levels of metabolism of the particular drug.

Not all insect development related to forensic casework is influenced by drugs. Some species' life cycles are unaffected if they consumed tissue intoxicated with a particular drug. Such is the case for flunitrazepam (Rohypnol®) the so-called date rape drug. Baia (2016) investigating the life cycle of a Brazilian population of *Chrysomya megacephala* (Fabricius) found little effect of this drug if supplied in food at intoxication levels of 4, 8, 16, and 32 $ng\,g^{-1}$, on development time or mortality. Although the drug did affect pupal weight and the weight of the resultant adults.

The decomposition stage at which insects start to feed on a drug-intoxicated body, and where on the body they feed, is important. Early colonisers such as scuttle flies, blowflies and flesh flies, can provide a rich source of information on the drugs present as they feed on the soft tissue which will not be greatly decomposed. To overcome any limitations of the analytical techniques, larval samples should be taken from a number of internal organs, or the head region. Species that feed later in the succession on drier tissue, such as dermestids, may be a less useful source of information if data on both drug identity and concentration are required. Shed larval skins and faecal waste from later colonisers may provide a drug concentration value that is less than was actually present in the body. Nonetheless, puparial cases and the pupae themselves are an alternative source of information if no corpse tissue is available for analysis. Puparia, in particular, can remain in a suitable state of preservation for a long period. Techniques such as fluorescence polarisation immunoassay, can be successful, provided that more than 10 $mg\,g^{-1}$ of puparial tissue is available.

The presence, speed of colonisation and degree of development of insects on a decomposing corpse can be influenced by substances other than drugs. Perpetrators often try to influence colonisation by disguising the existence of a corpse using substances commonly found in homes or offices. The items include herbicides such as paraquat, perfume, bleach, hydrochloric acid, caustic soda and gasoline besides more obviously insecticides and insect repellants such as *N*,*N*-diethyl-3-methylbenzamide (DEET) and citronella (a substance often used in mosquito repellents or for insect control during barbecues). Petroleum spirit, perfume and mosquito repellent (citronella), caused a mean delay of several days in the appearance of the first dipteran species as did the organophosphate Malathion, which is often used to kill lice. Malathion exposure caused no morphological change to *Megaselia scalaris* in adults or larvae. It did affect the eggs and pupae, increasing the length of the larval stage. In contrast, bleach and perfume had little effect on insect development time, although low concentrations of caustic soda and citronella increased development time. In such instances PMImin estimation will be affected. Other readily available substances such as gasoline, hydrochloric acid and insecticides such as permethrin, tend to kill insects outright.

17.7 Molecular Analysis and Insect Identification

Analysis of insect genetic material is a valuable tool in forensic entomology in a number of ways. Firstly, it is an important means of determining the species of larvae on a corpse. Flesh fly larvae, for example, are often difficult to identify from their morphological features without considerable expertise and access to a museum collection of reliably identified specimens. Species identification can often be achieved using DNA from larval specimens. Secondly it is possible to identify the host or corpse using DNA from the insect crop or gut content or some other aspect of an insect's behaviour. For example, successful identification of a missing corpse was achieved by using the cases of the case-making clothes moth and webbing clothes moth from a crime scene. Both these species had incorporated hair from the victim into their cases and hence were a source of the victim's DNA.

Molecular approaches require species-specific primers in order to amplify DNA specific to an insect family such as the Calliphoridae, Sarcophagidae or members of the Coleoptera. Once DNA has been extracted, a number of analytical techniques can be used to examine insect genetic material from crime scenes. The techniques include polymerase chain reaction-restriction fragment length polymorphism (PCR-RFLP), random amplification of polymorphic DNA (RAPD) and PCR-targeting mitochondrial genes such as the mitochondrial subunits of cytochrome oxidase I (COI), cytochrome B, 16S ribosomal RNA or microsatellites. Insect mitochondrial DNA (mtDNA) subunit cytochrome oxidase I (COI) has been chosen for analysis because it is the biggest of the cytochrome units and the protein sequence combines variable and highly conserved regions. Furthermore, mtDNA is resistant to degradation and can be rapidly analysed, thereby providing an efficient means of distinguishing between insect species.

A DNA barcoding system using cytochrome oxidase I (COI) has gained in significance. Such barcodes are intended to provide a standardised means of genetic identification of insects which is rapid and reliable as a means of species determination. The standard used is the 658 base pair region of the gene encoding the first subunit of mitochondrial cytochrome oxidase (COI) of the fruit fly *Drosophila yakuba* Burla. Using DNA barcoding has recently enabled identification of the first record of adult and third instar larvae of the picture-winged fly Physiphora *alceae* (Preyssler) on the well-decomposed corpse of a decapitated female in an urban district of the city of Milan, Northern Italy (Tuccia 2021). In this instance DNA barcoding and phylogenetic analysis were combined with identification using morphological characteristics. The methodology highlighted the forensic value of this species of Ulidiidae which, whilst recently found on animal carcasses, is more usually found associated with dung, and decomposing vegetable matter.

However, there are some limitations to the COI barcoding approach since it is a single-locus identification system. To overcome this analysis of a supplementary nuclear marker system alongside COI has been proposed; nuclear ribosomal expansion segments (rDNA) are considered as a possible option for species identification. In addition, inclusion of the names of expert taxonomists who identified the original specimens morphologically, which were used to provide the original genetic material in the database, has been mooted.

Aging specimens collected from bodies by molecular means is a further benefit of these techniques. Throughout the insect life cycle different genes are expressed at differing points in time. Such age-dependent differential gene expressions are potentially useful especially in combination with General Additive Models (GAM) to increase the precision of such age estimates. The use of real time PCR (RT-PCR) for differential gene expression measurement has been successful for determining the age of blowfly pupae, an otherwise difficult procedure.

17.7.1 DNA Analysis of Insect Gut Content

Comparison of DNA from tissue in the crop, (part of the gut) of insects which have recently been actively feeding on a corpse, with DNA known to come from the missing person may provide conclusive evidence of their previous presence at a location. To this end human short tandem repeat (STR) analysis has been successfully used to confirm associations between fly larvae and a particular corpse. Because the crop contains digestive enzymes the tissue of a newly dead corpse will be degraded over time; equally the tissue itself may be decomposing dependent upon the stage of decomposition of the host corpse on which the fly larvae are feeding. The analysis is not as likely to provide conclusive evidence with older post feeding larvae, starved maggots with empty crops or those with crops of 1 mm length or less. DNA from larvae starved after feeding on the corpse can be successfully extracted for up to four days. Unfortunately, if the larvae then feed on a different food source the original DNA is masked which impedes interpretation of the results beyond a two-day window.

17.8 Developments in Insect Identification Techniques

New methods have been developed or introduced into forensic contexts to provide objective insect identification. Some are based upon insect morphological characteristics such as frons and head widths, larval cephalopharyngeal skeletons (Figure 17.4) and wing

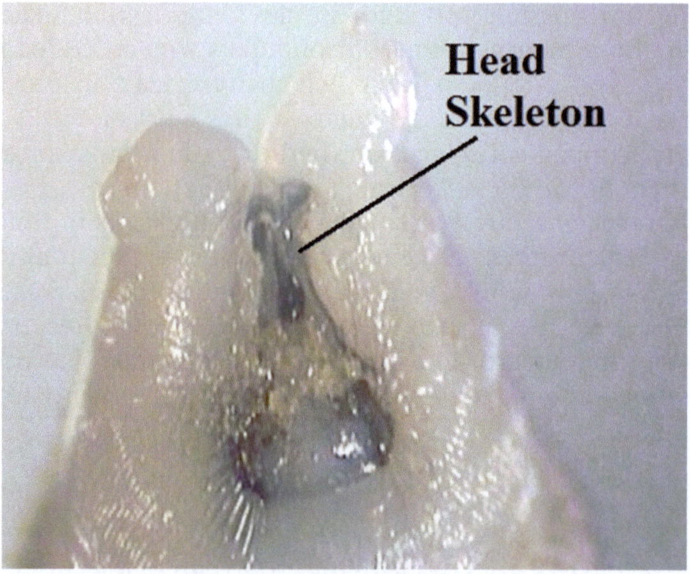

Figure 17.4 Cephalopharyngeal skeleton - mouthparts of the larva which have value as a means of species identification. Reproduced from ref. 2 with permission from John Wiley & Sons, Copyright 2012.

morphology. Confocal laser scanning microscopy (CLSM) for example, has been used successfully to identify first instar larvae of forensic interest using cephalopharyngeal skeletons and 3-D reconstruction. Other techniques take advantage of aspects of mathematical analysis. Variation and covariation of particular physical features has been used to assess differences in species or developmental stage shape, especially where it can be based on particular and consistent geometric coordinates present in an organism's morphology. Examples of such coordinates ('landmarks') which have been used are the intersections between wing margins and specific wing veins and also between various wing veins themselves. The method, called Geometric Morphometrics, can use outline approaches or landmark approaches. Using landmark approaches, the morphology of wings is the most utilized aspect of the body for such discrimination in many orders of insects including Diptera. The technique in forensic entomology can provide a means of discrimination between species as adults and also their sexual dimorphism and for the visualization of larval growth markers. Geometric Morphometrics can also be useful for interpreting the effects of drug consumption in relevant species, where the drug has a shape or size effect on the immature stages.

Spectroscopic methods such as near-infrared spectroscopy (NIRS) and medium-infrared spectroscopy (MIRS) are effective for discriminating between both species and life stages. Jales (2021) reviewed the use, application and importance of infrared spectroscopy in species identification, larval age determination and to distinguish the presence of drugs in immature insect life stages. NIRS successfully revealed the presence of the drug flunitrazepam in *Lucilia eximia* (Wiedemann) larvae for example. These authors tabulated the instances of successful use of these techniques highlighting their many advantages such as their low-cost, rapidity, non-destruction of the samples and accuracy. MIRS use, for example, enabled *Calliphora vicina*, *Lucilia sericata* and *Musca domestica* larvae to be identified with between 82.5 and 100% certainty.

Aging adult and larval insects can be difficult. The age of immature stages is usually based on their morphological features and restricted to determination of which instar is present. No morphological differences exist with which to age adult insects. However, hydrocarbons are present at all insect life stages from egg to adult. Aging of adult blowflies has been achieved as the total amount of the hydrocarbons increase as the fly ages. For example, a profile of over 130 different hydrocarbons making up the non-polar fraction of surface lipids secreted by 5-day-old adults of the screwworm *Cochliomyia hominivorax* (Coquerel) has been shown to change over time. Similarly, hydrocarbons provide a means of distinguishing puparia of recent *Hydrotaea aenescens* Wiedemann cohorts from those which are historic. Hydrocarbons have also proved a useful means of distinction between populations of post-emergence adults from different geographic origins.

Scanning electron microscope and energy dispersive X-ray spectroscopy (SEM/EDX) have been effective for identifying the nature of explosives using tissue from the gut content of insects feeding on the remains of bodies killed in an explosion. The value of the technique is confirmed by the results of analyzing third instar larvae (the life stage before pupation) of *Lucilia sericata* previously fed on TNT (trinitrotoluene) contaminated liver. The by-products and by-components of TNT increased the speed of the life cycle of this species by up to 41%.

17.9 Insects and Fire

There has been much discussion about the ability and conditions under which colonising insects might withstand burning. A rapid response to a fire by the fire brigade, if the fire was noted shortly after it has been started, can provide a fixed point for determining time since death accurately, even if all of the insects have died in the fire. The larval instar feeding on a body at the point when the fire was set can be determined under such circumstances. This is not true where the fire is not extinguished rapidly, such as when cars or buildings burn themselves out. In this situation a fixed end point for the fire may be unknown and the level of probability that the larvae withstood the fire and continued to develop becomes of significance. The amount of heat generated in the fire, level of protection that the insects receive from the body and the presence of blankets or other forms of insulation, will dictate whether or not insect remains are present. Burnt vestiges including fire debris can retain entomological evidence which provides a source of evidence of whether or not the victims died prior to the fire setting.

The attractiveness of burnt tissue to flies as an egg-laying site has been questioned and the issue of a possible delay in colonisation of the bodies raised. The state of the body and level of burning will dictate whether or not it is attractive to fly species, based on the odours released during and after burning. Evidence and unpublished observations suggest that burnt flesh is more attractive to both *Lucilia sericata* and *Calliphora vicina* than unburnt flesh. Therefore, colonisation of the areas of burnt flesh by these species is more likely to take place first. In contrast *Chrysomya megacephala* and *Chrysomya rufifacies* (Macquart) will delay oviposition if the corpse is strongly burned. Although both oviposition and larval development rate is unaffected in these species if the corpse exhibits minimal or minor burns.

The availability of cracks and flesh suitable as an oviposition site and for subsequent larval feeding and development will dictate whether and where a burnt corpse is colonized. The condition of the flesh may also determine speed of subsequent development. Such aspects of the state of the burned corpse have to be considered and may require examination by the forensic entomologist in person, rather than use of photographic evidence to provide an opinion and PMImin determination.

17.10 Animal Death and Cruelty Investigation

In addition to providing evidence relating to legal cases involving humans, forensic entomology techniques are applicable to cases of endangered species trafficking and poaching as well as domestic animal and wildlife mistreatment and neglect. Depending upon the country concerned law-enforcement officers, crime scene investigators, veterinary scientists and in some countries members of national wildlife forensic laboratories collect and appropriately interpret evidential information for particular cases relating to wildlife and domestic animal crimes.

In the UK the Animal Welfare Act 2006 and in England the Welfare of Farmed Animals (England) Regulations 2007 require that animal welfare is paramount especially in relation to farm and domestic animals. Flies are known to lay eggs on the bodies of pigs, sheep, deer, dogs, rabbits and also poultry – a condition known as fly strike. If this is neglected, the body of such animals can become infested by maggots very rapidly. The predominant species initiating these forms of infestation or strike are from the genera *Lucilia*, *Calliphora*, *Chrysomya* and *Phormia*. They are most likely to initially infest the perineal region of the body, being attracted by the odour of faeces or urine. Flies such as the house fly may then secondarily colonise any areas of lesion or dermatitis caused by the original flies. The infested animal will be living with the effect of maggots burrowing into its body and consuming its flesh (myiasis) (Figure 17.5) and may die directly as a result of infestation. Humane euthanasia of such a live infested animal may subsequently be necessary because the animal was beyond treatment. Such abuse is punishable by law.

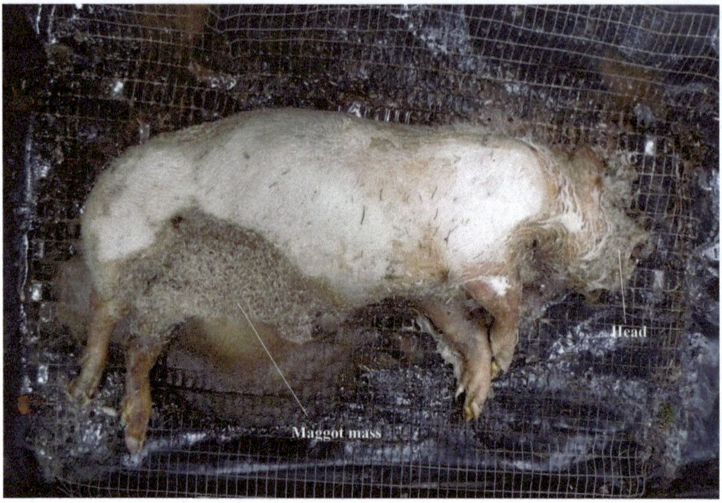

Figure 17.5 Dead pig illustrating the effects of myiasis if left.

Determining the time of death is paramount in cases where a domestic or wild animal is found dead, where the corpse of a wild animal is found with obvious evidence of the removal of visceral body parts, or where the dead animal is an endangered species. Poaching of endangered wildlife in countries on the African and Asian continents both to obtain live animals and animal products along with the movement of such products all fall under the remit of the Convention on International Trade in Endangered Species of Wild Fauna and Flora (CITES) as does trafficking of endangered species. (Appendix 1 of the convention – species amongst the most endangered and threatened with extinction – lists 69 species plus 7 sub-species of invertebrates. Amongst these are scorpions, stag beetles and scarab beetles as well as a number of species of butterflies and spiders.) Many of the countries at the forefront of attempts to control poaching and the illegal trafficking of wildlife and their products, have developed teams skilled in forensic entomology methods to estimate time of animal death, catch poachers and prevent further loss of wildlife.

17.11 Modelling Time Since Death (Post-mortem Interval)

The models most frequently used for determining time since death are the linear models based, between upper and lower temperature threshold limits, upon the direct relationship between temperature and time for the insect to develop. These models are called the accumulated degree hour (ADH) and accumulated degree day (ADD) models and are also commonly used in integrated pest management (IPM) strategies for controlling insect pest infestation on crops. The ADD and ADH models require that the corrected temperature values between the two thresholds are multiplied by a series of set time periods and the totals accumulated until they give a measure of the physiological energy budget needed to reach the life stage of the oldest specimens discovered on the corpse.

The egg stage is often the life stage of shortest duration. The presence of eggs alone is frequently considered to indicate death within a day to only a few days. However, it is important to consider whereabouts on the body the eggs were laid and where the larvae are located. If it is inside some of the body orifices, then the temperature could be greater than ambient and this must be corrected for in any calculations. Larvae tend, by the second instar, to gather together in groups. These maggot masses raise the group temperature above ambient and so can increase the rate of larval growth. (A similar heat effect may arise due to adult and larval Silphid beetles' feeding strategies. However fewer beetle life cycle data at fixed temperatures are available to be able to use the method.) The temperature effects of maggot masses *etc.* should be considered when calculating ADH or ADD – thermal summation methods.

The larval stage is the predominant insect life cycle stage used to calculate the post-mortem interval (PMI). Use of locally-generated species development rate data at a fixed experimental temperature is the most accurate means of determining the PMI for species in a particular geographic location. This information is used alongside an estimate of the temperatures at the scene before the corpse was discovered. Temperature correction factors are calculated, based on comparable temperature recordings at the crime scene in relation to records from recognised meteorological stations nearby. The temperatures at the crime scene while the body lay undiscovered, can then be back-calculated using corrected temperatures taking into consideration any microclimatic differences between crime scene and meteorological station.

Charabidze (2019) identified variation in the influence on larval growth of microclimate temperature at the crime scene as a potential source of contention. They concluded temperature measurement approaches at the scene, relative to that taken by the available weather stations, was a critical factor when trying to derive accurate data from which to calculate the temperature at the crime scene between death of the individual and the time their corpse was discovered. Differences in weather station recording strategies may exist which are not immediately obvious from the raw temperature figures. For example, differences in how and at what height above ground temperature measurements are taken; differences in how the average temperatures per hour or day are calculated. In addition, features at the crime scene may exist which would influence the temperature experienced by the insects on the body such as shading from trees, the influence of wind, the effects of concealment or location of the body in a building. The researchers outlined a qualitative checklist of factors which influenced the temperatures experienced by the larval life stages on the body by elevating, lowering temperatures or ameliorating their effects. Their proposed approach was to start the assessment for temperature effects on larval growth from the corpse itself, expanding the investigators interest outwards to take account of local features such as wind and sun exposure, or open windows in houses.

Such a checklist has been suggested for inclusion in the standard forensic entomology protocol for assessing a scene as it would better account for temperature effects on insect growth and provide more reliable temperature data from which to calculate ADH or ADD for the period before the body was found. In turn this would enable a more reliable estimation of the PMI. They also suggested modifications such as putting temperature recorders on any crime scene evidence bags containing larvae whilst they were being transported to the laboratory and stored in refrigerators, or during the period when larvae were kept in growth cabinets to complete their life cycle to the adult stage, to provide knowledge of the effects on larval growth.

In addition to using growth rates of the larvae, larval weight and larval length has been used to calculate PMI. From these data Isomegalen and Isomorphen graphs have been prepared. Isomorphen graphs relate growth to a particular instar at a particular stage. Such graphs have been produced for a small but growing number of species, notably the blowflies *Lucilia sericata* and *Calliphora vicina*, *Chrysomya albiceps* (Wiedemann), *Chrysomya megacephala*, *Chrysomya chloropyga* (Wiedemann) and *Chrysomya putoria* (Wiedemann), *Protophormia terraenovae* (Robineau-Desvoidy), *Hemipyrellia ligurriens* (Wiedemann), *Sarcophaga (Liopygia) argyrostoma* (Robineau-Desvoidy), the Asian species *Aldrichina grahami* (Aldrich), the Fannid *Fannia pusio* (Wiedemann), staphylinid *Creophilus maxillosus* and the silphid *Thanatophilus micans* (Fabricius).

Each line of the Isomorphen graph represents morphological change in the life cycle of the specific species which has been experimentally investigated at set temperatures. Areas between the lines reflect identical morphological stages within the sequence. This form of graph is valuable where puparia or post-feeding larvae have been recovered from a crime scene in which the temperature fluctuates within very narrow limits – usually this is an indoor crime scene. To use the graph effectively the specimens should have been kept under constant environmental conditions reminiscent of the temperature of the crime scene until they either pupariate, or if puparia were collected, the adults emerge. The age of the specimens can be deduced retrospectively using the graphical representations of the growth rate at the temperature of the crime scene (Figure 17.6).

Isomegalen graphs depict the length of maggots at a range of different temperatures in relation to their age in days. The graph allows the age of the maggot of a specific species in the crime scene to be read directly off the graph if the temperature is constant. Where the temperature is not constant it is impossible to determine the age range accurately. Therefore, both these methods of PMImin detection are restricted to specific conditions at crime scenes.

The use of larval width has been proposed as an alternative to length for the determination of the PMI. The measurement of body

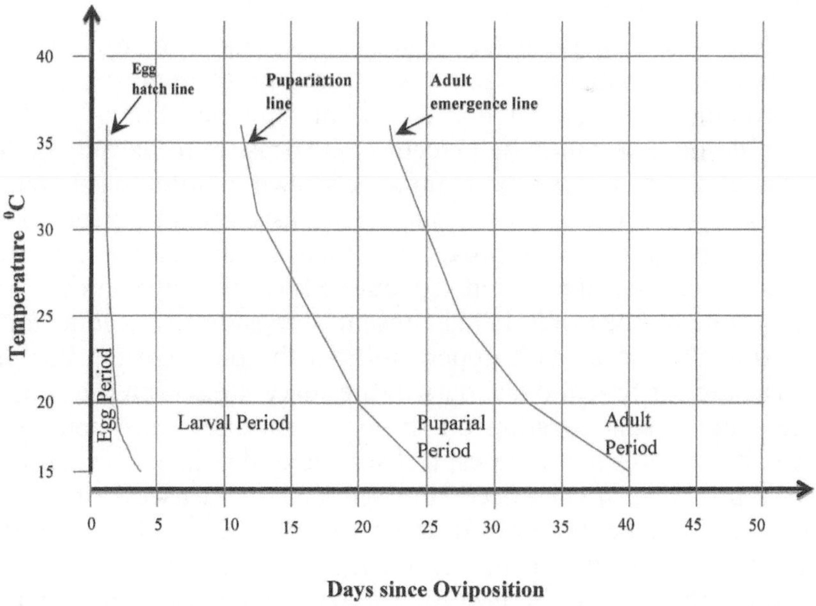

Figure 17.6 Example of an artificial Isomorphen graph of an unspecified insect.

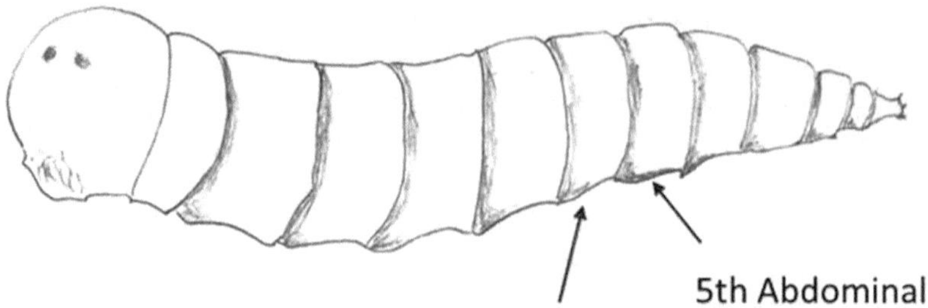

Figure 17.7 Diagram of a third instar larva showing the fifth and sixth abdominal segments: arrows indicate the positions for larval width measurement.

width between the ventral and dorsal surfaces at the junction of the fifth and sixth larval abdominal segments, as shown in Figure 17.7, is considered comparable with body length for predicting the age of the larva. Such measurements can be translated into length with 95% accuracy. The method is advantageous where the larvae have been damaged or are poorly preserved and accurate measurement is difficult.

A further means of determining post-mortem interval, especially for corpses advanced in decomposition, is to use insect succession. Background knowledge of a particular geographic region is required and within that area, the necrophagous species likely to be associated with each of any different habitat types present. Field and laboratory background studies on when and in what sequence, individual species would be expected to be present on a corpse and their life cycles at a range of fixed temperatures must be known in order to generate a post-mortem interval from the succession assemblage present on a corpse. The succession approach is based on the presence of two chosen species in a succession sequence whose presence on the body overlap; one species only being present alongside the second for a short time. The lower and upper limits of the post-mortem interval estimate are the first and last days in the succession when the chosen species are present together on the body. An alternative approach, developed in Hawaiian studies, utilises the end-point of two chosen species presence in the succession sequence. In this case the PMI time period bracket (lower and upper PMI estimate) is based on the point in succession where an early colonizing species becomes absent (providing the start of the PMI time bracket) compared with the last time of presence of a second more enduring species (the end of the PMI estimated period). Data on the development times for each of the colonising insect species' development stages (egg stage, first instar, second instar and so on) at various temperatures should either be known or determined. This is because continued insect existence on a corpse is affected by many variables; corpse mass, temperature and season amongst them.

17.11.1 Reducing Uncertainty of Time Since Death

Laboratory studies on species development are being validated through field studies for particular species. They will increase the accuracy of data on colonisation, development times, succession and pre-appearance intervals *i.e.*, the time before the insect species first colonised a body (the PAI). Matuszewski (2017) developed a technique which provides a measure of error. This uses the pre-appearance periods (PAI) of two chosen pointers (insect species), one recorded on the corpse in the oldest life stage available and one anticipated based on succession but was not yet present. The advantage of the method being that PAI for some species can be determined by methods based on temperature thereby avoiding localised growth effects on the insect life cycle. The post-mortem interval range is defined by a lower PMI estimate which is demarcated by PAI of the chosen pointer species currently present on the body. Whilst an upper PMI estimate is allocated using the PAI of a second pointer species that would be expected next in the succession sequence. Choice of second pointer species depends on which species appears in the succession on a body, later than the first pointer chosen. The PAI of the chosen pointer species is based on the literature. An inventory of such pointer (definitive) species whose development stage durations are known along with the effects of temperature on their speed of development, has been created for the rural environments of central Europe by Matuszewski.

From data based on research, it is possible to provide error rates for the PMI estimate; calculation of the midpoint of the PMI upper and lower estimate range provides a confidence interval. Depending upon the pointer species chosen it may be possible to estimate reliable PMI up to 3 months after the individual's death. Matuszewski concluded that by presenting the data as an interval surrounding the point estimate the method delivers a robust interval estimate featuring a single error rate *i.e.*, this delimiting helps overcome error sources in the individual components of the calculation included in PAI estimation and the determination of upper and lower PMI. A major disadvantage of the method is lack of background data for a large enough range of species. This may be the result of inconspicuous life stages in some species or an inability to distinguish between immature members of some orders, such as beetle larvae. Where suitable background information is absent the use of additional but lesser dominant species, which do have sufficient data attached to their colonisation times and behaviour, can assist.

Similar disadvantages of methodology pertain in forensic entomology application generally. These issues are overcome by: use of professional sampling methods and internationally standardised protocols by trained personnel; ensuring the data used to predict the post-mortem intervals are consistent with the conditions pertaining in the crime scene; data gathering on average monthly or seasonal pre-appearance intervals for the life stages of further necrophagous species; ensuring the analysis is based on the initial colonisation of a pointer species and not further generations of that species, or those which have over-wintered in an immature stage and recolonised a body; greater studies of succession in different countries and under varying conditions.

17.12 Forensic Entomology – Developments and Influences

Forensic entomology practices and procedures have been developed and refined as a result of worldwide discussions and the work of a number of researchers both forensic entomologists and in the wider forensic community. The development of a protocol by the European Association for Forensic Entomology (EAFE) to help maintain at least minimum standard operating procedures in forensic entomology has been influential in providing a means to ensure risk of error is minimised and that courts are better able to interpret and compare entomologists' findings. It has helped standardise good practice across all the forensic disciplines in which scientists collect entomological samples.

Since the production of that protocol, key legal cases and the deliberations of the U.S. National Research Council in 2009 have influenced the practice of forensic scientists including forensic entomologists not just in the United States but also in Europe and beyond. They have led to the production of significant scientific papers, and engendered discussion by entomologists within a range of professional scientific societies. These discussions have coloured the research which has since been conducted in an attempt to further refine forensic entomology procedures and amongst other things better determine the accuracy and validity of the estimates of post-mortem interval. A broadening range of arthropods which include the activities of common scorpionflies (Mecoptera), hoverflies (Syrphidae), nocturnal moth flies (Psychodidae) and noctuid moth larvae have been shown to be of value for forensic entomology case work. Research on the variation in development and behaviour of necrophagous species in different geographic areas has been initiated or furthered in a number of countries including African and South American countries. As a result, the body of information available to forensic entomologists with which to interpret entomological evidence at the crime scene has become greatly enhanced and its rigour greatly increased.

Further Reading

J. Hodeček and P. Jakubec, *Int. J. Leg. Med.*, 2022, **136**, 923.

F. Tuccia, G. Giordani, C. Cattaneo, D. Mazzarelli and S. Vanin, *Eur. Zool. J.*, 2021, **88**, 1071.

J. T. Jales, T. M. Barbosa, J. R. de Medeiros, L. A. S. de Lima, K. M. G. de Lima and R. A. Gama, *J. Forensic Sci.*, 2021, **66**, 2080.

D. Charabidze and V. Hedouin, *Int. J. Leg. Med.*, 2019, **133**, 633.

S. Matuszewski, *Int. J. Leg. Med.,* 2017, **131**, 877.

References

1. D. Gennard, *Forensic Entomology: An Introduction*, John Wiley and Sons Ltd, 1st edn, 2007.
2. D. Gennard, *Forensic Entomology: An Introduction*, John Wiley and Sons Ltd, 2nd edn, 2012.

18 Forensic Examination of Documents and Handwriting

Anthony Stockton

Stockton Pugh, The Catalyst, York Science Park, Heslington, York, UK
Email: info@stocktonpugh.co.uk

18.1 Introduction

Due to the uniqueness of handwriting, it was one the earliest forms of identification of the individual.

The forensic scientist specialising in the examination of documents and handwriting will have a number of areas of expertise. These areas include the identification of handwriting and signatures, a knowledge of modern office printers, the composition of inks, papers and the materials from which documents are produced, together with techniques such as electrostatic detection of impressions (ESDA) and imaging processes which allow the origin and history of documents to be studied. Training and experience coupled with the use of equipment, observational skills and logical thinking all assist the Forensic Document Examiner when addressing issues raised in a case.

18.1.1 Qualifications and Training

Whilst not completely essential, most forensic document examiners possess a science degree or science qualification, normally a Bachelor of Science degree with Honours. It is generally acknowledged that the

Crime Scene to Court: The Essentials of Forensic Science, 5th Edition
Edited by Niamh NicDaeid and Peter C. White

Published by the Royal Society of Chemistry, www.rsc.org

best form of training is at least a two-year apprenticeship alongside highly experienced document examiners before the examiner handles their own casework. The trainee should follow a formal training programme which aims to cover the different areas they will encounter in real casework.

18.1.2 Types of Cases Undertaken

The forensic examination of documents is not an academic exercise. It is a practicably based discipline, aimed at addressing the issues in the case in order to assist the court in arriving at a decision.

Some areas where the assistance of a document examiner is required include criminal prosecution casework, criminal defence casework (normally legal aided), Coroners Court, cases involving civil litigation/arbitration, tribunals, hearings and internal investigations.

A host of different offences and legal matters require the expertise of the document examiner. These include:

- Fraud and forgery, cheques, mortgages, banking documents and tax evasion
- Money laundering, drugs offences
- Violent crime, murder, rape
- Counter-terrorism
- Identity theft, counterfeiting
- Malicious communications, blackmail
- Civil litigation
- Internal investigations, professional standards and health and safety

The service is routinely used by police forces, HMRC, DWP and other government departments as well as lawyers, companies, courts and overseas institutions.

As such the forensic document examiner is exposed to thousands of different documents cutting across nearly all crimes and investigations for a large range of different submission authorities in the UK and overseas.

18.1.3 Choosing an Expert

When choosing an expert, it is still very much up to the individual lawyer or client to ascertain that the expert is competent and credible. There is no formal register of experts in the UK.

The properly trained and qualified forensic document examiner should, however, be able to demonstrate a solid scientific background at appropriate degree level, a period of training in an established forensic science laboratory or similar, possession of a fully equipped laboratory and a record of continued active participation in research and development in the field. Other areas to enquire about is the expert's length of service, whether they work alongside any other experts and if their work is peer reviewed. Most experts are members of a professional body such as the Chartered Society of Forensic Sciences (CSFS) or the American Society of Questioned Document Examiners (ASQDE). The expert should, if possible, undertake a range of casework which may include a mix of criminal prosecution/defence casework and civil casework as well as providing training to other experts in the field.

Most importantly however, the expert should be prepared to appear in court as an expert witness and understand and comply with their role.

18.2 Examinations

18.2.1 Why Do We Examine Documents?

There are many reasons why documents are examined in a forensic context.

- Is a document a forgery?
- Who wrote/signed the documents?
- Has the document been altered in any way?
- Have pages been removed and others inserted?
- Are documents which ought to be from different sources actually from the same source?
- What is the history of the documents?

18.2.2 Types of Examinations Undertaken

Various examinations are undertaken to address the above including:

Handwriting and signature comparisons/alterations to documents/recovery and interpretation of indented impressions examinations/detection of counterfeit documents/printing and typescript examinations/paper and ink comparison/stamp mark comparison/miscellaneous examinations including relative dating of documents, sequencing of entries, reconstruction of documents and decipherment of obliterated or faded entries.

18.2.3 Equipment Used for Examination

Most of the techniques employed in forensic document examination have been developed to extract as much information as possible from the document without damaging or altering it in any way. This is usually a prerequisite of any proposed examination since a disputed document is of considerable value – indeed, its very existence may itself be evidence.

The forensic document examination laboratory should have the following basic equipment:

(1) Good lighting sources including daylight and use of a darkened room.
(2) Hand lens and stereo zoom microscopy allowing magnification of $\times 5$ to $\times 40$.
(3) Infrared, ultraviolet, high intensity and transmitted light sources for studying inks, alterations, latent marks and for overlaying images accurately.
(4) Electrostatic detection apparatus (ESDA) or equivalent for the study of impressions.
(5) Oblique light and transmitted light sources *e.g.* light box.
(6) Methods of recording visual results by either photographic or computer image capture techniques, as well as facilities for preparing demonstration materials for presentation in court.

More sophisticated laboratories carrying out other aspects of forensic science may also have available advanced techniques for ink analysis and electronic imaging as well as databases of background literature, handwriting, typewriting, transmitting terminal identifiers of facsimile machines and ink data *etc.*

Whilst the use of sophisticated equipment can be of use in some cases to provide analytical results, it should not prevent the examiner from interpreting the findings in the context of the case based on their expertise and ability.

It may well be for example that the forensic document examiner will be asked to study a particular aspect of a document such as a disputed signature. This should not deter the experienced examiner from checking that other aspects of the document under examination are consistent. A disputed signature on the final page of a multipage document may be identified as genuine. However, if there are unobtrusive differences between the paper and type style of a signature page and that of the previous pages, this may give a considerably different view of the origin of the signature and authenticity of the document as a whole.

18.3 Identification of Handwriting

18.3.1 Uniqueness of Handwriting

> *"By my life, this is my lady's hand. These be her very c's, her u's, and her t's, and thus makes she her great P's. It is in contempt of question her hand" – William Shakespeare, Twelfth Night (Act 2, Scene 5)*

Believed to have been written around 1601–1602 as a twelfth night's entertainment for the close of the Christmas season we see Shakespeare commenting on the uniqueness of handwriting features in his play.

With the exception of fingerprinting and DNA profiling, the majority of forensic science sets out to establish links between individuals and places or objects. The identification of handwriting is one of the few forensic sciences that addresses the identity of the individual and can be a powerful tool.

Most adults can recognise the handwriting of their immediate family and close friends, for example, when a birthday card arrives in the post. A pattern recognition process is taking place based on familiarity of form much in the same way that we identify a person's face. A handwriting expert does not have that familiarity at their disposal and needs to rely on a range of known writings of the individual to make scientific comparisons in order to determine the degree of match or dissimilarity. It is an axiom that, given sufficient amounts, no two individuals write exactly alike. Every character of the alphabet, both block capital and cursive, and the numerals can be constructed in a number of different ways. Each person's handwriting therefore displays a particular combination of character forms which gives that handwriting much of its individuality.

18.3.2 Development of Handwriting

The basic shapes and construction of handwritings are taught in school. In some countries handwriting systems are adhered to rigidly and in the UK, children especially in the same school, follow a particular set format for development. Handwriting is a major activity for children and is a learned process involving a complex series of neural connections in the brain. Studies have shown that certain forms of brain injury have significant effect on the person's writing capability. A stroke can often mean that the person needs be to taught to write again. Because different nationalities teach handwriting in different ways, it is sometimes possible to determine where a person was taught to write. Gender however cannot be determined from handwriting.

Individual handwriting features begin to be introduced during adolescence when the young person begins to experiment with the appearance of handwriting and mimics attractive features from sources other than the standard taught systems. The handwriting of any individual tends to attain maturity in early adulthood and remains consistent in shape, structure and proportion over the years until changes are introduced as a result of old age.

18.3.3 Construction of Character Forms

Handwritten characters, and indeed figures, can differ in their construction, shape and relative proportions; of these, construction is probably the most important feature used to distinguish between handwritings. Some of the block capital forms, such as 'E', 'G', 'H' and 'K', can be written in several distinctly different forms (Figure 18.1).

Different constructions of other characters are more subtle. For instance, the point of entry and exit of the character 'O' or the numeral '8' can be of crucial importance in distinguishing two different handwritings. The shape of the individual character form is also an important consideration; characters may be generally angular or rounded in their formation.

Both the internal proportions of the individual character forms and their relative proportions will be taken into consideration. An individual may consistently introduce a large form of a particular character into handwriting. Similarly, only part of a character may be relatively large, such as the bowl of the 'P' or 'R'. Equally important are such features as the point and direction at which a crossbar is made across a vertical in such characters as 'H' and 'T'.

Frequently, the detailed construction of character forms cannot be determined accurately without magnification of the pen lines. This is particularly the case where it is necessary to determine the order and direction of movement of the pen lines. In some circumstances, the direction of movement of the pen can be used to determine if the writer is left- or right-handed. There are a number of different features of pen lines which can be used to determine the direction in which the

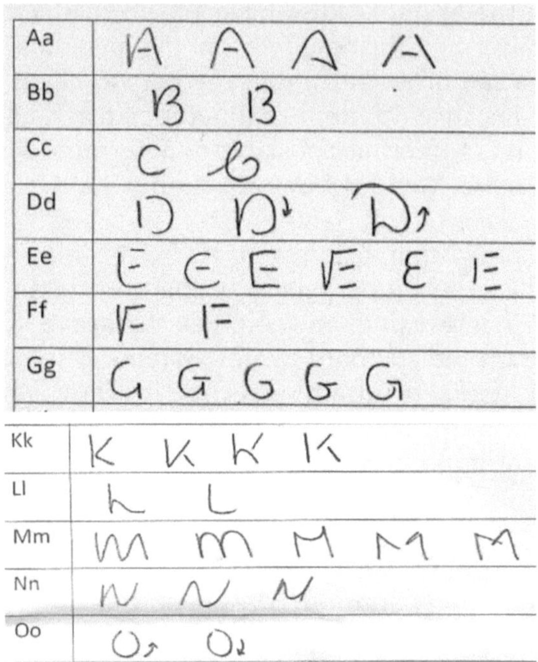

Figure 18.1 Some variations in character design.

pen is travelling. Ballpoint pen lines show features which are particularly useful in determining direction. They frequently show striations which follow the direction of curves and in addition the pen often deposits ink after a change in the direction of the pen line. Lightening of pressure at the endings of strokes indicates the pen leaving the surface of the paper fluently, whereas a definite spot of ink on the paper can indicate the position of the beginning of the pen stroke.

The striations always move outwards on a curve indicating the direction of the pen line. Other writing instruments produce pen lines that are more difficult to interpret.

The use of a good stereoscopic microscope and proper lighting is essential to study the character forms in detail. It is an essential part of an examiners training that they are able to identify and interpret striations in ball point pens. Inadequate exposure to these features can result in errors of interpretation and therefore authorship.

18.3.4 Natural Variation

Very few people write every character in the same manner on every single occasion. All handwritings will exhibit some natural variation, most often in the shape and proportions of characters but also occasionally in structure. It is essential that sufficient handwritings are available for examination to enable this range of natural variation to be determined. If insufficient handwritings are available for examination there will always be the problem of not knowing if any differences have arisen because they were written by different people or merely the result of natural variation in one person's handwriting.

18.3.5 Comparison of Writing

18.3.5.1 Comparison Material

The choice of comparison handwritings may well determine the scope of the final analysis. The best possible handwritings for an examination are those made in the course of day-to-day business as close in time as possible to the writings in question. This is particularly important when the writings of young or elderly people are under consideration. Specimen writings produced specifically for the purpose of comparison can also sometimes be of assistance in forensic examinations; it is however difficult to produce entirely natural handwriting under such circumstances. Furthermore, writings made all on one occasion will not show the normal range of variation that will occur from day-to-day. Sufficient comparison material must be examined in order to establish the full range of natural variation in the person's handwriting under examination. If only specimen handwritings are available, great care must be taken in their production to limit the possibility of copying the questioned document or including unnatural features. This is most easily achieved by dictating passages at intervals with breaks in between and removing the specimens from sight as they are produced.

18.3.5.2 Comparison Process

It is essential that handwritings are compared 'like with like'; a character 'a' with a character 'a', the character 'B' with a character 'B'. Similarly, cursive writing must be compared with cursive writing, block capitals with block capitals and signatures written in the same names.

Among the features which may be considered are elements of the writing such as abbreviation; alignment; arrangement, formatting, and positioning; capitalisation; connectedness and disconnectedness; cross strokes and dots, diacritics and punctuation; direction of strokes; disguise; embellishments; formation; freedom of execution; handedness; legibility; line quality; method of production; pen hold and pen position; overall pressure and patterns of pressure emphasis; proportion; simplification; size; skill; slant or slope; spacing; speed; initial, connecting, and terminal strokes; system; tremor; type of writing; and range of variation. Other features such as lifts, stops and hesitations of the writing instrument; patching and retouching; slow, drawn quality of the line; unnatural tremor; and standard lines of various forms are evaluated when present.

Normally an examination of the reference material is undertaken both visually and using magnification to determine, where possible, the fluency and range of variation to be found in the individuals handwriting style. A visual, side by side, comparison of the reference material with the questioned material is carried out and where possible, a comparison is made between the overall style of the writing, the fluency and line quality of the writing, the construction and formation of the characters and the shapes, relative proportions and connectivity of components.

18.3.6 Factors Affecting the Appearance of Writing

The appearance of handwriting can be altered by the circumstances under which it was made as well as the health of the writer. The consumption of alcohol is known to result in increased size and curvature of handwriting and other drugs, such as cannabis, may mimic this. The handwritings of elderly and infirm individuals often show uncertain pen lines and trailing strokes.

18.3.6.1 Alcohol

Larger more spaced out lettering, less legible, uncoordinated pen movements, re-touching and extraneous marks and positioning of writing may be affected. The detailed letter constructions are however likely to remain unchanged.

18.3.6.2 Drugs

The effect on handwriting is dependent on the drugs being used and personal sensitivity. Some drugs may improve writer skill by lessening tremor, for example in Parkinson's disease. Some other drugs *e.g.* local anaesthetic at the dentist may reduce writer skill, fluency and co-ordination. Cannabis will slow the writers' movements and legibility may suffer but only after high doses. It is therefore important for the examiner to remember to ask about drugs/medication and to obtain samples of writing before and after if possible.

18.3.6.3 Writing Position/Writing Surface

Use of a marker pen on walls or other non-paper surfaces can result in some variation in letter shapes and some disconnected letterforms. The formation of graffiti is another specialised area of forensic document examination requiring caution of interpretation.

A lack of fluency may seem apparent if writing or signatures have been produced on an uneven surface. It is therefore important to examine the reverse of the document.

A lack of coordination in writing movements may occur when writing is made in a car/taxi/train *etc.* and a lack of controlled movements observed when writing is produced on a tablet *e.g.* signing for packages. For many examples where conditions affect the overall appearance of the writing, the letter designs should not suffer greatly.

18.3.6.4 Age/Illness/Mood

Lack of coordinated pen movements result in "spidery" writing as we age. Parkinson's disease can result in tremor of the pen line and sometimes "micrographia" (very small writing). In Alzheimers disease there can be progressive loss of control, angular formations, abrupt changes in direction, and repetitions at the start. Other factors which may affect the appearance of writing include arthritis/stroke and use of the unaccustomed hand. Finally, emotional stress, feeling nervous and motivation to write are other factors to be considered when assessing the features in a person's writing.

18.3.7 Disguised Handwriting

Probably the most talked about and interesting area of forensic document examination is disguised handwriting. Unlike fingerprints and DNA, handwriting is a dynamic, moving image which can be altered to attempt to avoid detection. The process often begins with the attempted deception of the "Valentine's Card" at an early age. From the author's experience, some people are very good at disguising their writing and others are unable to change it at all without considerable thought and effort.

Such handwritings are often detected by their inconsistency: the slope of a piece of handwriting may vary, or unusual construction of character forms may appear. Often some characters are formed in eccentric ways, or features such as loops and curls are added.

18.3.8 Simulated Handwriting

An attempt can be made to simulate another person's handwriting. Any attempt at simulation is a compromise between accuracy and fluency. To simulate another person's handwriting it is necessary to suppress one's own natural handwriting characteristics and adopt those of the person whose handwriting is the target of simulation. This is a difficult process and one that cannot be maintained over long periods. Inevitably the simulated handwriting will show characteristics of both the simulator and the target handwriting, with the balance changing as time proceeds. The simulation will also be identifiable by differences in detail. The simulator may achieve a superficial similarity to the target handwriting, but it is unlikely that the detailed structure of the individual character forms will be correctly perceived or reproduced.

18.3.9 Non-roman Script

The principles of consistency in construction and natural variation are the same for all handwritings whether they are written in Roman, Arabic, Chinese or any other script. A forensic document examiner trained in the study of UK handwritings will be able to detect significant differences between handwritings in different languages and scripts but will be handicapped by not being familiar with the expected range of variation for any particular problem. In such circumstances the forensic document examiner will be cautious in attributing significance to similarities and differences.

18.3.10 Expression of Handwriting Conclusions

Having completed the comparison of two handwritings, the forensic document examiner will assemble the observations of differences and similarities in the detailed structure, shape and proportions of the component character forms of the handwritings, and from these draw the appropriate conclusions. The forensic document examiner's conclusions are based not only on the number of similarities but also on their quality. The variations in construction of character forms appear in different frequencies in the population. For instance, in the UK population the anticlockwise 'o' formation is far more common than the clockwise 'o' type form, which is normally associated with left-handed writers.

In all handwriting comparisons the presence of differences is of profound significance. There will always be similarities between handwritings because there are a finite number of character forms that can be used. However, the presence of even a single, consistent difference between handwritings must be explained since it can be a strong indicator of different authorship. Where there is a sufficient quantity of both questioned and genuine comparison handwritings, and they match in all respects without any significant differences being detected, then a strong conclusion of authorship can be given.

There are, however, many occasions when the quantity of handwriting, either questioned or genuine, is restricted. There may also be clear indications that one or other of the handwritings under examination is not natural or has been altered by an outside influence. In these circumstances, conclusions reached should be reflective of the limitations imposed on the comparison.

Because handwriting is to an extent a voluntary act and can be consciously changed, it is unusual to be able to say with total confidence that a person was not the author of a particular piece of handwriting.

The reliability of handwriting opinions was challenged in the US courts in 1995 and temporarily likened to a technical skill rather than a science. This challenge galvanized the forensic document examiner community in the USA into an overhaul of background research, validation of procedures in handwriting examinations and proficiency testing of examiners. The reliability of evidence in this field has been put on a firmer footing through the work of Dr Moshe Kam who, with the backing of the FBI, established error rates for the process of handwriting comparisons and demonstrated that forensic document examiners have identifiable skill in this field.

18.3.11 Copy Documents

Forensic document examiners are often asked to examine copies of documents. Ideally, the examination of handwriting and signatures requires an analysis of the fine detail of the handwriting including stroke direction and order, crossings between strokes, and pressure. Depending on the quality of the image, these details are often lost during the copying process and hence the conclusions that can be drawn from the examination of copies will be restricted. However, where the amount of handwriting under examination is large and its image clear, it is possible to determine sufficient characteristics of the handwriting for strong, and in some cases, very positive conclusions to be reached.

Documents presented in copy form will not show all of the finer detail of the writing present on the original. As such it is not possible to:

(a) Examine the documents using microscopy to see if the writing/signatures are in original form in "wet" ink.
(b) Examine the writing/signatures using microscopy to determine the detailed pen path ("formation") of the characters and signatures.
(c) Examine the detailed fluency and line quality of the writing/signatures using microscopy in order to confirm if there are any undue pen lifts or hesitations within the writing line.
(d) Test the paper surface using specialised lighting for indented or pencil guidelines associated with the tracing process.
(e) Examine the paper surface to see if documents have been written or signed whilst resting on the pages below or if other, extraneous, indented impressions are present.
(f) Examine the reverse of the document to determine the relative pressure of the writing.
(g) Examine the reverse of the document to see if the writing surface has affected the appearance or formation of the writing/signature.
(h) Examine the inks to see if similar or different inks have been used to write or sign each page within a document.

18.3.12 Graphology

Graphology is the analysis of handwriting to determine someone's personality traits. It does not have a scientific basis and is not the same as forensic handwriting examination which seeks to address authorship by way of scientific comparison.

18.4 Examination of Signatures

18.4.1 Introduction

Signatures are very specialised pieces of handwriting and specific problems are involved in their identification. Our signature is the piece of handwriting which we all use most frequently. Even individuals who write infrequently can produce a consistent signature. Because we use our signature so frequently it becomes more or less unconsciously produced each time it is written. The examination of signatures is essentially a pattern recognition process and as a result the nationality of the person whose signature is under investigation is of little consequence. The same rules apply to signatures written with Roman, Arabic or Chinese character forms or indeed any other script. A significant number of signatures made by individuals around the world contain no character forms but are merely a personalised design.

The greatest problem in determining the authenticity of a signature lies in the small amount of comparable material that is contained within any signature. This factor, combined with the natural variation which is inherent with any signature, makes signature authentication one of the more testing areas of forensic science.

It is not unusual for people to state that their signature changes each time they write it. Forensic Document Examiners are aware of this and furthermore, understand that if two signatures are identical then at least one of them cannot be an original signature written in ink.

Because of these difficulties, it is essential that the original signatures are examined if they are available. The amount of available information contained within a signature is so restricted that every comparable feature must be used. Fluency is difficult to be assessed accurately from a photocopy, nor can pen lifts or guide lines be detected. Conclusions on signatures can be drawn from the examination of copies but the limitations must be pointed out in the report.

The basis of signature identification is very much the same as that for handwriting. However, since signatures are used for personal identification they are frequently the targets of forgery. Considerable effort will be expended in attempts to simulate another person's signature. This can be achieved in a number of different ways.

18.4.2 Simulation

There are a variety of ways in which an individual will attempt to simulate someone's signature. In the majority of cases the individual will have a model of a genuine signature they will use to create the forgery. If they do not have a model and have to guess what the signature looks like, inevitably their attempt will bear no resemblance to the genuine signature style. The various means of producing a simulated signature include:

18.4.2.1 Tracing

If a document bearing a genuine signature is placed on a window or a sheet of glass over a light source and overlaid by another document, it is possible to trace the outline of the genuine signature on to the forged document. The tracing, often made in pencil, can then be inked over. The resulting forgery, although superficially similar to the genuine signature, will be detectable by its lack of fluency as the simulator laboriously follows the line of the original signature, and by mistakes in the detailed construction of the character forms. These will be superficially similar to those in the genuine signature but the simulator is generally only interested in obtaining a pictorial, passable simulation and will not necessarily use the correct number or order of strokes to achieve this.

Guide lines may be left on the document and even if only fragments remain these can be detected by viewing the signature under specialised lighting conditions. Pencil lines remain opaque when viewed in the infrared region of the spectrum, whereas many inks can be transparent at these wavelengths. Guide lines may also be in the form of indented impressions on the target document. These are produced by placing a genuine signature on top of the target document. The genuine signature is written over heavily so that impressions are transferred to the document beneath. These are then inked in to produce the simulation. The impressions can be readily detected in light shone at a shallow oblique angle to the document. In addition, the signature is likely to lack fluency and contain mistakes in a similar manner to the direct tracing.

18.4.2.2 Freehand Simulation

Probably the most encountered way of simulating a signature is where the individual attempts to copy the signature without any artificial means but merely creates their best image, known as freehand simulation.

This may be the most fluent way of simulating a signature however the simulator still needs to achieve a compromise between accuracy and fluency. Unusual hesitations and pen lifts in the pen line will occur whilst the simulator pauses before embarking on the next section. The simulation is detected by its superficial similarity yet differences in the fluency, detailed shape, structure and proportions of its component character forms.

18.4.2.3 Cut and Paste Simulation

Due to the complex motor control that is required to produce a signature, no two signatures are ever to be found so alike that they are superimposable on one another. It is to be expected that genuine signatures will exhibit a range of natural variation between them.

One common method employed by the fraudster is to take a genuine signature from a document and transpose it on to another to create a montage. The montage, by its nature will be a copy document. The method employed can be affected as follows:

Transposing handwriting/signatures from one document to another:

(1) It is possible using a variety of techniques to make an image of a genuine original signature or pieces of writing and to transpose these images onto another document to create a montage. The entries can be "cut" from separate areas within the document.
(2) Owing to the advent of computer technology it is possible to now execute this task electronically. However, the process of creating a montage document using "parts" from other documents has been historically carried out manually using scissors and glue, a process which can be undertaken by anyone dexterous enough to undertake the task.
(3) Once photocopied or scanned/printed, the new document will appear to bear genuine entries but will by its nature be a copy document. The signature/writing on the montage/composite document will not be in original form written with a pen and ink.
(4) There are a number of techniques which can be employed to carry out this process which include:
Mechanical Process
    (a) A genuine document bearing a signature/writing is photocopied or printed.
    (b) The copied signature/writing is cut out of the document using scissors.
    (c) The cut-out is then glued or fixed onto another document.
    (d) The resultant montage is then photocopied or scanned to produce a document which appears to bear a genuine signature/writing.

Electronic Process

(e) A genuine document bearing a signature/writing is scanned using a scanner.
(f) The scanned image is imported into a computer.
(g) Using widely available computer software the signature/writing on the document is selected by the user. Different areas can be selected.
(h) The selected signature/writing is then "cut" or "copied" electronically and "pasted" onto another document in an area(s) selected by the user.
(i) The resultant montage is then saved electronically as another document and can be printed or emailed.

(5) It is also possible to create a transparency of an image using a photocopier/scanner onto an acetate sheet. The sheet bearing the image can then be superimposed onto a document and copied/scanned to create a montage.
(6) There are variations to the general processes outlined but the basic method is relatively easy to execute by someone with a basic knowledge of how to use a photocopier and/or scanner and computer.
(7) Whilst the technique above describes signatures/writing, it can also be employed on typing/stamps or any other part to a document.

18.4.3 Authorship of Simulation

Simulated signatures are not natural writings. To produce a simulated signature the forger's own natural characteristics of writing must be suppressed and those of the person whose signature is the target of forgery adopted. Hence it is not normally possible to identify the author of a simulated signature by comparing it with natural handwriting.

18.4.4 Self-forgery

Certain signatures are written by individuals with intent to deny them at a later date. These signatures are generally different from normal natural signatures in some very obvious feature, often the form of initials or slope, but the detailed structures of the less obtrusive features are unchanged. A change in the length of the signature or the obvious omission of a part of the signature can also be a sign of self-forgery.

18.4.5 Vulnerable Signatures

Signatures which are short and simple and which contain a number of natural pen lifts are vulnerable since they reduce the requirement of maintaining fluency while executing unfamiliar and possibly complex character forms. Signatures that naturally demonstrate a wide range of variation are also vulnerable since it is difficult to assess if differences have arisen merely as a result of this variation or because of simulation by another person.

Signatures of elderly people often naturally contain many of the features which are the hallmark of simulation, such as lack of pen control, undue variation and poorly formed character forms. These signatures are, therefore, particularly vulnerable. In some cases, it is not always possible for the expert to offer an opinion but this in itself can be of assistance to the court.

18.4.6 Guided Hand Signatures

An infirm, elderly or partially sighted individual may be assisted by another person in signing documents. Guided hand or assisted hand signatures are a particularly difficult area for the forensic document examiner. The characteristics of such a signature may be a mixture of those of the signator and the assistant.

18.4.7 Comparison Material

The requirement for adequate comparison material in signature examinations is similar to that required in all handwriting comparisons. However, it is worth stating again that like can only be compared with like – in other words, signatures are only comparable with other signatures in the same name. 'Smith' cannot be compared with 'Jones', nor can signatures generally be compared with day-to-day handwritings. There is no ideal comparison material however a sample that consists of genuine signatures on between twelve and twenty documents signed in the course of day-to-day business close in time to the date of the signature in question should be adequate in many cases. It is sometimes useful to see the person's passport or identity card in case the signature on those documents, which are often reduced in size, has been used as the model for the simulation(s).

It is also useful for the individual to sign their name a number of times (for example 6 samples on five separate sheets of paper) as these can be used to verify the course of business signature and can be helpful in the comparison process as "known" or "witnessed" examples.

18.5 Examination of Copy Documents

Copies of documents can be produced in a number of ways, *e.g.* simple photocopying or electronic scanning. The limitations of examining copies have already been explained. However, it is often the case that a copy is the only evidence of a pre-existing original. The forensic document examiner is then charged with the task of determining from the copy if it was indeed prepared from an authentic contemporaneous original document or if the copy has been recently manufactured or manipulated to misrepresent the facts.

The copy may be a montage produced, either physically or electronically, partly from genuine documents. The simplest demonstration of this is the image of a genuine signature transferred by photocopying on to a fraudulent document as described above.

It may be important to demonstrate the origin of a particular photocopy document. There are two types of mark which appear on photocopies that can be used to do this: trash marks and drum or mechanism marks. Trash marks are transient and produced by dust particles or debris on the glass surface of the photocopier. These will appear as a series of dots or marks on the resultant copy. The same configuration of trash marks on a number of documents would indicate that they have a common origin. However, it is important to remember that if a copy bearing trash marks is copied, the daughter copies will also bear the same configuration of marks. Marks made by the drum or the mechanism of the photocopier will persist for a longer period and can be used to identify a specific machine. Drum marks do not necessarily appear in the same position on every page but can be detected by the constant intervals at which they appear. The interval is directly related to the circumference of the drum and not the dimensions of the document being copied.

The development of colour photocopiers raised concerns as to their possible use in counterfeiting. Many photocopier manufactures (and also colour laser printer manufacturers) incorporate a code of yellow dots into every copy produced. These codes may assist in identifying the brand of copier, model, serial number and, in some cases, the date of a particular copy. They are often referred to as the Machine Identification Code.

18.6 Printing and Typewriting

18.6.1 Modern Office Technology

The replacement of the old-style manual typewriter with fixed type bars, first by electric typewriters containing golf-balls and print wheels, then by dot-matrix printers and today by the ubiquitous ink jet or laser printer, has completely revolutionised the forensic examination of typescript in the last 40 years.

18.6.2 Word Processors

The term 'word processor' actually refers to the computer program used by the keyboard operator to prepare a computer file of the document which is to be printed. The printing of the document from the computer can be made using a number of different types of machine, most commonly laser printers and ink jet printers or occasionally dot-matrix printers. The appearance of the printed characters that finally appear on the paper is to a large extent governed by the computer program.

Changes in software operating systems can be very important in dating computer-generated documents. The introduction of the Times New Roman typeface in the early 1990s by Microsoft, and its subsequent replacement as a default typeface by Calibri in 2007, are particular markers which have been important in the authentication of documents. It is, however, very difficult to distinguish between the myriad of type styles available, some of which may differ in only very subtle features.

18.6.3 Laser Printers

The laser printer works very much on the same principle as a photocopier. It contains a photosensitive drum which becomes charged as the laser light hits it, producing an 'electrostatic negative' of whatever image is to be printed. The drum is sprinkled with negatively charged toner which clings to the charged areas originally scanned by the laser beam. A sheet of paper is passed under the rotating drum and given a charge greater than on the drum, allowing transfer of the toner from the drum to the paper where it is then fused in place by heat or pressure. Images produced by laser printers are generally of very high quality. They are also very reproducible, and it is extremely difficult to distinguish between the work of two different laser printers. However, it has been established that, on occasions, drum faults may occur on laser printers in the same way as they do on photocopiers. As the laser drum is often smaller than in a photocopier and the paper wrapped around the drum three times, any marks are often seen three times on a single sheet, around 94 mm apart. It is important to be aware that this is one mark repeated three times and not three separate marks on the drum.

As difficult as it is to distinguish the work of one photocopier from another, it is similarly difficult to identify any document as having been produced by a particular computer/printer system. The number of fonts available and their versatility is immense.

18.6.4 Ink Jet Printers

In ink jet printing technology, a digital image is created by spraying droplets of ink onto paper and plastic substrates. Ink jet printers are one of the most commonly used type of printer and range from small inexpensive consumer models to expensive professional machines.

The main difference between the laser and the ink jet printer is the state of the ink. When the ink from the ink jet printer hits the paper it spreads into the fibres and gives the print a slightly ragged appearance. The fused appearance of the toner of the laser printing is clearly different from the ink jet image using microscopy.

Although the work of ink jet printers can readily be identified it is as difficult to identify a particular machine or differentiate between the work of several machines as it is with laser printer copy.

18.6.5 Typewriters

Typewritten documents are only occasionally encountered in casework and are sometimes used by someone writing a hate mail correspondence. However, they can be encountered more commonly on documents that were produced before 1990 when other methods of production were not readily available. Both fixed bar and single element typewriters that allowed interchanging of typestyles and sizes, frequently produced features that varied with usage and could be used to identify the origin of an individual document and to link documents together to show a common source.

18.7 Origin and History of Documents

18.7.1 Authenticity of Documents

There are occasions when documents used in criminal activities or in the course of civil litigation are either created specifically for the purposes of deception or altered for similar ends. For example, entries in diaries, or occasionally whole diaries, are prepared to show that a particular sequence of events occurred in the past. Similarly, files can be reconstructed and correspondence backdated.

The forensic document examiner is often asked to assist in determining the authenticity of such documents and entries. The dating of documents and writings is a difficult forensic science problem. Since there are no reliable techniques available in the UK which allow the absolute dating of inks, it is necessary to use other features of documents to determine where and when they were created and their history.

Take for example a diary which purports to represent a chronological record of events, yet it is suspected of having been bought and completed recently.

If the diary is a genuine chronological record of events, we may expect to find some of the following features:

- The spine is worn or creased in a number of places, or even torn.
- The outside cover has scuff marks or dirt on it, particularly the back cover.
- The fabric bookmark may be creased, worn and beginning to come away from the spine. Its location in the diary may be significant?
- There is writing on the inside cover, important names/numbers or important dates. There may be other extraneous writings on the pages outside the main calendar *e.g.* in the notes page or on personal pages.
- The used pages are "well thumbed" and show crease marks when viewed under oblique lighting. Examination of the diary from the side, when closed, will show which pages have not been turned or written on. They will lie flatter than the used pages.
- There may be loose scraps of paper or business cards in the diary or post-it notes.
- A variety of inks will have been used. If the diary is old there may be variable discolouration of the inks, particularly near the edges of the page where light has penetrated.

- The writing will vary in terms of speed and rhythm and there may be a variety of styles evident (*e.g.* arrows, brackets, date styles, asterisks, underlinings, punctuation).
- There should be crossings out and other alterations present. Correction fluid may have been used.
- There may be "doodles" or areas where pens have been tested for colour or to see if they are working. There may be areas where pens have run out.
- There may be areas where additions have been made. For example, at the tops or bottoms or pages or lines squashed into text in a different spacing or using different ink.
- Some crossed out appointments may appear later or earlier in the diary.
- There may be a variety in the amount of writing on each page; some pages may be blank, others full.
- There may be yellowing of the edges to varying degrees, depending on storage.
- The diary will not smell new.
- The "unused" pages may have one or two future appointments on them.
- The "unused" pages may have been "thumbed" but not to the same extent as the "used" pages.

Alternatively, if the diary is new and has been written up recently on one occasion, we may well find it to be clean and does not look "used". The writing is neat and consistent, although a number of pens may have been used.

The presence and absence of those features observed in the genuine diary should be taken in to account when addressing the authenticity of a diary.

18.7.2 Examination of Inks

The ink used to write particular entries on a document may be of interest for several reasons. Principally the forensic document examiner will be interested in establishing if inks are similar or different. In this way it is possible to determine if entries have been added to or altered.

The examination of inks on questioned documents is generally confined to non-destructive comparative analysis. Inks that appear similar to the unaided eye can be very different when viewed in the infrared region of the spectrum. Similarly, illuminating an ink with high-energy light at one wavelength can promote it to release light energy at different wavelengths. The degree to which this occurs depends on the chemical nature of the ink. Several instruments have been developed which allow inks to be viewed under a wide range of different lighting conditions to enhance some of the less visible effects. The fact that many inks can be rendered transparent in the higher wavelengths of the infrared region of the spectrum is an essential tool in the examination of questioned signatures. Pencil markings remain opaque at these wavelengths and can easily be detected under the transparent ink. Micro-spectrophotometry and Raman spectroscopy are other largely non-destructive techniques which can prove useful in distinguishing between inks which cannot be distinguished using more traditional infrared and fluorescence techniques.

On some occasions further destructive techniques can be employed to compare inks more rigorously. In fact, this destructive analysis can be carried out using relatively tiny amounts of ink which are carefully removed from within the pen line of a single character form. Thin-layer chromatography (TLC) or high performance liquid chromatography (HPLC) of these ink samples is used to separate and compare the dye components.

Most of the ink analysis carried out in the UK is comparative. If parts of a document can be demonstrated to have been written using two different inks, then clearly two pens have been used. However, the opposite is not equally true – if the inks used in two writings are indistinguishable they could have been written either using the same pen or pens containing similar inks. Since most ink dye manufacturers do not change basic ink compositions very often, there are likely to be several million similar pens available for any one type of ink.

Some high-end pen manufacturers can incorporate "markers" in the ink for a particular pen to identify the important individual who owns the pen.

18.7.3 Examination of Paper

The size, thickness, density, colour and finish of papers can be compared to determine if they are from a similar or different origin. Optical brighteners are incorporated into some papers and give differing reactions when viewed in ultraviolet light.

The origin of a paper is most accurately identified by its watermark. Some paper manufacturers periodically introduce changes into watermarks. Repairs to the dandy roll used in the manufacturing process to produce the watermark may be identifiable in subsequent production runs. However, not all manufacturers keep detailed records of changes made over the years and most paper in normal circulation does not bear watermarks.

18.7.4 Development of Handwriting and Signatures Over Time

As noted previously the handwriting and signatures of individuals vary over the years. If sufficient examples of genuine handwritings and signatures are available to demonstrate the different stages of development, questioned writing and signatures can be matched to a particular period. This type of examination is particularly applicable to the writings of young and elderly people.

18.7.5 Impressions

Impressions of handwriting may be left on a document as a result of writing on another document which at the time of writing was overlying the first one. These impressions are rarely visible to the naked eye and need to be detected using specialised equipment and techniques.

Deep impressions can be detected using light directed at a low oblique angle to the document. In this way the surface of the document is thrown into relief and the impressions can often be read.

Fainter impressions can be detected using the technique of electrostatic detection (ESDA). This technique was developed jointly by the London College of Printing and the Metropolitan Police Forensic Science Laboratory in London in 1979 and revolutionised forensic document examination. The long and tedious hours piecing together fragments of indented impressions in dimly lit rooms were replaced with a relatively fast and highly sensitive technique which could develop impressions made through several layers of paper. Such impressions are too faint to be detected by any other methods. It is thought that the ESDA process detects areas where the paper fibres have been deformed by the movement of the writing implement or by two documents moving against each other.

The technique is not always successful – impressions cannot be detected on documents which have been treated with solvents. It is therefore imperative that any document which will eventually be examined for fingerprints should first be submitted to the forensic document laboratory before treatment with fingerprint reagents. A number of reports have been made of 'secondary impressions'. These impressions are caused not by the action of writing but by storing two documents together such that the fairly deep embossing of actual handwriting on one document causes detectable impression marks on the second document. Such 'secondary impressions' can be very confusing, particularly if they appear on the same document as those impressions caused by the act of writing. Whilst the presence of impressions is positive evidence of handwriting made on an overlying document the absence of impressions cannot be taken as positive evidence that the event did not occur. The ESDA response is variable and can be affected by ambient humidity and the conditions under which a document has been stored.

The ESDA technique can sometimes be used in the determination of the order of writings. The ESDA 'lift' bears not only black traces caused by the development of the impressions but also traces caused by the ink lines present on the document. On occasion it is possible to determine which came first, the impressions or the ink line, however considerable caution needs to be expressed when interpreting the findings and the technique can only be used when the ESDA lift is of good quality and there are a substantial number of intersections which are clear.

Although ESDA has now been in routine use in forensic document laboratories for over forty years the underlying mechanisms of the process are still not fully understood and the ESDA response can be variable. However, it is one of the most powerful weapons in the armoury of the forensic document examiner.

18.7.6 Folds, Creases and Tears

The physical condition of a document may betray much of its history. The arrangement of folds and creases in a document may indicate if it was folded into an envelope or folded in conjunction with other documents. Torn edges of a document may be matched to those of another document, particularly if they were torn together from the same pad. It is also possible to reconstruct documents by studying the mechanical fit of the torn edges, even if the documents have been shredded.

18.7.7 Staples and Punch Holes

The act of stapling one document to another causes physical damage to the document as the staple ends are punched through. If the staples are removed the staple holes remain obvious and can also be seen after the document has been scanned or photocopied. Further damage can be caused to the back of a document where the ends of the staples are forced in, often distorting the paper between the staple holes. The relative size and the positioning of staple holes can be used to identify documents which have been stored together.

The study of staples is also important in considering if sections of documents have been removed, such as in notebooks. Punch holes also cause irreparable changes to a document. Punch holes become worn and distorted over a period of time as a result of movement of documents against the holding prongs of a file or clip. Often documents stored together for a period of time will show a similar pattern of damage to the punch hole.

18.7.8 Erasures, Obliterations and Additions

The act of erasing an entry from a document, be it typewritten, handwritten or printed, will damage the document. The extent and nature of the damage will be different depending on whether the erasure was carried out by mechanical or chemical means.

In general, mechanical removal of an entry will physically damage the document. Using an eraser to remove entries results in the paper fibres of the document being disturbed. This disturbance, if large enough, may be detected using obliquely directed light which throws the surface of the document into relief. Remnants of the pencil can be detected using specialized lighting techniques.

Chemical erasure involves the use of a solvent to remove an ink entry. This may be comprehensive and leave no trace visible to the unaided eye. The effect of the solvent on the document, however, can be detected by viewing it in ultraviolet light or under conditions which promote infrared fluorescence.

Traces of entries removed by erasure can be enhanced by viewing the document under different optical conditions. Pencil entries are opaque in the infrared region of the spectrum and can therefore be enhanced using an instrument which combines this lighting facility with powerful magnifying capabilities. Similarly, traces of an ink which has fluorescent properties can be enhanced by promoting the fluorescence and magnifying the image. Some of these techniques can be useful in enhancing entries on documents that have aged.

Obliterated entries are detected by separating the original entry from the obliterating material. This can be achieved non-destructively by exploiting the optical characteristics of inks and correcting fluids.

A black ink entry obliterated with another black ink can be recovered if the inks react differently, for instance, in infrared light. Entries obliterated with white correcting fluids can be recovered using a combination of lighting conditions including viewing from the back of the document. Today's modern range of instrumentation in forensic document laboratories allows images of such features to be captured electronically and manipulated to give a clear view of the original obliterated or erased entry.

The detection of added ink entries again involves the non-visible properties of the inks. Anyone making a fraudulent addition to a document will make an effort to choose an ink of similar colour. However, these visually similar inks may well have very different properties in infrared light and can easily be demonstrated as having been made with different pens.

18.8 Stamp and Seal Identification

Examination of stamp impressions on documents such as passports and commercial company documents are frequently carried out in laboratories in many parts of the world. In parts of the Far East a stamp is often associated with a handwritten signature collectively referred to as person's "Chop".

Materials used to produce stamps vary both in quality and type. Pre-inked stamps are very common. Such stamps contain ink reservoirs that can be used to produce up to one thousand impressions. Stamps may exhibit manufacturing defects and also defects introduced as a result of wear and tear which can assist in their identification.

18.9 Printed Documents

Most practicing forensic document examiners will have a working knowledge of printing techniques sufficient to identify the process by which a document or parts of a document have been produced. However, some fields of printing required a highly specialised knowledge.

High-value documents, *e.g.* bank drafts and identity documents such as passports, are specially printed using methods to deter alterations or counterfeiting. Documents can be printed on papers which are sensitive to the application of chemicals or mechanical erasures and which change colour. Coloured or fluorescent fibres can be added to the paper. Security backgrounds can be produced with fugitive inks which run on application of a liquid, and complex patterns of printing can be produced which are difficult to imitate. Sections of documents can be printed using inks with special spectral properties or which change colour under different lighting conditions. The techniques used for studying alterations to documents can also be used to detect breaches of a document's security.

A knowledge of printing processes is frequently needed in the identification of counterfeits. The counterfeit must be compared directly with the genuine article, be it a passport, cheque or bank note. The forensic document examiner needs to be able to identify the differences in printing between the products to show differences in their mode of production.

18.10 Procedures, Protocols and Quality Assurance

Forensic document examiners should be all following the same, broadly agreed, standards for the discipline.

Whilst it is innovative to try out different methods, particularly using new instrumentation, the examiner must be mindful that any new techniques should be validated and accepted within the scientific community.

In forensic document examination, as in any forensic science, certain criteria need to be applied to ensure a correct interpretation of the evidence.

Any document, either questioned or of known origin, submitted to a forensic document laboratory needs to be handled properly and securely to ensure it retains its integrity throughout the examination. Documents need to be properly identified, labelled and recorded. The appropriate testing must take place and this should always be the most rigorous available. The results of examinations and tests should be recorded clearly.

The forensic document examiner's report should be comprehensive, setting out the examinations employed, the results and the conclusions drawn. It must be clear and understandable to the reader, be it a member of the jury or a High Court Judge.

Finally, it is critical that the Forensic Document Examiner is impartial to proceedings and that results are provided on an independent basis, having consideration to any possible bias that may occur. Always look out for the unexpected, consider all the other possible alternatives and never assume any case is straightforward, as it never is.

Further Reading

V. Aginsky, Ink Aging Testing – Do Preceding Indentation Examinations Affect Ink Ageing Parameters? *J. Am. Soc. Quest. Doc. Exam.*, 2014, **17**, 49–63.

D. M. Ellen, *The Scientific Examination of Documents*, CRC Press, Boca Raton, 3rd edn, 2006.

W. R. Harrison, *Suspect Documents*, Sweet & Maxwell, London, 1958.

R. A. Huber and A. M. Headrick, *Handwriting Identification Facts and Fundamentals*, CRC Press, Boca Raton, 1999.

J. S. Kelly and B. S. Lindblom, *Scientific Examination of Questioned Documents*, CRC Press, Boca Raton, 2nd edn, 2006.

A. S. Osborne, *Questioned Documents*, Boyd, Albany, NY, 1929.

Proceedings of 2nd European Academy of Forensic Science Meeting (European Document Experts Working Group), Cracow, 12–16 September 2000 (papers by C. Neumann and W. D. Mazella, and by E. Wagner).

J. S. Kelly, *Forensic Examination of Rubber Stamps*, Charles C Thomas, Springfield, Illinois, 2002.

19 The Courts and Expert Evidence

Ann Priston

The Chartered Society of Forensic Sciences, 19A Mount Parade, Harrogate, North Yorkshire, HG11BX, UK
Email: annpriston@gmail.com

19.1 Introduction

The prosecution of offences in England and Wales is an extremely convoluted process riddled with exceptions. Hence only the procedures that apply generally will be described here.

The law of England and Wales has built up gradually over the centuries in a variety of different ways referred to as sources of law. Historically the most important of these is the unwritten law that was developed by custom and the decision of judges. This is called common law. As parliament became more powerful, laws began to be created by legislation and Acts of Parliament became the new sources of law. This source is called statute law. European law has recently made a considerable impact but judicial decisions are still important in interpreting parliamentary law.

The decision to investigate an offender normally rests with the police. Once started, the decision to prosecute becomes one for the Crown Prosecution Service (CPS). This is a separate government department, headed by The Director of Public Prosecutions and is responsible for the conduct of prosecutions in England and Wales.

Crime Scene to Court: The Essentials of Forensic Science, 5th Edition
Edited by Niamh NicDaeid and Peter C. White

Published by the Royal Society of Chemistry, www.rsc.org

Sometimes the CPS is consulted before proceedings are begun. Private individuals can also bring prosecutions (although this is rare) as can companies and large organisations.

19.2 Offences and the Courts

19.2.1 Offences

Since 1977 offences have been classified depending on how they are dealt with and which of two courts will try them. This classification is called 'by mode of trial'. The offences are classified as Summary, Indictable and Either Way and the courts which hear criminal trials are the magistrates' court and the Crown Court.

The least serious offences are called 'summary' offences and are always tried in the magistrates' court. This is an actual determination of the case and not a preliminary enquiry. These include most driving offences, common assault, public disorder, *etc.*

The most serious are called 'indictable only' offences and must be tried at the Crown Court before a judge and jury. These include murder, robbery, rape, terrorism offences and others.

The majority of offences are middle range and called 'either way' offences. These are indictable offences of a slightly less serious nature and, as the name implies, can be tried in either court. They include theft, actual bodily harm, deception, *etc.* A complex procedure exists to decide the progress of either way offences.

Thus classification of offences as indictable, summary or either way broadly reflects a distinction between serious and minor crime, and these together with their penalties are laid down in the 1977 Criminal Law Act (as amended by subsequent legislation).

19.2.2 The Courts

Courts are broadly divided into superior courts (Supreme Court, Court of Appeal, High Court, Crown Court, Restrictive Practice Court, *etc.*) and inferior courts (magistrates', youth and county courts). Most criminal trials are dealt with in the magistrates' court and the Crown Court.

19.2.2.1 Magistrates' Courts

The most important of the inferior courts with respect to criminal cases are the magistrates' courts. All cases start in the magistrates' court and 97% are dealt with there. With respect to crime the court's functions include (i) the issuing of summonses, granting of arrest and search warrants, granting of bail; and (ii) dealing with mode of trial, sentencing and summary offences.

The judges in these courts are part-time magistrates called Justices of the Peace (JPs) or lay magistrates. There are over 30 000 lay magistrates who sit as a panel of three, with only one person (the chair person) acting as the spokesperson. They have no legal qualifications but do undergo an initial period of training and ongoing updating on sentencing guidelines as the law changes. A legally qualified adviser is present to advise them on the law and procedure. Magistrates are addressed as 'Your Worship' (or 'Sir/Madam'). Lawyers appearing before them are not wigged and gowned. There are also approximately 100 legally qualified salaried magistrates called District Judges in some large cities. These are full-time and sit alone. Because they are legally trained they do not have the need of a legal adviser. District Judges are addressed as 'Sir' or 'Madam'.

The sentencing powers of the magistrate are strictly limited. They can impose fines of up to £5000 (for either way offences), make community orders and impose a maximum sentence of 6 months custody or twelve months for two or more triable either way offences. Where they feel they do not have sufficient powers for sentence, they can send the case to the Crown Court for sentencing by a Circuit Judge.

Dress in the magistrate's courts is relatively informal in that there are no robes.

19.2.2.2 Youth Court

Proceedings involving juveniles, children between 10 and 14 years old, and young persons between 14 and 18 years, normally take place in a youth court which is a type of magistrates' court specially modified for dealing with young people. The court is composed of either a bench of lay magistrates (JPs) or a District Judge who have specialist training and are qualified in dealing with young offenders. Normally three lay magistrates sit including a chairman and it is usual to have both sexes represented on the panel.

19.2.2.3 Coroner's Court

The Coroner's Court is principally concerned with inquests into cases where there is cause for suspecting violent or unnatural death, sudden death with unknown cause or death within prison. It also has jurisdiction over treasure trove and all outbreaks of fire (in London). Experts in fire investigation and toxicology are more likely than others to give evidence at a Coroner's Court where the system is inquisitorial. The Coroner usually sits alone to hear inquests, but may empanel a jury. Although an inferior court attendance, this court takes precedence over even the Central Criminal Court because its formation as the King's Court in the 12th century predates the other courts. The Coroner is addressed as 'Sir' or 'Madam'.

19.2.2.4 Crown Court

Since the 1971 Courts Act all trials of indictable offences committed in England, Wales and Northern Ireland take place at the Crown Court (as discussed in more detail in Section 19.5.2). In addition, its functions include hearing appeals following summary conviction in magistrates' courts and sentencing persons committed for sentence following summary conviction.

To hear trials on indictment the court is composed of a High Court Judge, a Circuit Judge or a Recorder, who sit with a jury of 12. When hearing appeals from magistrates' courts the judge sits with at least two (sometimes one) and not more than four Justices of the Peace. Then the decision is by majority and the judge has the casting vote if necessary.

19.2.2.5 High Court

The High Court has both civil and criminal jurisdiction. There are three divisions of the High Court: the King's Bench Division; the Family Division and the Chancery Division. The only division with criminal jurisdiction is the King's Bench Division and judges from this Division sit in the Crown Court to try the most serious crimes.

19.2.2.6 Court of Appeal

There are two divisions of the Appeal Court; The Criminal Division and the Civil Division. The Criminal Division hears appeals from decisions of the Crown Court. The court consists of Lords Justices of Appeal, the Lord Chief Justice (its president) High Court Judges and sometimes Circuit Judges. It is normal for two High Court Judges to sit either with the Lord Chief Justice or with one Lord Justice.

19.2.2.7 The Supreme Court

This is the final court of appeal for criminal cases in England, Wales and Northern Ireland and was formerly the appellate division of the House of Lords. It was established in 2009 and considers appeals on points of law and public importance. It has both civil and criminal jurisdiction.

19.2.2.8 Scottish Courts

Although concentrating on the English and Welsh legal systems here, the structure of the courts in Scotland is somewhat different, although the principles of justice and the responsibilities of the forensic scientist remain essentially the same.

Justice of the Peace Courts are replacing District Courts which form the lowest tier of courts in Scotland. Here lay justices sit to deal with the least serious criminal matters. The principal local court is the Sheriff Court which has both criminal and civil jurisdiction. A sheriff, who is legally qualified and in some ways may be likened to a District Judge, may sit alone (Summary Procedure), or with a jury of 15 members (Solemn Procedure) depending on the seriousness of the crime to be tried. A jury can return a verdict of 'guilty', 'not guilty' or 'not proven', provided that eight of the jurors agree. A defendant cannot elect for a trial by jury. In 2023, the decision was made to remove the not proven verdict and reduce jury numbers from 15 to 12.

The High Court of Judiciary is the most senior court and hears the most serious offences. It sits with a jury of 15 in Edinburgh and other major towns. These courts also hear appeals from the lower courts. The equivalent of the coroner's inquest is a fatal accident enquiry, which is presided over by a sheriff. In court all judges and sheriffs are addressed and 'My Lord/Lady'; Justices of the Peace are addressed as 'Your Honour'.

Prosecution in the Scottish courts is undertaken by the Crown Office Procurator Fiscal Service (COPFS), led by the Lord Advocate. Procurator Fiscals (solicitors), operate through local offices in a manner which is similar to that of the Crown Prosecution Service in England and Wales. The procurator fiscal is most often involved in the investigative phase of a criminal case from the start. They often attend forensic strategy meetings where investigative decisions are discussed between police, forensic scientists and other agencies (as appropriate) and where the Procurator Fiscal can ask that further investigative work be carried out. The Crown Office also instructs advocates and advocate deputes (barristers) to present the Crown's case in the Courts.

19.3 The Legal Profession and the Judiciary

19.3.1 The Legal Profession

Having briefly mentioned the lay magistrates and the District Judges who sit in the magistrates' courts there now follows some discussion about the rest of the legal profession. It is a distinctive feature of the English legal system that the profession is divided into two branches; solicitors and barristers.

19.3.1.1 Solicitors

No unqualified person may act as a solicitor. Before being admitted as a solicitor a person must usually have attended a course of legal education and completed a training contract. However, the training for solicitors is currently being reviewed. Their name must appear on the Solicitors Roll kept by the Law Society, who must issue a practising certificate each year.

The major duty of a solicitor doing criminal work is to ensure that the client is properly advised as to the legal position (which may involve consultation with a barrister); they also prepare cases for trial by seeing witnesses. All solicitors have full rights of audience in the magistrates' court and have a limited right in the Crown Court. Some solicitors called Solicitor Advocates have been granted higher rights of audience by the Law Society and may appear in the Crown Court before a judge and jury. When they do, they are robed but do not wear wigs.

19.3.1.2 Barristers

A barrister, also referred to as 'counsel' is a person called to the Bar by one of the four Inns of Court (Inner Temple, Middle Temple, Lincoln's Inn or Gray's Inn). For this they must be 21 years of age, have completed the requisite number of qualifying sessions as a student in their Inn (normally 12) and passed the relevant examination. In order to become a barrister a person must be called to the Bar by the Benchers (senior members) of their Inn and must serve an apprenticeship called a pupillage, usually for two periods of six months each. Unless and until they become King's Counsel a barrister is termed a junior: they wear a gown of 'stuff' and is a member of the Outer Bar. After a minimum of ten years they are eligible to apply to the Lord Chancellor to become King's Counsel (KC). They wear a silk gown, are termed a 'silk', and are members of the Inner Bar. They then undertake less routine work and rarely appear in court without a junior to assist them. Most judicial appointments are made from amongst these ranks. Generally, a barrister may not act unless first instructed by a solicitor. Apart from advocacy in court, they may give advice on points of law and draft proceedings and other documents such as conveyances and wills, *etc.*

19.3.1.3 Law Officers

The Law Officers, legal advisers to the Crown, are the Attorney General and Solicitor General. Both are political appointments. The Attorney General is the government's chief legal adviser and as head of the English Bar has precedence over all King's Counsel. In certain circumstances the Attorney General has the power to stop a criminal prosecution. The Solicitor General's position is basically that of deputy to the Attorney General.

The Director of Public Prosecutions (DPP), though not strictly a Law Officer, does work under the supervision of the Attorney General. They are senior barristers or solicitors, appointed by the Attorney General, whose duties are to institute and have the conduct of most criminal proceedings. The Prosecution of Offences Act 1985, established a Crown Prosecution Service (CPS) with the DPP as its head, formally answerable to the Attorney General, and thence to Parliament. Barristers and solicitors employed by the CPS are known as Crown Prosecutors.

19.3.2 The Judiciary

Another distinguishing feature of the English judicial system is that there is no separate judicial profession. Judges (other than lay magistrates) are selected from experienced members of the Bar and solicitors, and it is from amongst the ranks of KC's that the Judges of the High Court are usually selected.

19.3.2.1 Circuit Judges and Recorders

As a result of the 1971 Courts Act, all county court and Crown Court judges became Circuit Judges. This was a new, full-time appointment. Barristers and solicitors of ten years standing first sit as part-time judges and they are called Recorders. Recorders have the same powers as Circuit Judges although they do not normally try cases of the most serious nature, *e.g.* not murder and they are addressed as 'Your Honour'.

At the Central Criminal Court in Old Bailey (commonly referred to as 'The Old Bailey') the most senior judge is The Recorder of London. The next most senior judge is The Common Sergeant. These are resident City of London appointments and the judges do not sit in other courts. All judges here (and at other 1st tier courts) are addressed as 'My Lord'. The Recorder of London normally wears a scarlet gown with black velvet facings, but for certain occasions he wears a purple gown. A rosette can be seen on the back of the gown which is a symbol of the wig-bag. The Common Sergeant, next in importance, wears a black tufted gown. Circuit Judges wear black gowns with purple edging and a scarlet sash which indicates they are trying a criminal case (a purple sash indicates a civil trial).

Judges of the Crown Court usually sit with a jury, unless hearing appeals from the magistrate's court when they sit with two Justices of the Peace.

19.3.2.2 High Court Judges

These are an elite band of full-time judges, often known as Red Judges because of their scarlet robes, over which is worn a black scarf secured by a black belt. Over the scarf lies a scarlet band which passes diagonally from the right shoulder to the left front. During autumn and winter the robe is trimmed with ermine, and for spring and summer with slate coloured silk. Upon appointment to the bench the new judge of the High Court is knighted, known by the title 'The Honourable Mr Justice Smith', is referred to as 'His Lordship' and addressed in court as 'My Lord' (appropriately changed for females). He is written as 'Smith J'.

All High Court judges dealing with criminal cases are members of the King's Bench Division of the High Court. They sit in the Crown Court to hear serious cases on indictment and may also sit with the Lord Chief Justice or a Lord Justice of Appeal in the Criminal Division of the Court of Appeal.

Experts giving evidence in the most serious cases are likely to encounter these judges.

19.3.2.3 Lords Justices of Appeal

The judges of the Court of Appeal are called Lords Justices of Appeal. These are appointed from High Court judges. On appointment they become members of the Privy Council and carry the title 'The Right Honourable Sir John Smith' or 'The Right Honourable Lord Justice Smith'. They are referred to as 'Lord Justice Smith' and written as 'Smith LJ'. In court they are addressed as 'My Lord'. They sit either with the Lord Chief Justice and a High Court judge (King's Bench Division) or more often with two King's Bench Division judges.

19.3.2.4 Justices of the Supreme Court

The judicial role of the House of Lords ceased with the creation of the Supreme Court in 2009 and the Lords of Appeal in Ordinary (known as Law Lords) who formerly sat in The House of Lords took the title of Justices of the Supreme Court. There are 12 Justices of the Supreme Court who sit, unrobed, to consider points of law of general public importance.

19.4 Evidence

19.4.1 Laws of Evidence

All litigation involves the court making findings of fact and applying the law to those facts. Therefore, it is not allowed to accept just any evidence of those facts but must observe a large number of rules about them. These rules make up the laws of evidence.

For historical reasons, many of the rules of evidence are rules of exclusion. Until quite recently in our history, lay justices and jurymen were the only judges of fact, and being completely without experience in evaluating witnesses and evidence it was assumed that they would overvalue certain types of misleadingly attractive evidence which a trained person would handle with caution. The law in fact overcompensated for this assumption by not allowing them to hear the suspect evidence at all. These rules are not relaxed even though the modern jury person will most probably know at least something about the burden and standard of proof before they ever get into the jury box.

The law of evidence is particularly active on the criminal side of the law because of the presence of the accused person and the need to protect his special position. It is chiefly concerned with which facts may be proved (*i.e.* types of evidence – relevance and admissibility) and how they may be proved, (*i.e.* by testimony of witnesses, documentary or real evidence).

19.4.2 Types of Evidence

The object of evidence is to prove a fact and for evidence to be acceptable to a court it must be both relevant and admissible. However, not all relevant facts are admissible, for example hearsay evidence may be relevant but inadmissible. Hearsay, or 'second hand' evidence and most opinion evidence are generally excluded as a means of proof, for example a witness may testify only as to facts that they have directly perceived. Hearsay consists of evidence of what someone said or wrote, for example evidence of the contents of a policeman's notebook containing the statement of a witness to an accident is hearsay evidence of the accident. Conduct may also be hearsay, for example if, in answer to a question a person shook his head or pointed, evidence of this conduct would be hearsay.

The opinion of one whom the court regards as expert is admissible in wider circumstances than the opinion of an ordinary witness. Expert opinion is limited to matters involving specialist knowledge but the expert witness may draw inference from the facts. This will be enlarged on later in the chapter.

19.4.3 Witnesses

19.4.3.1 Ordinary Witnesses

One way in which a fact may be proved in law is by the testimony of witnesses. In trials at the Crown Court, attendance is usually voluntary (*i.e.* the police or defence solicitors notify the witness as to when they should attend court). However, attendance may be compelled by the issue of a witness summons.

19.4.3.2 Character Witnesses

Where the character in question is that of the accused the fact that they are of good character is usually relevant and admissible. However, evidence of bad character is admissible only with the leave of the court if the court considers that its probative value outweighs its prejudicial effect. After conviction and before sentence, it is common for the court to receive evidence of the accused's character and antecedents in order to arrive at the appropriate sentence.

19.4.3.3 Expert Witnesses

Many cases involve the giving of expert evidence; it is into this category that forensic scientists fall. The distinction between an ordinary (or lay) witness lies in the fact that a lay witness may only testify to relevant facts within their knowledge or expertise, whereas an expert witness (called skilled witnesses in Scotland), may offer their opinion. Opinion has been defined as 'an inference drawn or to be drawn from observed and communicable data.' The expert must usually be either qualified or experienced but their qualifications may be challenged by cross-examination. In English law there is no particular standard with which an expert witness must comply but the court (judge) must be satisfied that the fact which the expert's opinion is tendered to prove is something which may be the subject of expertise, *e.g.* fingerprints, ballistics, *etc.* The Criminal Procedure Rules and the Criminal Practice Directions place an express duty on expert witnesses to actively assist the court in fulfilling its obligations.

19.4.3.4 Witness Statements

Before 1967, witnesses had to appear in person to give their evidence at the lower (magistrates') court and this wasted a lot of time where the evidence was not in dispute at that stage. The Criminal Justice Act of 1967 enabled the magistrates' court to commit a defendant for trial **without** consideration of the evidence providing certain criteria were fulfilled. Those were:

(1) the statement must bear the witness' full name, professional address and occupation,
(2) each page must be signed and the date of birth available and, most importantly,
(3) that the certificate at the beginning must be signed.

This certificate gives the statement the full force of a sworn testimony and renders the witness liable to a charge of perjury in exactly the same way as if an oath had been sworn. In other words, it allowed for statements to be admitted as evidence of the same value as oral evidence providing they were in the form required by the Act *i.e.* they could be read to the court.

19.5 The Legal Process

Since 2005 the conduct of the trial, both before and during the hearing, is determined by the Criminal Procedure Rules. The rules apply in all magistrates' courts, the Crown Court and the Court of Appeal (Criminal Division) in England and Wales.

19.5.1 Pre-trial

Committal procedures for cases which are to go to the Crown Court were abolished and have been replaced by transfer proceedings under Section 51 of the Crime and Disorder Act 1998 and take place as soon as possible after the first appearance in court.

Under Part 3 of the Criminal Procedure Rules 2020, Plea and Trial and Preparation Hearings (PTPH) will take place 21 days (if the defendant is in custody) and 28 days (if on bail) after the case is transferred from the magistrates' court. For not guilty pleas the defendant will be arraigned, the trial date set, the trial issues identified if possible, the timetable for pre-trial preparation provided and provision made for any further Case Management hearing if necessary.

There are now four stages to pre-trial preparation.

(1) The prosecution must serve most of the material on which they rely 50 days (if the defendant is in custody) and 70 days (if on bail) after the case is transferred. If the defendant has identified issues not admitted in the Streamlined Forensic Report (SFR) full forensic science statements can be ordered at this stage.
(2) The defence must serve their Defence Statement 28 days after stage 1. This includes any expert evidence they are relying on.
(3) The prosecution must respond 14 days (in custody) or 28 days (on bail) after stage 2. This also includes ordering full forensic science statements if appropriate.
(4) The defence to provide final material or make applications arising out of prosecution disclosure, including any expert evidence to be relied upon that could not be served by Stage 2.

All parties are obliged to complete the PTPH form, the objective of which is to ensure a speedy and effective trial. The total time a person may be held in custody awaiting trial (custody time limits) is 182 days and if exceeded may lead to a defendant being granted bail.

19.5.1.1 Disclosure

In order to ensure a fair trial, the Criminal Procedure and Investigations Act 1996 requires all material gathered during the course of the prosecution to be disclosed to the defence if it either undermines the prosecution case and/or assists the case for the defence, unless it is deemed sensitive material.

19.5.2 Trial at Crown Court

England and Wales adopt an adversarial system of dispute resolution. This means there are two competing 'sides' to each case, one put forward by the prosecution and the other put forward by the defence with each cross examining the others witnesses. Because of the presumption of innocence the obligation of proving the facts falls upon the prosecution. This is called the 'burden of proof'.

The standard to which these facts must be proved by the prosecution is 'beyond reasonable doubt'. The current preferred phrase to be used in directing juries (and magistrates) is that "they must be satisfied so that they are sure". Where the burden of proving a fact rests with the defence in a criminal case this is to the civil standard of proof 'on the balance of probability', *i.e.* more likely than not.

19.5.2.1 Court Procedure

On the day fixed for the trial the defendant will enter the dock, they will be asked their name and assuming they are the person named on the indictment it will be read to them. This is called the arraignment.

Trials take place before a jury of twelve persons, chosen by ballot from among those on jury service. Before they are sworn in, the Clerk of the Court invites the accused to challenge any of them if they object to their presence. There are a limited number of challenges for good cause (which satisfies the trial judge *e.g.* he knows a juror *etc.*). Equally, the Crown can only challenge with reason.

The trial begins with the prosecuting counsel making an opening speech which outlines the allegations against the defendant and the evidence they propose to call in support of their case. Prosecution witnesses will then be called, the victim usually being the first witness. As a general rule, all witnesses must give evidence on oath, which requires them to swear on their preferred holy book that the evidence given to the court is true. Alternatively, a witness may affirm.

Witnesses will be subjected to 'examination-in-chief' by prosecuting counsel, 'cross-examination' by defending counsel and possibly 're-examination' by prosecuting counsel. Examination-in-chief is always carried out by the party calling the witness. They must not ask leading questions (*i.e.* those framed in such a way as to suggest the answer), however leading questions may be asked during cross-examination. If there is more than one defendant, cross-examination of prosecution witnesses will normally be taken in the order in which the defendants' names appear on the indictment. There may be some further questions from the prosecution by way of re-examination but this can only be to develop an issue raised by the defence, which the prosecution could not have foreseen. When all the witnesses for the prosecution have been examined prosecuting counsel will close their case.

Defence witnesses are then called and if the defendant is being called it is usual for them to give evidence first. All defence witnesses are subject to an examination-in-chief, a cross-examination and a re-examination in the normal way. Once the case for the defence has been closed the prosecution may be allowed to call rebutting evidence but this is rare. Defence has the right to cross-examine a witness called in rebuttal.

The judge has the right to recall witnesses or even to call witnesses whom neither side has called. Closing speeches then follow, each side addressing the jury (unless the defendant is unrepresented and either gives no evidence or is the only person giving evidence; then the prosecution has no right to a speech). Prosecuting counsel speaks first, followed by counsel for the defence.

The judge sums up and invites the jury to consider their verdict. Initially the judge will not direct them as to majority verdicts but encourage them to reach a unanimous verdict. The Crown Court cannot accept a majority verdict unless the jury has had at least two hours to consider it. In practice 2 hours and 10 minutes is allowed. If the jury cannot agree in this time the judge should send them back to 'try again'. If they still disagree a majority verdict of ten out of eleven or twelve, or nine out of ten is possible. When a verdict of guilty is given the foreman must state the number of jurors agreeing. This is not so with a verdict of not guilty. If the defendant is found guilty there will be an enquiry into their antecedents and/or a plea in mitigation. The judge will usually refer the case for reports before sentencing. If the verdict is one of not guilty the defendant is free to go. Where the jury cannot agree a re-trial may be ordered.

19.6 Standards

At all stages of any practitioner's work, be it in the laboratory or at a crime scene, quality standards must apply and must be adhered to. Most large providers of scientific evidence have subscribed to an assessment of quality standards for many years. Within individual laboratories scientists will be trained in every aspect of the work to meet the very high standards imposed upon them by their own internal training programmes.

However, in 2008, following criticism and concerns about some quality issues, the post of Forensic Science Regulator was created to ensure that appropriate standards are met by all providers of forensic science to the Criminal Justice System. This included the adequate acknowledgement of and dealing with quality failures. The Regulator is assisted by an Advisory Council and a number of specialist advisory groups and has issued guidance and Codes of Practice in many disciplines. These are all available on the Forensic Science Regulator's website. The Forensic Science Regulator was provided with statutory powers in 2022 which came into force in October 2023.

19.6.1 Accreditation

Accreditation is a means of assessing the technical processes and working practices of an organisation. The United Kingdom Accreditation Service (UKAS) is the national accreditation body recognised by government to assess organisations against internationally agreed standards.

The Regulator's recommendation is that all providers of forensic services should be assessed by UKAS to an appropriate standard, this being a way to ensure the consistency of the information given to the courts. The international standard with which laboratories have to comply and which demonstrates technical competence to produce accurate data is ISO 17025. As well as looking at procedures and technical processes this standard also regularly reviews the effectiveness of the organisation's quality management systems and is the single most important standard for laboratories producing data of all kinds.

19.6.2 Competence

Apart from organisational accreditation of the laboratories supplying scientific services, the courts also need to know that the individual giving the expert opinion is competent to do so. The Criminal Procedure Rules 2020 and accompanying Criminal Practice Directions 2023 give the courts guidance on ways to assess an expert's reliability and credibility with the scientific arena, including any potential flaws in the expert's opinion.

The forensic scientist's own professional body, The Chartered Society of Forensic Sciences, issues subject specific Certificates of

Professional Competence to individual members who have satisfied criteria set by the Quality Committee of the Faculty of Sciences of Staffordshire University. Details can be found on the Society's website.

19.6.3 Cognitive Bias

Bias exists in many categories and in all disciplines. Case workers need to know the circumstances surrounding any offence submitted for examination but it is essential that practitioners ensure that they guard against having their judgement influenced either by limited and one-sided information supplied to them (contextual bias) or by requests to confirm a test rather than make an independent assessment (confirmation bias). They must be made aware, through training and continual assessment, of the risks associated with information which could potentially influence their judgement and final outcome.

This applies to police officers instructing scene goers, checkers within a laboratory, experts evaluating data and anyone preparing a witness statement.

19.7 Expert Evidence

Both the court and the public have expectations of an expert and in turn the expert has certain duties to them. However, because of the very special position of the expert within the trial process there may be a greater emphasis or importance attached to their evidence. Apart from by cross-examination, one of the safeguards against this under common law is that evidence must satisfy the admissibility test; that is the witness must be impartial, the evidence must be relevant, it must be outside the knowledge of the jury and the expert witness must be competent to give that opinion.

Prompted by the Government's Select Committee report *Forensic Science on Trial* the Law Commission sought to apply further safeguards in their report *Expert Evidence in Criminal Proceedings in England and Wales*, the key recommendations of which were:

(a) to introduce a statutory admissibility test, to be applied in appropriate cases, whereby expert opinion is admissible in criminal proceedings only if it is sufficiently reliable to be admitted ("the reliability test"), and

(b) to provide, by statute, judges with a single list of generic factors to help them apply the reliability test.

However, the government did not legislate to this effect, maintaining that there was adequate provision within the Criminal Procedure Rules.

The Law Commission has acknowledged that there is nothing at common law which prevents the court from making the same inquiries of expert evidence that were the subject of their recommendations. Therefore, in determining the reliability of an opinion Criminal Practice Directions 2023, part 71.2 details the factors which the court may take into account in determining the reliability of expert opinion, and include:

(a) the extent and quality of the data on which the expert opinion is based,
(b) the validity of the methodology employed by the expert,
(c) if the expert's opinion relies on an inference from any findings, whether the opinion properly explains how safe or unsafe the inference is (whether by reference to statistical significance or in other appropriate terms),
(d) if the expert's opinion relies on the results of the use of any method (for instance, a test, measurement or survey), whether the opinion takes proper account of matters, such as the degree of precision or margin of uncertainty, affecting the accuracy or reliability of those results,
(e) the extent to which any material upon which the expert's opinion is based has been reviewed by others with relevant expertise (for instance, in peer-reviewed publications), and the views of those others on that material,
(f) the extent to which the expert's opinion is based on material falling outside the expert's own field of expertise,
(g) the completeness of the information which was available to the expert, and whether the expert took account of all relevant information in arriving at the opinion (including information as to the context of any facts to which the opinion relates),
(h) if there is a range of expert opinion on the matter in question, where in the range the expert's own opinion lies and whether the expert's preference has been properly explained, and
(i) whether the expert's methods followed established practice in the field and, if they did not, whether the reason for the divergence has been properly explained.

19.7.1 The Duties of an Expert

These are contained within the Criminal Procedure Rules 2020, Part 19 of which relates to expert witnesses in particular. Rule 19.2 now requires an expert witness, as part of his or her duty to the court to actively assist the court in complying with directions or informing the court of any significant failure. For example, providing a time by when a report must be served or warning of substantial delay in the preparation of a report. The Rules states that an expert's evidence must be objective, unbiased and within the expert's area or areas of expertise and this duty overrides any obligation to the person from whom the expert receives instructions or by whom the expert is paid. The expert must notify all parties if their opinion changes. At all times the parties must have a mind to the overriding obligation to ensure that criminal cases are dealt with justly. *The Crown Prosecution Service: Guidance on Expert Evidence* has been prepared to assist experts in understanding their duties. This provides a clear explanation of the Rules and all the relevant authorities and is essential reading for any expert giving evidence in court.

19.7.2 Expert Reports

Experts for the defence are not obliged to comply with the Criminal Procedure Rules although they do have to adhere to the Regulator's Codes of Practice and of course legal aid criteria. Crown experts on the other hand present their findings in one of two ways which are governed by strict guidelines as described below.

19.7.2.1 Streamlined Forensic Reports

Forensic evidence is often complex, time consuming and costly to produce and in the past has been routinely requested without establishing whether or not it is required. This wastes significant forensic resources in both time and costs and produces delays which prove unhelpful to the courts when trying to set trial dates. In an attempt to speed up the criminal justice process, reduce cost and ensure fairness for the defendant, scientific findings can, in some cases, be presented in stages. This process is called Streamlined Forensic Reporting (SFR) and enables investigators, scientists, prosecutors, defence and the courts to comply with the Criminal Procedure Rules. It is a two-stage process, the purpose of which is to deliver forensic evidence proportionate to the needs of the case.

On an SFR 1 form the initial key scientific findings (results and conclusions) are summarised and should be accompanied by photographs where possible. The SFR process puts defendants in a more informed position and allows them to participate and comment on forensic evidence much earlier in the process. The objective of the SFR is to enable early identification of the issues so that guilty pleas can be entered at the first opportunity.

If the defence takes issue with any aspect of the report they can ask for an evaluative report which is expressed on the form SFR 2. Thus through the case management process, any areas of dispute are identified and addressed.

In complex cases, or when the defence requests, following service of an SFR, a full statement can be produced.

19.7.2.2 Full Witness Statements

When full witness statements are provided experts should recognise the need for clarity and accuracy and should communicate effectively. Historically a number of factors contributed to so called 'miscarriages of justice' and scientists were accused of being 'sparing with the truth'.

In an effort to overcome some of these failings an agreement with the Crown Prosecution Service and the Association of Chief Police Officers have resulted in certain standards being adopted to ensure that reports contain sufficient certainty and selectivity to make the report informative and clear to lay persons and that conclusions should be reasonably terse (courts do not want a scientific treatise).

CPR Rule 19.4 states that expert reports should include:

(1) details of the experts academic and professional qualifications, and relevant experience and accreditation,
(2) details of the information the expert has received, the matter on which their opinion is sought, the materials provided and all the information that is material to the opinion they have expressed,
(3) information on who has carried out the examinations and whether under the experts supervision or not,
(4) where there is a range of opinion this should be summarised and the expert's own opinion explained, and
(5) finally, it should include a statement to the effect that in the event that his or her opinion changes, the parties/court will be informed.

To comply with these requirements, guidelines provide the basic framework upon which scientists can construct their statements. Therefore statements prepared for the prosecution would normally include the following:

19.7.2.2.1 Introduction

This should describe the academic and professional qualifications in full and indicate the experience of the witness. This will help to provide the court with the means to assess the witness's credibility as an expert. It is important also to alert the reader if assistants have been used.

Thereafter, headings as shown below are recommended to help guide the reader to the appropriate parts of the statement.

19.7.2.2.2 Case Circumstances/Information Received

Relay succinctly the information that has been given (bullets are ideal). Only the information that has a bearing on the examination should be supplied. Care is required with the wording, as the prosecution version is a non-proven allegation.

Any statements that have been provided should be included here. Where there has been an attendance at a scene the time, date, others attending, *etc.*, then these must be mentioned. Also any exhibit taken by the scientist must be given the scientists' exhibit number and identified.

19.7.2.2.3 Receipt of Items

Item numbers and dates of everything that comes into the expert's possession should be mentioned. If any item is not examined, then this should be recorded separately. The integrity of the packaging should also be referred to in this section.

19.7.2.2.4 Purpose of the Examination

Care needs to be taken to ensure that the purpose of the examination is addressed from an unbiased standpoint (*i.e.* whether or not X was present). There should be no attempt to address the question – 'did the person commit the crime?' – that is a job for the jury.

19.7.2.2.5 Technical Note

This should be as brief as is consistent with accuracy. If macros are used they must be relevant to the case. It may also be put into an appendix.

19.7.2.2.6 Examination/Results

Dealing with items as they appear in the exhibit list can make for easy cross referencing but can be repetitive. Items may be better grouped by location, defendant or, in the case of drugs, by drug or class. Common names, more easily recognised by the layman, should be included along with the chemical name. Generally the use of jargon, and technical terminology should be avoided, as should abbreviations and the excessive use of identification numbers.

19.7.2.2.7 Interpretation/Conclusion

The conclusions should be reasonably terse (courts do not want a scientific treatise). Usually, the Bayesian approach which addresses two alternative propositions is adopted. Generally, this deals with source level or activity level and a weighting is given to the strength of support for one proposition or another. Offence level is normally the province of the jury.

19.7.2.2.8 Disclosure Schedule

In the course of any scientific examination inevitably there will be material generated that ceases to be of relevance as the work progresses. There will be conversations held, emails, telephone calls, decisions made. All these must be accurately recorded and, if not specifically mentioned in the statement, must be disclosed as unused material. The police and CPS require the expert to reveal to them all their unused material in the form of an index appended to their statements.

In summary, therefore, proper disclosure will require the expert to:

(1) retain all the material (including documentation) generated in the course of the examination,
(2) record everything at all stages,
(3) reveal everything that has been recorded.

Experts are fully expected to comply with these requirements and failure to do so can have serious consequences. The *Guidance Booklet for Experts-Disclosure: Experts' Evidence, Case management and Unused Material* has been prepared to assist with understanding these obligations.

19.7.3 The Expert in Court

Giving evidence is a vitally important part of the court-going scientist's job. Their appearance as an expert entitles them to limited privileges in court, and also exposes them to more severe assessment as a witness, not just on the word of their evidence but on their dress, demeanour and conduct. During their (usually) brief appearance in the witness box the forensic expert must give an impression of competence and honesty. The average juror will never have seen a forensic scientist before but will certainly have an idea of how they expect them to be and so appearance and demeanour are important aspects of the expert's presentation.

The following information may be of help to experts attending a court for the first time and for the reader to understand the procedures.

19.7.3.1 Magistrates Court

These days courts sit all the year round, usually from 10.30 am to 4 pm (although this does vary), with a lunch break between 1 pm and 2 pm. Magistrates' courts also sit on Saturday mornings but only to hear cases where the defendant is in custody.

It is unusual for a magistrate's court hearing to call an expert witness but when they do the trial procedure is the same as for the Crown Court. Remember that a JP is addressed as 'Your Worship' and District Judges as 'Sir/Madam'.

19.7.3.2 Crown Court

On arrival at the court the expert witness should consult the lists for trials, which are usually very clearly posted to locate the correct courtroom easily. Arriving in plenty of time avoids getting flustered and worrying the officer unnecessarily. If the court is in session, the witness should see the usher so that they can let the officer know that the witness has arrived and then wait outside the court room until being called. Any conversation should only be with the officer in the case or to prosecuting counsel after being introduced to them. If the evidence is particularly complicated or needs any explaining, and there has been no previous case conference, then the witness should try to speak to counsel before giving evidence.

19.7.3.3 In Court

When called, the witness should not enter cluttered up with handbag, coat, umbrella, *etc.* However, it is advisable to carry the file containing case notes rather than keeping them in a briefcase.

After being escorted to the witness box by the usher they will ask the witness whether they would like to swear or affirm. Either way they should make sure to read from the card. It is easy and embarrassing to get it wrong. If choosing to take an oath the witness will be asked which holy book they wish to swear on.

> The oath is "*I swear by Almighty God (or whichever holy book your faith dictates) that the evidence which I shall give shall be the truth, the whole truth and nothing but the truth*".
>
> The affirmation is "*I do solemnly, sincerely and truly declare and affirm that the evidence I shall give shall be the truth, the whole truth and nothing but the truth*".

For the entire time giving evidence, remarks should be addressed to the judge and jury (usually on the opposite side of the courtroom) using the correct form of address. At the Central Criminal Court (and any 1st tier centre) circuit judges are addressed as 'My Lord' (or Your Lordship). Lady Judges are called 'My Lady' (or your Ladyship). At most Crown Courts Circuit Judges and Recorders are called 'Your Honour' regardless of sex. The correct form of address is usually written on a card on the sill of the witness box.

There are several ways that counsel may take the expert through their evidence-in-chief. They may take them word for word through their statement or they may select certain items and gloss over the rest. Where there is a written a summary, counsel may ignore the bulk of the statement and take the expert only through this. Whichever form of examination is used, the expert may refer to their notes providing they were made at or near the time of the examination *i.e.*, they were contemporaneous. However, this has to be established in court, so the expert has to be prepared to answer questions about the circumstances in which the notes were made, most importantly the time they were made in relation to the examination. If they were made by an assistant and checked and signed by the expert, or were made under their direction, supervision or dictation, they are perfectly admissible.

The witness should place their notes on the sill of the witness box and not continually shuffle the pages. Counsel should always ask the judge's permission for the witness to refer to them but the right to use the notes is rarely challenged. The court may ask to examine the notes. The defence or the judge and jury has the right to see them if they wish, and the defence has the right to apply to have them made an exhibit if there is information in them useful to their case, or damaging to the prosecution's case. This right is not normally exercised, particularly as the disclosure regime (Section 19.5.1.1) means that if there is any issue in relation to the notes, they should have been disclosed to the defence before the trial.

The expert witness is not normally expected to produce exhibits, but may be asked to display them. If asked, the witness should hold them up to the judge and the jury and point out the areas under discussion clearly.

The judge will probably be writing down the evidence in longhand so the witness should speak slowly and also keep an eye on the judge's pen. Questions should be answered slowly and the expert should take care not to blind the court with science, this will only earn a rebuke from the judge. Remember that in all cases the expert is there to be helpful. Answers should be directed at the jury and there should be no engagement in a dialogue with counsel even though it is counsel who is asking the questions.

During the cross-examination it is defence counsel's job to test the evidence. The expert witness should never trespass beyond the bounds of their own skill and competence. If unable to answer a question they should say so. If not sure of the question, ask for it to be repeated. Do not try to be clever and never argue with counsel.

If there are several defendants each may have their own counsel who will question the expert in turn. Prosecuting counsel may re-examine, but this is only just to clarify points that have been raised during cross-examination. The judge may question an expert witness at the end of the re-examination or at any time during giving evidence.

At the end of the cross-examination or after their re-examination the prosecuting counsel will usually say 'No further questions' and ask the judge if they have any questions. The judge or counsel will then usually remember to release the witness.

If the hearing adjourns for lunch, *etc.*, in the middle of giving evidence, the witness should not speak about the case to anyone and will probably be warned by the judge not to do this. It is better to lunch alone.

After giving evidence, the expert witness may sit in the court and listen to the other witnesses; the usher will tell them where to sit. Sometimes the expert can be asked to sit in court and advise counsel when cross-examining defence experts and sometimes an expert witness is entitled to sit in and hear evidence before giving theirs, though this rarely happens. For either reason the witness will sit in the well of the court behind counsel. The prosecution expert should not speak to defence experts about their evidence unless asked to do so.

19.7.4 Defence Experts

Many forensic scientists will carry out work for both prosecution and defence during the course of their professional lives. The over-riding duty to assist the court applies to Crown and defence witnesses alike. An expert for the defence is likely to want access to material and notes in order to carry out their evaluation of the scientific findings and conclusions.

Defence experts will need to apply to the police for access to exhibits but can only take possession of them in exceptional circumstances. Most of the time where a defence expert is called, the prosecution team are given ample warning and indeed Part 19.6 of the Criminal Procedure Rules encourages experts to meet before the trial to discuss areas of agreement.

During the trial the defence expert usually sits with counsel to give advice during the prosecution expert's evidence. Similarly, prosecution counsel is likely to require the presence of their expert if the defence expert gives oral evidence.

Further Reading

The Criminal Procedure Rules 2020 and Criminal Practice Directions, 2023, https://www.gov.uk/guidance/rules-and-practice-directions-2020.

CPS Guidance Booklet for Experts on Disclosure, Unused Material and Case Management, 2029, https://www.cps.gov.uk/legal-guidance/cps-guidance-experts-disclosure-unused-material-and-case-management.

The Crown Prosecution Service: Guidance on Expert Evidence, 2022, https://www.cps.gov.uk/legal-guidance/expert-evidence.

Subject Index